3	4	5
00	Clouds generally forming or developing during past hour	Visibility reduced by smoke	Haze
10	Lightning visible, no thunder heard	Percipitation within sight, but NOT reaching the ground	Precipitation within sight, reaching the ground but distant from station
20	Rain and snow (NOT falling as showers) during past hour, but NOT at time of observation	Freezing drizzle or freezing rain (NOT falling as showers) during past hour, but NOT at time of observation	 Showers of rain during past hour, but NOT at time of observation
30	Severe dust storm or sand storm, has decreased during past hour	 Severe dust storm or sand storm, no appreciable change during past hour	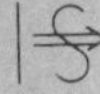 Severe dust storm or sand storm, has increased during past hour
40	Fog, sky NOT discernible, has become thinner during past hour	Fog, sky discernible, no appreciable change during past hour	Fog, sky NOT discernible, no appreciable change during past hour
50	Intermittent drizzle (NOT freezing), moderate at time of observation	Intermittent drizzle (NOT freezing), thick at time of observation	Continuous drizzle (NOT freezing), thick at time of observation
60	Continuous rain (NOT freezing), moderate at time of observation	Intermittent rain (NOT freezing), heavy at time of observation	 Continuous rain (NOT freezing), heavy at time of observation
70	Continuous fall of snowflakes, moderate at time of observation	Intermittent fall of snowflakes, heavy at time of observation	 Continuous fall of snowflakes, heavy at time of observation
80	Slight shower(s) of rain and snow mixed	 Moderate or heavy shower(s) of rain and snow mixed	 Slight snow shower(s)
90	Slight snow or rain and snow mixed or hail at time of observation, thunderstorm during past hour, but not at time of observation	Moderate or heavy snow, or rain and snow mixed or hail at time of observation; thunderstorm during past hour, but NOT at time of observation	 Slight or moderate thunderstorm without hail, but with rain and or snow at time of observation

(Continued on back endpapers)

METEOROLOGY

SECOND EDITION

METEOROLOGY

THE ATMOSPHERE IN ACTION

JOE R. EAGLEMAN
University of Kansas

Wadsworth Publishing Company
Belmont, California
A Division of Wadsworth, Inc.

Acquisition Editor: Thomas P. Nerney
Production Editor: Deborah O. McDaniel
Design and Cover: Cynthia Bassett
Print Buyer: Barbara Britton
Copy Editor: Carol L. Dondrea
Technical Illustrators: Joan Carol
Salinda Tyson
Cover painting: J.M.W. Turner,
Waves Breaking Against the Wind,
ca. 1835. Tate Gallery, London.
Paging Dummy: Virginia Mickelson

Printed in the United States of America

1 2 3 4 5 6 7 8 9 10—89 88 87 86 85

ISBN 0-534-03352-0

Library of Congress Cataloging in Publication Data
Eagleman, Joe R.
Meteorology, the atmosphere in action

Includes index.
1. Meteorology. I. Title.
QC861.2.E16 1985 551.5 84-13021
ISBN 0-534-03352-0

Photo Credits

Frontispiece National Oceanic and Atmospheric Administration (NOAA)
Part One NOAA
Part Two National Center for Atmospheric Research (NCAR)
Part Three Henry Lansford/NCAR
Part Four NOAA
Chapter One NCAR
Special Topic 1 Dave Baumhefner/NCAR
Chapter Two NCAR
Chapter Three NOAA
Chapter Four NCAR
Chapter Five NCAR
Chapter Six NCAR
Special Topic 8 NCAR
Chapter Seven NOAA
Chapter Eight NOAA
Chapter Nine Dave Baumhefner/NCAR
Chapter Ten Gene Moore
Chapter Eleven Gene Moore
Chapter Twelve NCAR
Chapter Thirteen NOAA
Chapter Fourteen NOAA
Chapter Fifteen NCAR

CONTENTS

PREFACE

Numerous students are attracted to introductory meteorology courses, often intending to take only one course to satisfy their intellectual curiosity and complete basic science requirements. This book is designed for such students, who generally have little physical science background. For this reason, mathematics is not emphasized or required, but appropriate equations are included in Appendix A.

Students (including meteorology majors) who take an introductory course expect to learn about weather forecasting and are interested in understanding more about various atmospheric storms. They are well aware of the existence of thunderstorms, tornadoes, lightning, hail, blizzards, and hurricanes, yet these are the very topics that frequently receive little attention in introductory meteorology texts. This book contains chapters on these subjects designed not only to give the student fuller appreciation of our atmospheric environment but also to provide guidance for dealing with the hazards associated with atmospheric storms.

Synoptic weather maps are presented in Chapter 2 to give students an immediate appreciation of the nature of current weather data processing and forecasting. Forecasting is the topic that beginning students typically associate with meteorology, since weather forecasts have already affected their activities in many ways. The book's approach allows current weather conditions and synoptic maps from the facsimile receiver or morning newspaper to be used more effectively in lectures or laboratories as the atmosphere and the weather that it generates are studied throughout the remainder of the course. However, the chapter is set up in such a way that instructors who prefer to cover these topics later can do so.

Another unique feature of the text is that it stresses the jet stream's role in governing surface weather and climate. Since the discovery of the jet stream only a few decades ago, it has become increasingly clear that it has tremendous influence on most midlatitude weather events. This text emphasizes the relationship between the jet stream and observed surface weather features in a manner that should be informative to the undergraduate student.

The book is organized into four parts. Part 1 gives a general survey of the atmosphere. Part 2 is devoted to the fundamentals of weather and climate. This part includes a discussion of radiation, winds, stability, and the raindrop formation process, a topic sometimes slighted in textbooks, even though it is fundamental to introductory meteorology. Part 3 covers atmospheric storms in a comprehensive way, and Part 4 includes selected topics of current and continuing interest, such as weather modification, atmospheric optical effects, and air pollution.

The book is designed for maximum flexibility. Within the section on atmospheric storms, an instructor might emphasize only storms of particular local importance. Similarly, Chapters 12 through 15, on applications and selected topics, may be taught in any order.

The numerous illustrations are designed to provide visual impact for the material discussed in each chapter. The figure captions reinforce concepts developed in the text and provide the student with maximum opportunity for efficient learning.

A list of technical words and concepts is included at the end of each chapter. Questions are also included with each chapter for use as study aids or classroom exercises. A combined index and glossary helps overcome the problem of terminology which faces all students in a new discipline.

Metric units are used throughout the book, with English equivalent units given in parentheses in many cases. Conversion factors are provided in Appendix B.

The helpful suggestions of many who used the first edition were appreciated. Also, the comments of several critical reviewers were very helpful in giving direction to revisions included in this edition. These reviewers include: Wilford A. Bladen, University of Kentucky; Carl R. Chelius, Pennsylvania State University; Robert L. Clark, United States Naval Academy; Russell L. DeSouza, Millersville State College; Ken Mackay, San Jose State University; G. L. Reynolds, Southern Illinois University; Colin E. Thorn, University of Illinois; and Gerald F. Watson, North Carolina State University.

P A R T O N E

INTRODUCTION

CHAPTER ONE

SURVEY OF THE ATMOSPHERE

INTRODUCTION

The atmosphere is an amazingly thin shell of air surrounding the earth, as can be seen by the layer of clouds in Figure 1-1. More than half the atmosphere is contained in a layer extending upward only 6 km (3.7 miles) from the surface of the earth, yet life on earth would be impossible without it. The atmosphere is a mixture of compressible gases that exhibit pressure and volume changes as the temperature varies. The winds resulting from these changes may bring forth a damaging storm or cleanse the air of harmful contaminants. It is amazing to think that such different phenomena as the beautiful sunset, the towering thunderhead, the terrifying winds within a tornado, the peaceful ocean breeze, lightning, and hailstones are all generated within the atmosphere from nothing but air, water vapor, minute suspended particles, and sunlight.

This chapter provides an introduction and overview of the atmosphere and the weather generated within it. A historical setting is provided through a discussion of ideas and discoveries that can be considered milestones in meteorology. Basic features of the atmosphere are introduced that may help promote understanding in later chapters.

Everyone is affected by the weather. Such diverse activities as industrial production and the world's food supply in any given year are strongly dependent on the weather. Some agricultural areas have reduced their dependence on water from the atmosphere by using irrigation systems, but most of the world's agriculture still remains tied to atmospheric processes.

Throughout the world, all outdoor, and much indoor, activity is influenced in one way or another by the weather. An example is the frigid winter of 1977, which not only closed many schools but also put more than two million people in the United States out of work because the natural gas needed for industry had to be used instead for heating. January of 1978 and 1979 were also much colder and snowier than normal, as traveling atmospheric storms repeatedly produced blizzard conditions across the central and eastern United States. These winters were the harshest of the twentieth century thus far, breaking many long-standing weather records. Figure 1-2 shows one of the problems caused by the extra snowfall in New York.

Some atmospheric processes are more predictable than others. Twenty-four-hour temperature forecasts, for example, are generally quite good. Also, it is usually possible to predict, with reasonable accuracy, the general areas of precipitation for the next day. However, it is much more difficult to predict the amount of precipitation, in spite of its importance. The current understanding of atmospheric processes is not sufficient to allow us to predict whether an area will get 10 or 20 cm of rain, although quantitative snow forecasts are slightly better.

One of the exciting things about studying meteorology is that many frontiers still exist. While it may be frustrating sometimes to realize how little is known about some of the important behavioral characteristics of our atmosphere, it is at the same time encouraging to know that there are still new frontiers to be conquered.

FIGURE 1-1 Clouds covering Florida and the southeastern United States give some indication of the thickness of the atmospheric layer in this Gemini V satellite photograph taken on August 23, 1965. (NASA photograph.)

METEOROLOGY—OLD AND NEW

The first recorded ideas and speculations on meteorology were made by early Greek philosophers. At the forefront of these was Aristotle, who, in his book *Meteorologica*, written about 340 B.C., gave in-depth explanations of many atmospheric phenomena. At that time, hail, rain, and all other substances that were suspended in air or fell from the sky were called meteors. Since much of Aristotle's writing in *Meteorologica* dealt with weather phenomena, the title gave rise to the name meteorology for the study of the atmosphere and weather. This designation has persisted to the present time.

Meteorology is therefore the study of the atmosphere and all the processes and phenomena that result in our particular weather. The atmosphere and weather are not only interesting to study but are more important than many people realize. Our moods, for example, are frequently geared to the weather since it is generally much easier to be happy on a sunny day than on a very gloomy, overcast day.

Modern studies of the atmosphere and the processes that occur there may also be called atmospheric science. This emphasizes that the study of the atmosphere is a science closely related to the other physical sciences. In fact, advanced studies of meteorology or atmospheric science may require an understanding of such sciences as physics, mathematics, chemistry, computer science, oceanography, geology, geography, and hydrology. However, the elementary facts concerning the atmosphere and weather are relatively simple to understand and can be mastered without advanced training in these related sciences.

FIGURE 1-2 Heavy snow causes many transportation problems, as shown here in Adams, New York, February 2, 1977. (Photograph by Dr. Ken Dewey, NOAA.)

A proper understanding of the atmosphere and the weather it generates is necessary for complete enjoyment of life as well as for the ability to deal effectively with the many weather-related facets of our environment. Atmospheric research contributes to improvements in the quality of life through numerous applications in air quality, world food supply, air travel, space flight, hydrology, architecture, engineering, and many other areas. However, much remains to be learned about the atmosphere. The weather cannot be forecast perfectly because both the solution methods used to solve the equations describing atmospheric processes and the observations describing the atmosphere are imperfect. Improvements in forecasting are coming, though, as modern technological developments add tremendous quantities of new data on the atmosphere, and provide the tools—computers, microfilms, and so on—for handling it efficiently.

MILESTONES IN METEOROLOGY

Aristotle's ideas in *Meteorologica* predominated for almost 2000 years. This Athenian philosopher suggested that the sun drew two kinds of exhalations from the earth. One was a cool moist material that was the source of rain. The other, according to Aristotle, was a hot, dry kind of smoke that was the natural substance of wind. Together the two made

FIGURE 1-3 Athenians used this octagonal Tower of the Winds in the first century B.C. The pointer on top showed the prevailing wind direction and also pointed to one of eight personified climatic conditions. Sundials on the walls showed the time of day on sunny days. (From J. Stuart and N. Revett, *The Antiquities of Athens, 1762.* Courtesy of the British Library.)

air, which converted into clouds when the temperature changed.

After clouds had formed, Aristotle suggested, the forcible ejection of rain in large quantities produced storms. He thought winds ejected from clouds burned to produce lightning. The thunder was produced when the winds ejected from one cloud bumped into surrounding clouds.

One of Aristotle's students, Theophrastus, published an elementary book of weather forecasting called the *Book of Signs.* It contained 80 signs of rain, 45 of wind, and 50 of storms. According to Theophrastus, "The plainest sign is that which is to be observed in the morning, when, before the sun rises, the sky appears to be reddened over; and it indicates rain." There is some basis for this rule since greater humidity in the air causes the sky to be red before sunrise. Many other signs, however, have proved to have no basis in fact.

The ideas of Aristotle and Theophrastus could not be tested until instruments were developed. This did not occur until the seventeenth century, although rain gauges consisting of clay bowls were in use in India during Aristotle's time and an elaborate wind vane was built on top of Athen's Acropolis in 100 B.C. (Fig. 1-3). No other instruments were available until Galileo Galilei demonstrated in 1600 that he could warm a flask with his hands, then invert it into a reservoir of water and observe the water flowing up into the flask as the air cooled and contracted. This was the first thermometer, although it was subject to pressure as well as temperature variations in this first stage of development.

Evangelista Torricelli, who followed Galileo as philosopher and mathematician to the Grand Duke of Florence, invented the barometer in 1644. Torricelli and others at the Academy of Experiments in Florence delighted in disproving Aristotle's ideas. Aristotle had written that a vacuum was impossible anywhere in the universe and that light waves could not travel through a vacuum. Torricelli's efforts to produce a vacuum resulted in his invention of the mercury barometer. By filling a tube with mercury and inverting it into a vessel containing mercury he developed a vacuum at the top of the tube and also proved that light could pass through it. He correctly explained that the mercury was held in the tube below the vacuum by the weight of air.

A short time later, in 1646, French philosopher Blaise Pascal suggested that the mercury barometer could be used to measure atmospheric pressure changes. This idea was verified when a barometer was carried up a mountain in France. Pascal correctly explained that air can be compressed, just like a huge pile of feathers, with the air on the bottom—closer to sea level—being denser than air higher up. The name Pascal is now used as a metric unit of pressure.

Less than thirty years later, in 1667, British physicist Robert Hooke, developed the first anemometer to measure wind speed. For the first time it was possible to quantify atmospheric temperature, pressure, and wind.

Further progress was made with the discovery of the fundamental gas laws. Robert Boyle developed the first gas laws, relating pressure and volume changes, in 1662. The effect of temperature on the volume of a gas was discovered by Jacques Charles in the next century (1787). These discoveries led to a greater understanding of the gaseous atmosphere.

During the 1700s the instruments required to measure atmospheric properties were improved and more weather-related discoveries were made. The Fahrenheit scale, for example, was developed in 1714, and the Celsius scale in 1736. Also, important contributions to meteorology were made during this time by Benjamin Franklin. Besides conducting his well-known lightning studies, Franklin became the first to document that storms moved toward the northeast in midlatitudes of the Northern Hemisphere. On October 21, 1743, Franklin wanted to watch a 9:00 P.M. eclipse of the moon, but a violent storm clouded the skies. He was surprised to learn a few days later that the sky over Boston had been clear during the eclipse, but had become stormy the following day. He began to study reports of the weather more carefully and made his discovery about the movement of storms.

Observations made by Franklin during the night of the eclipse were a source of frustration to him, since he observed that surface winds came from the northeast in a direction opposite to that needed to explain the storm's apparent movement from the southwest. He wrote to a friend that the pattern of winds in the storm must have caused this, but he could not quite put together the rotation of storms. Almost a century later William Redfield discovered that storms are actually giant whirlwinds. Redfield spent the autumn of 1821 walking across Connecticut, Long Island, Rhode Island, and Massachusetts to observe the aftermath of a strong hurricane that had swept through in September. As he mapped out tree fall directions he was amazed to find that the passage of a huge whirlwind would explain the observed damage. He spent ten years examining the logs from ships that had been caught in storms and published a detailed anatomy of storms in 1831.

His new idea of circular winds in large storms was greeted with great skepticism by some scientists, particularly James Espy. Espy argued that air rushed directly toward a low pressure center and that rising air currents near that center were explained by the fact that the air was heated. He further suggested that clouds formed as the air cooled and that the release of heat from the condensing moisture was what produced violent storms. Espy and Redfield bickered in print for the rest of their days and never realized that the best parts of both their theories were correct.

During the twentieth century the character of the atmosphere is being discovered in much more detail. In 1902 atmospheric layers were given the names troposphere and stratosphere by Teisserence de Bort. In 1903 Vilhelm Bjerknes, who was then a 41-year-old professor of physics at the University of Stockholm, decided to change his career and study the weather. He intended to apply known physical laws to the state of the atmosphere at any particular time in order to calculate a future state of the atmosphere. He worked at the Bergen Geophysical Institute in Norway and organized Norway's weather service. While his primary goal was not satisfactorily achieved until the invention of computers in recent years, Vilhelm and his son Jacob made major contributions to meteorology. In 1918 they discovered that most of the weather of midlatitudes results from the interaction of warm and cold air masses and not from simple pressure changes. They borrowed the word *front* from the combat terminology of World War I as they became involved in forecasting the weather for military operations. The idea of weather fronts and the association of fronts with the life cycle of a cyclonic storm system represented a major breakthrough in understanding and forecasting weather.

In 1939 Carl-Gustaf Rossby was appointed assistant chief of the U.S. Weather Bureau. Rossby had studied briefly in the Bergen school of meteorology, where he had immediately become interested in the winds aloft and had developed a dynamic theory of long waves in the atmosphere. The meandering stream of air he described as being at a height of about 12 km soon became known as the jet stream. The discovery of this major feature of the atmosphere was another significant advance in the understanding of weather systems.

In 1948, only a short time after the jet stream was discovered, cloud seeding methods were developed from experiments by Vincent Schaefer. Schaefer's experiments showed that dry ice caused supercooled water drops to crystallize into many ice crystals. Bernard Vonnegut discovered a short time later that silver iodide crystals also induced the formation of ice crystals in a supercooled atmosphere. These experiments were the basis for most current weather modification efforts.

The development of radar, satellites, and rockets in the twentieth century has allowed in-depth exploration of the atmosphere. Weather radar was developed in 1935, and the first upper-level exploration of the atmosphere with rockets was conducted in 1946. The first manmade satellite was launched by the USSR on October 4, 1957. On January 31, 1958, the United States launched its first satellite Explorer I. The first meteorological satellite TIROS I was launched by the United States on April 1, 1960.

It has been only during this century that weather science has approached the development we know today. In 1922 L. F. Richardson attempted, although the effort was unsuccessful, to predict the weather by numerical process involving mathematical equations. Only with the development of computers have these methods been successfully used in making quantitative forecasts.

Communication systems and technological developments have also helped advance meteorology to its present state. The impact of modern communications is apparent when past weather events are considered such as the 1925 tornado that traveled through Missouri, Indiana, and Illinois killing over 600 people. Because of the lack of communication systems and tornado forecasts no prior warning of the tornado was given. The first tornado forecast was issued in 1948 by Fawbush and Miller for Tinker Air Force Base in Oklahoma. As radio, and later television, came into widespread use, they became very important in disseminating current weather information to the public.

The paths of all hurricanes originating over oceans have been successfully observed and documented only since 1960, when the TIROS satellite was launched. Prior to the advent of satellites only sketchy information was available on storms over tropical oceans with the result that ships were frequently caught in them and many lives were lost. Today, modern meteorological satellites photograph the whole earth several times per hour and reveal these storms as soon as their spiraling winds develop.

Even with modern electronic and technical devices that photograph, measure, and analyze the atmosphere, humans are still limited as much by their incomplete understanding of atmospheric behavior as by the incomplete measurements of the state of the atmosphere at any one time. If these obstacles can be overcome it will be possible to advance the science of meteorology beyond our present comprehension. The weather is so influential in determining so many aspects of our existence that it is awesome to think of the potential outcome of understanding and controlling the weather, and possibly bending it to our particular desires.

COMPOSITION OF THE ATMOSPHERE

The **atmosphere,** which is so essential for all life forms on earth, is a mixture of many gases. The major constituents are nitrogen and oxygen. The third most abundant gas, argon, comprises less than 1% of the total, while the next, carbon dioxide, is present in even smaller quantities (Table 1-1). Numerous other gases constitute the remainder of the atmosphere.

TABLE 1-1

The composition of air

Gas	Parts per million	Percent by volume
Air with no water vapor		
1. Nitrogen	780,840	78.08
2. Oxygen	209,460	20.95
3. Argon	9,340	0.93
4. Carbon Dioxide	330	0.03
5. Neon	18	0.0018
6. Helium	5.2	0.00052
7. Methane	1.4	0.00014
8. Krypton	1.0	0.00010
9. Nitrous Oxide	0.5	0.00005
10. Hydrogen	0.5	0.00005
11. Ozone	0.07	0.000007
12. Xenon	0.09	0.000009
Average air		
1. Nitrogen	769,000	76.9
2. Oxygen	207,000	20.7
3. Water Vapor	14,000	1.4
4. Other Gases	10,000	1.0

The relative concentrations of the major atmospheric gases, nitrogen and oxygen, as well as argon, carbon dioxide, neon, and helium are uniform for a considerable distance above the earth. Thus the name **homosphere** is applied to this layer. Above 80 km is the **heterosphere,** where the gases are more stratified according to their weights. The most variable component of the unpolluted homosphere is water vapor, which varies from almost 0% to 4% by volume near the surface. This atmospheric gas decreases rapidly with height and is confined almost exclusively to the lower 16 km. If a sample of air were collected near the earth's surface, the amount of water vapor present would be small in comparison to the amount of oxygen, but atmospheric water, like oxygen, is very significant. The three major components of ordinary air near the surface are, therefore, nitrogen (76.9%) and oxygen (20.7%), with the next largest component being water vapor (1.4%).

Many gases in the atmosphere are capable of chemical reactions. Some of those present in trace amounts may form combinations that are considered to be pollutants. In many urban areas these and other potentially harmful gases, including nitrogen oxides, sulfur dioxide, carbon monoxide, methane, ozone, and ammonia, are monitored by State Health Departments or the Environmental Protection Agency.

Atmospheric gases in the upper homosphere and lower heterosphere exist in their ionized states. Therefore, this region is also called the **ionosphere.** Radio waves can travel around the world because they are reflected back down to earth by the ionosphere. Charged particles or bursts of energy from the sun, however, may interfere with radio wave propagation and cause fadeouts. These charged particles are drawn toward the magnetic poles of the earth and produce the beautiful aurora displays that occur within 30° latitude from the poles. Different colors are produced as the charged particles interact with the various atmospheric gases.

CHARACTERISTICS OF THE ATMOSPHERE

Our atmosphere can be so bland that we are unaware of its presence, or it can be so violent that we are literally blown off our feet, soaked with rain, bombarded by hailstones, or covered with snow. The basic differences in its behavior are determined by the distribution of sunlight, temperature, pressure, and moisture. These elements dictate the amount and type of wind and precipitation. In this survey chapter we will introduce some of the basic characteristics of each of the separate atmospheric elements; a discussion of the interrelationships among temperature, pressure, wind, and moisture will be reserved for later chapters.

Vertical Temperature Distribution Temperature is defined as the degree of hotness or coldness of a substance and is actually a measure of the internal energy of the substance. It is common knowledge that a hot-air balloon will rise, and that cold air engulfs the top of a mountain. The vertical change in atmospheric temperature is a critical factor in many weather-producing processes. Numerous measurements have been made to provide accurate information on the average temperature distribution through the atmosphere as shown in Figure 1-4. The global average air temperature decreases very rapidly with altitude above the surface, with the freezing level at a height of only 2.3 km (7000 ft). We must consider the source of heat for the atmosphere to explain these observations. The atmosphere is almost transparent to solar radiation. This makes the earth's surface the primary absorber of heat from the sun. Some of this heat from the earth's surface is then transferred into the atmosphere, resulting in warming upward from the surface through the lowest layer of the atmosphere. This increases the average temperature of the air near the earth to 15°C while just above, at 11 km, a very chilling −56°C exists.

As you can see in Figure 1-4, the air is warmed in the upper atmosphere to a temperature approaching 0°C at 50 km. This warming is caused by the absorption of the shorter wavelengths of sunlight (ultraviolet radiation) by a special form of oxygen called **ozone.** Ozone is a gas that is made up of three oxygen atoms per molecule (O_3) instead of two, (O_2), as found near the earth's surface. Ozone is a very unstable gas that constantly forms and decays. Its formation is related to the dissociation of O_2, among other factors. The absorption of ultraviolet radiation by ozone is important to us, because without it we would get many more severe sunburns.

Ozone is so effective in absorbing ultraviolet radiation that more heating occurs at the very top of the ozone layer, near 50 km, than lower in the atmosphere, where ozone is more concentrated. This effect is similar to squirting a strong stream of water into a thick forest, where the first layer of trees stops the water. Our concern for the ozone layer has led to the elimination of Freon from compression sprays, such as deodorants, since research indicated that fluorine from the Freon might reach the ozone layer and decrease the amount of ozone formation.

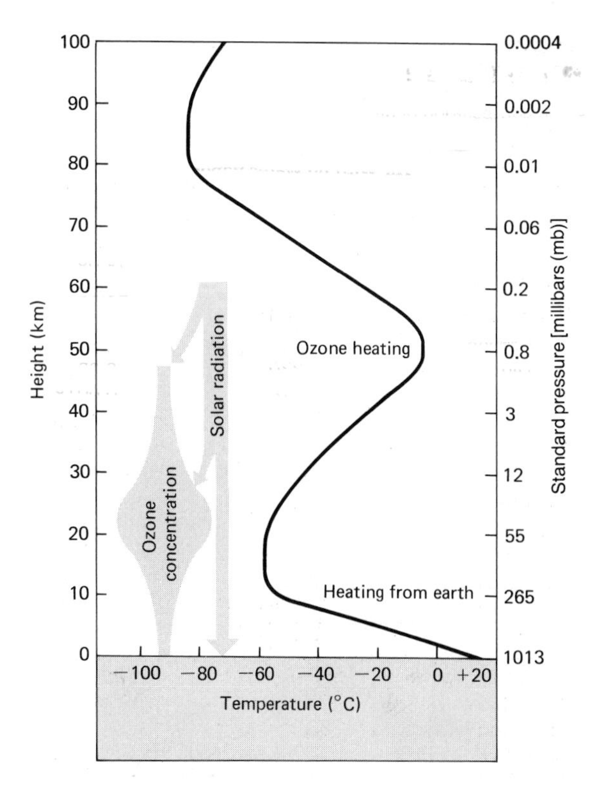

FIGURE 1-4 Variations in temperature with height through the lower 100 km of the atmosphere. The vertical temperature profile results from surface heating and ozone absorption of sunlight at 50 km.

The vertical temperature profile up to 50 km is thus developed from heat absorption at two different locations by very different absorbing agents, the earth's surface and ozone.

Above 50 km there is nothing to absorb energy from the sun so the temperature decreases with height. Eventually, above 90 km, the temperature again increases. However, the atmosphere is so thin at this altitude that a shaded thermometer exposed there could not measure the high internal energy of the air molecules because they are so sparse. A satellite located at this height, therefore, becomes very hot on the side that absorbs sunlight and very cold on the other, shaded side.

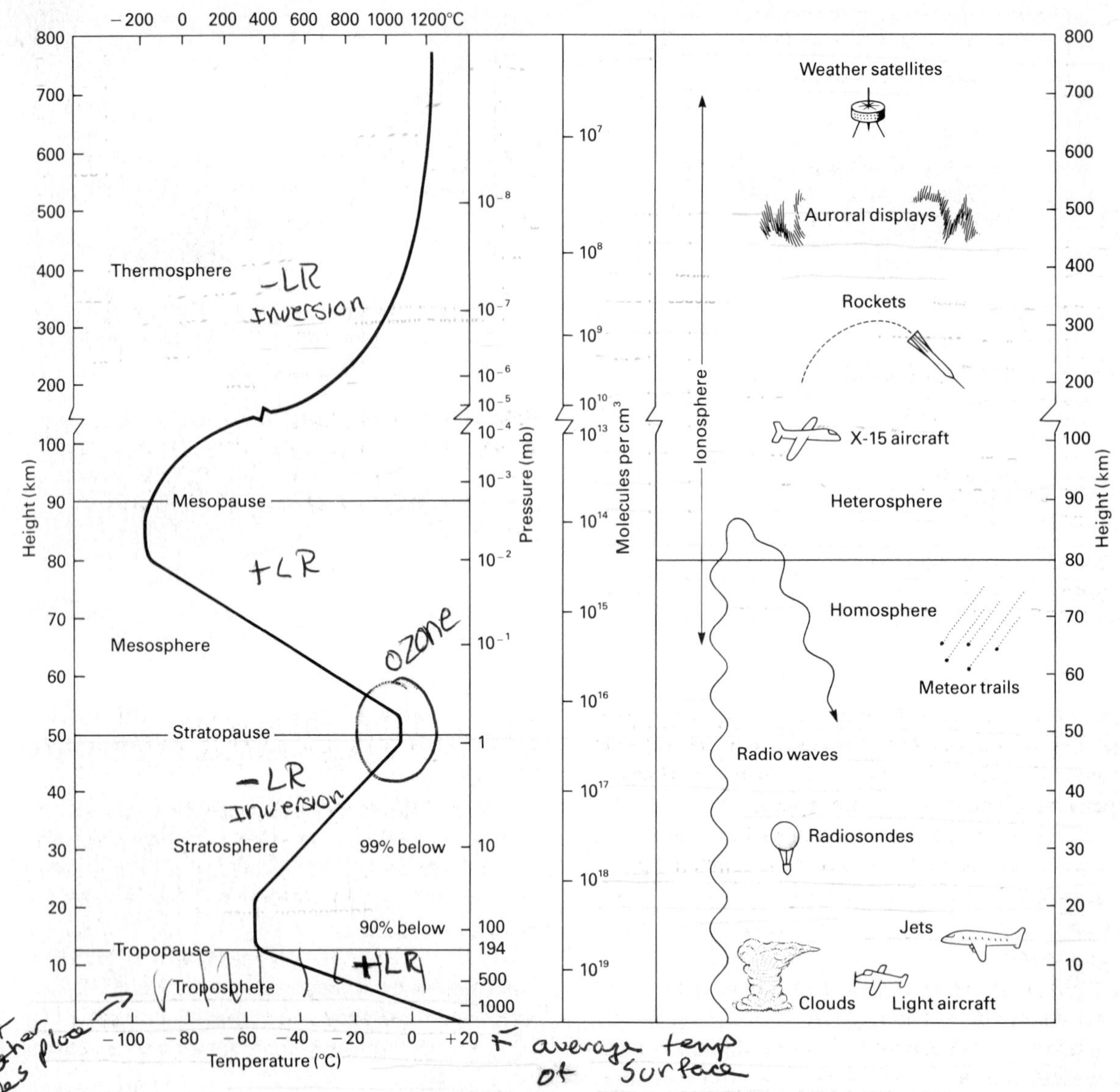

FIGURE 1-5 Atmospheric layers are determined by the temperature profile and composition of the atmosphere. Some atmospheric phenomena and observational systems are shown for various regions of the atmosphere.

The temperature profile is used as a basis for giving names to the atmospheric shells (Fig. 1-5). The lowest and most important atmospheric shell for weather processes and for human existence is the **troposphere,** where the temperature decreases with height. The region above the troposphere is called the **stratosphere** since it is generally stratified, with little mixing compared to the lowest layer of the atmosphere. The stratosphere is thicker than the troposphere, extending from 12

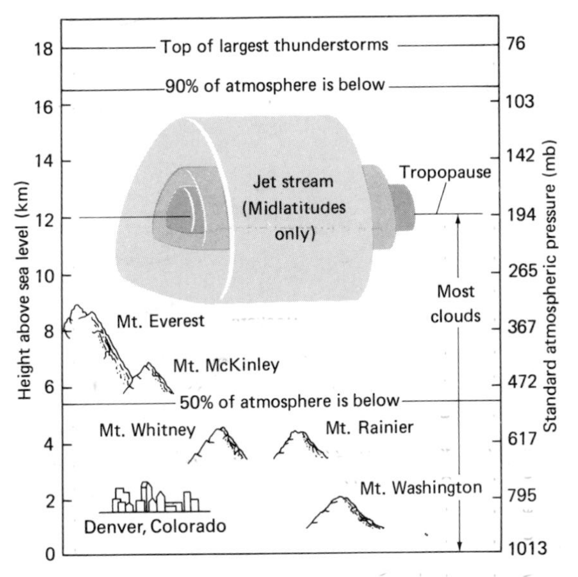

FIGURE 1-6 Height of the tropopause and jet stream in relation to the top of a large thunderstorm, Mount Everest, and other geographic features.

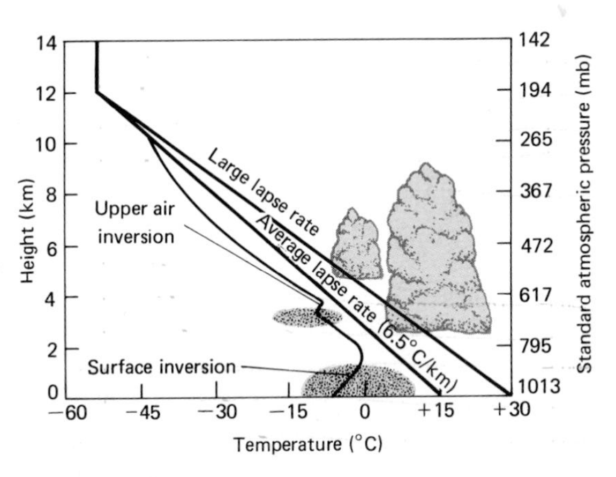

FIGURE 1-7 Comparison of large lapse rates, which result in vertical cloud development, to lesser lapse rates and inversions, which cause the atmosphere to become stratified and allow the concentration of pollutants.

km to the top of the ozone layer at 50 km. Above the stratosphere is the **mesosphere,** which extends upward to 90 km. Above 90 km is the **thermosphere,** which reaches outer space.

The tops of some of the layers have their own names: The top of the troposphere is called the tropopause; the top of the stratosphere is the stratopause; and the top of the mesosphere is the mesopause.

The most dynamic layer of the atmosphere is the troposphere, which contains 80% of the total mass of the atmosphere, because of surface heating and decreasing temperature with altitude. The thickness of the troposphere in comparison to some atmospheric and geographical features, including the tallest mountain on earth, Mt. Everest (8.8 km), which extends over halfway through the troposphere, is shown in Figure 1-6. Large thunderstorms penetrate the tropopause and form a mushroom- or anvil-shaped top as the rising air spreads out in the lower stratosphere. They cannot grow far into the stratosphere because of the uniform temperature structure of that layer, which dampens future growth into it. Most weather activity, therefore, is confined to the lowest layer of the atmosphere.

The vertical variation of temperature in the atmosphere is so important that it is given a specific name: **lapse rate.** A normal lapse rate occurs when the temperature decreases with altitude, while a temperature increase with height, or a negative lapse rate, is known as an **inversion.** The temperature decreases through the troposphere from an average surface temperature of 15°C to −56°C at a height of 11 km, resulting in an average lapse rate of 6.5°C/km.

Inversions can develop near the surface or in the upper atmosphere (Fig. 1-7) and exist through any layer in the atmosphere where the temperature increases with height. The existence of an inversion gives the atmosphere very different characteristics because cold, heavier air is present beneath warmer air. This produces an atmosphere with little mixing; thus an inversion contributes to more concentrated pollution levels near the ground. Such increases in temperature with height also produce stratification of the atmosphere, while large decreases in temperature with height contribute to clouds that develop vertically because of the presence of vertical air currents in the lower atmosphere.

Vertical Pressure Distribution The force exerted by the gases that comprise the atmosphere on a unit area of surface is the **atmospheric pressure**. The force arises from the activity of the molecules in the gases, which is determined by air temperature and by the pull of gravity (the molecules have a small amount of mass). We do not normally think of the atmosphere as having mass until we hold our hand out the window of a moving automobile, for example, or see damages caused by strong winds, but the mass of the atmosphere in 1 m^3 of air is about 1 kg near the surface. The standard pressure at sea level, resulting from the weight of air above, is 1013.25 mb (14.7 lb./in.^{2}). The milibar (mb) is a metric unit of pressure equal to 1000 dyn/cm^2 (dyn = dyne) and is the unit of pressure commonly used in meteorology.

1mb=

A pressure of 500 mb can be measured at a height of about 5.5 km, indicating that this atmospheric layer includes the lower half of the atmosphere since this pressure is about half the surface pressure (Fig. 1-8). Ninety percent of the mass of the atmosphere (100 mb level) is below 16.5 km. Since atmospheric pressure requires the presence of gaseous molecules with mass, the pressure decreases quite rapidly with altitude as the pull of gravity concentrates more of the mass of the atmosphere close to the earth's surface. Because of this characteristic we say that the atmosphere is compressible and has greater **density** near the surface. The density of the atmosphere is defined as its mass per unit volume and is about 1 kg/m^3 near the earth.

Pilots of aircraft with unpressurized cabins may be affected by the lower atmospheric pressure at greater altitudes. At a sea-level pressure of 1000 mb a person is accustomed to breathing oxygen at 21% of the total atmospheric pressure or 210 mb, since **Dalton's law** of partial pressures tells us that each separate atmospheric gas contributes to the total atmospheric pressure in proportion to the percentage of its concentration. Thus oxygen pressure decreases with height in proportion to the total atmospheric pressure decrease. Supplemental oxygen is needed above 3 km as the oxygen pressure decreases from its surface value of 210 mb to below 150 mb. A pilot gaining altitude, or making a prolonged high-altitude flight without supplemental oxygen or a pressurized cabin will notice a feeling of exhaustion followed by impairment of vision and perhaps unconsciousness.

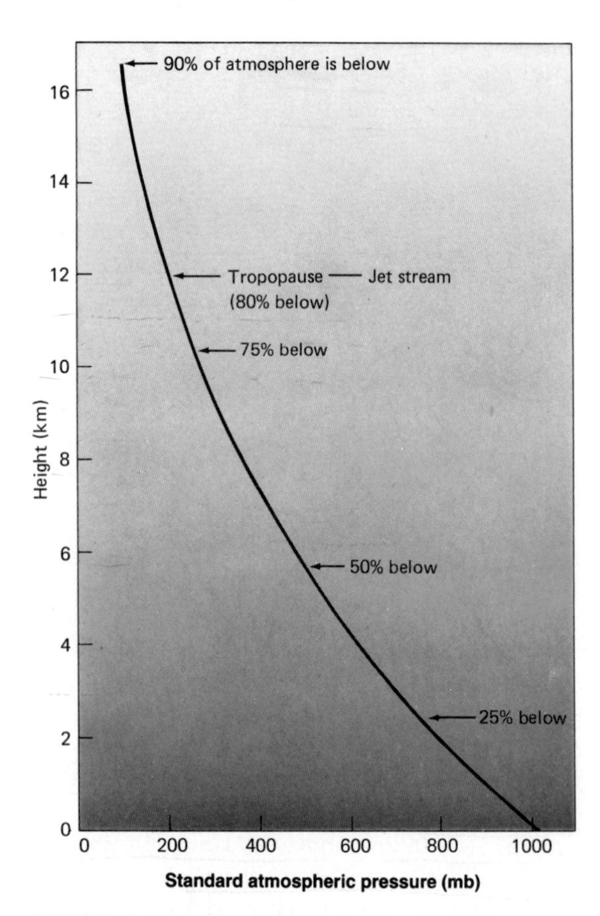

FIGURE 1-8 Pressure decreases rapidly with height in the lower atmosphere. Ninety percent of the atmosphere is below 16.5 km.

Horizontal variations in atmospheric pressure also occur continually. They are a prime factor in

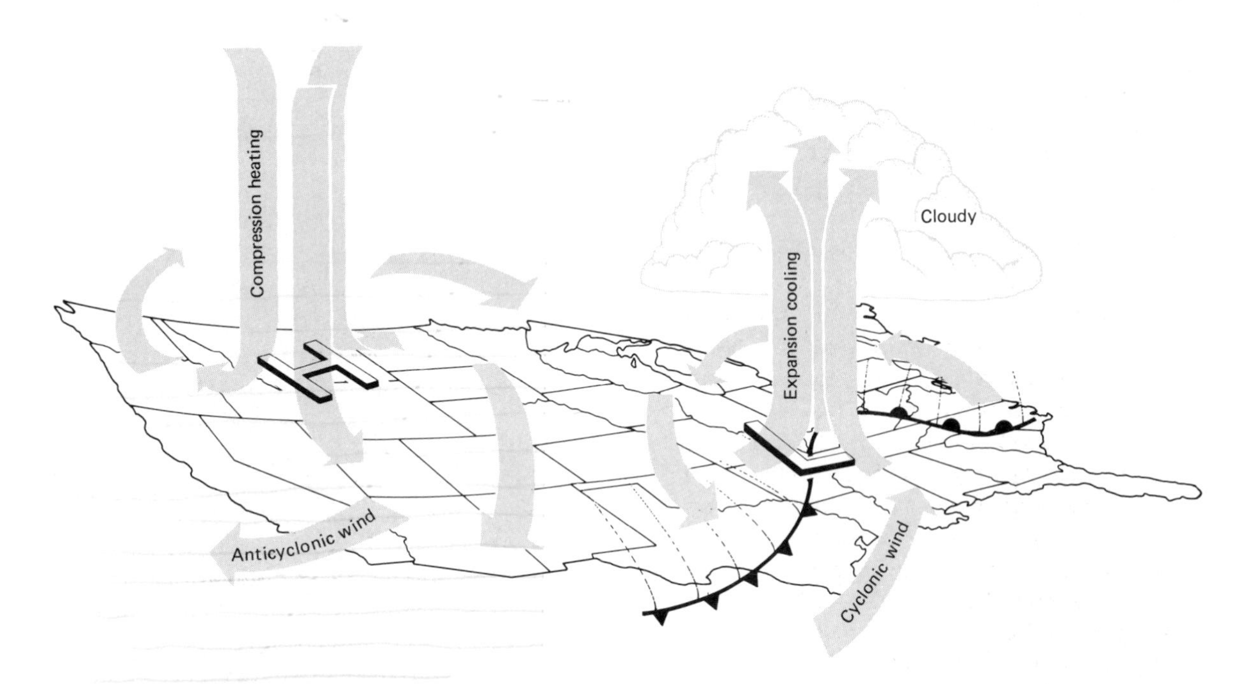

FIGURE 1-9 A surface high-pressure system is characterized by subsiding air above it and outflow at the surface, giving rise to clear skies. A surface low-pressure system has air spiraling in toward the lower pressure and rising above the low-pressure center; this leads to cooling as the air expands into lower pressure, with resulting cloud formation.

determining winds and the particular type of weather on any given day.

Introduction to Atmospheric Winds One of the distinguishing features of the atmosphere is its dynamic nature. It is always in motion, never achieving the equilibrium state that would exist if the sea-level pressure were 1013 mb all over the earth. Unequal heating from sunlight and an uneven distribution of atmospheric mass create traveling high- and low-pressure systems. The surface pressure, adjusted to sea level and disregarding hurricanes and tornadoes, varies by about 5% (960 to 1050 mb) as traveling high- and low-pressure systems develop and move through the midlatitudes. Wind is generated because of these horizontal pressure differences, but does not blow directly from high to low pressure because the earth's rotation also affects the wind direction, as will be discussed in more detail in Chapter 4, which deals with winds.

You can estimate the general location of low pressure from the surface wind direction. In 1857 Buys Ballot determined that if you stand in the northern hemisphere, with your back to the wind, the lower pressure will be to your left and slightly in front of you. The **Buys-Ballot law** has proven true, in general, because air always blows counterclockwise around a low-pressure system in the northern hemisphere (opposite in the southern hemisphere).

The counterclockwise winds around a low-pressure system blow slightly toward the center of low pressure (Fig. 1-9). This inflow of surface air forces some of the air upward above the surface low-pressure center, resulting in expansion cooling as the air rises into lower pressure. This produces the

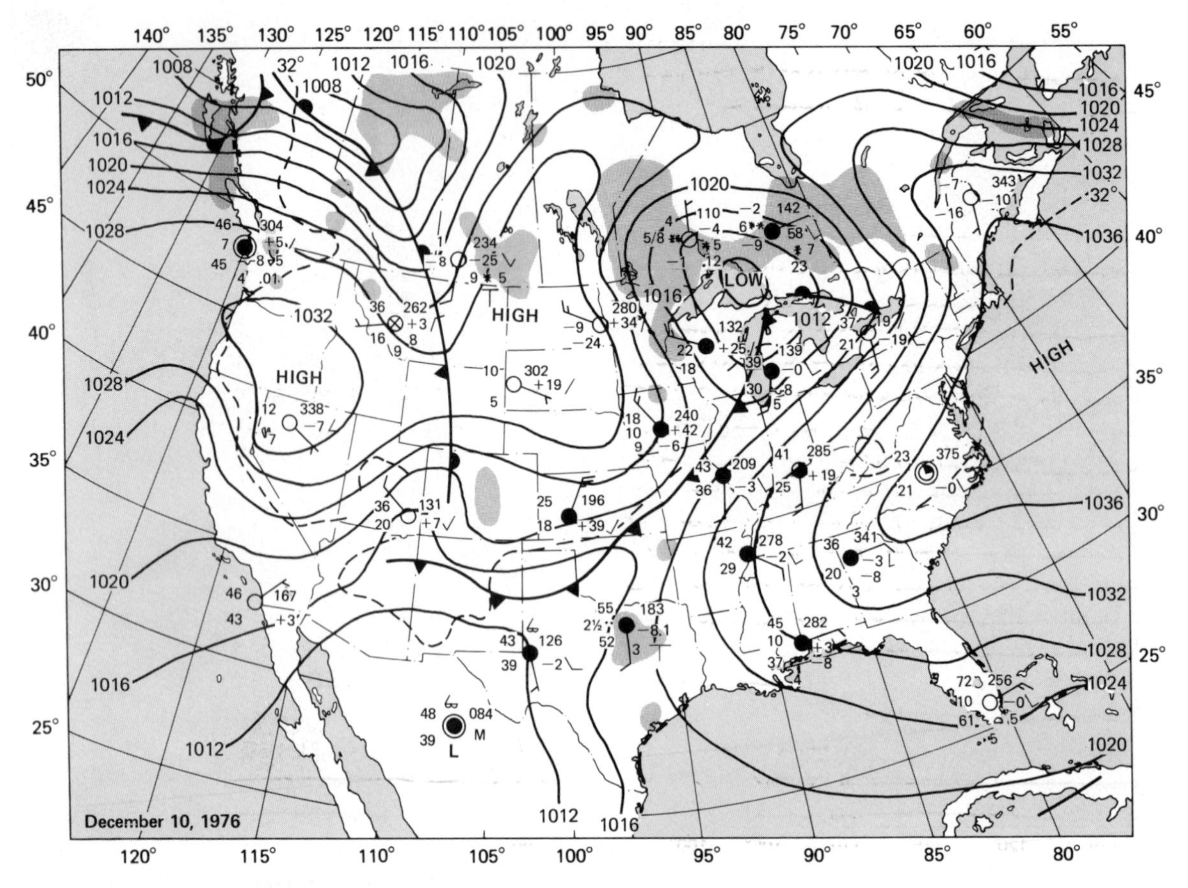

FIGURE 1-10 The wind blows counterclockwise or cyclonically around an area of low pressure and slightly in toward its center. The winds are anticyclonic or clockwise around areas of higher pressure and slightly away from the center of high pressure. This weather map is for December 10, 1976.

cloudy weather that is typical of low-pressure systems.

An area of lower pressure is called a **cyclone** in both the northern and southern hemispheres, and air flowing around it is said to be flowing cyclonically. A high-pressure system is called an **anticyclone,** and air flows around it in a clockwise or anticyclonic direction. Figure 1-10 shows a surface cyclone and anticyclone, with associated winds. The winds are slightly across the isobars (lines of constant atmospheric pressure), with some flow outward from the high-pressure center and toward the low-pressure center.

In the upper troposphere, the wind blows anticyclonically around high-pressure areas and cyclonically around lows (Fig. 1-11). The cyclonic bends in the air streams, called **troughs,** occur around areas of lower pressure, while anticyclonic bends, called **ridges,** occur around high-pressure areas. Important areas of lower pressure are frequently characterized by a trough without a circular low-pressure pattern, as can be observed over

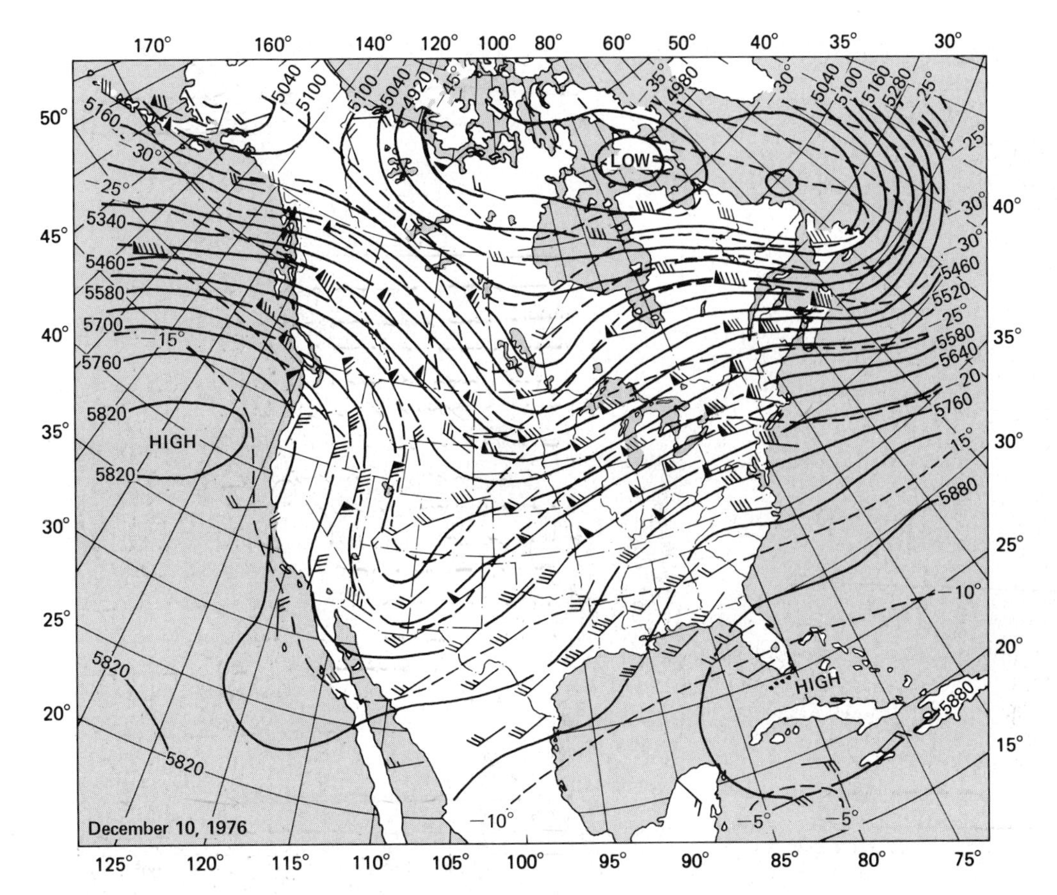

FIGURE 1-11 This weather chart was constructed from the measured heights, in meters, to the 500-mb pressure level for December 10, 1976. It shows wind above the surface blowing along the lines of constant height (pressure), and blowing cyclonically or counterclockwise around areas of low pressure, anticyclonically or clockwise around areas of high pressure.

western Colorado in Figure 1-11. At 500 mb, the winds blow along the height contours with very little flow from higher to lower pressure; this is in contrast to the surface wind, where the effect of friction with the earth's surface alters the direction.

Wind speed increases with altitude above the earth's surface. In setting up a system to tap the energy of the wind, one should build the necessary rotating blades on towers or locate them at higher elevations in order to take advantage of this wind characteristic. If the surface wind is from the southwest in midlatitudes, it frequently changes to the west at higher altitudes. The wind speed continues to increase up to the region of the **jet stream,** a stream of air circling the earth at a height of about 12 km (40,000 ft).

Jet stream speeds of over 550 km/h have been measured. In addition to the effects the jet stream has on surface weather, these strong winds frequently interfere with air traffic. If an aircraft must fly against the wind, time will be lost; with tail winds, the aircraft will arrive ahead of schedule, in

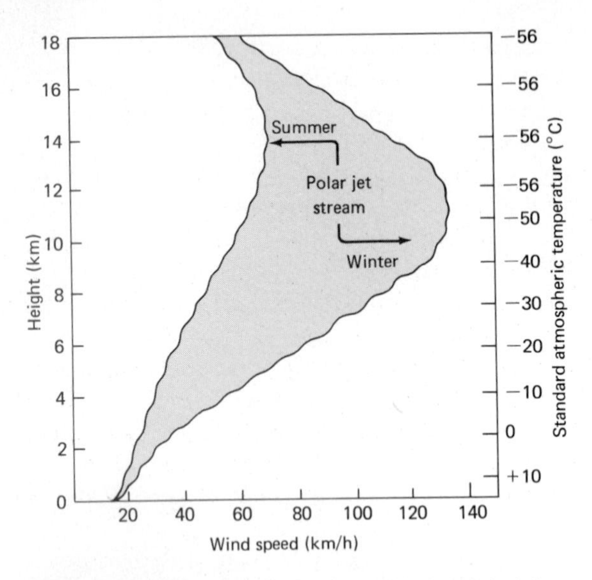

FIGURE 1-12 Variations in wind speed with height above midlatitudes during the summer and winter seasons. The polar jet stream attains its highest velocity during the wintertime, as the temperature gradient in the northern hemisphere is the largest.

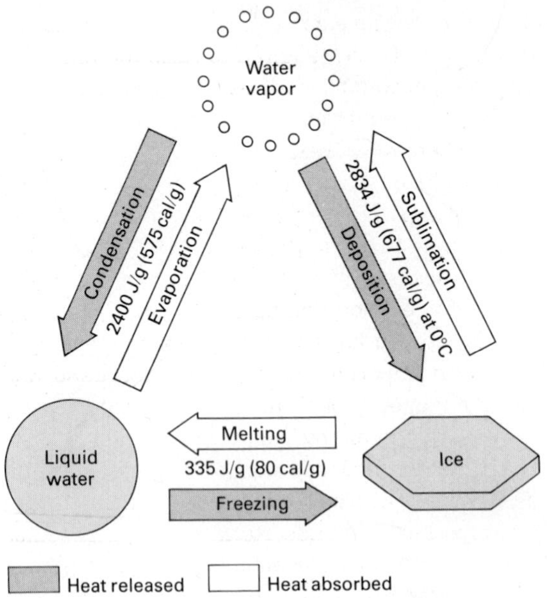

FIGURE 1-13 The amount of latent energy involved in various transformations of water among vapor, liquid, and solid states.

either case disrupting passenger arrival plans. The average speed of the jet stream is 130 km/h during the wintertime (Fig. 1-12), but only about 65 km/h in the summertime. The height of the jet stream and tropopause increases during the summer as sunlight warms the lower atmosphere and expands the air.

Introduction to Atmospheric Moisture The atmosphere serves as an important reservoir of our fresh-water resources. Water is the only substance that occurs in the atmosphere in all three forms: gas, liquid, and solid. Water in the gaseous state gives the atmosphere properties that you might not expect it to have. For example, a parcel of moist air containing invisible water vapor is actually lighter than a similar parcel of dry air. The molecular weight of the combined gases that form dry air is 28.9 compared to a molecular weight of 18 for water vapor. Thus, water evaporating into the atmosphere acts somewhat similarly to helium that is being added to the atmosphere; that is, the moist air is buoyant and will float upward more easily than dry air. This factor is important in explaining the occurrence of tornadoes along the boundary separating warm, moist air from warm, dry air.

Whether water takes the form of an invisible gas, solid, or liquid depends on its temperature and pressure environment, with the temperature ordinarily being the primary factor. Energy is required or released if water changes from one form to another (Fig. 1-13). The internal energy of water is greatest in vapor form and least in solid form. Large quantities of heat are released to the atmosphere as cloud droplets condense [2400 J/g (575 cal/g)] and as snowflakes form [2834 J/g (677 cal/g)]. These same quantities of heat were previously supplied to transform the liquid water to vapor (evaporation) and to transform ice to water vapor without passing through the liquid stage (**sublimation**). Since this heat is held within the water vapor, it is called **latent heat** or latent energy. The quantity of heat can be expressed in calories, defined as the amount of heat required to raise the temperature of 1 g of water 1°C, or as the number of joules (J) of energy required to produce 1 cal of heat. Conversion factors for these and other units are given in Appendix B.

Latent heat is a major factor in many weather processes. Condensing water vapor supplies heat to the surrounding air and makes it more buoyant; conversely, evaporation uses heat and is a cooling process, as you have probably noticed when you sweat. An important part of the generating and sustaining energy of a hurricane is the latent heat supplied to the air by condensing cloud droplets within the dense clouds surrounding the eye (center) of the storm. The latent heat of condensation is also important in thunderstorm development since air within the cloud becomes lighter as additional heat is added from condensing water droplets.

The quantity of water vapor in the atmosphere is very important in determining the region of cloud formation and atmospheric drying conditions at the surface, as well as a variety of other events. Either the **specific humidity** or the **mixing ratio** can be used to express the actual quantity of water vapor in the atmosphere, although they must be calculated from some other measurable quantity. The specific humidity is defined as the mass of water vapor divided by the mass of moist air and may be expressed as g/g or g/kg. The mixing ratio is similar since it is the mass of water vapor per unit mass of dry air.

If the amount of water vapor is increased sufficiently, the air will reach a state known as **saturation.** This represents the maximum amount of water vapor that can exist in air of a particular temperature. If the amount of vapor is increased further, or if the air is cooled, **condensation** of the excess water vapor into liquid will occur, or **deposition** as ice will occur on impurities that are usually present in the atmosphere.

The water **vapor pressure,** or portion of the total atmospheric pressure due to water vapor, may also be used to express water content, since Dalton's law of partial pressures applies to water vapor as well as oxygen, as noted earlier. The saturation vapor pressure is the pressure exerted by water vapor when the air is saturated. The typical contribution of water vapor to the total atmospheric pressure is about 15 mb.

The **relative humidity** is a common expression of humidity that gives the percentage of saturation of the atmosphere. It does not describe the actual amount of moisture in the air, since that may not change; but the relative humidity does change as the temperature changes. The relative humidity can be calculated from the actual water vapor content of the atmosphere (specific humidity, mixing ratio, or vapor pressure) divided by the amount of water vapor at saturation. This fraction is then multiplied by 100 to give the relative humidity in percent.

The moisture supply of the atmosphere is depleted as clouds form and precipitation falls to the ground. Since it is replenished by evaporation of water from oceans and humid continental areas, the concentration of water vapor is greatest near the earth's surface and decreases with altitude in the atmosphere. Colder temperatures higher in the atmosphere influence water vapor concentrations since cold air can retain only small quantities of water vapor before becoming saturated. The seasonal distribution of water vapor at the ground also shows the effects of cold temperatures—during January the amount of water vapor is reduced considerably (Fig. 1-14). The specific humidity decreases rapidly with height, with the quantity of water vapor in the atmosphere becoming extremely small at altitudes above 6 km.

AIR MASSES

The concept of **air masses** is helpful in understanding the nature of the atmosphere. The main atmospheric properties that determine the particular type of air mass are temperature and humidity. Air covering a large region that has a uniform temperature and humidity represents a single air mass.

The **source region** determines the type of air mass since the basic temperature and humidity properties are acquired there. Air masses that originate over continents are drier than those that

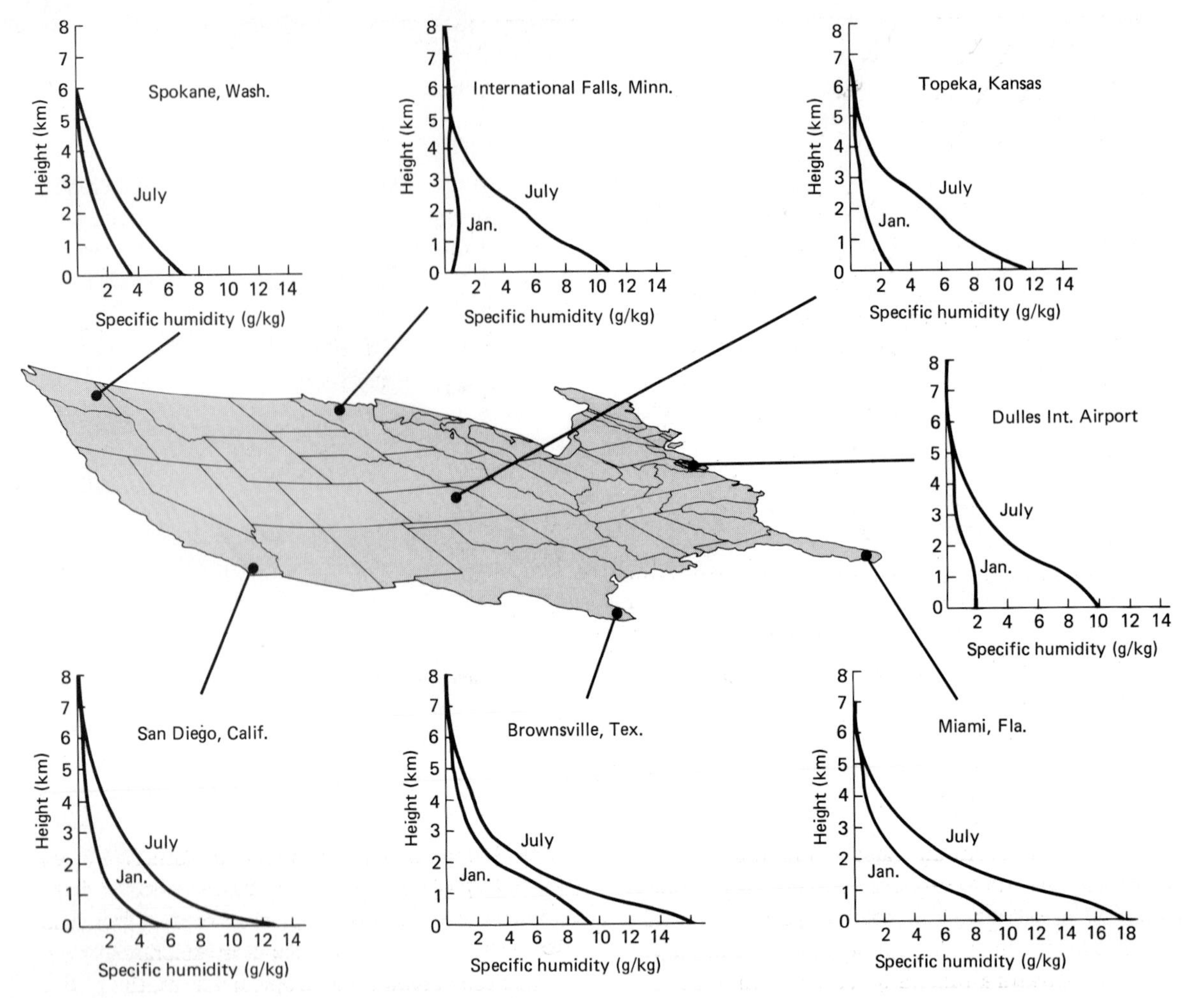

FIGURE 1-14 The vertical distribution of specific humidity during January and July for several different locations.

originate over oceans. Air masses originating north of midlatitudes are cold while those originating southward are warm.

Air masses are classified according to temperature and humidity characteristics acquired from the source region. Uppercase letters are used to indicate air mass temperature: A for arctic air masses, P for polar air masses, T for tropical air masses, and E for equatorial air masses. Moisture characteristics of air masses are commonly indicated by lowercase letters: m for maritime air (originating over oceans) and c for continental air (originating over land). Therefore, mT (maritime tropical) air masses are warm and humid while cP (continental polar) air masses are cold and dry.

Two additional symbols are used less frequently. These are the symbols N for a modified air mass and I for an indifferent air mass. A modified air mass is one that has been altered by radiation, or other factors, after it has moved out of its source

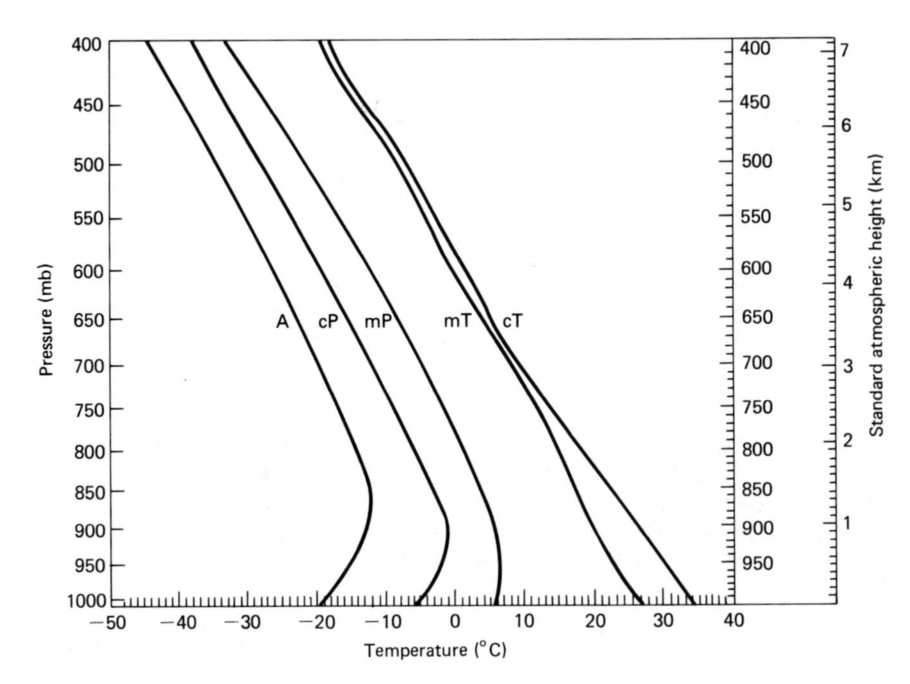

FIGURE 1-15 Typical temperature profiles for various air masses from the surface to 400 mb.

region. An indifferent air mass is still in its source region and has characteristics similar to the underlying surface. Thus air in polar regions would be designated indifferent after it became cold and dry.

The temperature and humidity characteristics of different air masses extend for considerable distances above the ground. A comparison of these characteristics of different air masses is shown in Fig. 1-15. The temperature variations from the surface to 400 mb show that large differences exist for the various air masses. The specific humidity of the different air masses is also quite different, with the cold air containing very little water at the surface or in the upper atmosphere. Maritime tropical air is much warmer and more humid than continental polar air at all levels in the upper atmosphere. For this reason, a wedge of continental polar air advancing beneath maritime tropical air creates a contrast in temperature not only at the surface but in the air above as well.

Major air masses that affect the United States (Fig. 1-16) are continental polar air masses from the north and maritime tropical air masses from the Gulf of Mexico, with occasional invasions of arctic air masses, continental tropical air masses, and maritime polar air masses. Three air masses frequently involved in severe thunderstorm outbreaks are continental polar, maritime tropical, and continental tropical. As the air circulates around a large midlatitude cyclone, the spiraling air converges toward the center of the lower pressure and creates sharp boundaries between different air masses.

FRONTS

When two different air masses meet, a weather front is formed at the boundary. The boundary is frequently quite sharp, with very little mixing of the two different air masses. If the cold air is advancing, the cold dense air pushes underneath the warm air mass and a **cold front** develops. If the warm air mass is advancing to replace cold air, a **warm front** is created with different characteristics from a cold front. If viewed from the side (Fig. 1-17), a cold front would appear as the leading edge of a thin wedge of cold air advancing beneath warmer air, with an average slope of 1 km vertically for every 100 km horizontally. The cold air is denser than the warmer air and pushes it upward. Clouds with vertical growth (cumulonimbus) are common near a cold front because of the rapid uplift of warm air.

A warm front viewed from the side (Fig. 1-18) would appear as an even thinner wedge of cold air being pushed along as the leading air in the warm air mass glides up over it. The typical slope of the warm front is 1 km vertically for every 200 km horizontally. Since the slope is less than that for a cold front, the weather associated with a warm front is less violent. Clouds of a more stratified nature (nimbostratus), with steady soaking rains, are commonly associated with warm fronts. The average warm front moves along at a lesser speed (25 km/h) than a cold front (35 km/h). The smaller slope of the warm front causes the clouds and weather to be spread over a larger area and hence to last for a longer time.

A shift in wind direction accompanies the passing of both a warm and cold front. As a warm front passes a location, the winds normally shift from the southeast to the southwest; a cold front passing a location is frequently accompanied by a wind shift from the southwest to the northwest, as can be seen on the weather map shown in Figure 1-10. These shifts in wind direction along with temperature changes may alert a farmer that a change in air mass has occurred, while such data plotted on a weather map are used by meteorologists to locate weather fronts and to record their movement.

The weather accompanying a warm front consists of light intensity rainfall lasting from several hours to a few days. After the rains cease the weather is mild, unlike the weather that arrives following the passage of cold fronts, which is cold. In

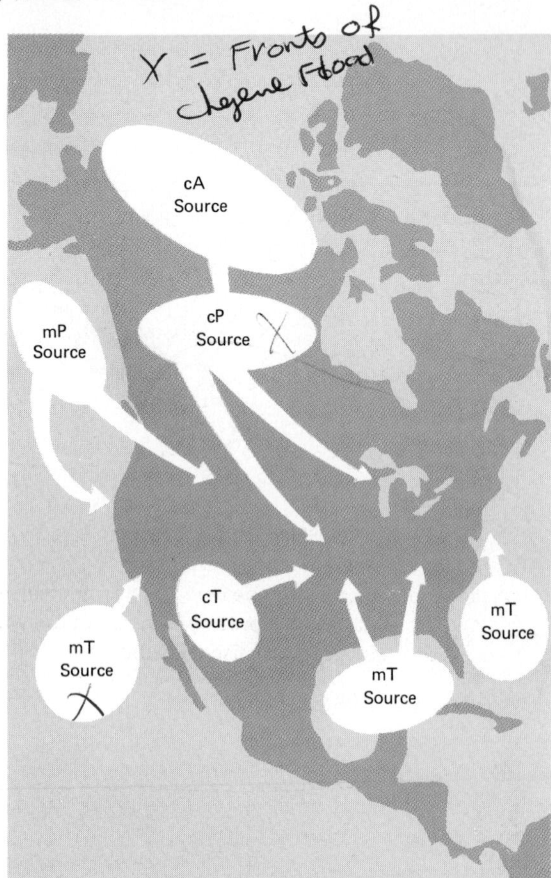

FIGURE 1-16 Air masses that affect the United States are shown in their source regions, with arrows indicating common paths.

a day or so the cold front may arrive bringing thunderstorms and higher intensity rainfall.

Since a cold front normally travels faster than a warm front, a part of the cold front near the midlatitude cyclone center frequently passes the warm front, creating an **occluded front** (Fig. 1-19). When this occurs, the warm humid air south of the midlatitude cyclone is lifted off the ground and may continue to release precipitation for some time.

Another type of frontal system is a **stationary front.** Mountain ranges or specific weather patterns, shown in Figure 1-20, can cause either cold or warm fronts to stall and become stationary.

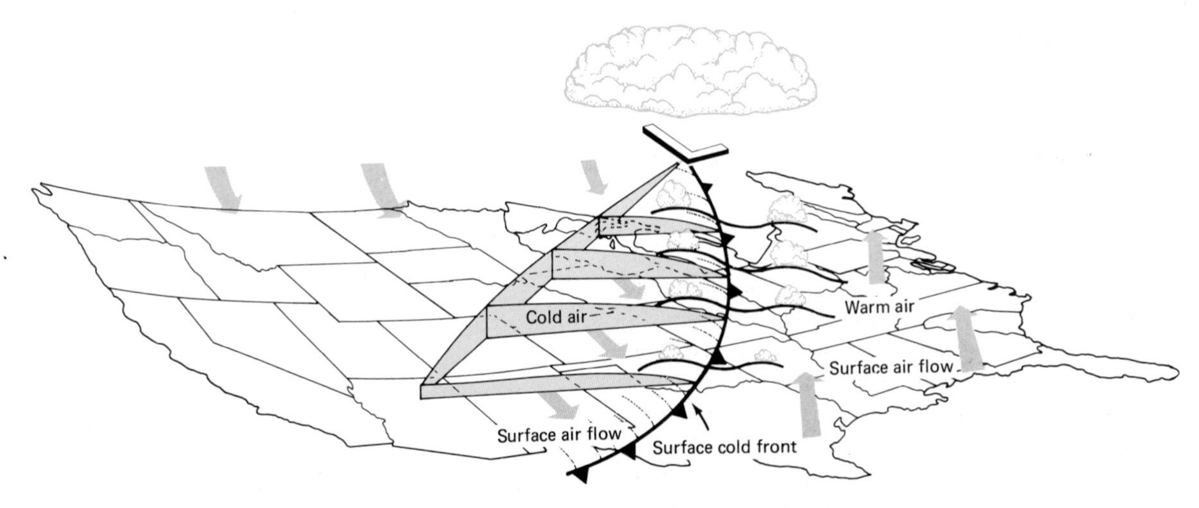

FIGURE 1-17 A surface cold front is the leading edge of a very thin wedge of cold air moving southward and eastward around a center of low pressure. Clouds and precipitation occur along the cold front as the warm air is lifted, or ahead of the cold front as prefrontal waves propagate through the warm air mass. The rising air above the center of low pressure also contributes to cloud formation.

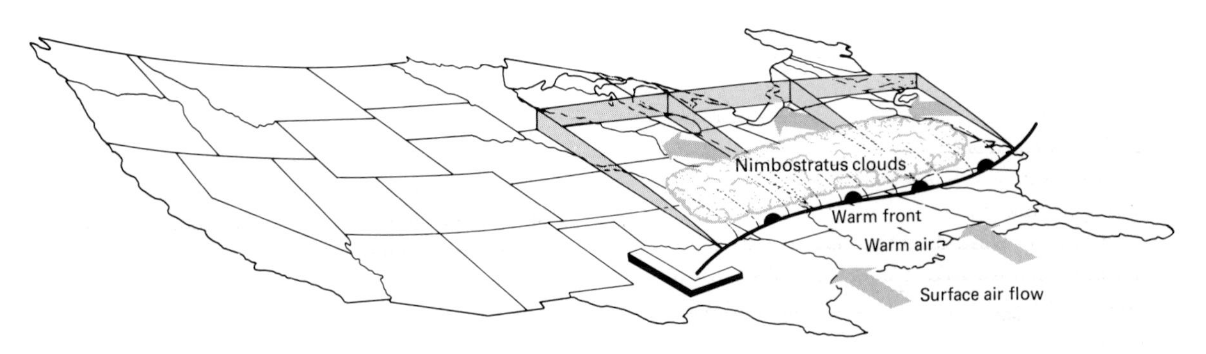

FIGURE 1-18 A warm front is the leading edge of a warm air mass, which is rising over cooler air. Such fronts typically occur eastward from the center of a low-pressure area. The warm air rises gradually over the warm front producing stratified clouds.

Stalling occurs when a weather front moves into a parallel position to the jet stream, thus losing its source of energy for movement and becoming a stationary front. If the front is oriented east-west and the surface pressure patterns provide strong southerly airflow south of the stationary front with northerly flow north of the front, rainfall may continue for several days over the same area and cause severe flooding. Precipitation continues as long as the front exists, with the warm humid air from the south flowing up over the wedge of cold air to condense into clouds and rain.

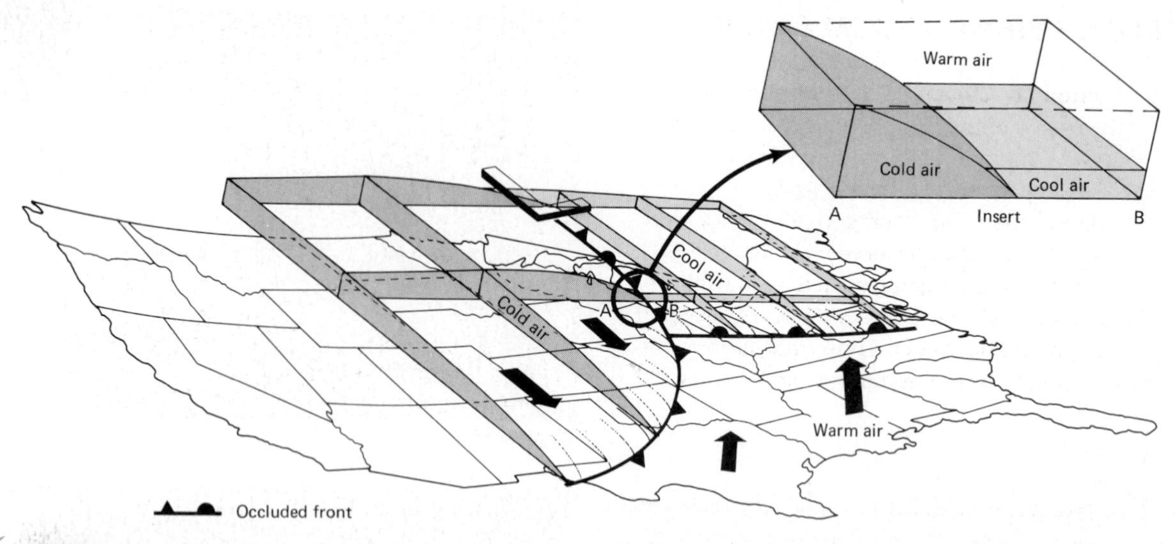

FIGURE 1-19 An occluded front forms as a cold front overtakes a warm front, lifting the warm air above the surface. An occluded front may accompany more mature midlatitude cyclones.

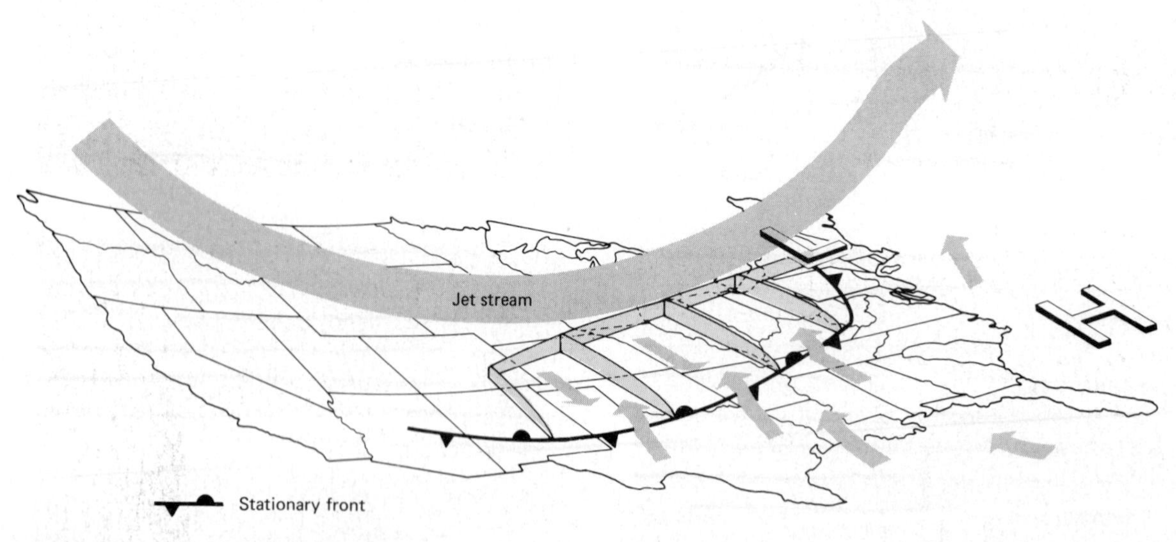

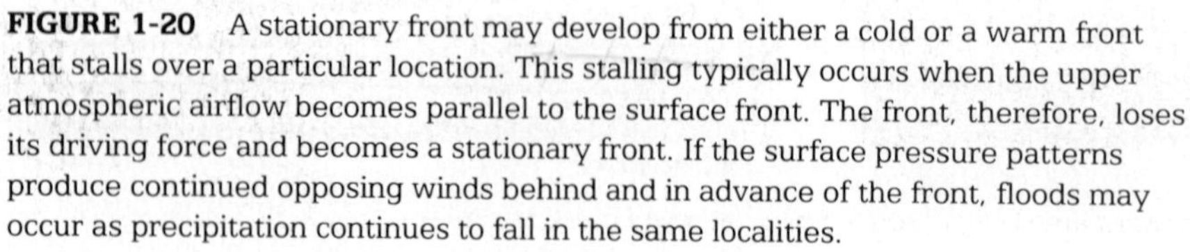

FIGURE 1-20 A stationary front may develop from either a cold or a warm front that stalls over a particular location. This stalling typically occurs when the upper atmospheric airflow becomes parallel to the surface front. The front, therefore, loses its driving force and becomes a stationary front. If the surface pressure patterns produce continued opposing winds behind and in advance of the front, floods may occur as precipitation continues to fall in the same localities.

CLOUDS VIEWED FROM SPACE

Introduction to Clouds Day-to-day changes in the appearance of the sky are an important aspect of our lives. The amount and type of cloud cover determines the amount of solar heating and acts as a constantly changing backdrop for all outdoor scenes. To the serious observer, clouds also provide clues to processes going on in the atmosphere.

For classification purposes clouds are grouped into ten *genera* (Fig. 1-21). Those that form in the upper troposphere are composed of ice crystals.

Cirrocumulus

Cirrus

Altostratus

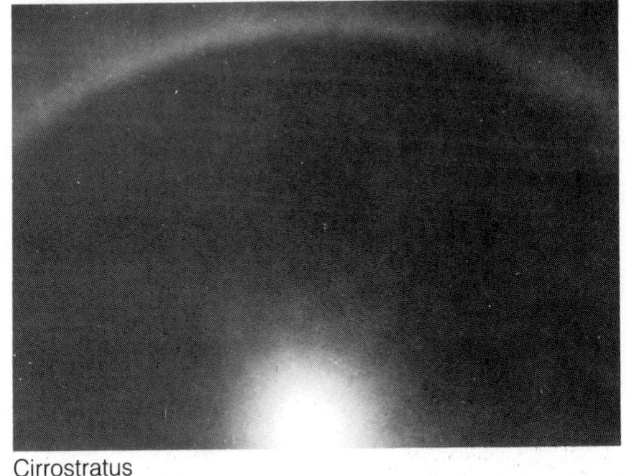

Cirrostratus

Altocumulus

FIGURE 1-21 The ten cloud genera are shown here in order, according to their various heights. The tenth genus, cumulonimbus, however, may extend vertically through the whole troposphere. (Photographs courtesy of NOAA.)

Nimbostratus

Cumulus

Stratus

Cumulonimbus

Stratocumulus

Such high clouds (6–14 km) are called cirrus, cirrocumulus, and cirrostratus. The middle clouds (2–7 km) are altocumulus and altostratus, and are frequently composed of supercooled liquid water droplets mixed with ice crystals. Low clouds (below 2 km) are nimbostratus, stratocumulus, and stratus and are composed of liquid water droplets. The remaining two genera, cumulus and cumulonimbus, grow vertically and may contain water droplets below the freezing level and ice crystals at higher levels.

The study of clouds and their relationship to weather on earth has become more interesting since the development of satellites. A Skylab photograph (Fig. 1-22) shows part of Missouri and the

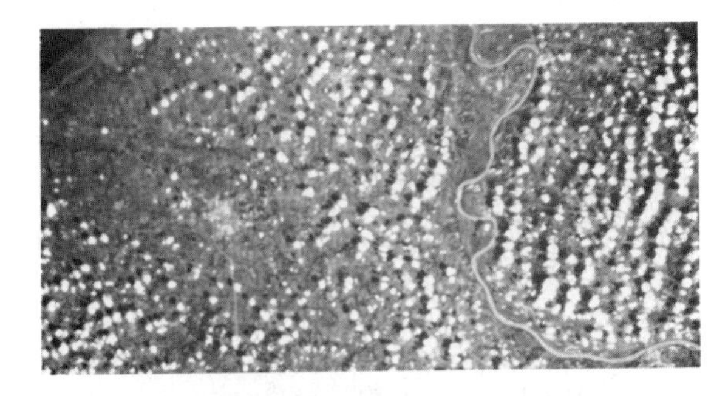

FIGURE 1-22 Skylab photograph showing scattered cumulus clouds over Missouri on September 18, 1973. Note the lack of clouds over the Missouri River. (NASA photograph.)

Missouri River, where scattered cumulus clouds were located over the surrounding land but not over the river or the river valley. This indicates that small differences in the temperature of the earth's surface (either warmer or colder) are very important in cloud formation.

Photographs from the NOAA (National Oceanic and Atmospheric Administration), ATS (Applications Technology Satellites), or SMS (Synchronous Meteorological Satellite) satellites, which are orbiting at much higher altitudes, are more useful for observing general cloud patterns. Figure 1-23 shows much of the northern hemisphere, with as-

FIGURE 1-23 These ATS photographs were taken on June 8, 1974, at 10:50 and 14:36 CST (Central Standard Time). Atmospheric vortices of various size are clearly visible in these photographs. Much smaller but very intense atmospheric vortices (tornadoes) are not visible from such photographs but were occurring over Oklahoma and Kansas during this time period. The thunderstorms that produced the tornadoes, however, can be seen on these photographs. (Photographs courtesy of NOAA.)

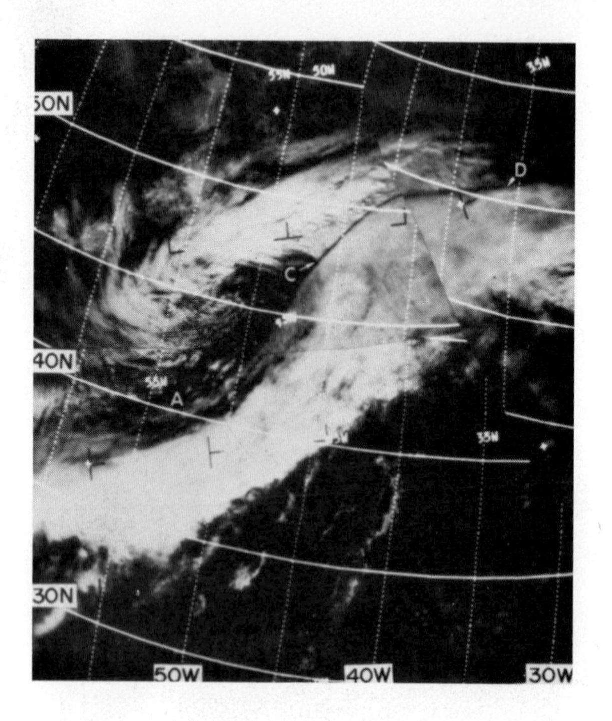

FIGURE 1-24 Midlatitude cyclones appear as comma-shaped cloud patterns, as in this photograph taken by a meteorological satellite, ESSA 3, at 9:02 CST, October 23, 1967. The air is spiraling counterclockwise and slightly toward the center of lowest pressure. Midlatitude cyclones are developed by the jet stream, which can also be identified in this photograph. An extensive cirrus cloud shield often forms along the tropical side of a jet stream core. This edge lies parallel to the jet core and within 150 km of it. This higher cirrus shield frequently casts a shadow on the upper surface of a lower cloud deck, as in this photograph. With the right sun angle it shows up as a shadow marking the tropical side of the jet stream core (also as in this photograph), where the shadow extends from C to D. (From *Direct Transmission Users Guide,* National Environmental Satellite Center, 1969, NOAA.)

sociated cloud formations. Cloud patterns are much larger in area and more homogeneous in pattern than was anticipated before the use of satellites. Traveling midlatitude cyclones are apparent in satellite photographs as large, swirling atmospheric storms.

Space photographs may show storms as small as individual thunderstorms, but they are generally more appropriate for studies of larger storms such as hurricanes and midlatitude cyclones. The appearance from space of a midlatitude cyclone is shown in Figure 1-24. The center of the low-pressure area is near the top of the comma-shaped cloud formation, with the tail of the comma corresponding to the cold front at the surface.

The relationship between surface weather and cloud patterns photographed from a satellite is shown in Figure 1-25. The cloud cover is denser over the surface fronts. Upper atmosphere wave patterns are also related to cloud formation, as shown in this figure. Clouds are more frequent from the trough to the ridge with the eastern edge of the cloud cover frequently indicating the presence of an upper atmospheric ridge.

A hurricane is a smaller vortex storm than a midlatitude cyclone and its appearance from space is more symmetrical. The rain bands spiraling in toward the central core of the hurricane, as well as other characteristics of hurricanes, can be seen on photographs from space (Fig. 1-26). Techniques (which will be described later) are being developed for using satellite photographs to obtain specific information on the nature of the pressure and winds inside a particular hurricane.

Figure 1-27 shows a view from one of the Gemini spacecrafts that includes several individual thunderstorms near Florida. Note the anvils at the top of the largest thunderstorms that are blown downwind. The top of a single thunderstorm viewed from one of the Apollo spacecraft is shown in Figure 1-28. This thunderstorm occurred in the equatorial region over Africa, where winds at the top of the storm were light, giving it the appearance of a huge bubble of rising air.

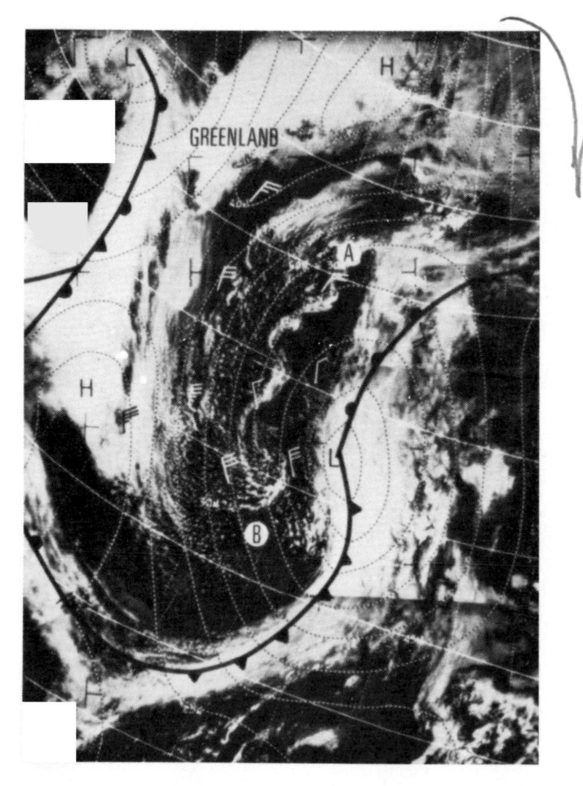

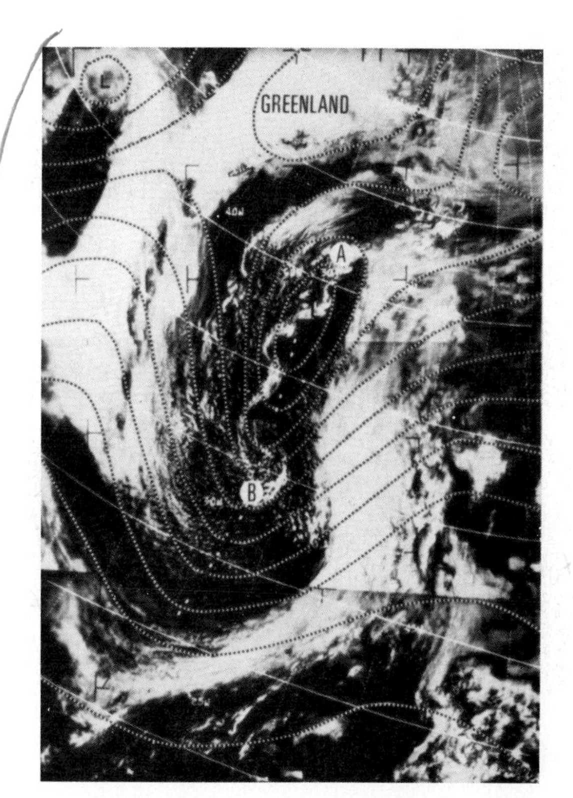

FIGURE 1-25 The nature of cloud patterns provides considerable information on surface weather conditions, as in this photograph taken at 7:52 CST, June 14, 1969, from the meteorological satellite ESSA 9. On the left, the superimposed frontal systems, isobars, and wind directions show the direct relationship between the cloud patterns and these atmospheric conditions. This midlatitude cyclone is in an early stage of development and does not show the extreme comma shape that was shown by Figure 1-24. The streamlines of airflow at 500 mb are superimposed on the same ESSA 9 photograph on the right. A 500-mb ridge can often be located since it corresponds to the eastern edge of cloud bands in the northern hemisphere. Cloud bands and storm activities are usually between the trough and ridge. (From *Applications of Meteorological Satellite Data in Analysis and Forecasting,* Supplement No. 2, National Environmental Satellite Service, 1973, NOAA.)

FIGURE 1-26 This Apollo 9 photograph clearly shows the spiraling bands of clouds associated with a decaying tropical storm. The extensive cirrus cloud shield that ordinarily covers more vigorous storm activity has disappeared, revealing the lower clouds and the spiraling nature of the associated winds. (NASA photograph.)

FIGURE 1-27 This Gemini 5 photograph taken on August 22, 1965, shows large individual thunderstorms over the ocean off the coast of Florida and much more extensive cloud development over the Florida peninsula. (NASA photograph.)

FIGURE 1-28 This Apollo 9 photograph shows the top of a large equatorial thunderstorm as it penetrates into the upper atmosphere under conditions of very light upper atmospheric winds. (NASA photograph.)

SUMMARY

The word *meteorology* comes from the writings of Aristotle several hundred years before the birth of Christ. Modern studies of the atmosphere are called meteorology or atmospheric science. A few of the important milestones in meteorology are given in this chapter, such as the development of meteorological instruments and some of the scientific discoveries that contribute to modern studies.

It is interesting to note that meteorological studies have been revolutionized with the development of computers that can handle large quantities of data and satellites that are designed specifically for meteorological observations. The coming of television in the 1950s has increased the impact of the field of meteorology on society.

An important characteristic of the atmosphere is its temperature profile. Most clouds and weather phenomena occur in the lowest atmospheric layer, called the troposphere, where the temperature decreases with height. This temperature decrease occurs because the absorption of sunlight by the earth's surface heats it. Heat is then transferred upward into the lower atmosphere. The stratosphere is also warmed near 50 km from the absorption of radiation by ozone molecules.

The atmosphere is a mixture of gases, with the three most abundant constituents being nitrogen, oxygen, and water vapor. The water content of the atmosphere varies up to 4% and can exist in all three different states: vapor, liquid, or solid. Latent energy is involved when water vapor changes from one state to another.

Standard atmospheric pressure at sea level is 1013 mb, but frequently varies up to 5%. Winds are related to pressure differently at the surface than in the upper atmosphere. Surface winds blow counterclockwise around low-pressure systems and slightly in toward them; they blow clockwise and slightly outward from high-pressure systems. In the upper atmosphere winds blow along the height contours with very little flow from higher to lower pressure. The average speed of the jet stream is greater during the winter than during the summer.

The amount of water vapor in the atmosphere is very important to weather processes. The specific humidity, mixing ratio, or vapor pressure can be used to specify the amount of water vapor in the atmosphere. The amount of water vapor required to produce saturation of the air increases for warmer temperatures. The percentage of saturation of the atmosphere is expressed by the relative humidity.

The concept of air masses is important in considering the general distribution of large masses of air differing in temperature and humidity. Common air masses are maritime polar, maritime tropical, continental polar, and continental tropical. A cold front occurs at the leading edge of a cold air mass as it replaces warmer air, while a warm front occurs at the leading edge of a warm air mass.

Clouds viewed from satellites provide information on total cloud cover as well as on specific cloud formations. The jet stream may be accompanied by high clouds extending for many thousands of kilometers, while midlatitude cyclones are frequently comma shaped. Satellites have been very useful in identifying hurricanes over the oceans, where observations were previously very inadequate.

STUDY AIDS

1. List the resource materials for all weather phenomena.

2. Describe as many economic, physical, or other effects of weather that you can think of.

3. Explain the reasons for the observed temperature profile from the surface to 100 km.

4. Describe some of the basic differences in the character of the troposphere and stratosphere.

5. If the atmospheric temperature near the earth's surface on a particular day were 25°C and an average atmospheric lapse rate existed, calculate the temperature that would be observed at 5.5 km above the surface.

6. Describe the differences in wind direction and speed at the surface and in the upper troposphere.

7. Explain two ways that air may become lighter and more buoyant.

8. What is latent heat?

9. Keep a record of the type of air masses and frontal passages in your area for two weeks.

10. Explain some of the ways that satellite photographs may be used in weather studies.

TERMINOLOGY EXERCISE

Consult the glossary for more information on any of the following terms, used in Chapter 1, that are still unfamiliar to you.

Atmosphere
Homosphere
Heterosphere
Ionosphere
Ozone
Troposphere
Stratosphere
Mesophere
Thermosphere
Lapse rate
Inversion
Atmospheric pressure
Density
Dalton's law
Buys-Ballot law
Cyclone
Anticyclone
Troughs
Ridges
Jet stream
Sublimation
Latent heat
Specific humidity
Mixing ratio
Saturation
Condensation
Deposition
Vapor pressure
Relative humidity
Air masses
Source region
Cold front
Warm front
Occluded front
Stationary front

THOUGHT QUESTIONS

1. Try to imagine not having any of the current meteorological instruments. Suggest new ways of measuring temperature, humidity, and pressure. Does this give you any insight into the obstacles faced by people who lived several centuries ago?

2. Do you think the composition of the atmosphere is changing? If so, how?

3. If water vapor is lighter than air, why doesn't it escape to outer space and diminish the total supply of water on earth?

4. Why do air masses form weather fronts instead of thoroughly mixing at their boundary regions?

5. Describe ways in which satellite photographs can be useful for weather studies.

SPECIAL TOPIC 1

UNDER THE WEATHER

Some of the effects of weather on people can be very subtle. Evidence is accumulating, for example, that some people are sensitive to the large concentrations of positive ions produced by thunderstorms. The inhalation of positive ions increases the production of serotonin—a powerful hormone that acts in the midbrain on sleep, nerves, and mood—and thus people who are sensitive to positive ions may experience headaches, irritability, and sleeplessness as large thunderstorms move overhead.

Another effect of positive ions may be the increased production of adrenaline. This may cause a person to experience a burst of energy, but the excessive adrenaline eventually leads to exhaustion.

A person's skin is affected by the relative humidity. It expands when the humidity is high and contracts when the humidity is low. Since scar tissue has less water content and is less flexible, low humidity may cause large scars to become painful.

Still another effect of weather on some people arises because broken bones that have healed may still be sensitive to atmospheric pressure changes. Also, persons affected by arthritis may experience a swelling of the joints and pain when atmospheric pressure decreases. It is not surprising, then, that the expression, "I'm feeling under the weather today," is very common.

SPECIAL TOPIC 2

ATMOSPHERES OF OTHER PLANETS

The nature of the atmosphere and climate of other planets has continued to stir the imagination of earthlings. The space program of the United States has included major efforts to answer some of the questions about the atmospheres of some of the other planets. Data from the Pioneer, Mariner, and Viking spacecraft have provided detailed observations from other planets for the first time. Observations from the Pioneer spacecraft, for example, indicate that the Great Red Spot of Jupiter is similar to a hurricane on Earth except that its lifetime is several decades and its size is several times larger than the Earth.

Since the ozone layer is important in providing an environment suitable for animal life, the presence of such a layer on other planets is also of interest. Spacecraft observations made by Mariner instruments detected ozone near the winter pole of Mars but could not detect it elsewhere on the planet.

The composition of the martian atmosphere is primarily carbon dioxide (95%) as shown in Table S2-1. Nitrogen and argon are also present in concentrations greater than 1%. The martian atmosphere is quite thin by earth standards, with a surface pressure of only about 6 mb. The atmosphere of Venus, in comparison, contains much more mass than the Earth's atmosphere, with a surface pressure 90 times that of Earth. The atmosphere of Venus, like that of Mars, is composed largely of carbon dioxide.

The winds on Mars cause major dust storms because of the lack of liquid water to hold the dust particles together. The dust has been identified as silicates and is carried to altitudes of 50 km above the surface. Sandblasting at the surface is apparently the main erosion mechanism.

The polar ice caps on Mars have been a source of fascination since they were first observed from Earth by telescope. The north polar ice cap is larger than the southern one, and a dark band surrounds it as it shrinks in the spring. This was previously thought to result from the melting of water ice. Viking measurements have shown, however, that the temperatures are too low for water to exist as liquid, indicating that the caps are composed of carbon dioxide ice.

Clouds in the martian atmosphere (Fig. S2-1) are composed of dust, water ice, or carbon dioxide ice. Carbon dioxide clouds apparently predominate in polar regions, water ice clouds in midlatitudes, and dust clouds elsewhere. The clouds on Venus, however, are composed of sulfuric acid.

All the information obtained from other planets indicates that a trip to Mars or Venus would not be nearly as romantic as science fiction has made it appear. The tremendously large atmospheric pressure on Venus with scorching temperatures beneath the sulfuric acid clouds would not seem to make the planet an inviting place to live or take a vacation, while the very low atmospheric pressure on Mars with frigid temperatures and choking concentrations of dust make this a very unfriendly environment beneath the dry ice clouds.

TABLE S2-1

Comparison of the atmospheres and other physical properties of Venus, Earth, and Mars (after NASA SP-334)

Physical property	Venus	Earth	Mars
Atmospheric composition (%)			
Carbon dioxide	97	0.03	95
Nitrogen	2	78.09	3
Argon	—	0.9	1.5
Water vapor	0.0001	1–4	0.01–0.1
Oxygen	—	20.9	0.15
Carbon monoxide	0.0002	trace	0.6
Distance from sun (millions of km)	107.8	149.6	227.8
Number of earth days per year	225	365.25	686.98
Inclination of axis of rotation (deg)	10	23.45	24.94
Day length (h)	243	24	24.62
Surface atmospheric pressure (mb)	90,000	1013	6.0
Average temperature (°C)	450	15	−62
Clouds	Sulfuric acid	Water, ice	Carbon dioxide, water

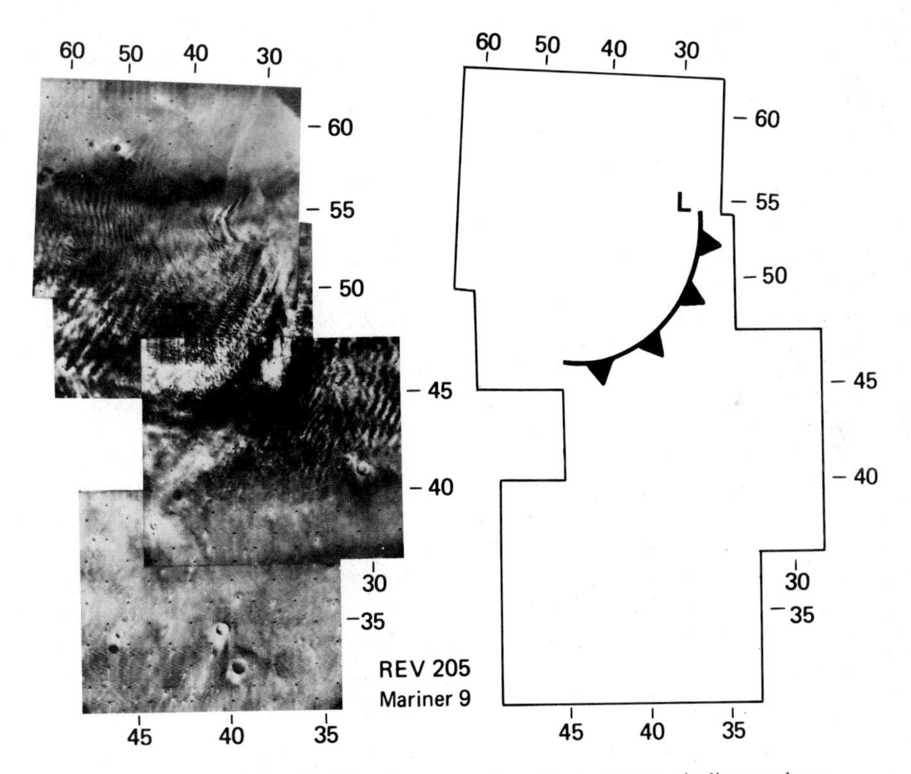

FIGURE S2-1 Cloud cover within the Martian atmosphere that seems to indicate a low-pressure system and cold front. (Photograph courtesy of Jet Propulsion Laboratory.)

PART TWO

FUNDAMENTALS OF WEATHER AND CLIMATE

CHAPTER TWO
WEATHER OBSERVATIONS, SYNOPTIC CHARTS, AND FORECASTS

INTRODUCTION

The most important practical application of meteorology is weather forecasting. Although it is not the only important application, weather forecasts are thoroughly intertwined with our everyday activities. In addition to this application of forecasting, many industries are willing to pay private meteorologists for forecasts related to their special situation, whether it be a commodity exchange, construction company, or one of a large number of other weather-sensitive industries.

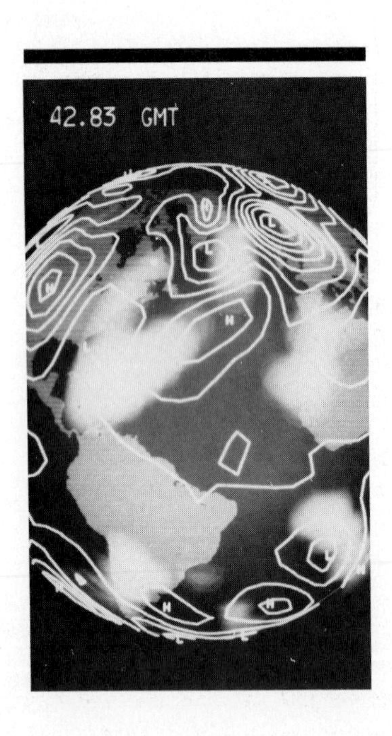

WEATHER OBSERVATIONS

Surface Data In order to understand the atmosphere and to forecast future weather conditions, many different measurements must be obtained. Observations of the state of the atmosphere are made at a large number of appropriately located stations by the National Oceanic and Atmospheric Administration (NOAA). More than 200 synoptic and basic weather stations in the United States provide weather data (Fig. 2-1). The term **synoptic** means that meteorological data are obtained over a large area at a specified instant of time. Some of the weather stations are less than 50 km apart, while others are almost 200 km from any other. Observations of atmospheric temperature, humidity, pressure, precipitation, wind and sky cover, type of clouds, and visibility are recorded.

The synoptic observation program provides measurements at the world standard synoptic times, 0000, 0600, 1200, and 1800 GMT, with reports transmitted throughout the world (Fig. 2-2).

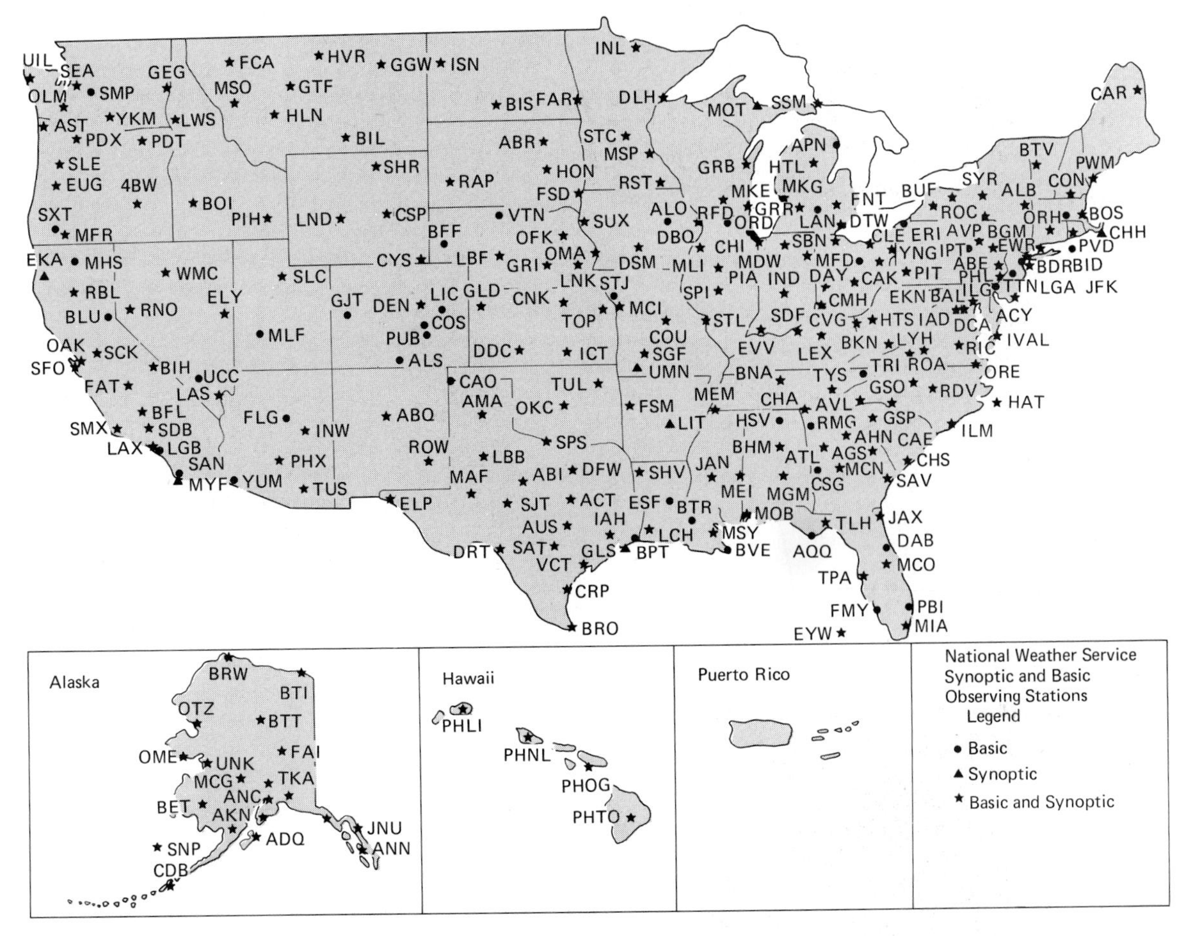

FIGURE 2-1 Locations of all the NOAA synoptic and basic weather stations where meteorological data are collected to serve forecast programs, and to provide data for international exchange as well as for all other uses. Observations on the amount of sky cover, type of clouds, wind, visibility, weather, temperature, dew point, and pressure are made at world standard synoptic times, 0000, 0600, 1200, and 1800 GMT. The basic observing stations also provide data at hourly intervals and include cloud height and altimeter settings for aircraft operations. (From *Operations of The National Weather Service,* NOAA.)

Greenwich Mean Time (GMT) is converted by subtracting five hours to obtain EST, six hours to obtain CST, and so on. The time may also be labeled Zulu (Z), which is another designation of Greenwich Mean Time or World Standard Time. The basic observation program provides data at hourly intervals with special reports of significant weather changes in between. These special reports contain the same atmospheric measurements as those from the synoptic stations, but also include cloud heights and altimeter settings for aircraft.

In addition to the synoptic and basic weather stations, the United States has numerous aviation weather and cooperative weather observing stations. A single state, for example, may have more than 300 such stations providing detailed coverage

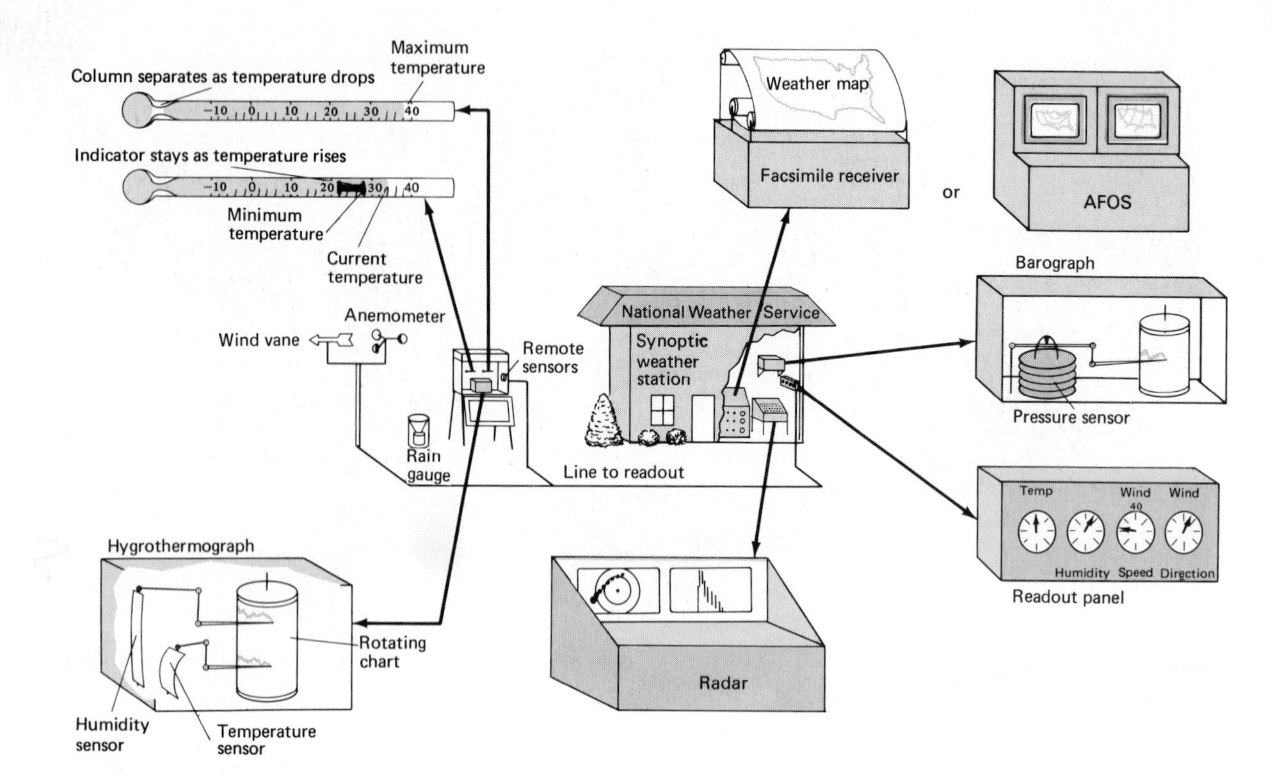

FIGURE 2-2 The instruments in a synoptic weather station include a maximum and minimum thermometer, anemometer, and wind vane, and recording instruments such as a hygrothermograph for temperature and humidity, and a barograph for pressure. Remote sensors may also be exposed in an outdoor shelter with readouts inside the station. Other pieces of equipment include a radar system, an AFOS (Automation of Field Operations and Services) computer-based system for sending and receiving data and weather-map receiving equipment.

of certain atmospheric conditions. However, the data gathered at these stations is not as useful as that from the synoptic stations, since most of them obtain temperature and precipitation data only twice daily instead of hourly or continuously. Observations at such stations are sent daily to a nearby synoptic weather station to be transferred to the National Meteorological or Climatic Centers. These numerous land stations are supplemented by meteorological measurements from ships, ocean buoys, and data from other countries to gain a more complete picture of synoptic weather conditions.

Upper Atmospheric Data Information on the upper atmosphere comes from another network of stations. Certain synoptic and basic weather stations are also designated upper air stations. In the

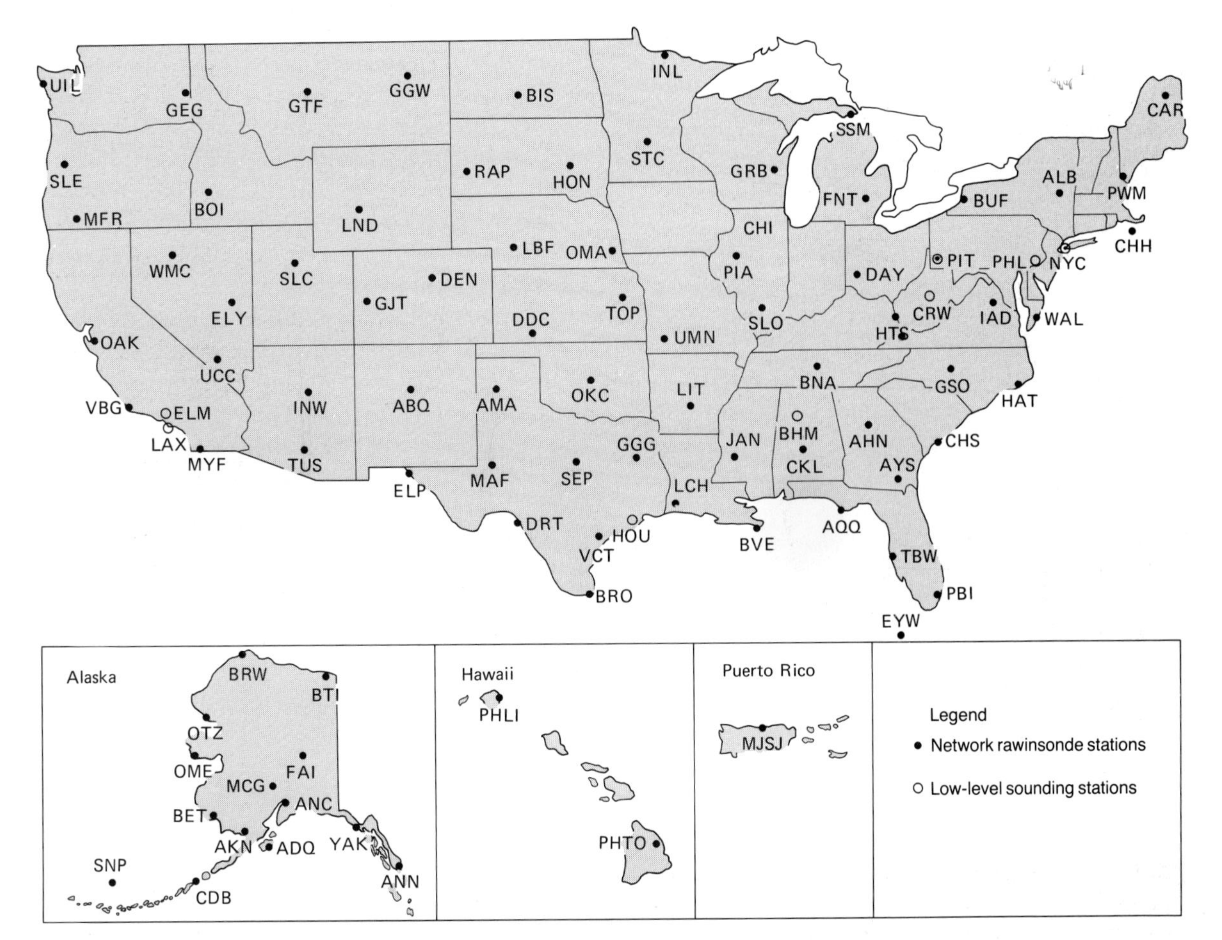

FIGURE 2-3 The location of rawinsonde stations, where observations of pressure, temperature, moisture, and winds aloft are made. These stations provide information on the third dimension (various heights) to supplement the horizontal distribution of measurements of the state of the atmosphere at 0000 and 1200 GMT. (From *Operations of the National Weather Service*, NOAA.)

United States, there are only 93 such stations (Fig. 2-3), thus the distance between measurement sites for upper air data is much greater than for surface weather data. Each upper air station sends up a balloon twice each day at the world standard times of 0000 and 1200 GMT with an instrument package that measures temperature, humidity, and pressure. These data are then transmitted back to the weather station.

The instrument package that measures and transmits the upper air data is called a **radiosonde** (Fig. 2-4). The helium-filled balloon carries the instruments to a height of perhaps 30 km, where the balloon bursts and the instruments are carried back to the ground by parachute. If the package is tracked by optical or other means, the wind speeds at various levels are also obtained. The instrument system is then called a **rawinsonde.**

FIGURE 2-4 The radiosonde consists of temperature, humidity, and pressure sensors that gather data, and a radio to transmit the information back to the weather station. The instruments are carried by balloon from the surface through the atmosphere. (Photograph courtesy of NOAA.)

Data Handling Data from all the synoptic and basic weather stations go by high speed communication lines directly to the **National Meteorological Center** in Suitland, Maryland, where they are used to develop current weather maps, also called **synoptic charts,** and forecasts. Within 90 minutes, analyzed surface weather maps based on the combined data go out from the National Meteorological Center to be received by weather stations, TV or radio stations, universities, and other owners of facsimile weather map receivers or other data receiving and display systems. Data from the synoptic and basic stations are also used for making 12-, 24-, 36-, and 48-hour forecasts, as well as 3- and 5-day outlooks that go to this same group of users.

The **National Climatic Center,** located in Asheville, North Carolina, has the responsibility of storing and publishing weather data. The most detailed weather information is published as *Local Climatological Data,* which contains observations made every 3 h at the synoptic and basic weather stations. Daily temperature and precipitation for these and many other weather stations are published in *Climatological Data,* a monthly publication for each state. Daily temperature and precipitation records for each of the synoptic and basic stations are also published in *Climatological Data, National Summary,* at monthly intervals. Weather maps based on 1200 GMT observations are published weekly as *Daily Weather Maps.*

In spite of the enormous quantity of weather data, the coverage provided by the 200-station network is still not fine enough for forecasting local weather events. If the distance between weather stations is 150 km, for example, no one can make accurate forecasts for weather disturbances much smaller than this; yet, as we all know, the weather can vary considerably between two towns only 10 km apart.

Providing data on the initial state of the atmosphere, against which changes can be measured, is the major purpose of these numerous weather stations.

ANALYSIS OF DATA

The next step after making observations is to analyze and interpret them. This means assembling and displaying the data in a matter that reveals the weather-producing features of the atmosphere. One way to accomplish this is to construct a three-dimensional display of the atmospheric data (Fig. 2-5) that represents several horizontal slices through the atmosphere. For such a display, mete-

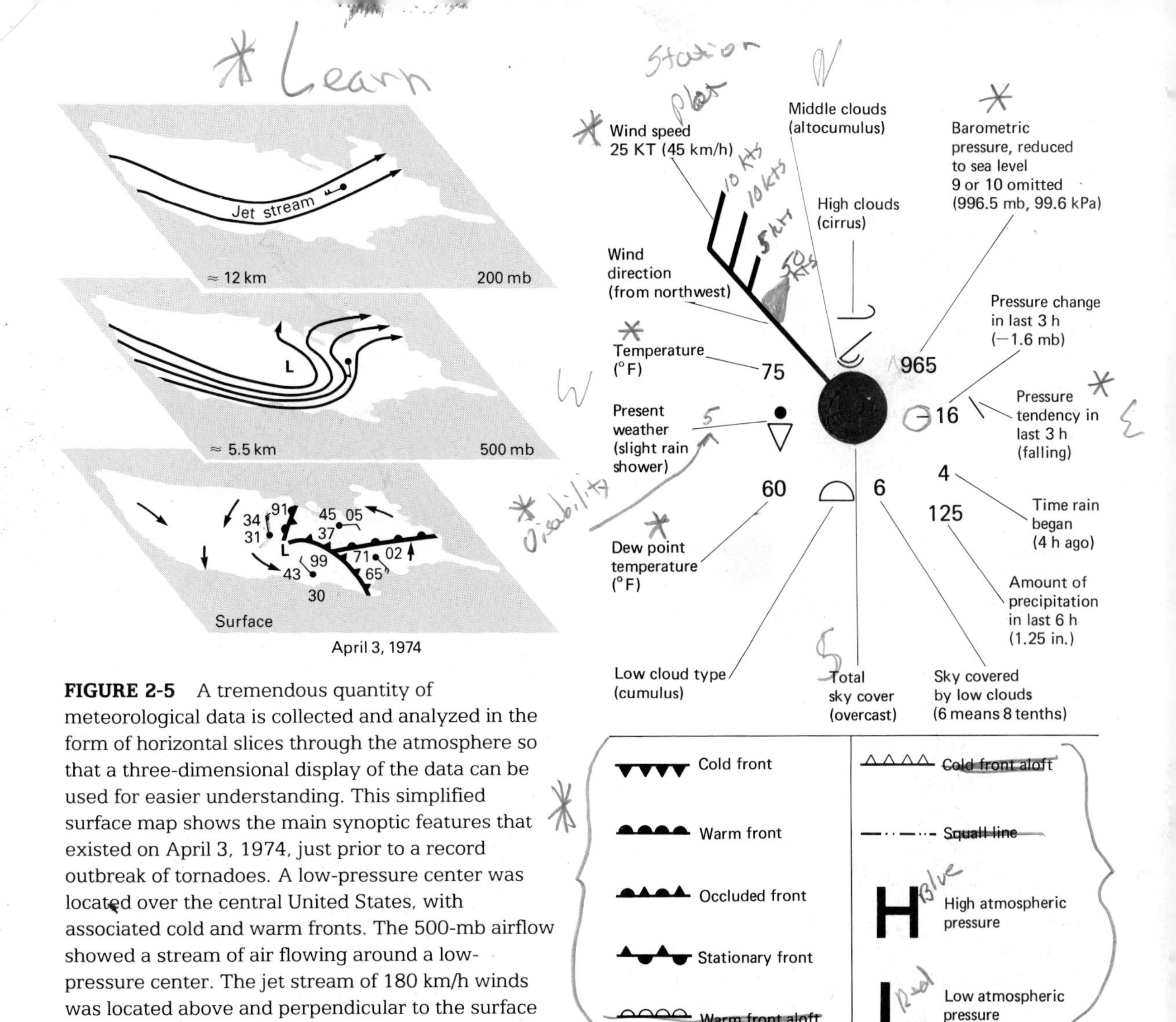

FIGURE 2-5 A tremendous quantity of meteorological data is collected and analyzed in the form of horizontal slices through the atmosphere so that a three-dimensional display of the data can be used for easier understanding. This simplified surface map shows the main synoptic features that existed on April 3, 1974, just prior to a record outbreak of tornadoes. A low-pressure center was located over the central United States, with associated cold and warm fronts. The 500-mb airflow showed a stream of air flowing around a low-pressure center. The jet stream of 180 km/h winds was located above and perpendicular to the surface frontal system.

FIGURE 2-6 Surface weather data are plotted on the synoptic map according to this model. The amount of sky cover is shown in the circle indicating the observation point on the map. The various other weather data are plotted in specific positions around the circle. Symbols for various types of fronts, squall lines, and atmospheric pressure are also shown.

orologists obviously need information not only for the earth's surface but also for higher layers of the atmosphere.

Making Synoptic Charts The horizontal distribution of the surface weather variables is displayed by creating a synoptic weather map for the surface by plotting the observed data according to a consistent format. The station plotting model is shown in Figure 2-6 and in more detail in Appendix C. The temperature is plotted in the upper left, current weather in the left center, and the dew point temperature in the lower left from a circle that indicates the station location. The atmospheric sea-level pressure is plotted in the upper right, the **pressure tendency** (change in pressure in the last three hours) in the right center location, and the amount of precipitation in the lower right. The wind is plotted as a flag so that a south wind of 35 km/h (20 knots), for example, will extend toward the bottom of the map with two flags (one for each 10 knots). When such information is plotted for all weather stations, the geographic distribution of the weather variables can be determined.

The vertical dimension of pressure variations is provided by measurements from the atmosphere's upper region, birthplace of some of the winds that influence surface weather. The most used upper atmospheric level is the 500-mb pressure, usually located about 5500 m (18,000 ft) above sea level, but ranging from 4900 to 5900 m.

In order to display the upper atmospheric pressure distribution, the height in meters of the 500-mb pressure is used rather than the pressure at a constant height. For a given air temperature, a low altitude of the 500-mb pressure level means the pressure is also low at that height (Fig. 2-7). The hydrostatic equation (Appendix A) tells us that for a constant density of air, changes in pressure are directly related to changes in height. Thus, the height of the 500-mb pressure is plotted and used to express pressure variations. A constant pressure level can be considered as a surface comparable to a topographic surface, with the mountains corresponding to higher atmospheric pressure and the valleys to lower atmospheric pressure.

Since the jet stream usually travels at the 200-mb level, averaging 12 km (7.2 mi) above sea level, the height of this constant pressure surface is also used. At least two others, the 850- and 700-mb levels are plotted as well, with average heights of 1500 and 3000 m, respectively.

After the surface and upper air data are plotted, they must be analyzed. Several different analyses of the surface data are completed, including isobaric, frontal, and isallobaric analyses, in that order.

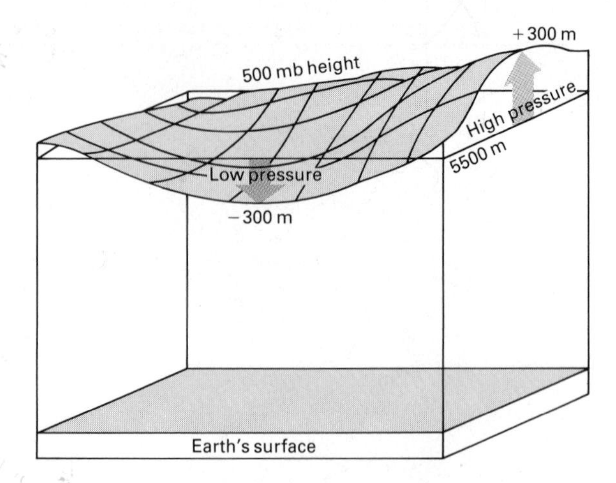

FIGURE 2-7 The distance up to the 500-mb pressure level is used to show horizontal pressure variations instead of using actual pressure differences at a particular height. If the distance to the 500-mb level is less than at another location, the pressure at that level is also less in direct proportion to the height difference.

Isobaric Analysis An **isobaric analysis** consists of drawing lines of constant pressure on the surface synoptic chart by considering the plotted sea-level pressure at each of the weather stations. Lines of constant pressure (isobars) are drawn by starting with a certain pressure, for example, 1000 mb, and drawing a line corresponding to this pressure between all plotted pressure values. On one side of this line would be a pressure lower than 1000 mb and on the other side a pressure higher than this value. When the 1000-mb line has been completed, isobars are drawn for 1004, 1008, 996, 992 mb, and so on, for every 4-mb interval above and below 1000 mb up to the highest pressure and down to the lowest pressure on the synoptic chart. When the entire isobaric map has been completed, areas of high and low pressure will be revealed, such as those in Figure 2-8, and these, in turn, give general information on the weather in their corresponding locations.

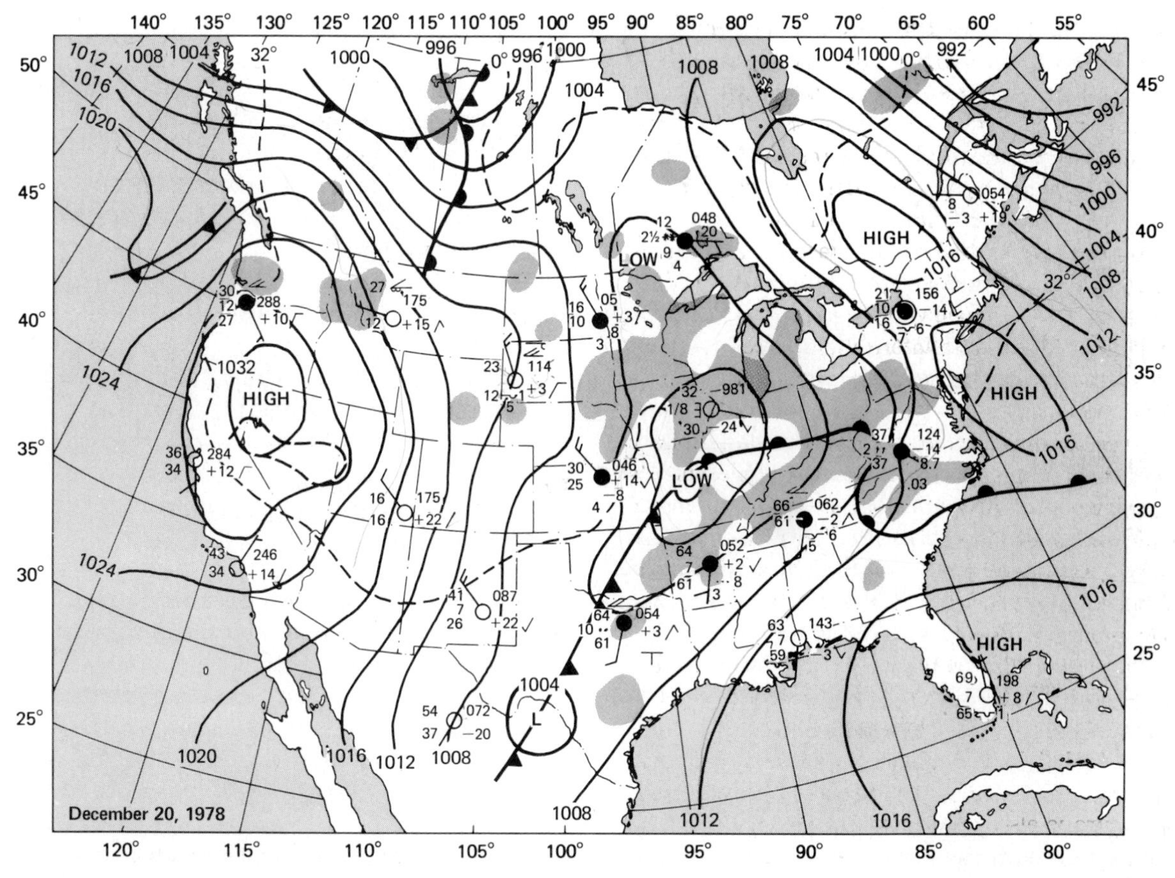

FIGURE 2-8 The surface synoptic map for December 20, 1978. A center of low pressure was located in the central United States, with a cold front extending for a great distance toward the southwest. A high-pressure system was located over the western United States. On this date much of the eastern United States experienced cloudy and rainy weather, while the western United States enjoyed fair weather. Such a synoptic map is drawn by plotting specific weather information for the synoptic observing network.

Frontal Analysis The second step, a **frontal analysis,** consists of locating cold fronts, warm fronts, and stationary fronts on the synoptic weather map. Weather fronts can be located in several ways: one obvious method is to look for contrasts in temperature. The region southward from a low-pressure center should be inspected for a cold front, as a cold air mass is usually brought southward behind (to the west of) a low-pressure center. The region to the east and southeast of a low-pressure center should be inspected for a warm front, which may also be located by means of a temperature contrast between two neighboring weather stations.

Another indication of a weather front is a **wind shear line.** If the winds are from the southwest in one location and from the northwest at a nearby station, the opposing winds exhibit shear and the presence of a cold front is indicated. If a line can be drawn separating a general geographic area with east winds from another with southwest winds, a warm front is indicated. Such a wind shift line, as well as a distinct temperature contrast, exists across a well-developed weather front, and either characteristic may be used to locate the position of the front (Fig. 2-8). Since weather fronts are frequently weak, both characteristics are usually used in developing a complete frontal analysis.

Isallobaric Analysis A third type of synoptic analysis, the **isallobaric analysis,** is also quite useful. The pressure change within the last three hours is plotted on the surface weather map as a basis for the isallobaric analysis, which consists of drawing contours representing lines of constant pressure change (isallobars). This analysis then shows locations where the pressure is changing most rapidly. This characteristic of the atmosphere is important; for example, if the pressure is dropping, a low-pressure center is approaching. If the pressure is rising, fair weather is indicated with a coming high-pressure center.

The pressure tendency is plotted as a number (1.0 mb is plotted as 10) preceded by a plus or minus sign, which indicates whether the pressure has increased or decreased over the last three hours. The number is followed by a line that may slant up, down, or be a combination of these to show the current direction of pressure change. If the line has the appearance of a check mark, it indicates that the pressure dropped during the first part then rose during the last part of the 3-hour period.

The isallobaric analysis is begun in a manner similar to the isobaric analysis: a certain pressure tendency, −10, for example, is chosen and a line is drawn between stations so that those with greater pressure tendency are separated from those with a smaller pressure tendency. This isolates the area where the pressure is dropping fastest. If this area is near an existing low pressure, it indicates that the center of low pressure is moving in that direction. As we shall see, the movement of low-pressure systems is an important part of weather forecasting. In very simple terms, a **low-pressure system** is an area of low atmospheric pressure where less air is above the surface at that point. If the pressure is lower at the surface, the surrounding air will circle in toward the center of lower pressure. This causes the air to rise above the low pressure and generate cloudy weather with precipitation. In contrast, a **high-pressure system** contains subsiding air that leads to clear skies and warm temperatures.

After the surface analysis has been completed and the highs and lows have been located along with the fronts and areas of greatest pressure change, the main characteristics of the surface weather have been specified for the particular time of these observations. However, it is impossible to forecast the surface weather for any future time without information on the characteristics of the atmosphere above the surface.

Upper Air Analysis and Use After the data have been plotted for each upper air station showing the temperature, height of the 500-mb pressure, and wind direction and speed, lines of constant height, called isohypses, are drawn connecting the points of equal 500-mb heights, just as isobars were drawn on the surface map. Isohypses are drawn for each 60-m (5400, 5460, and so on) or each 200-ft interval (18,100, 18,300, and so on) until the upper air synoptic chart is completed. Since the winds above the surface blow along the lines of constant height, which are also lines of constant pressure, if a **streamline analysis** were performed, by connecting stations with lines oriented along the wind direction flags, a result similar to the **isohypse analysis** would be obtained. Thus **streamlines,** corresponding to the path a parcel of air would follow as it moved in the atmosphere, show airflow around low-pressure areas with large cyclonic bends and around high-pressure areas with anticyclonic bends, as indicated by the 500-mb heights (Fig. 2-9).

The weather associated with a 500-mb ridge (anticyclonic curvature) is quite different from that for a trough. Locations under a trough are more likely to have precipitation and cloudy weather along with the entire region from the trough to the ridge (Fig. 2-10). Fair weather is much more likely

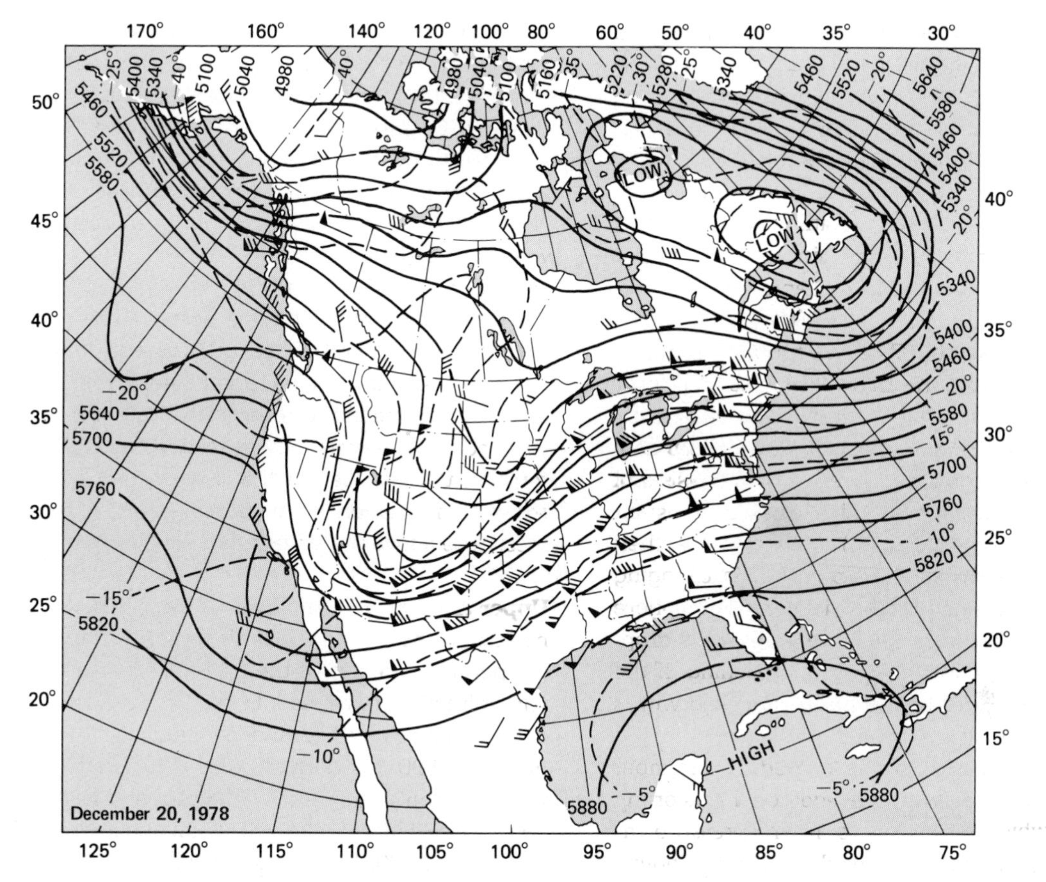

FIGURE 2-9 The 500-mb map for December 20, 1978, corresponding to Figure 2-8. The 500-mb height contours (in meters) and winds flow around the areas of higher and lower pressure.

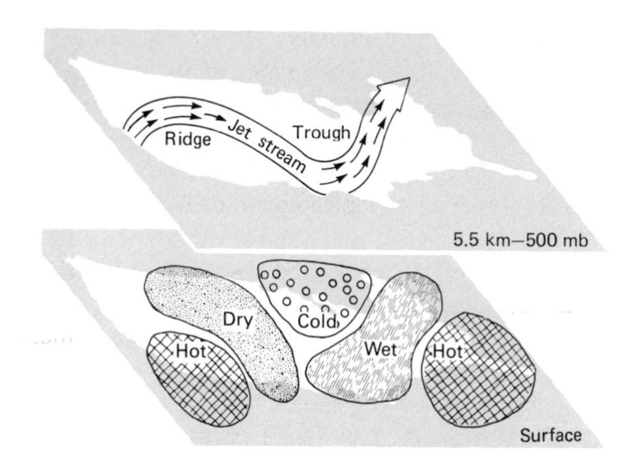

FIGURE 2-10 Upper atmospheric conditions are very influential in determining surface weather patterns. Regions located beneath southwesterly upper air streams are generally wetter than normal, while locations beneath northwesterly upper air streams are typically drier than normal. The general temperature distribution is also related to the upper air stream. Surface weather is related to specific regions of the upper air stream because many atmospheric storms originate only in certain portions of the upper air patterns.

under the ridge and in the whole region downwind from the ridge to the next trough.

The horizontal flow of air above the surface within the layer from 500 to 200 mb (5.5 to 12 km) is very influential in determining the surface weather. Abnormally dry weather, such as occurred in the central United States in 1980, is typically associated with an upper atmospheric ridge formed as the jet stream meanders in a general direction from west to east above midlatitudes. The jet stream is such a large stream of air that the 500-mb winds are ordinarily in a similar direction to the winds at 200 mb, where the center of the jet stream with highest wind speeds is usually located. Because of this the airflow at 500 mb can normally be used to locate ridges and troughs associated with the jet stream. Major changes in surface weather occur as the jet stream patterns and 500-mb patterns change. Just as droughts occur in locations from the ridge to the trough, above-normal amounts of precipitation are likely at surface locations between the trough and ridge.

WEATHER FORECASTS

Once the meteorological data have been obtained and various surface and upper air analyses have been performed, the next step is predicting the future weather. This is accomplished by using one or more of the three different types of forecasts: *persistence, meteorological,* and *climatological* forecasts.

The persistence forecast is good for only a very short time, usually six hours or less. Persistence forecasts are based simply on past history. The current location of a low-pressure center is noted as well as its past movement. The same intensity, speed, and direction as in the past is used for future projection. The persistence forecast is simply a projection of the past history of the storm into the future. Since no factors are considered that may change the course or intensity of the storm, the forecast is frequently inaccurate.

Meteorological forecasts are made from various rules and equations using computers as well as the experience of forecasters to develop an improved forecast that may be for six hours or as much as five days in the future. This is the type of weather forecast or prognosis that is most common.

The third type, the climatological forecast, is based on upper air patterns, cycles, statistics, or sunspots and is used to give variations in the weather over an extended period of time. These result in a general forecast of future weather conditions a month, year, or several years away. The meteorological and climatological forecasts are important activities of the National Weather Service and many professional meteorologists. Therefore, they will be considered further.

Meteorological Forecasts Meteorological forecasts are the most accurate of the various types of forecasts and are of primary concern since they are made for 12 hours to 48 hours in the future, with less-detailed predictions made for three, four, and five days in the future. A good forecast can normally be developed for 6 to 12 hours by an experienced forecaster simply by inspecting the surface and the upper air synoptic charts. An experienced forecaster has seen most of the different types of storms and is aware of the associated upper air patterns. Most surface weather systems are related to the upper air streams that guide the storms below them. For this reason, weather systems in midlatitudes come from the west or southwest. A simple forecasting rule for projecting the center of a low-pressure system, located beneath the jet stream, is to use a forward speed of 50% of the speed of the 500-mb winds, since these can be expected to steer the surface low-pressure system along in the same direction as that of the jet stream.

Numerical weather prediction is a technique used by the National Meteorological Center to make a more accurate forecast. This technique involves the use of mathematical equations that relate the temperature of the air, wind, speed, pressure, and other atmospheric variables. The mathematical equations relating these variables are used by starting with the known state of the atmosphere, for example, the wind, temperature, and pressure. The equations then project these atmospheric variables into the future ten minutes, for example. The state of the atmosphere ten minutes after the measurements is then used to project another ten minutes into the future and so on (Fig. 2-11). Numerical weather forecasts require high speed computers, since a 24-hour prediction for North America requires half a billion calculations based on several complex equations.

The numerical model of the atmosphere used by the National Meteorological Center is called the limited fine mesh (LFM) model. LFM charts are prepared in four panels (Fig. 2-12) for 12, 24, 36, and 48 hours in the future. Each of the four panels shows two variables, one with solid lines and the other with dashed lines. These are 500-mb heights (solid) and vorticity; surface pressure (solid) and 1000- to 500-mb thickness; 700-mb heights (solid) and relative humidity; and the quantitative precipitation forecast (solid) and 700-mb vertical velocity (positive is upward). Some of these variables are probably unfamiliar to you, and require explanation. **Vorticity** is a word used to describe the amount of circular motion in the atmosphere. A tendency of the winds for circular motion near the jet stream occurs in the same way as circular eddies occur in a deep river flowing downhill. Cyclonic vorticity is denoted by positive numbers and anticyclonic vorticity by negative numbers. Numbers greater than 16×10^{-5} radians per second are considered to be strong circular motion. Another variable, the 1000- to 500-mb thickness, is useful since the thickness between these pressure levels is related to the temperature of the air, with warm air causing a greater thickness.

All eight variables are extensively used by forecasters. If they were always perfectly accurate, the forecaster's job would be reduced to the task of reading a map. However, there is considerable room for the use of a forecaster's experience and knowledge because, for several reasons, such numerical models are not always completely accurate.

One type of inaccuracy arises because the equations required for a numerical forecast are simplified. This simplification causes some inaccuracy after a day or so of projection. Forecasts also lose

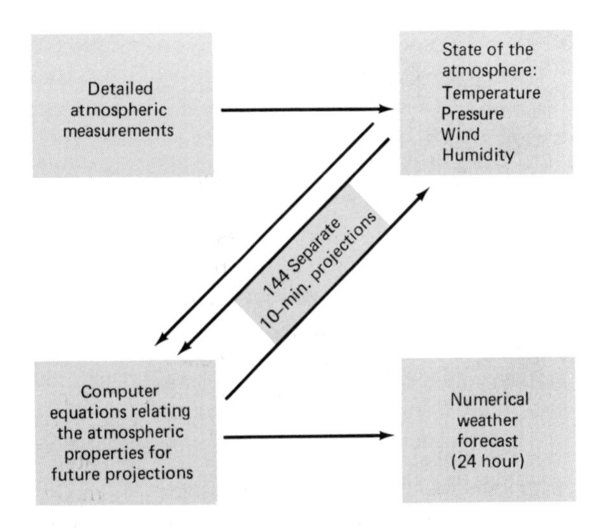

FIGURE 2-11 Numerical weather forecasts for 24 hours involve computer analyses of the equations describing the atmosphere, using measured data on the initial state of the atmosphere and many separate projections and evaluations of the state of the atmosphere.

some accuracy because of incomplete coverage of initial weather observations, since only data from the synoptic and basic weather stations are fed into the computers.

The distance between the synoptic weather stations is about 150 km so they can provide details on weather events that occur only within the **macroscale** (100 to 1000 km). A **mesoscale** (1 to 100 km) event such as a thunderstorm is much smaller than the distance between synoptic stations. Therefore, a single large thunderstorm 30 km across could not be expected to be forecast from the synoptic weather station coverage.

Meteorological forecasts are issued for a variety of weather elements such as the maximum and minimum temperatures, humidity, wind, cloud cover, and probability of precipitation. The **precipitation probability** represents the certainty of rainfall at a particular point (Fig. 2-13) and is forecast for 12-hour periods up to 48 hours in advance by

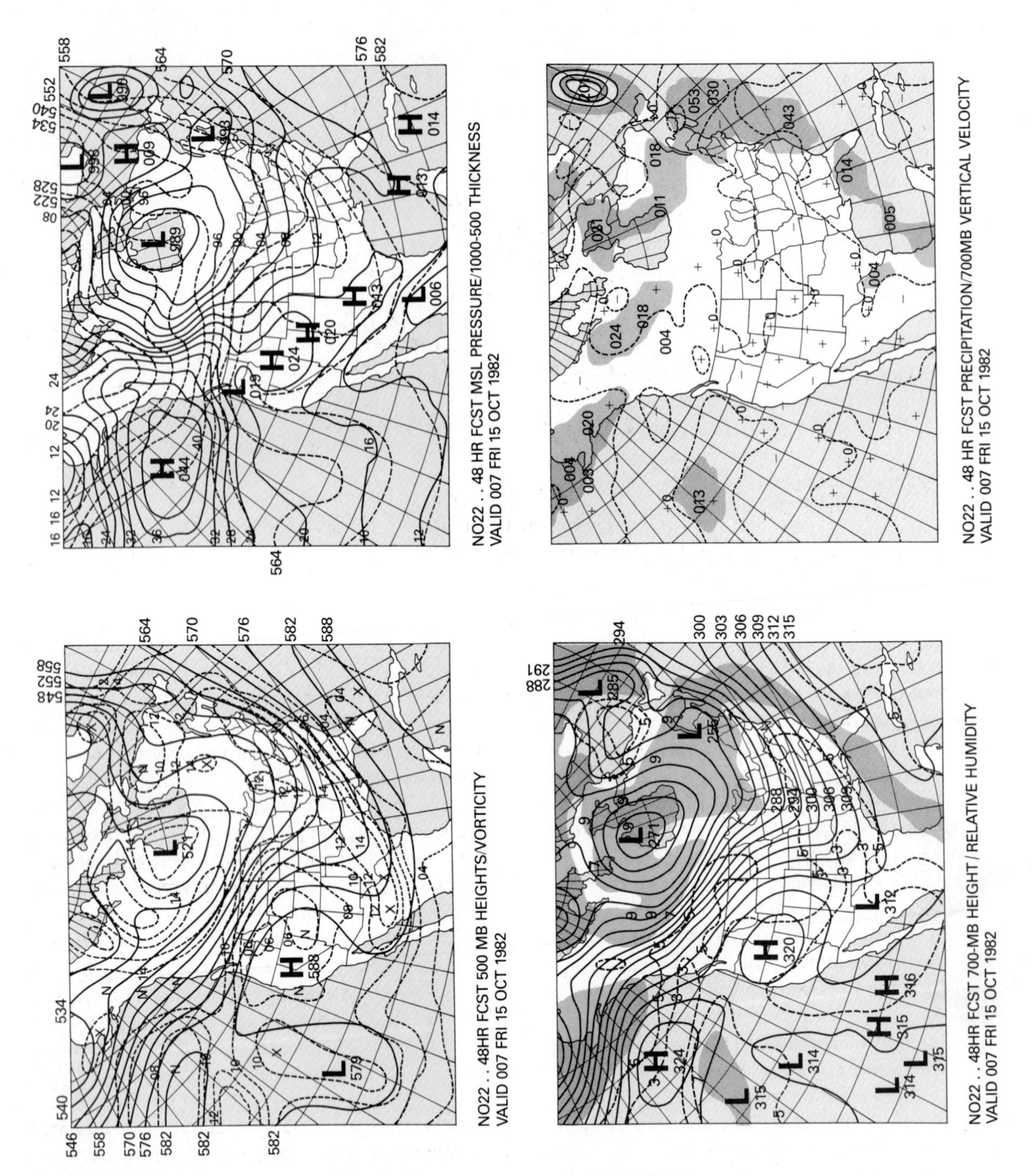

FIGURE 2-12 Numerical weather forecasts are made by the National Weather Service in the form of four-panel LFM charts. These are produced for 12-hour intervals up to 48 hours in the future. (NOAA.)

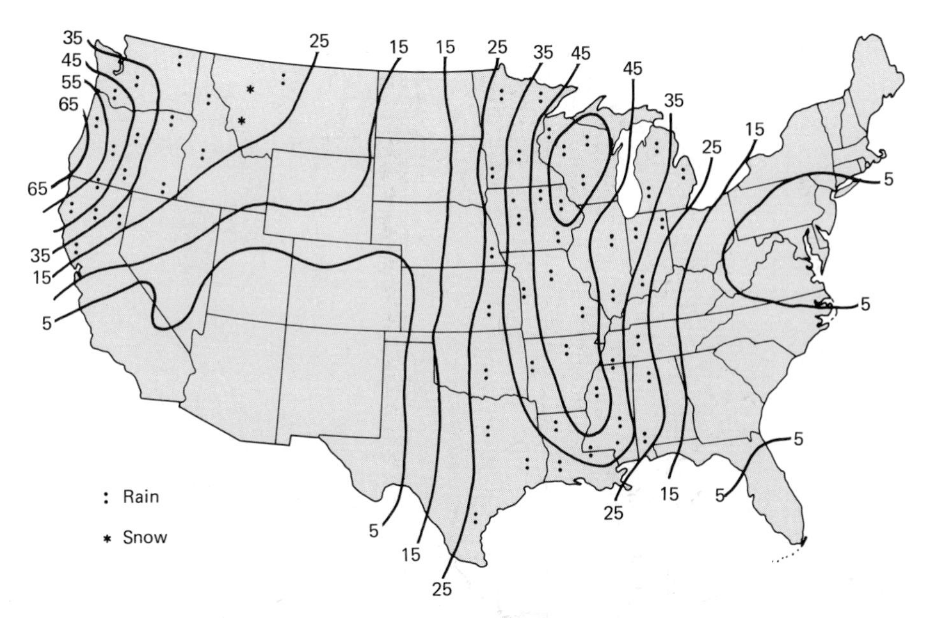

FIGURE 2-13 Precipitation probability forecast for 0000 GMT April 21, 1979, issued 48 hours before the date. Such forecasts as the type and distribution of expected rainfall are also prepared by the National Meteorological Center for 12, 24, and 36 hours.

the National Meteorological Center. If rainfall has occurred in the past on half the occasions when atmospheric conditions were similar to the present situation, the precipitation probability is 50%. Therefore, the precipitation probability is a useful way of making a forecast more representative of the nature of rainfall variability.

Meteorological forecasts issued for the longest period of time are the 3- and 5-day outlooks (Fig. 2-14). As you might guess, these are subject to more error than forecasts for shorter time periods. For this reason, the information contained in these forecasts is more general than in shorter forecasts.

Climatological Forecasts Climatological forecasts for longer periods of time, normally greater than a week, can be made in several different ways. These forecasts are less accurate in general because of the length of time involved. Thirty-day forecasts (called average monthly weather outlooks) are prepared by the Long Range Prediction Group of the National Meteorological Center. These give the expected precipitation, above or below normal, for the next 30 days, as well as the expected temperature, above or below normal (Fig. 2-15).

The procedures used in generating the predicted monthly temperature and precipitation patterns involve construction of charts of expected monthly upper air circulation around the northern hemisphere. The corresponding surface weather is then inferred from the upper air circulation. Although the analysis is based on meteorological observations from weather stations over most of the continents in the northern hemisphere, the incomplete coverage and extrapolation of upper air circulation to surface weather events results in considerable uncertainty. However, the outlooks usually provide a good indication of the general weather for at least a couple of weeks ahead.

Extended forecasts can also be developed from statistics of past weather patterns. Some indication

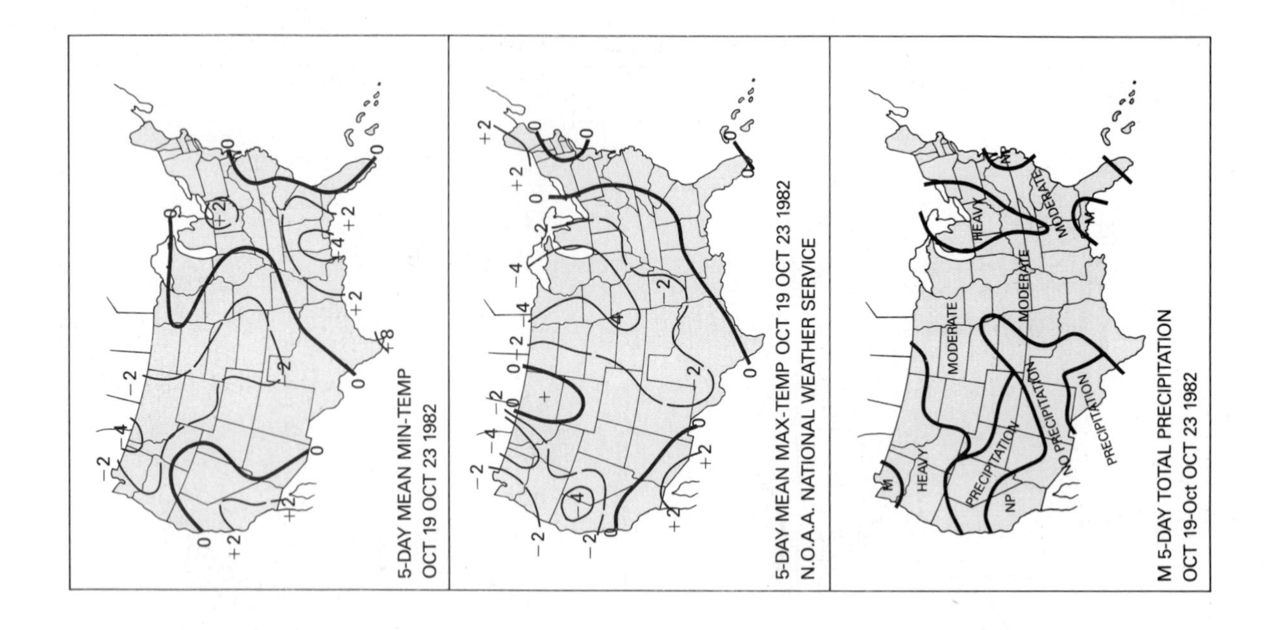

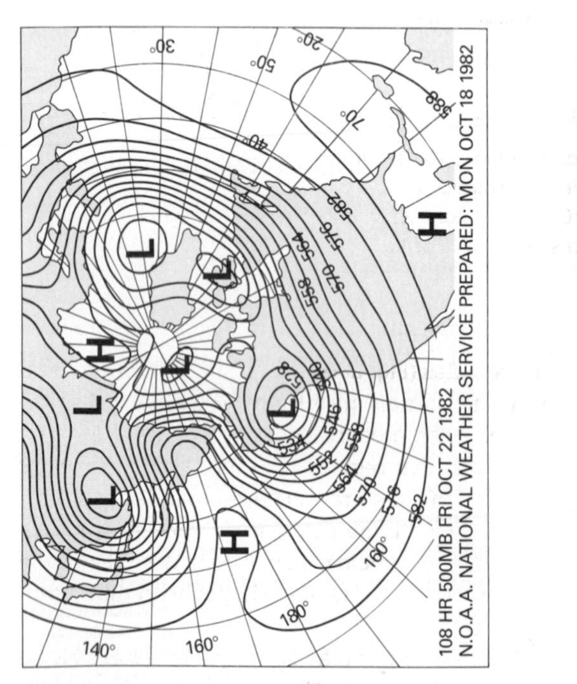

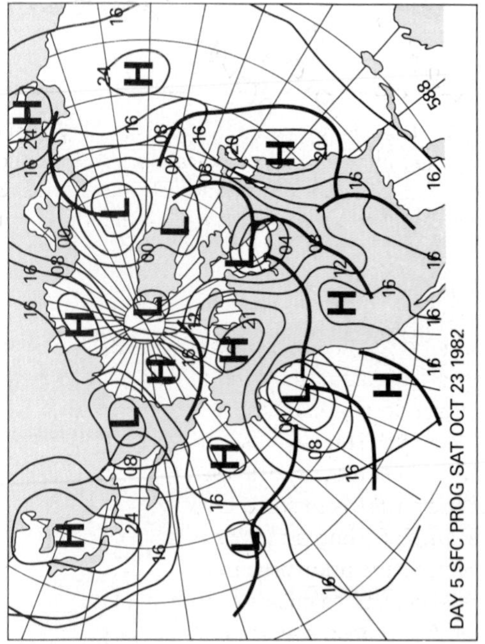

FIGURE 2-14 Five-day forecasts are made for the surface weather chart, 500-mb chart, maximum and minimum temperature, and precipitation. (NOAA.)

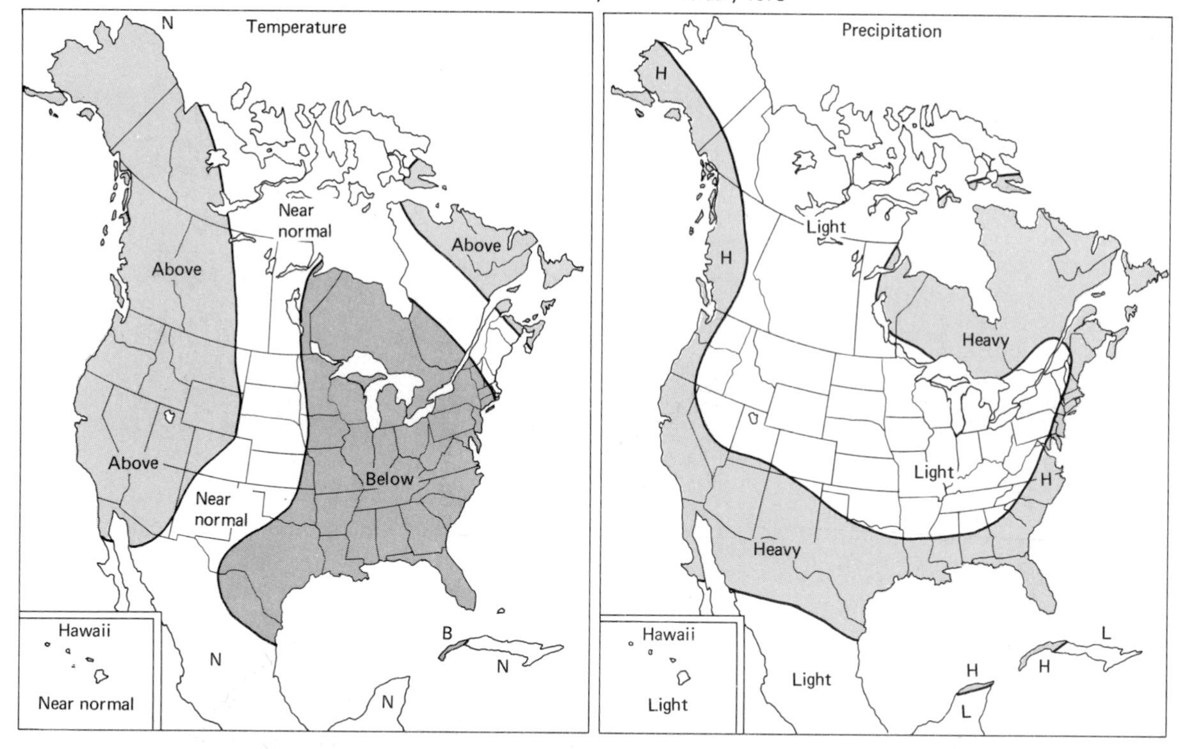

FIGURE 2-15 The average monthly outlook is a type of climatological forecast produced by the National Meteorological Center. It provides information on the expected temperature and precipitation during the next 30 days by denoting those regions likely to have above- or below-normal temperatures and heavy or light precipitation.

of the probability of precipitation can be obtained by comparing the current 500-mb airflow patterns and surface synoptic conditions with past meteorological records showing similar conditions. Enough information may be provided to predict the future climatic patterns to some degree.

Another way of using statistics is by compiling the data existing for prior precipitation over a 50- or 75-year period and calculating the probability of precipitation from these statistics. Thus the probability of receiving 25 mm (1 in) of rain in a 1-wk time interval for a specific time starting January 10, for example, can be computed. This can be completed for a large number of stations, giving information on the geographical distribution of these probabilities (Fig. 2-16). This type of information is useful for certain general considerations. For example, these probabilities may be useful for considering which regions are more likely to require irrigation of moisture-sensitive crops. They are accurate over a large number of years but any one particular year could not be expected to receive the specific amount of rainfall suggested by the probabilities.

Experimental climatological forecasts are based on correlations with **sunspots,** dark storms that occur on the sun at regular intervals of about 11 years, with circulation reversals of the sunspots after each interval, thus creating a 22-year cycle. An analysis performed in 1972 showed that the

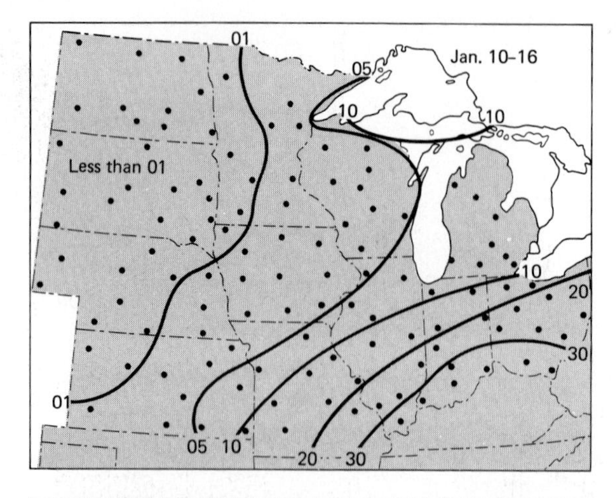

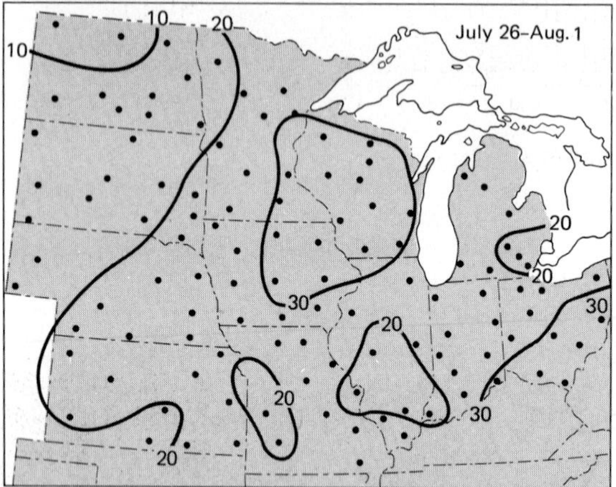

FIGURE 2-16 The probability (%) of receiving 1 in. or more precipitation in the north central United States during two different 1-week intervals. (From R. H. Shaw, G. L. Barker, and R. L. Dale, "Precipitation Probabilities," Bulletin 73, University of Missouri, 1960, courtesy of W. L. Decker.)

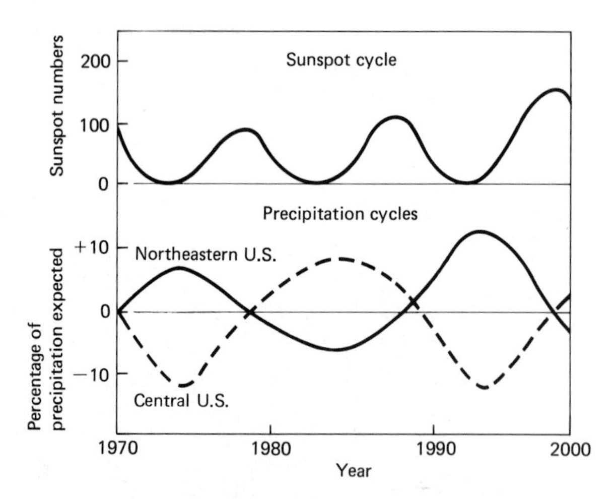

FIGURE 2-17 The variation in sunspot numbers and the amount of precipitation expected in the northeastern United States and central United States projected to the year 2000. (From James Marshall, "Precipitation and Sunspot Numbers," Dissertation, University of Kansas, 1972.)

central United States is likely to be dry when the northeastern part of the United States is having greater amounts of precipitation. The central United States was dry in the mid-1950s, the mid-1930s and in the teens. This 20-year cycle seems to be correlated with sunspot activity. The reasons for this are uncertain, but statistics indicate that alternating periods of minimum sunspot activity are correlated with dry weather in the central, and wet weather in the northeastern, United States. On the basis of these correlations, it was projected in 1972 that 1976 would be dry in the central United States (Fig. 2-17). This was, of course, the case. This particular long-range climatological forecast indicates that the mid-1980s will be wet in the central United States, with the northeastern United States having dry weather. Again, in the 1990s, drier weather and drought conditions can be expected to return to the Great Plains.

This type of climatological forecast is not as useful as it would be if the dry years occurred in a more uniform pattern. For example, Figure 2-18 shows the annual total precipitation at Manhattan, Kansas, from 1858 to 1980. It is apparent that the mid-1970s, mid-1950s, and mid-1930s were dry but there is no smooth transition into the wetter years in the 1940s and 1960s. Because of the extreme year-to-year variation, climatological forecasts of precipitation amounts lose much of their significance. They may be useful, however, for

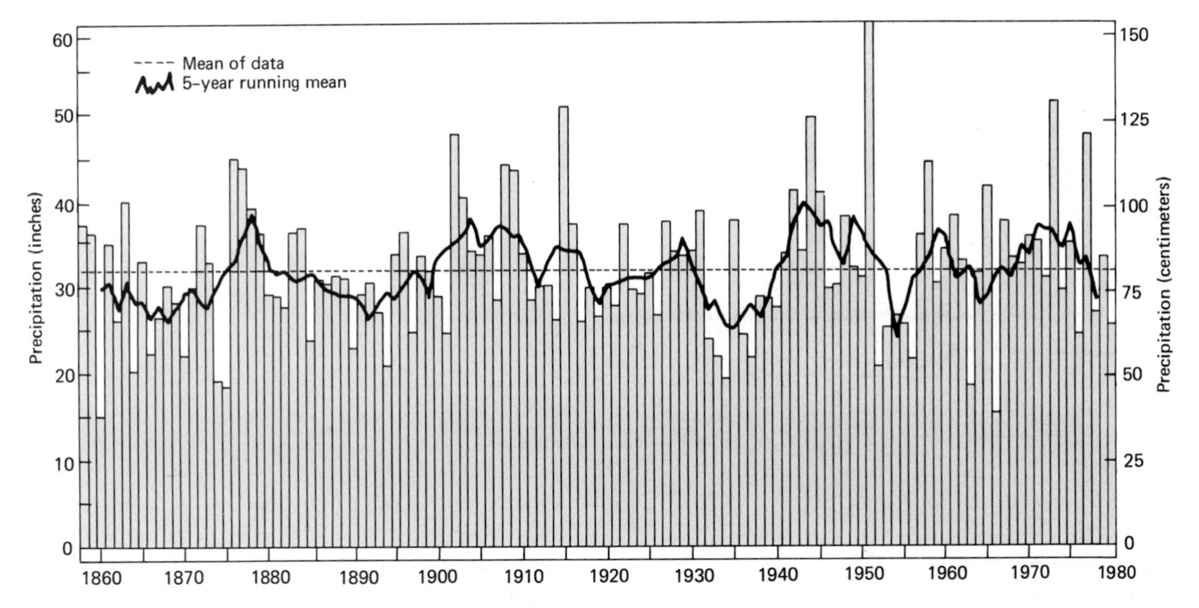

FIGURE 2-18 Variations in the annual precipitation recorded at Manhattan, Kansas, from 1858 to 1980. (Courtesy of L. Dean Bark.)

those applications where the 5-year average rainfall amount is of primary concern.

NATIONAL WEATHER FORECASTING CENTERS

Three different national centers are responsible for the various weather forecasts. The **National Meteorological Center** in Suitland, Maryland, is responsible for issuing meteorological forecasts, including the direction and speed of traveling high- and low-pressure centers, weather fronts, and associated weather events. Meteorologists there use computers to plot the maps, draw isobars, locate fronts, do an isallobaric analysis and upper air analysis, and make numerical weather forecasts. The completed maps then go over telegraph lines to all the facsimile machines over the United States that reproduce the maps as analyzed or forecast by the National Meteorological Center (Fig. 2-19).

FIGURE 2-19 A forecast chart is generated by a computer at the National Meteorological Center. The chart will be inspected by the meteorologist before it is sent by wire to facsimile map receivers located throughout the United States. (Photograph courtesy of NOAA.)

The **National Hurricane Center** in Miami, Florida, is responsible for locating tropical storms and predicting the time and location of landfall of hurricanes near the United States. Most hurricanes are first identified from satellite data, since cloud systems over the whole earth are photographed every few minutes. After they are located, the climatology of past hurricanes is used along with several numerical schemes to predict the future location of the storms.

The **National Severe Storms Forecast Center** in Kansas City, Missouri, is responsible for forecasting severe thunderstorms, including tornadoes, for the entire United States. Individual tornadoes cannot be forecast, but it is possible to specify general areas where the state of the atmosphere is appropriate for the development of this type of violent storm. Atmospheric conditions are watched, such as the humidity, temperature, and the upper air conditions, including the nature of the jet stream and the presence of a particular temperature profile.

ADVANCED TOOLS FOR WEATHER FORECASTING

Radar The development of **radar** has given the meteorologist "eyes" to see a storm that may be 300 km away. A radar set sends out short pulses of radio waves from a transmitter and, with the use of radio receivers, records the returned radio waves between transmitted pulses. Objects such as airplanes, raindrops, hail, and even smaller cloud droplets and ice crystals reflect or "echo" a small part of the emitted radio waves back to the radar antenna, where it lights a spot on a cathode ray tube or fluorescent screen similar to a television screen.

The use of radar for severe weather detection began with the observation of a hook-shaped echo from a tornado-producing thunderstorm by the Illinois State Water Survey in 1953. The radio waves reflected back to the radar antenna are usually displayed by a Plan Position Indicator (PPI). As the radar antenna rotates horizontally, a thin horizontal slice through a thunderstorm is displayed on the PPI screen (Fig. 2-20) as radio waves are returned by the precipitation and small droplets within the cloud. If the radar antenna is moved vertically, it can be used to obtain height information when the returned signal is displayed on a Range Height Indicator (RHI). The RHI screen displays a vertical slice through a thunderstorm or gives the height of an airplane, for example.

The Doppler effect, first described by Christian Doppler in 1842, allows a radar to determine the speed of any moving object that returns radio waves. It is used by police officers to check the speed of automobiles and by meteorologists to determine the speed of raindrops within a cloud. A Doppler radar transmits continuous radio waves instead of pulses. Objects that are moving toward or away from the radar antenna cause the reflected radio waves to have a different frequency or number of vibrations per second. This shift in frequency is then used to determine the speed of the object.

During the 1970s the technology was developed for training two or more Doppler radar sets on a thunderstorm to obtain the three-dimensional velocity of raindrops carried by air currents within the cloud. Such **dual Doppler radar** has provided new information on severe storm development and is likely to be used operationally by weather stations in the 1990s.

The National Weather Service has established a network of radar, Figure 2-21, that operates at wavelengths of 3 to 10 cm and provides important information on severe thunderstorms and hurricanes. Meteorologists use a hook-shaped echo as shown in Figure 2-20 to indicate that a thunderstorm has the right internal structure for the development of a tornado. The movement of such severe thunderstorms, as well as the approach of hurricanes, can be observed on the radar screen.

Satellites Several **meteorological satellites** have been placed in orbit specifically for atmospheric studies. The first of these was the TIROS (Television and Infra Red Observation Satellite) satellite launched in April 1960 (Fig. 2-22). Following the TIROS weather satellites, Nimbus, ESSA, ATS, NOAA, SMS, and GOES meteorological satellites have collected a tremendous number of photo-

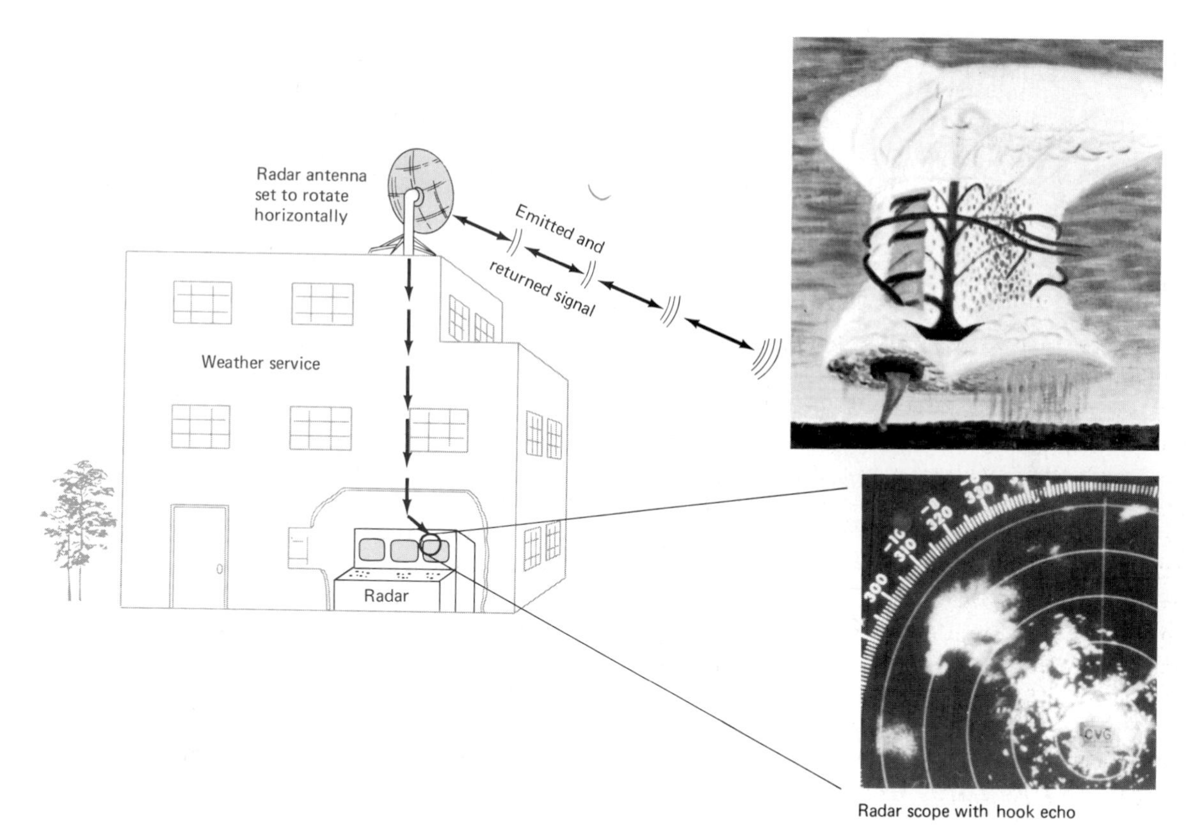

FIGURE 2-20 Weather radar are used for monitoring thunderstorm activity in a locality. The development of a hook-shaped radar echo such as that shown in this photograph indicates that a thunderstorm has the right internal structure for generating a tornado.

graphs from space. The ATS, SMS, and other similarly equipped satellites are located the greatest distance above the earth's surface at 35,877 km in geosynchronous orbits, taking 24 hours to make one revolution around the center of the earth and hence appearing to an earth observer to be over the same point on earth at all times—a "stationary" satellite.

These satellites send coded information from their earth-sensing instruments that is transformed into photographs in the receiving laboratory on earth. Meteorologists use the images taken from these satellites on many TV weather programs. Photographs are taken in the visible and infrared wavelengths and are useful in providing information on general cloud patterns, midlatitude cyclones (Fig. 2-23), hurricanes, and individual thunderstorms. These satellites are also used to transmit processed weather maps as well as photographs obtained by other satellites.

The manned space flights have also provided a variety of meteorological photographs and data. These include the Mercury, Gemini, Apollo, and Skylab space missions. In addition, the Landsat satellite provides repeated coverage of the earth's land. Satellite photographs have proven to be a valuable source of information to forecasters and research meteorologists.

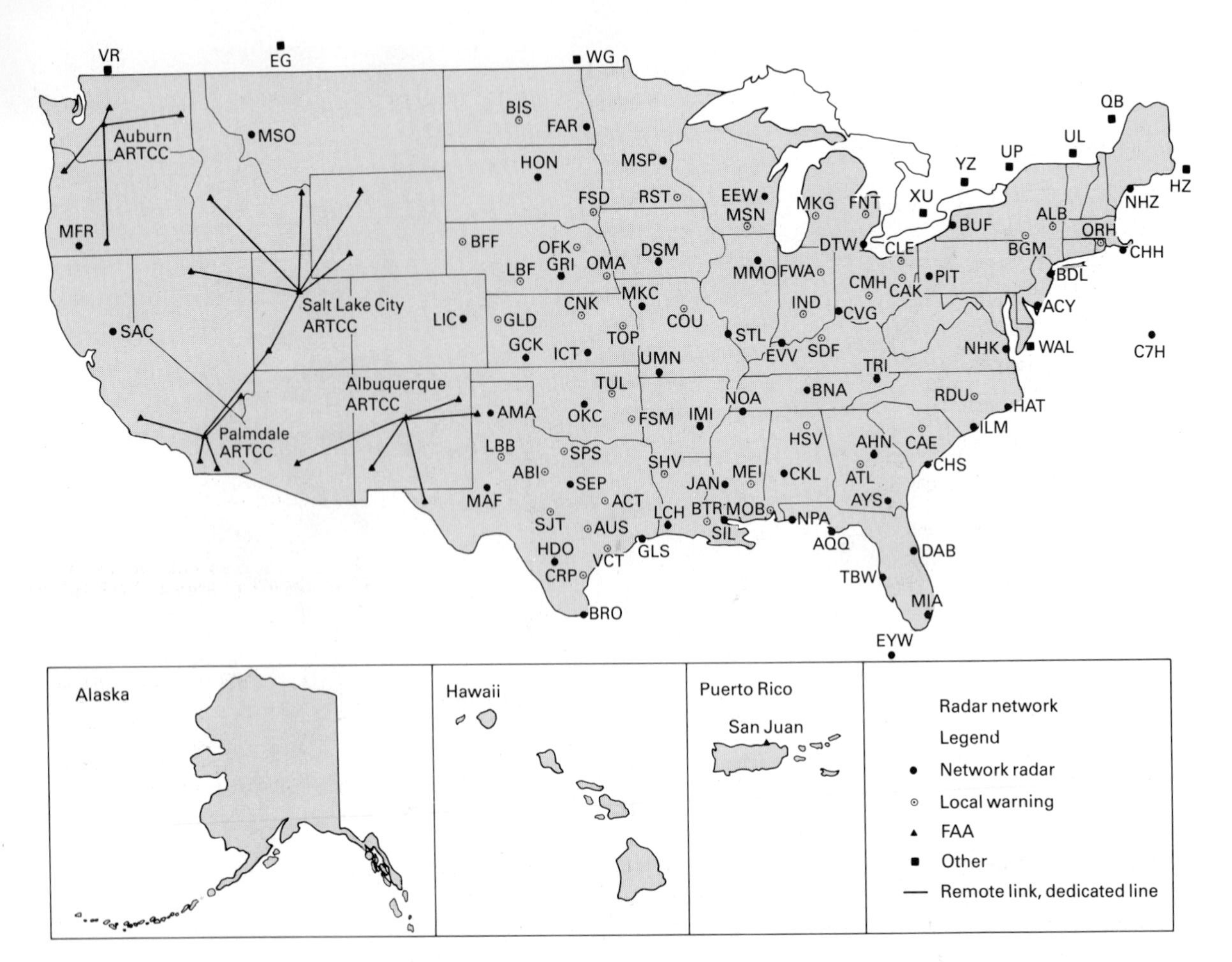

FIGURE 2-21 The location of weather radar systems used for monitoring severe weather. Coastal stations along the Gulf Coast and eastern United States are used in tracking hurricanes, with those in interior locations used for tracking severe thunderstorms. (From *Operations of the National Weather Service*, NOAA.)

Computers Modern **computers** can perform more than a million calculations in a single second. This tremendous capacity for speedy calculations has helped advance meteorology from an art to a science by allowing complex equations describing the atmosphere to be used with current atmospheric measurements to make much more precise weather forecasts.

Computers are also used to draw weather maps after calculations have been performed. In addition, they are a basic tool of the research meteorologist since routine and boring calculations can be done easily by simply writing a computer program to process millions of individual numbers. Since the computer is a machine, it can perform only the operations that are specified for it. Thus, the major limitation of computers is still the imperfect data and instructions supplied by the scientist.

Automatic Weather Facilities The National Oceanic and Atmospheric Administration is in the process of changing to completely electric high-speed data-handling equipment. The transition to these modern "weather stations of the future" is ex-

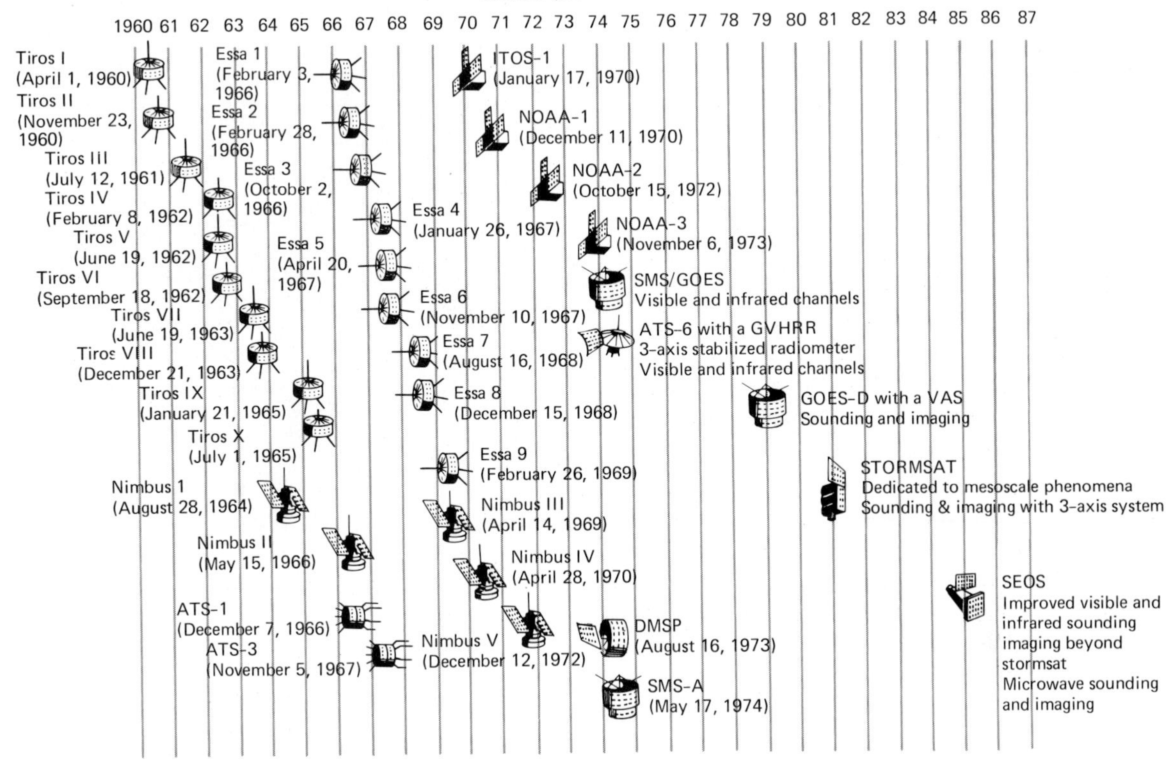

FIGURE 2-22 Various weather satellites have been placed in orbit beginning with TIROS I in 1960. Additional satellites with specialized sensors are being designed for launching within the next few years. (From *Bull. American Meteorological Society, 56,* No. 2 [1975] and Preprint Volume, 9th Conference on Severe Local Storms, American Meteorological Society, 1975.)

pected to be completed in the mid-1980s with the installation of more than 200 systems. The program is known as **AFOS**, an acronym for Automation of Field Operations and Services. This computer-based system frees meteorologists from much of the drudgery of data handling that was previously required. The program is designed to be helpful in making split-second decisions that provide warnings to people about the possibility of weather-related property damage and loss of life.

AFOS provides National Weather Service stations with high-speed data-handling and display capabilities by means of on-site minicomputers linked together in a nationwide network. Weather information from the minicomputers is displayed on TV-like screens instead of on paper, Figure 2-24. This all-electronic system eliminates the tedious process of tearing off and filing the huge volume of messages and maps previously received on teletypewriters and facsimile machines. In addition, a weather map can be produced on the TV-type displays in only about 15 seconds.

AFOS will be instituted at two levels. The first level was the automation of 52 Weather Service

Forecast Offices, four National Centers (the National Meteorological Center, the National Hurricane Center, the National Severe Storms Forecast Center, and the National Climatic Center), and 14 River Forecast Centers. These are linked by communication lines called the National Distribution Circuit.

The second level will be the extension of AFOS into each Forecast Office's area of responsibility (generally a state) by some degree of automation of the basic and synoptic weather stations. Messages between Forecast Offices and these satellite offices also will be relayed by high-speed communications.

Forecasters will have available a tremendous variety of weather maps and messages they can call up within seconds to prepare forecasts and warnings. To keep the data manageable, the minicomputers will be programmed to pull off the National Distribution Circuit only that data a Forecast Office wants. Information that has outlived its operational usefulness will be automatically purged.

The AFOS system will be enhanced by other automatic devices and systems in existence or being developed, such as automatic weather observing stations, digitized radar, and computer-assisted measurements of the upper air. These linkages will allow fast and frequent observations of changes taking place in the weather.

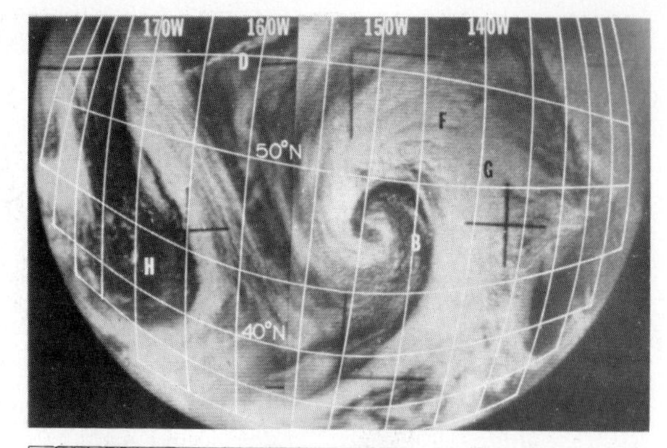

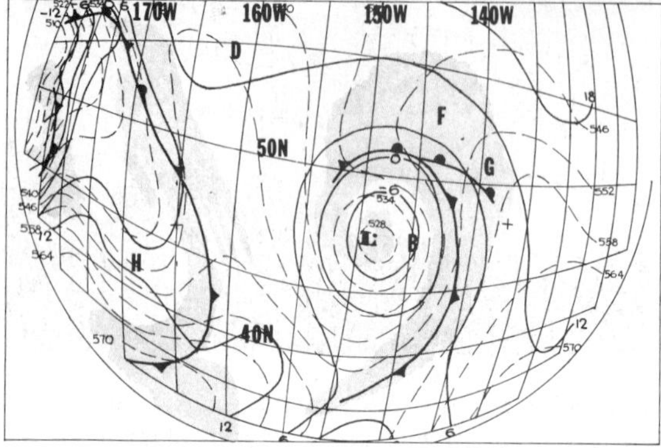

FIGURE 2-23 Satellite photographs give additional information on atmospheric storms, as shown in this comparison with the surface synoptic chart for the same date. The cloud cover associated with a mature frontal cyclone shows a rougher texture at F, where the jet stream crosses going northeastward, than at G. The inflow of clear air at B is affecting surface weather. (From *Direct Transmission System Users Guide,* National Environmental Satellite Center, 1969, NOAA.)

SUMMARY

Weather studies must begin with accurate observations of the current condition of the atmosphere including its temperature, humidity, pressure, precipitation, wind, and sky cover at various weather stations. After these data, as well as upper air measurements, have been obtained, various analyses are used to interpret the data. A surface synoptic chart, as well as upper air synoptic charts for the 850-, 700-, 500-, and 200-mb levels, are generated. This information on the current state of the atmosphere is then used to forecast the future weather.

Persistence weather forecasts are made by simply noting the past history of weather events and projecting these into the future. Meteorological forecasts made by considering various factors that may change the intensity or path of a particular weather system are more accurate. Meteorological forecasts of midlatitude cyclones with associated weather events are made by the National Meteo-

FIGURE 2-24 The new computerized weather data-handling network, AFOS, is used for displaying weather maps for daily forecasts and storm warnings. Minicomputers process and store large amounts of data for almost instant retrieval and display on TV-type consoles such as these in use at the National Meteorological Center. (Photograph courtesy of NOAA.)

rological Center, while hurricane forecasts are the responsibility of the National Hurricane Center, and tornado forecasts are the responsibility of the National Severe Storms Forecast Center.

A surface weather map analysis is performed by plotting detailed weather information and drawing isobars to locate the areas of high and low pressure. A frontal analysis is used to locate cold fronts, warm fronts, or stationary fronts on the surface weather map. An isallobaric analysis is used to show the areas of greatest pressure change. If the surface pressure is falling rapidly, this indicates that a low-pressure center is approaching with its associated weather. If the pressure is rising rapidly, then the approach of a high-pressure center is indicated, along with fair weather.

Upper air data is used to determine the regions of cyclonic or anticyclonic bends in the long waves of the atmosphere. Very different weather is associated with a trough, or counterclockwise curvature in the jet stream, than with a ridge. Long-range forecasts are less accurate than forecasts for shorter time periods and may be based on the upper atmospheric long-wave patterns, meteorological statistics, or variations in sunspot activity. The precision and utility of climatological forecasts are generally less than for meteorological forecasts made for shorter time periods.

Sophisticated tools are available to improve meteorological data and weather forecasts. Weather satellites provide photographs of cloud patterns and characteristics every few minutes. Modern computers are used by meteorologists for numerical weather prediction and for many other more specialized problems that involve handling the many numbers describing the atmosphere.

Facsimile weather map receivers and microcomputers are available commercially to obtain synoptic weather maps, forecast maps, and weather data transmitted from the National Meteorological Center. Many weather stations have radar equipment that is useful for tracking severe thunderstorms and hurricanes. During the 1980s, the National Weather Service will be automated with minicomputers linked together in a nationwide network. This will allow much more rapid transfer of weather data, maps, and warnings of value to the public.

STUDY AIDS

1. Refer to the detailed plotting model in Figure 2-6 and plot the following information obtained at a weather station: temperature 20°C, dew point temperature 18°C, winds east at 35 km/h (20 knots), pressure tendency 2 mb decrease, overcast sky with high cirrus clouds, 20 mm of rain occurred in the last 6 hours.

2. What publication would be used to obtain detailed weather information for every 3 hours from a synoptic weather station?

3. Where are the National Meteorological Centers located, and what specialized forecasts do they issue?

4. Describe some ways in which surface weather is related to upper atmospheric conditions.

5. Describe how you would perform an isobaric analysis on a surface weather map.

6. Describe how you would perform a frontal analysis on a surface weather map.

7. Describe how you would perform an isallobaric analysis on a surface weather map.

8. Compare persistence, meteorological, and climatological forecasts as to their accuracy and length of time for the forecast.

9. Describe the LFM charts and give some indication of how they can be used.

10. Discuss the use of radar and satellites in obtaining meteorological measurements and information.

11. How will automatic weather facilities benefit the public?

12. Speculate on ways to improve meteorological and climatological forecasts of the future.

TERMINOLOGY EXERCISE

Check the glossary for more information on any of the following terms used in Chapter 2 that are still unfamiliar to you. Many of these will be used or dedeveloped further in later chapters.

Synoptic
Greenwich Mean Time (GMT)
Radiosonde
Rawinsonde
National Meteorological Center
Synoptic charts
National Climatic Center
Pressure tendency
Isobaric analysis
Frontal analysis
Wind shear line
Isallobaric analysis
Low-pressure system
High-pressure system
Streamline analysis
Isohypse analysis
Streamlines
Numerical weather prediction
Vorticity
Macroscale
Mesoscale
Precipitation probability
Sunspots
National Meteorological Center
National Hurricane Center
National Severe Storms Forecast Center
Radar
Dual Doppler radar
Meteorological satellites
Computers
AFOS

THOUGHT QUESTIONS

1. Why is the use of a world standard time, such as Greenwich Mean Time, important in meteorological observations?

2. Why is it so important to consider synoptic maps for several different layers in the atmosphere, instead of simply using the surface chart?

3. How can some of the relationships between the jet stream and surface weather be used more effectively?

4. Do you feel that more emphasis should be placed on climatological forecasts? Explain.

5. Do you think that the gathering and dissemination of weather data should be the responsibility of the federal government or private industry?

6. Speculate on future changes in weather data-gathering techniques and dissemination programs.

SPECIAL TOPIC 3

NUMERICAL MODELS AT NMC

Modern weather forecasting at the National Meteorological Center (NMC) involves the use of mathematical equations that are built around the so-called primitive equations. These five time-dependent equations are programmed into the computer to predict the future state of the atmosphere. Two of the primitive equations are for the horizontal wind components. One equation involves the temperature, another the pressure, and the fifth includes the total water content of the atmosphere.

The numerical techniques based on these primitive equations (PE model) used at NMC were implemented operationally in June 1966. This model is the ancestor of two of the important current numerical weather prediction models. These are the limited area fine mesh (LFM) model and seven-layer hemispheric PE model.

The LFM model was introduced into routine operation in 1971. Computer limitations allowed it to be run only to 24 hours until 1974, when it became feasible to extend the LFM calculations to 48 hours. A further improvement came in 1977 when the grid spacing (distance between data points) used in the LFM calculations was reduced from 174 km at 45°N latitude to 116 km. The LFM model as used operationally by NMC calculates future atmospheric conditions for most of the North American continent.

The seven-layer hemispheric PE model is used to calculate future atmospheric conditions for the entire northern hemisphere. While the LFM model uses an equally spaced gridpoint analysis of wind, height, temperature, and humidity to specify the initial state of the atmosphere, the hemispheric PE model employs even more complicated techniques such as spectral analysis. This involves expressing the information in the form of a series of mathematical functions and then calculating the future state of the atmosphere.

Output products include maps showing the predicted state of the atmosphere 12, 24, 36, and 48 hours in the future. The LFM maps provide considerable detail on pressure, humidity, and winds expected over the next two days. The hemispheric PE model provides information on the expected future path of the jet stream and pressure patterns. In the same way that you do not have to understand everything about an automobile to use it, you can use the numerical weather forecasts for studying and understanding current and future atmospheric conditions before gaining an understanding of the exact nature of their generation.

CHAPTER THREE
RADIATION AND ATMOSPHERIC HEAT EXCHANGE

NATURE OF RADIATION

The sun, earth, and atmosphere constitute a gigantic heat engine. The sun radiates heat to earth and warms the tropical atmosphere more than the atmosphere at other locations. This pushes the atmosphere into action and causes what we call weather.

Tremendous quantities of energy are required to produce and sustain the winds within the atmosphere, especially in jet streams, tornadoes, hurricanes, and midlatitude cyclones. The mechanism for transferring energy from sunlight to many of the weather systems is not direct, but nevertheless, sunlight is the source of the energy. The general circulation and the major wind systems are driven by differences in the density of air that develop because of variations in the amount of solar energy absorbed in different locations. Energy from the major wind systems is then transferred to some of the smaller circulations.

The amount of radiation from the sun that reaches the earth's outer atmosphere is relatively constant. For this reason it is called the **solar constant;** the solar constant is 1380 W/m^2 (2 cal/cm^2/min). Since the earth is a rotating sphere it intercepts the same quantity of sunlight that a flat disk the same diameter would intercept if it were perpendicular to the sun's rays. Only half the earth receives sunlight at any one time—the other half is in darkness—but the earth's rotation allows the sunlight to be spread over the earth. Sunlight is intercepted by the earth's whole system: the atmosphere, cloud tops, and ground (depending on the amount of cloud cover). The amount of radiation received from the sun by the earth-atmosphere system is a staggering 180,000 trillion watts (J/sec).

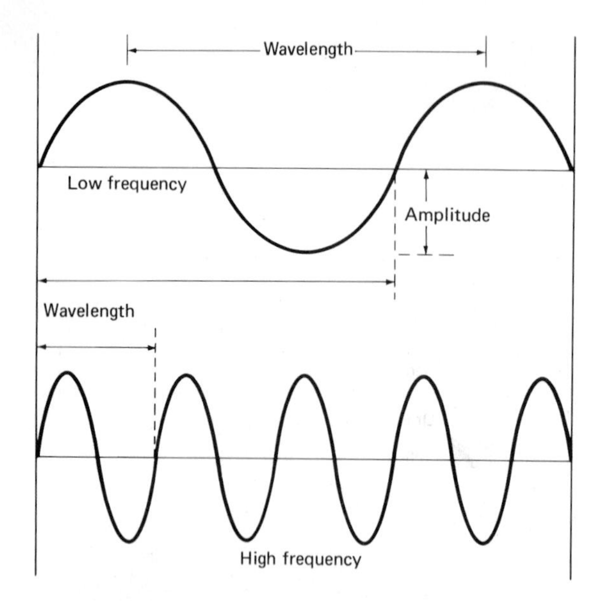

FIGURE 3-1 Radiation can be characterized by its wavelength or its frequency. The wavelength is measured from the crest of one wave to the crest of the next, or from center to center of vibration, as indicated. The amplitude gives the magnitude of the vibrations.

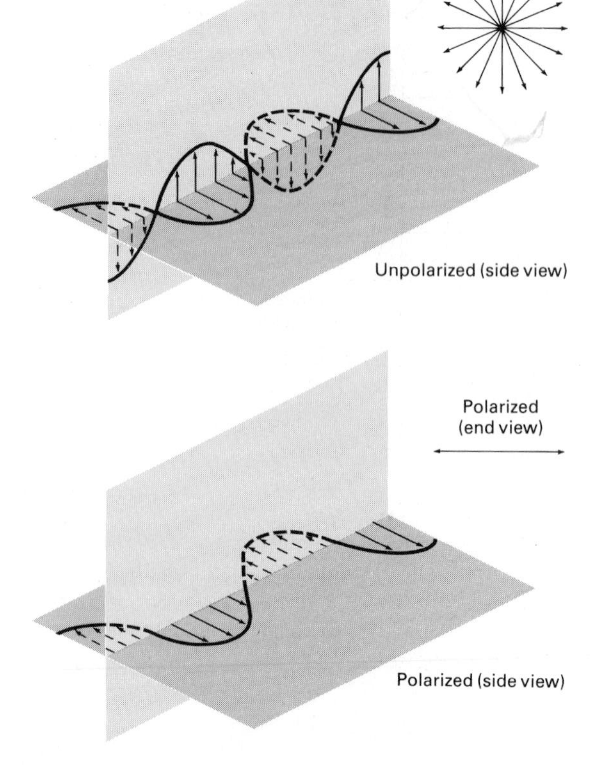

FIGURE 3-2 Radiation can be unpolarized and free to vibrate in every direction, or it can be polarized with vibrations in only one direction.

Other minor amounts of energy are available at the earth's surface from such sources as the heat flux from the hot interior of the earth (24 trillion watts) and radiation and reflected sunlight from the moon (2 trillion watts). Heat is also added to the atmosphere by the combustion of fuels, including oil, gas, coal, and wood, used as energy sources for factories or dwellings. The heat lost up chimneys, smokestacks, and through the walls, doors, and windows amounts to 0.001 trillion watt. The sum of all these minor heat sources is still about 10,000 times less than the energy received from the sun and would not support plant or animal life if these were the only heat sources.

The earth and the lower atmosphere are warmed because of the absorption of radiation from the sun. Because of the importance of this warming process we will now consider the nature of **radiation** in some detail. Radiation has very different characteristics depending on its wavelength. Radiation is propagated as waves, with some similarity to waves on water, with crests and troughs that remain at constant intervals as the radiation travels through space. The distance from crest to crest for a particular wave gives the wavelength of the radiant energy (Fig. 3-1). Energy emitted by the sun travels through space without changing its wavelength and reaches the earth in only eight minutes.

Radiation is normally free to vibrate in any direction: not only up and down, but horizontally and all other directions as well. This type of radiation is

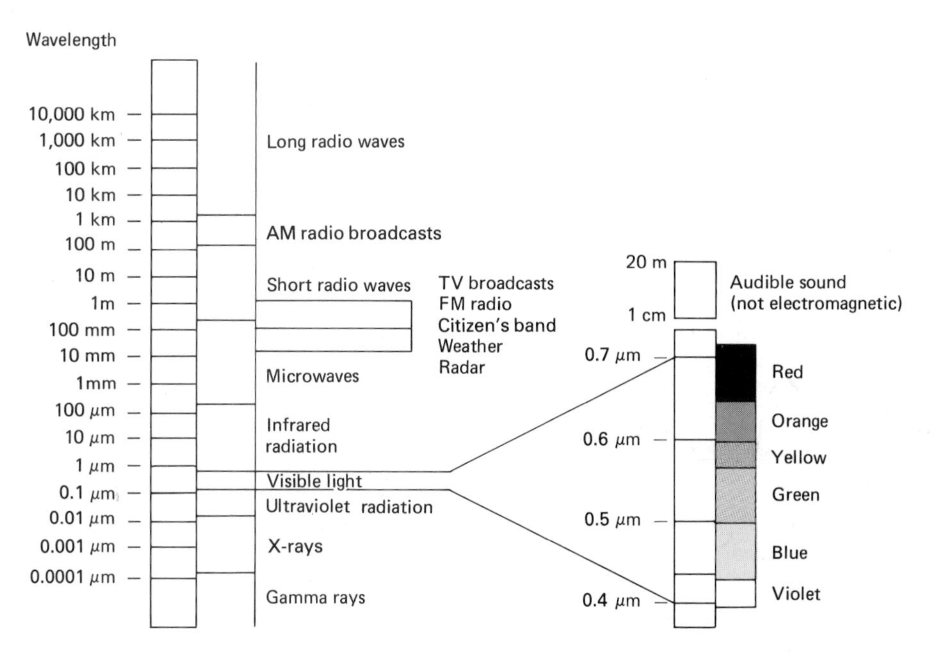

FIGURE 3-3 The electromagnetic spectrum characterizes radiation according to its wavelength or frequency. Radiation of different wavelengths has very different characteristics. Short wavelengths are very penetrating, longer wavelengths vibrate at the right frequency to be seen by human eyes, and still longer wavelengths exhibit other characteristics.

unpolarized light (Fig. 3-2). Polarized light vibrates in only one direction. Some of the light from the sun may become polarized, for example, as it bounces off the pavement. For this reason polarized sunglasses are effective in reducing glare. If the intense sunlight is vibrating horizontally, and the sunglasses allow only vertically vibrating light to pass through, the intense light is blocked. There are many other applications based on polarized light.

The **electromagnetic spectrum** includes all the different wavelengths of radiant energy (Fig. 3-3). Many different categories of wavelengths are given names. For example, the long wavelengths are called long radio waves. Their wavelengths are several hundred kilometers, with vibrations of 20 to 20,000 Hz (hertz, or cycles per second). Specific bands in the radio wavelengths have been designated by the Federal Communications Commission for broadcasts from radio stations, television stations, and citizen's band radios. Radio stations broadcast in the range of 1 km or less. Television and FM radios use wavelengths of a few centimeters. Forty channels are reserved for the operation of citizen's band radios. Weather radar operates at 3 or 10 cm wavelengths. These particular wavelengths are reflected back to the radar set by precipitation in clouds.

Our eyes are sensitive to a small portion of the electromagnetic spectrum called **visible light.** The visible part of the electromagnetic spectrum is a relatively narrow band from about 0.38 to 0.72 μm. All the various colors—violet, blue, green, orange, and red—have wavelengths between these limits. In order for our eyes to see a particular color they have to distinguish between very small differences in the wavelength of energy. The difference in wavelength between orange and green light, for example, is so small that it is amazing more people are not colorblind.

Our ears are sensitive to sound waves, which are not electromagnetic radiation but actual atmospheric vibrations with wavelengths from 1 cm to 20 cm. The actual range varies for different persons. Some animals such as dogs and bats can hear sound waves of much shorter wavelengths.

Radiation with slightly longer wavelengths than visible light is called **infrared radiation.** Wavelengths slightly shorter than those in the visible part of the spectrum are called **ultraviolet radiation.** The shorter wavelengths are x-rays and gamma rays. These have specific properties quite different from those of radiation with longer wavelengths. X-rays, for example, are used for penetrating flesh and visualizing bones. In general, a small change in the wavelength of radiation causes it to have very different properties.

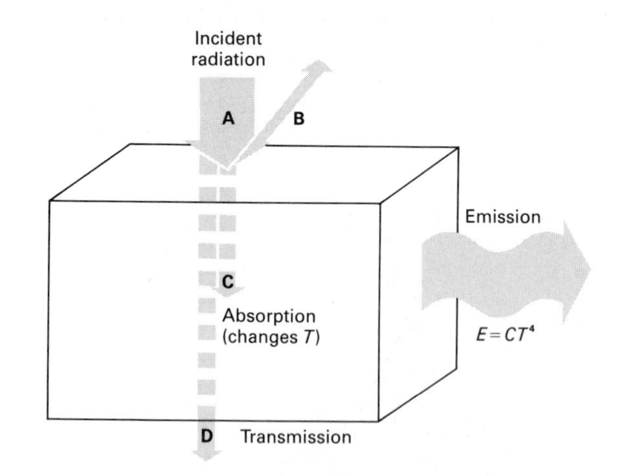

FIGURE 3-4 Incident radiation is reflected, absorbed, or transmitted. In addition to these energy processes, all objects emit some energy at a different wavelength according to their temperature, as specified by the Stefan-Boltzmann law (Appendix A).

Emission of Radiation Radiation is continuously emitted by all masses since their temperatures are above zero K. The characteristics of the emitted radiation are determined by the temperature of the emitting substance. The sun emits radiation at a tremendous rate because it is so hot; the earth emits less per unit area since it is much colder. The Stefan-Boltzmann law (see Appendix A) can be used to calculate the emitted radiation since the amount emitted from a perfect emitter is a function of the temperature. Both the earth and the sun are similar to perfect emitters, although the atmosphere is not. Many solid objects, such as the earth, emit the maximum amount of radiation that can be emitted by any substance at that particular temperature. This is called **blackbody emission.** Dull black objects are good emitters and also good absorbers of radiation. The sun is a very good emitter of radiation and approaches blackbody emission even though it does not appear black to us. Most gases, including those comprising the atmosphere emit and absorb particular wavelengths of radiation quite readily, but are transparent to other wavelengths.

Fate of Radiation When emitted radiation strikes any object or substance, there are only three possible results. First, some radiation may be reflected (Fig. 3-4). The amount reflected is determined by the basic properties of the substance intercepting the radiation. A specific quantity of energy is reflected compared to the quantity of energy striking the substance. In Figure 3-4 the quantity B/A gives this ratio, or the reflectivity. If the quantity is multiplied by 100, it becomes a percentage and is then called the **albedo.**

Some of the radiation that is not reflected may be absorbed. The quantity absorbed is the amount C/A in Figure 3-4, and is called the absorptivity.

If the energy is not reflected or absorbed, it is transmitted through the substance or object. The transmissivity is the quantity D/A, or the amount transmitted through, compared to the amount striking, the substance. The sum of the reflectivity, absorptivity, and transmissivity must equal 1.0 since the energy has to be reflected, absorbed, or transmitted.

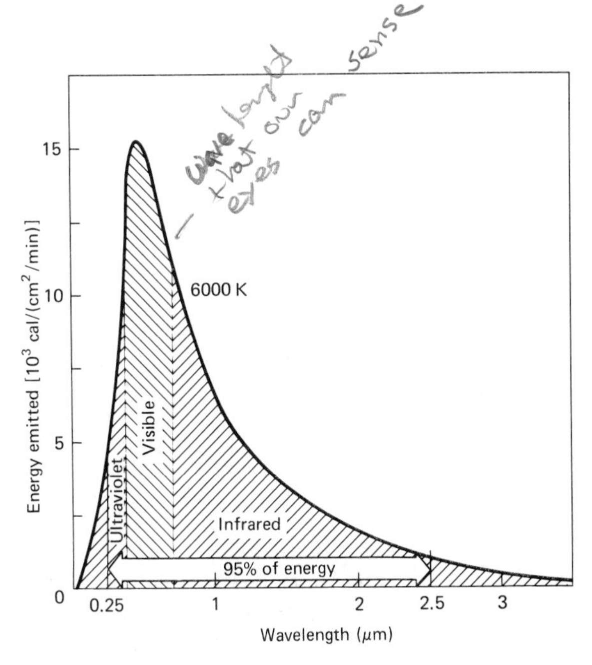

FIGURE 3-5 Sunlight is composed of ultraviolet, visible, and infrared radiation, with 95% of the energy between 0.25 and 2.5 μm.

SOLAR AND TERRESTRIAL RADIATION

The type of radiation emitted by the sun is determined by its very hot temperature. Ninety-five percent of the radiation produced by the sun has wavelengths between 0.25 and 2.5 μm, with much more radiation emitted in some wavelengths than in others. The relative amount of radiation emitted by the sun in different wavelengths is shown in Figure 3-5. Much of the solar radiation is emitted in the visible part of the spectrum, with more radiation of 0.47-μm wavelength (blue light) than of any other. Solar radiation also contains some infrared as well as ultraviolet radiation.

The ultraviolet wavelengths of solar radiation that are shorter than visible light are more effective in penetrating the skin and producing sunburn. The ozone layer extending from 25 to 50 km absorbs most of the ultraviolet wavelengths, while some selective absorption of other wavelengths also occurs in the atmosphere below the ozone layer.

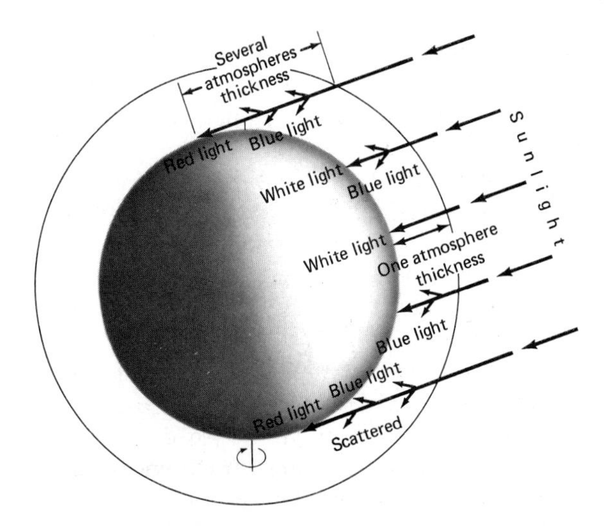

FIGURE 3-6 The shortest wavelengths of sunlight are scattered the most easily; the longer wavelengths, such as red light, form a larger percentage of sunlight as it travels through a greater atmospheric thickness.

It is easier for the longer wavelengths in the visible part of the spectrum to penetrate through the atmosphere. The sun looks red near the time of sunset or sunrise since the sunlight has a longer path through the earth's atmosphere and more of the shorter wavelengths are depleted (Fig. 3-6). The shorter wavelengths, which are scattered throughout the atmosphere, create the blue sky.

The earth, like the sun, continuously emits radiation, but the wavelengths are different from sunlight since the temperature of the earth is quite different. The sun's temperature of about 6000 K produces radiation in the visible part of the spectrum. The temperature of the earth's surface frequently varies between 250 and 300 K with an average temperature of 288 K. The resulting emission spectrum has about 95% of the radiation between 2.5 and 25 μm (Fig. 3-7). Radiation from the earth peaks at about 10 μm in the infrared part of the spectrum. A comparison of earth and sun radiation illustrates the general rule that warmer objects emit energy having shorter wavelengths. In fact, the wavelength corresponding to the maximum emission can be calculated from Wien's law (see Appendix A) if the temperature is known.

Radiation-Controlled Cycles Seasonal and daily temperature cycles are governed by the input of solar radiation. The amount of solar radiation available on earth varies because of the revolution and rotation of the earth. The earth receives the maximum amount of energy in the northern hemisphere at the time of the summer **solstice** because the northern hemisphere is tipped $23^1/_2°$ with respect to its orbital plane, toward the sun (Fig. 3-8). North of $23^1/_2°$ N latitude, the sun reaches its highest point above the horizon in the northern hemisphere during the summertime—at noon on June 21. At the time of the vernal **equinox** on March 21, the sun is overhead at noon at the equator. This occurs again on September 22 at the time of the autumnal equinox. The lowest noontime elevation of the sun in the northern hemisphere occurs on December 22, at the time of the winter solstice.

The elevation of the sun above the horizon can be calculated if the latitude is known. For example, at 40° N latitude, the height of the noontime sun at the time of the equinoxes is 90° minus 40° or 50°. To obtain the noontime elevation angle at the winter or summer solstice, simply add or subtract $23^1/_2°$, since this is the tilt of the earth's axis. For the summer solstice add $23^1/_2°$ to 50°, giving $73^1/_2°$. At 40° N latitude the noontime elevation angle on December 22 is 50° minus $23^1/_2°$ or $26^1/_2°$. This produces a large difference in the slant of the sun rays in summer and winter. If the sun rays are slanted, the radiation is spread over a much larger area. This, along with daylength, is the reason for large temperature differences between summer and winter.

The maximum temperature on any particular day does not occur at noon. Neither is the maximum heating during June in the summertime, since a temperature lag occurs on a daily and seasonal basis. On a clear day, radiation is received by the earth from sunrise to sunset while heat is lost by emission, convection, conduction, and evaporation (Fig. 3-9). The time of maximum daily temperature corresponds to the time in the afternoon when the amount of radiation received is equal to the total heat lost (about 3 P.M.). Heating continuously occurs until then, with net cooling after this time.

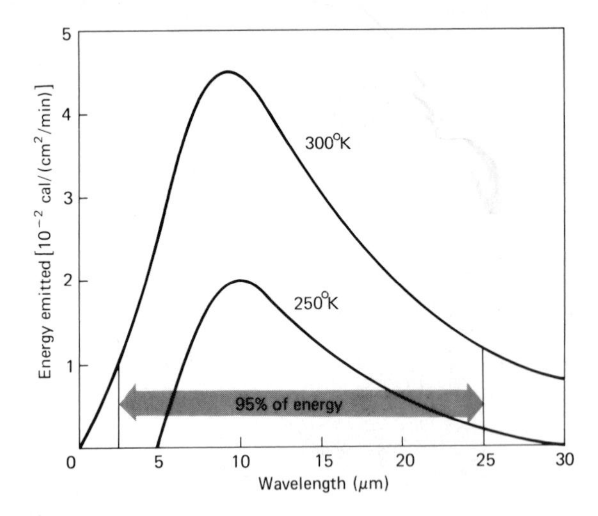

FIGURE 3-7 Terrestrial radiation is related to the temperature of the earth's surface. In general, 95% of terrestrial radiation is between 2.5 and 25 μm.

A similar seasonal lag in temperature occurs annually (Fig. 3-10). The maximum amount of radiation is received in late June in the northern hemisphere north of $23^1/_2°$ latitude since the sun is more directly overhead. Some radiation is also continuously emitted by the whole earth, and additional heat is lost by convection, conduction, and evaporation. About one month after the time of maximum heating, the earth begins to lose more heat than it receives. In midlatitudes this occurs in late July, so the hottest months (summer) are June, July, and August rather than May, June, and July. Similarly the coldest winter months are December, January, and February.

GLOBAL HEAT BUDGET

For the earth as a whole, some of the radiation traveling from the sun through space strikes cloud tops

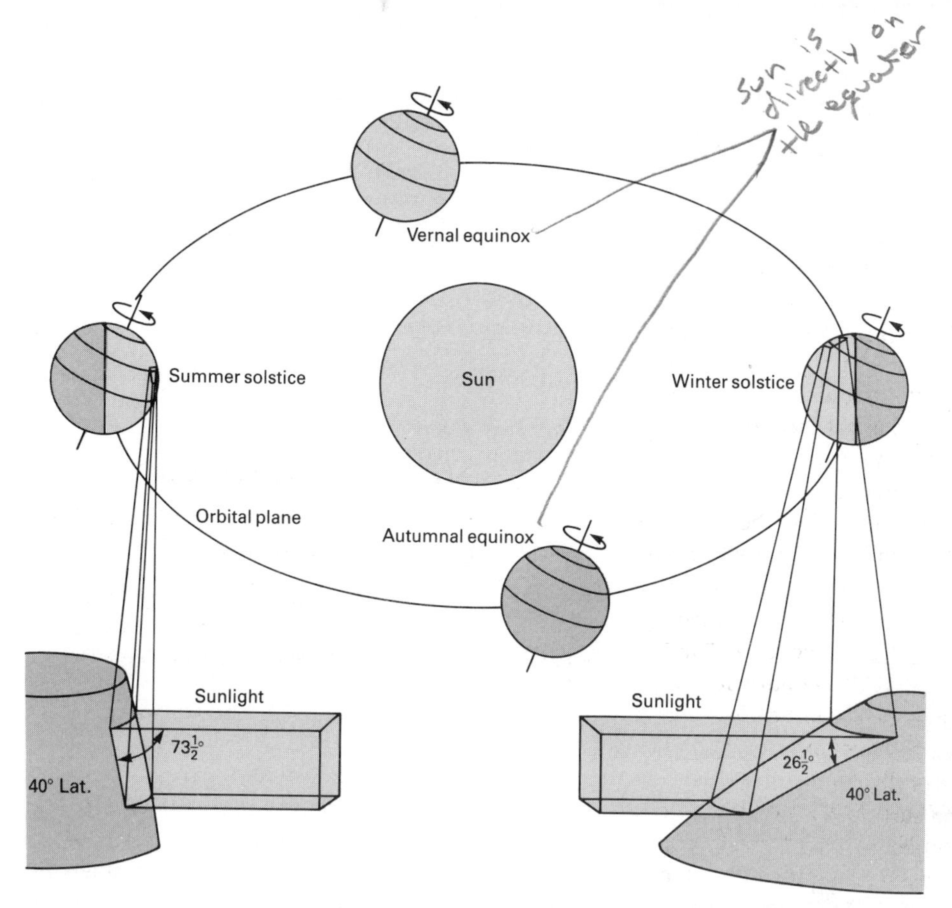

FIGURE 3-8 The summer solstice on June 21 or 22 is the time of year when the earth's axis is tipped $23^1/_2°$ toward the sun in the northern hemisphere. Since the sun's rays are more perpendicular to the earth's surface in the northern hemisphere, this corresponds to the time of maximum receipt of radiation. Six months later, at the winter solstice on December 22 or 23, the earth's axis is tipped $23^1/_2°$ away from the sun in the northern hemisphere, giving colder temperatures. During the vernal and autumnal equinoxes on March 20 or 21 and September 22 or 23, the sun is directly over the equator at noon.

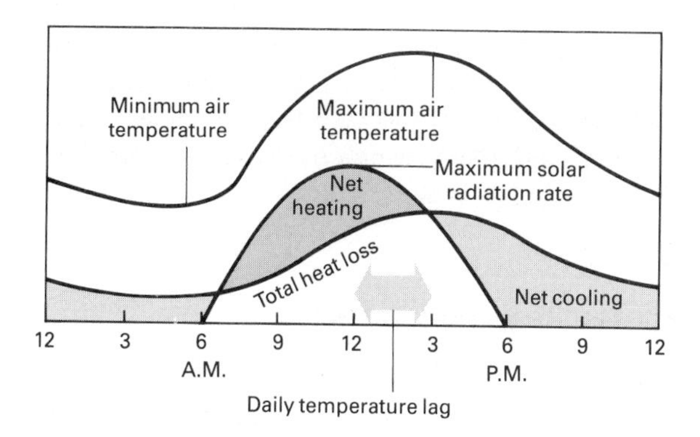

FIGURE 3-9 Daily temperatures do not coincide with the maximum radiation from the sun, but occur two or three hours later, since the earth continues to heat until the solar radiation is less than the total heat loss.

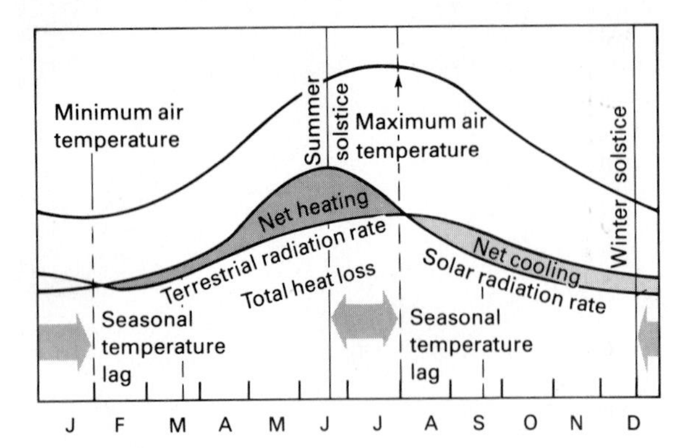

FIGURE 3-10 On a seasonal basis the maximum temperature does not occur on June 22 at the time of maximum solar radiation but occurs later, around August 1, as the solar radiation received just matches and then drops below the total heat loss.

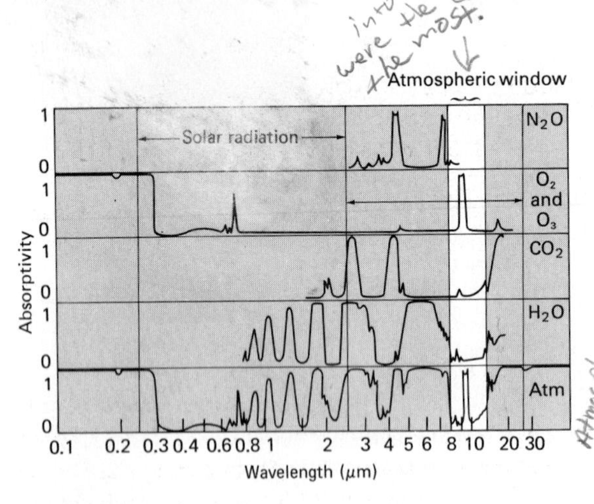

FIGURE 3-11 Absorption bands of various constituents of the atmosphere at various wavelengths, and the absorptivity of the atmosphere as a whole for both solar and terrestrial radiation. Note that the atmosphere is transparent to visible light (0.4–0.7 μm) and also to terrestrial radiation of 8–12 μm. (Adapted from R. G. Fleagle and J. A. Businger, *An Introduction to Atmospheric Physics*, New York: Academic Press, 1963.)

and is reflected; a small amount is absorbed by the atmosphere; and some reaches the surface where it is absorbed. The global heat budget can be evaluated in general terms by considering the partitioning of heat for the earth-atmosphere system. Part of the radiation budget consists of solar energy with terrestrial radiation comprising the rest. The total heat budget must include both the solar and terrestrial radiation since it is their combination that determines the overall heat balance for the earth-atmosphere system.

At any given time some parts of the atmosphere are much more transparent to solar radiation than others, depending on the amount of cloud cover. The cloudless atmosphere transmits more than 80% of the solar radiation through itself. Within the **atmospheric window,** from 8 to 12 μm, the atmosphere is transparent also to terrestrial radiation, allowing the radiation to pass through, and thus accounting for its name. The atmospheric window permits surface temperature measurements to be made from satellites because wavelengths within the atmospheric window are chosen and these can penetrate through the atmosphere.

Outside the atmospheric window **absorption bands** occur (Fig. 3-11). Gases in the atmosphere absorb radiation in specific wavelength bands and allow radiation with different wavelengths to go through. CO_2 has an absorption band for wavelengths slightly longer than 12 μm. For wavelengths shorter than 8 μm, absorption bands exist in the atmosphere because of the presence of water vapor, CO_2, oxygen, and ozone. Radiation with wavelengths less than about 0.3 μm are absorbed almost 100% by ozone. The atmospheric window exists because none of the atmospheric gases have strong absorption bands between 8 and 12 μm, except for a narrow absorption band of ozone.

Large differences exist in the way solar radiation is reflected by different materials (Table 3-1). Snow, for example, reflects as much as 90% of the sunlight that strikes it. For this reason sunglasses are needed if much time is spent in the snow during hours of bright sunlight. The reflectivity of snow also delays the warming of the earth and at-

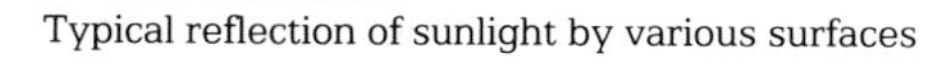

TABLE 3-1

Typical reflection of sunlight by various surfaces

Surface	Albedo
New snow	90%
Old snow	50%
Average cloud cover	50%
Light sand	40%
Light soil	25%
Concrete	25%
Green crops	20%
Green forests	15%
Dark soil	10%
Asphalt	8%
Water	8%

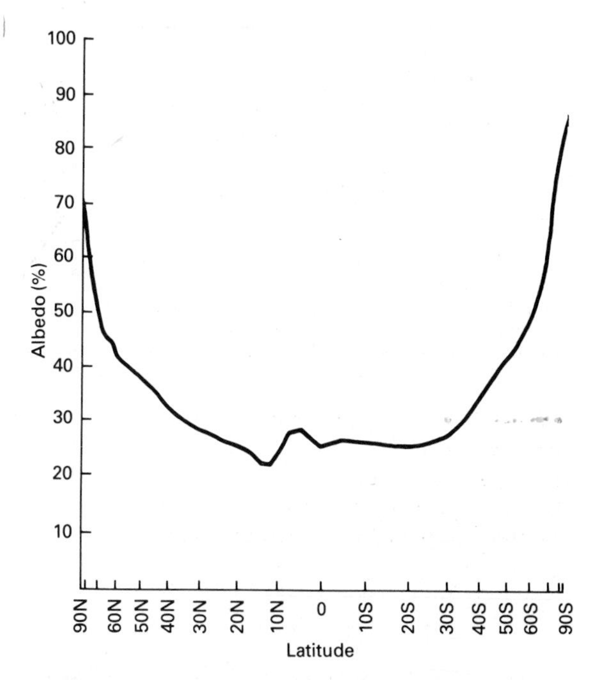

FIGURE 3-12 Measurements of the earth-atmosphere system reveal large variations in the albedo at different latitudes. Major increases in albedo are produced by snow cover and clouds. (Adapted from *Earth-Atmosphere Radiation Budget Analyses Derived from NOAA Satellite Data, June 1974–February 1978,* vol. 2, NOAA.)

mosphere after a snowfall. Winds blowing over snow are generally quite cold for the same reason. The reflective properties of snow can be demonstrated by placing a black cloth over a snow surface on a sunny day; the black cloth will absorb the sunlight and melt the snow beneath it. The snow surface and black cloth represent the extremes in albedos of different earth materials.

Some clouds are almost as reflective as snow. The most reflective clouds reflect more than 80% of the sunlight and transmit most of the rest. On the average, about 50% of the energy that strikes cloud tops is reflected; the exact amount varies with the thickness and type of cloud.

Most other materials such as grasslands, forests, and soils reflect in the range of 10 to 25% of sunlight. Water is an exception—on the average, only about 8% of the energy is reflected from water surfaces. The exact albedo of water depends on the angle at which the sunlight strikes its wavy surface. If the sun is near the horizon, most of the sunlight is reflected.

Large differences exist in the albedo of various latitudes (Fig. 3-12) because of different surface material. In general, cold regions reflect much more solar radiation than warm regions, which makes them even colder.

The presence of clouds is a major factor influencing the amount of energy at the surface. In the United States, the desert Southwest has few clouds and therefore has many more hours of sunshine than other locations. The driest part of the United States, it has about 4000 hours of sunshine per year on the average while the east coast has only about 2800 hours at the same latitude. If it were not for cloud cover, the amount of radiation received in various regions would depend only on the latitude. Variations in cloud cover (Fig. 3-13) are very important in modifying the radiation budget since they determine which areas receive direct sunlight at the ground.

The global heat budget can be evaluated by considering the percentages of radiation that are absorbed, transmitted, and reflected in various parts

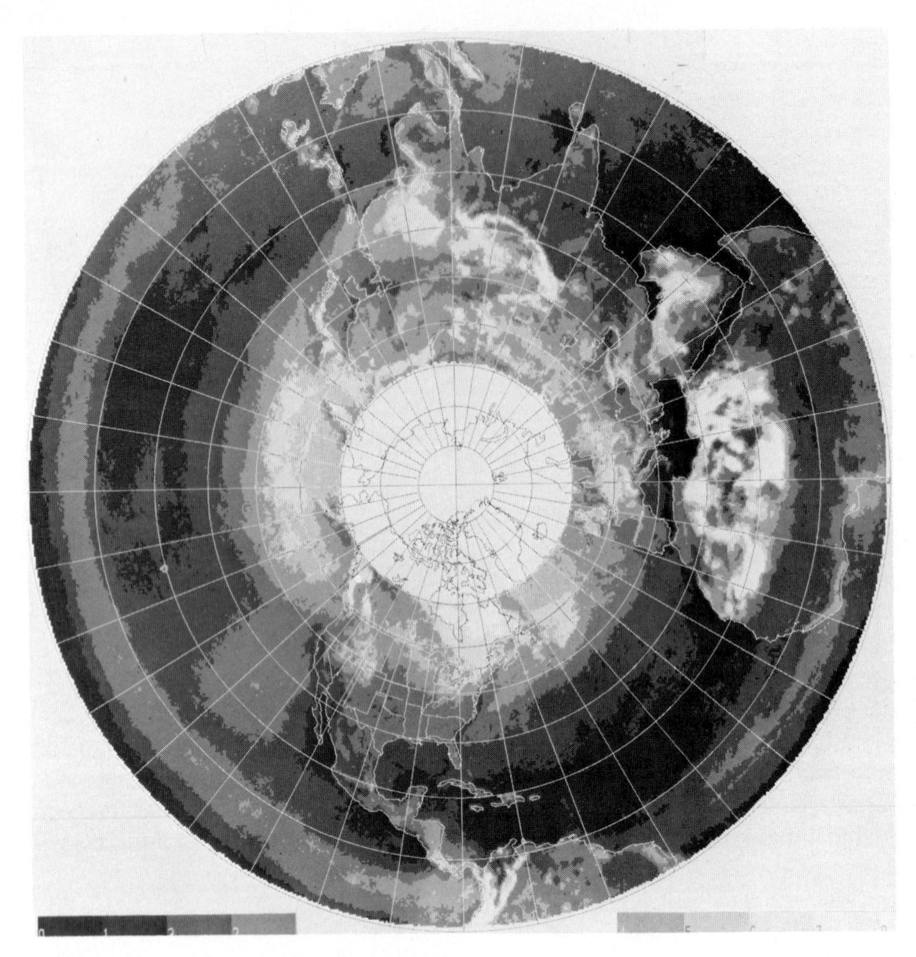

FIGURE 3-13 Average annual cloud cover as measured by satellites from 1967 to 1970. Note the cloud band at about 5° N latitude, which affects the albedo (as shown in Fig. 3-12), and also the dense cloud cover north of about 40° N latitude. (From *Global Atlas of Relative Cloud Cover 1967–70*, NOAA and USAF, 1971.)

of the atmosphere (Fig. 3-14). About 20% of the energy from the sun is absorbed by the atmosphere (approximately 17% by air and 3% by clouds). This is absorbed primarily by ozone and water vapor. Satellite data indicate that the amount of solar radiation reflected back to space is about 30%. This amount is composed of back-scattering by the atmosphere (6%), reflection off the cloud tops (20%), and reflection from the earth's surface (4%). Therefore, the average albedo of the earth-atmosphere system is about 30%. This leaves a deficit of 70% above the atmosphere, which is balanced by infrared radiation from the earth.

If the solar radiation is not absorbed by the atmosphere or reflected by the earth's surface, it is absorbed by the earth's surface. This amounts to about 50% of the solar radiation. The energy absorbed by the atmosphere is used primarily for emitting infrared radiation. The infrared radiation emitted by the atmosphere goes into space as well as down to the earth's surface. This loss of radiation by the atmosphere amounts to more than the

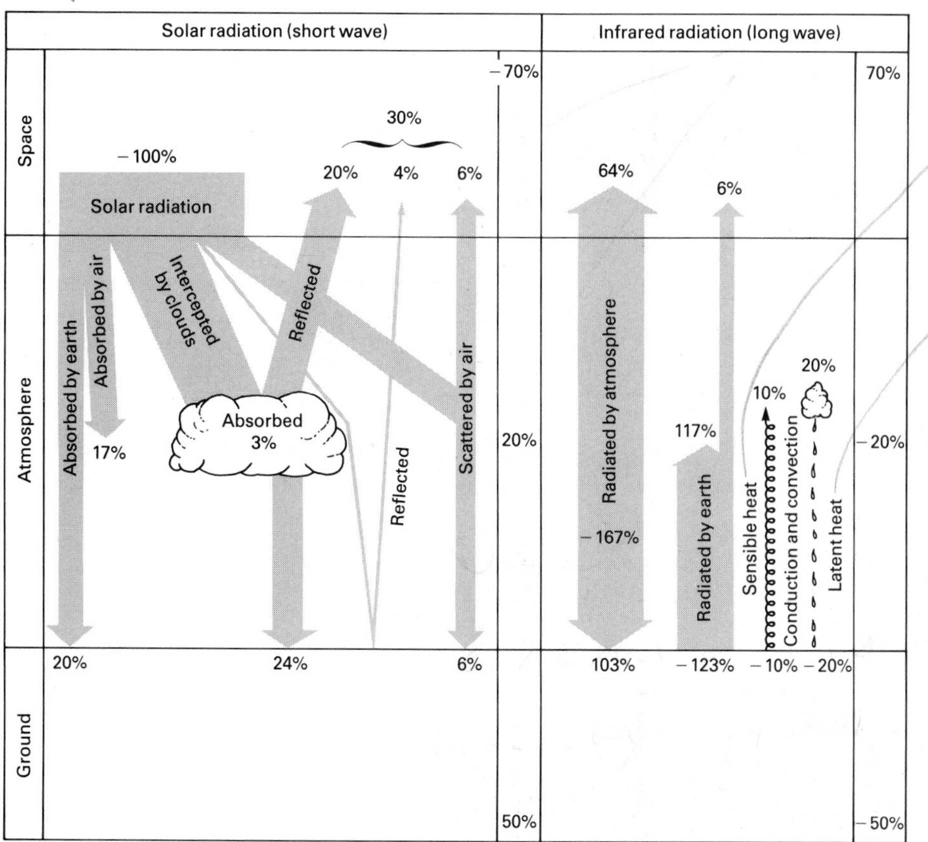

FIGURE 3-14 Global heat budget for solar and terrestrial radiation, showing the percentages of losses and gains to space within the atmosphere and at the earth's surface.

amount of sunlight absorbed by the atmosphere.

The atmosphere could not lose more radiation than it gains for very long without getting colder and colder unless other energy were available—and it is. Heat is transferred to the atmosphere from the earth's surface. Some energy is transferred by heat we can feel as the sunlight absorbed by the earth's surface warms the lower few meters of the atmosphere by conduction. Deeper layers of the atmosphere are affected as the earth's surface warms the atmosphere through convection and turbulent mixing. Considerable heat is also transferred upward as water evaporates from the earth's surface and condenses as clouds in the atmosphere, releasing latent heat in the process. The amount of radiation emitted by the atmosphere is so great that the absorbed radiation from the earth, sensible heat, and latent heat transfer to the atmosphere, as well as the absorbed sunlight, are all required to balance this heat loss.

If the global heat budget for the whole earth-atmosphere system did not balance, the earth

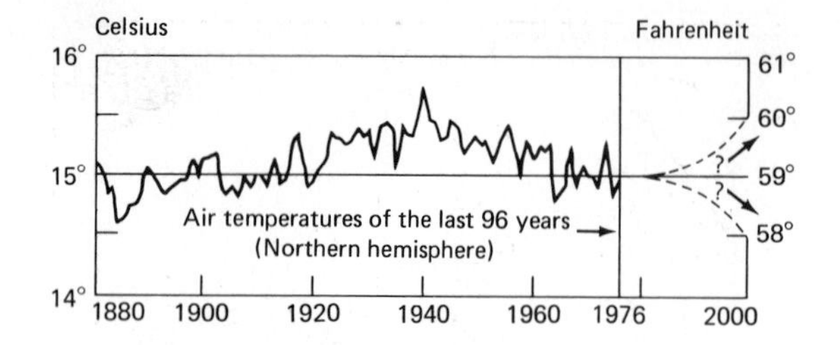

FIGURE 3-15 Long-term average temperatures of the northern hemisphere show variations during the period of time that measurements have been available. The temperature increased until 1940, and then began to decrease. (Adapted from a *National Geographic* drawing.)

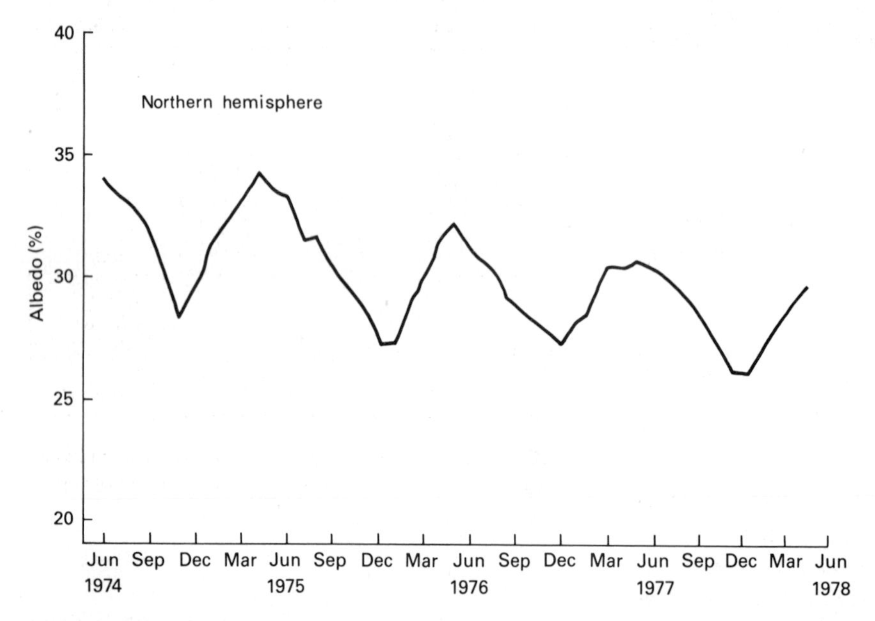

FIGURE 3-16 Time series of mean monthly albedo for the northern hemisphere from satellite observations. Note the gradual decrease in albedo for this time period. (From *Earth-Atmosphere Radiation Budget Analyses Derived from NOAA Satellite Data, June 1974–February 1978*, vol. 2, NOAA.)

would get colder or hotter over a long period of time. It is generally assumed that the heat budget balances for short periods of time. However, long-term temperature measurements indicate variations (see Fig. 3-15). The mean temperature of the northern hemisphere increased for several decades until about 1940 and then began a gradual decrease. It is not known with certainty whether this decrease will continue.

Changes in albedo are interesting, as they have an effect on temperature. Beginning in 1974, satellite measurements showed a gradual decrease through time. This short-term lowering of albedo would indicate a continuing warming trend if these few years shown in Figure 3-16 are representative of longer trends.

LOCAL ATMOSPHERIC HEATING

The temperature of the lower atmosphere is so important to life on earth, as well as to weather processes, that it deserves special attention. The air temperature within the lower atmosphere is related to the earth's surface temperature. Since the

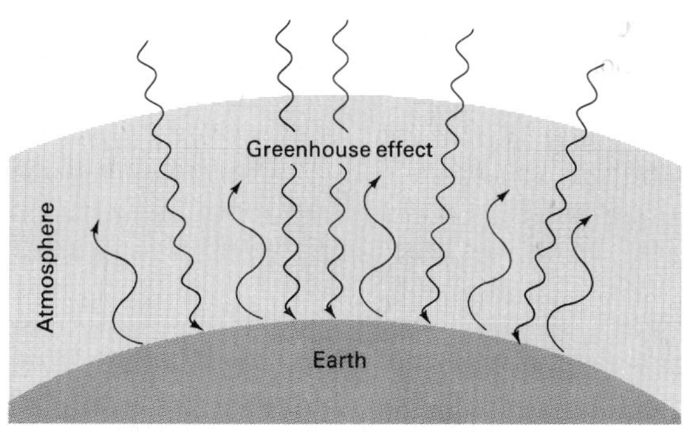

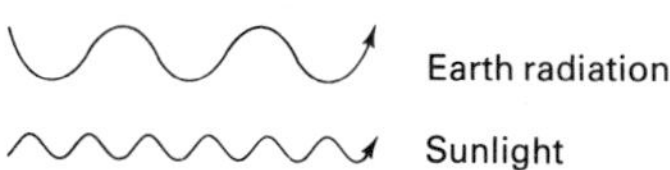

FIGURE 3-17 The greenhouse effect is important in determining the earth's equilibrium temperature. It arises because the atmosphere is more transparent to solar radiation than to terrestrial radiation.

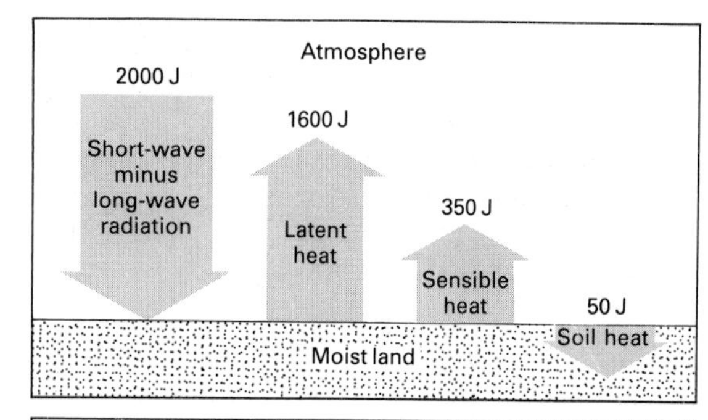

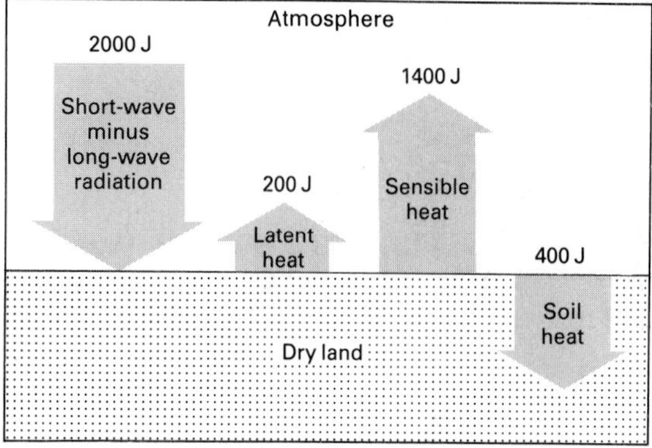

FIGURE 3-18 Local heat budget for a moist surface and for a dry surface, assuming 2000 J of available energy.

air is transparent to solar radiation in comparison to the ground, much of the atmospheric heating occurs as the ground absorbs energy in a very thin layer and transfers it upward by conduction, convection, radiation or evaporation. Molecular conduction from the ground to the atmosphere is very slow and affects only a very shallow atmospheric layer. Convection and turbulent mixing (interaction of parcels of air), along with radiation from the ground to the atmosphere, are responsible for warming most of the lower atmosphere.

The transfer of heat from the ground to the lower atmosphere is rapid, as demonstrated by the differences in air temperature between a sunny and cloudy day. The temperature on a cloudy day does not increase very much during the daylight hours as the cloud cover reflects most of the solar radiation and reduces the supply of heat at the ground. On a sunny day, the average diurnal temperature variation is about 15°C; on a cloudy day, this variation is only a few degrees. The absorption of solar radiation on a sunny day increases the temperature of the earth's surface rapidly. As it heats, the lower atmosphere is also warmed, by convection and radiation emitted from the ground. Infrared radiation is readily absorbed by the atmosphere, in contrast to shorter wavelength solar radiation, which is transparent to the atmosphere (Fig. 3-17). This absorbency of infrared radiation and transparency of solar radiation creates an effect called the **greenhouse effect,** which gets its name from its similarity to what goes on in a plant greenhouse. The primary absorbers of the atmosphere are carbon dioxide and water vapor.

The local heat budget is determined by the amount of radiation and its partitioning between the various processes that use energy at the earth's surface. Consider a typical summer day, in midlatitudes, where about 2000 J of energy are available per square centimeter of earth's surface. If it has rained recently, most of the energy is used for the latent heat of evaporation of water (Fig. 3-18). The evaporation of a 1-cm depth of water requires 2400 J, since this is the latent heat of evaporation, as described in Chapter 1. About 1600 of the 2000 J goes into the evaporation process, resulting in the evaporation of 0.67 cm of water. Almost all the remaining energy (350 J) goes into heating the atmosphere, with a small amount (50 J) going into warming the soil.

When the soil is dry, as in desert regions, energy cannot be used for the evaporation of water. Much more energy, about 1400 J, therefore goes into heating the atmosphere than is the case in wetter climates. A larger amount than in wetter climates, about 400 J, also goes into heating the soil, with only 200 J used for evaporating water. Because the radiation that would normally be used for the evaporation of water is used instead for warming the land and air above, the desert is very hot during the day. In spite of the daytime heat, however, deserts are colder than humid regions at night because less water vapor is present in the atmosphere to absorb the earth's radiation. This means that more infrared radiation is lost at night, resulting in faster cooling of the land and air near the ground. Thus, deserts are cooler than normal at night and hotter than normal during the day because of the local energy budget.

The extreme surface temperatures are interesting to note. The lowest temperature on record is −88°C (−127°F), measured in the Antarctic. The world record maximum temperature (58°C) was recorded in Libya in 1922. Within the contiguous United States, the lowest temperature on record is −57°C (−70°F) measured at Roger's Pass, Montana, in 1954. The lowest temperature measured in Alaska is only slightly lower, −60°C. The highest temperature measured in the United States was in Death Valley, California, where the temperature reached 57°C (134°F).

The maximum diurnal variation in air temperature occurs near the earth's surface because of the influence of the ground temperature on air temperature, with cooling at night and heating during the day. Throughout the troposphere the air temperature normally decreases with height. However, during the day, in the lower kilometer, the atmosphere warms rapidly as solar radiation is absorbed

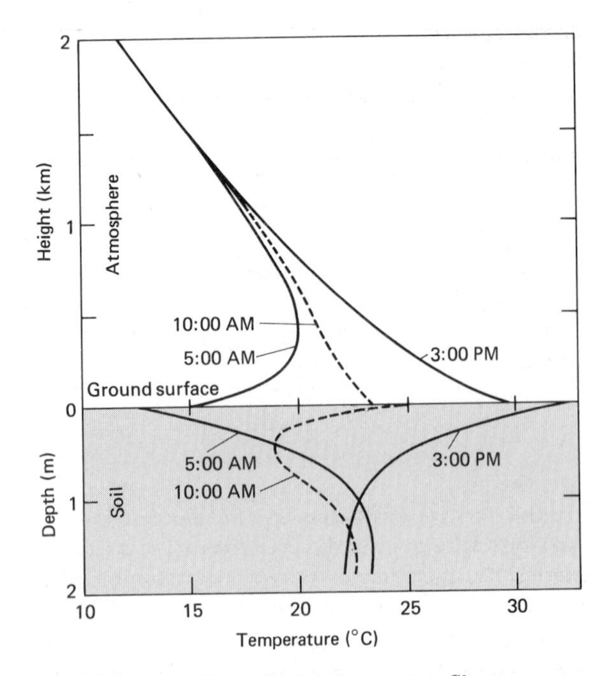

FIGURE 3-19 Typical temperature profiles throughout the day for the lower atmosphere and soil surface to a depth of 2 m.

(Fig. 3-19). During the night, a surface inversion develops through several meters or tens of meters in the lower atmosphere since the air above is cooled less rapidly. For this reason, measurements of the diurnal variation in temperature at various elevations in the atmosphere show the largest variations closest to the earth's surface.

Effects of Surface Material The particular type of earth surface material greatly influences the amount of heating that results from the interception of sunlight. Water is much more transparent to radiation than land, and radiation penetrates to a greater depth in the ocean (Table 3-2). Only about one-half of the solar energy is absorbed in the first 11 m of the ocean. Land surfaces absorb all the energy in the first few millimeters of soil. Therefore, there are large differences in the absorption of energy by oceans and continents: the oceans heat in a thick layer and continents heat in a very thin layer.

TABLE 3-2

Summary of characteristics of earth surface materials

Properties	Water	Snow or ice	Dry soil	Wet soil	Clear air
Reflection	8%	80%	15%	10%	7%
Absorption of 50% of the sunlight	11 m	5–12 cm	1 mm	1 mm	Many atmospheres
Heat capacity (cal/cm^3/deg)	1.0	0.50 (ice)	0.33	0.66	0.00024
Conduction of heat	Turbulent mixing (fast)	Molecular conduction (slow)	Molecular conduction (slow)	Molecular conduction (slow)	Turbulent convection (fast)

A separate influencing factor for heating rates is the large difference in the **heat capacity** of land and water. The heat capacity determines the quantity of energy required to warm a unit volume of material 1 degree. The ocean requires about three times as much energy for an equal temperature increase as the same volume of soil. Another way to express this characteristic is that water has a large heat storage capacity. One calorie of energy must be absorbed to raise the temperature of 1 cm^3 of water 1°C. The same quantity of heat will warm 1 cm^3 of soil about 3°C. Therefore the heat capacity favors much more rapid heating of land surfaces than oceans.

An even more important factor in the heating rates of oceans and continents is the type of mixing that occurs. Water is free to distribute heat by turbulent or convective mixing of warm and cold water, while soil transfers heat by molecular conduction. Molecular conduction is much slower than turbulent mixing of heat, and thus also contributes toward a faster heating of the top surface layer of the land.

All three factors—absorption characteristics, heat capacity, and mixing characteristics—favor the rapid warming of land surfaces and the great heat storage capacity of the oceans. For this reason continents are much hotter in summer, cool much faster in the fall, and become very cold in winter as compared to oceans. Hotter land during the summer means warmer, less dense air above, with resulting lower atmospheric pressure. During the winter, the air is denser over the continents, giving rise to higher atmospheric pressure.

On the local scale, differences in atmospheric heating develop because lakes, for example, absorb sunlight differently from land surfaces. Smaller differences in heating from sunlight occur because of variations in type of soil, moisture content of the soil, and vegetative cover. More solar radiation is frequently absorbed in one area than in another, eventually influencing the temperature of the air above. Since the atmosphere obeys the ideal gas law, warm air expands and becomes less dense while cold air contracts in volume and becomes denser. The density of air is directly related to pressure through the gas law. When pressure gradients develop, the air begins to move, as will be described in more detail in the next chapter. Thus solar heating sets up the mechanism that causes the movement of air, creating wind.

Redistribution of Heat Differences in solar heating occur globally as well as locally. Tropical regions absorb excess energy from the sun because of direct rays while polar regions absorb very little energy because of less-direct sun rays (Fig. 3-20). As a result of large inequities in atmospheric heating, the air is more expanded and less dense in tropical regions. Atmospheric density differences

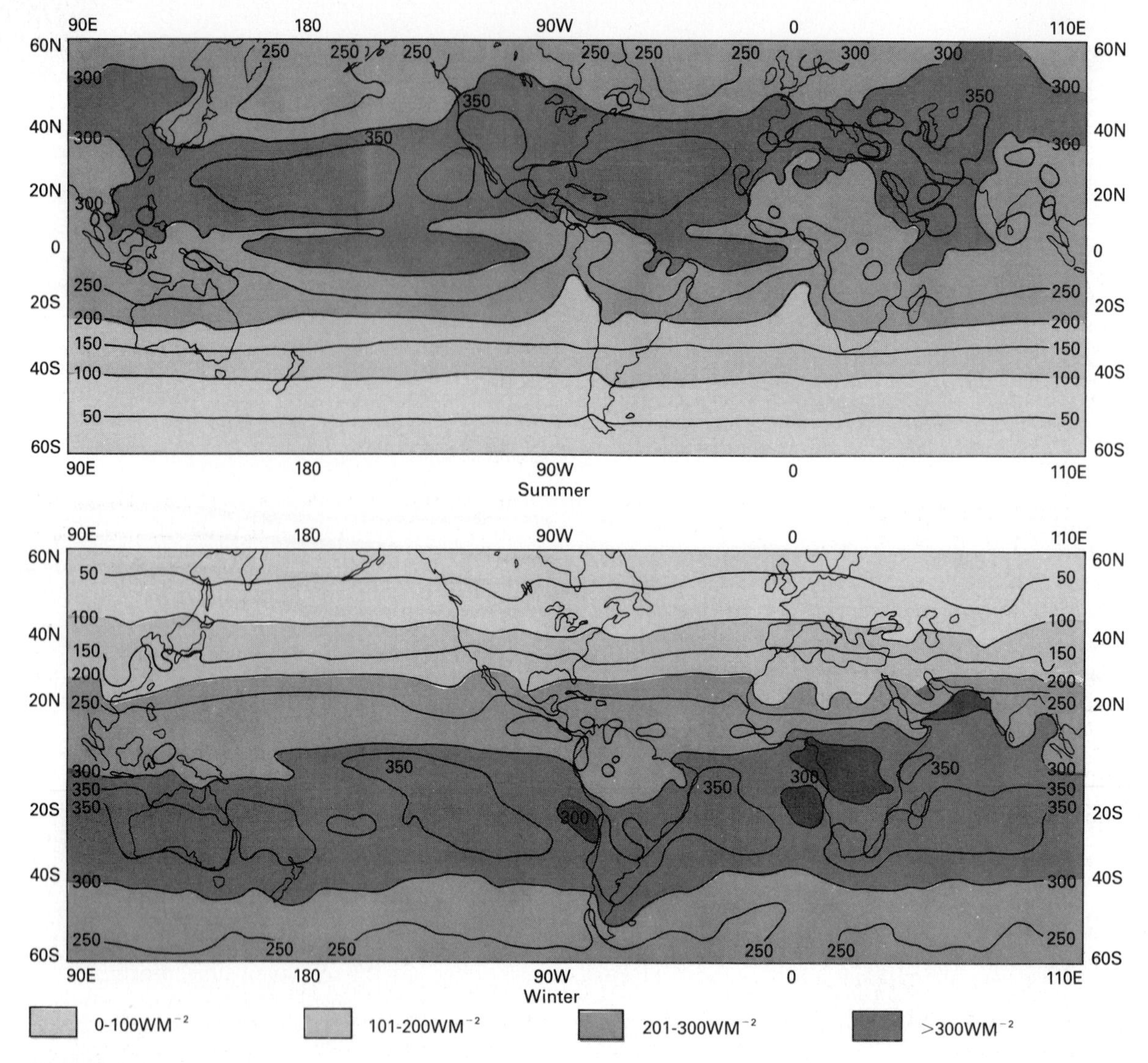

FIGURE 3-20 Amount of absorbed solar radiation varies with season and with location, as shown by satellite observations. (From NOAA Meteorological Satellite Laboratory, Washington, D.C.)

create pressure differences, which produce global wind systems.

A general relationship exists between solar radiation and larger atmospheric circulations such as the polar jet stream. After the jet stream is in motion, however, it is influential in redistributing heat and developing other circulations as it meanders through the atmosphere. The polar jet stream occurs at the boundary between equatorial and polar air. Surplus energy is absorbed by the earth-atmosphere system from the equator to about 30° N latitude with a deficiency of energy northward (Fig. 3-21). More than 4000 trillion watts must be transferred northward across 30° N latitude to maintain normal temperatures in the tropical and polar regions. The jet stream with its associated

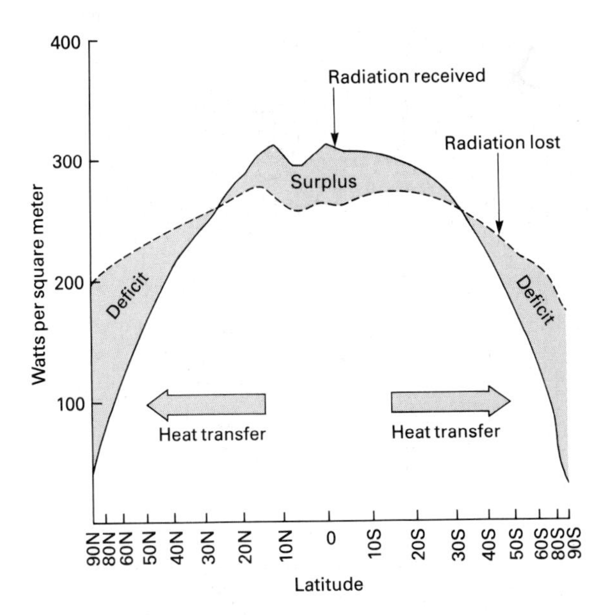

FIGURE 3-21 Comparison of the incoming solar radiation and outgoing terrestrial radiation in different latitudes, showing the resulting surplus and deficit of energy for different latitudes that results. The transfer of tremendous quantities of energy is required to maintain a balance. (Satellite data 1974–78 from NOAA Meteorological Satellite Laboratory, Washington, D.C.)

smaller storm systems is an important part of the earth's heat exchange system because of the high wind speeds and momentum that develop.

More heat must be transferred across midlatitudes than other regions, as shown in Figure 3-21. Heat is transported northward by latent energy and by large air masses of different temperature (Fig. 3-22). The jet stream occurs at the boundary between cold polar air and warm equatorial air. As the jet stream meanders through the atmosphere the amplitude (north-south variation) of the **long waves** frequently changes. As the amplitude increases, cold air masses are brought southward in the cyclonic bends of the jet stream, while hot air masses flow northward in the anticyclonic bends. As the jet stream returns to a straighter path around the earth, pools of cold air are displaced southward while pools of warm air are displaced northward. Tremendous quantities of heat are exchanged by this process, just as a large amount of heat is involved in the evaporation-condensation process.

Ocean currents are another important mechanism of redistributing heat. For example, the cold current off the California coast has a moderating effect on the air temperature of southern California. The warm northeasterly Gulf Stream along the East Coast greatly affects winter temperatures there.

SUMMARY

Solar radiation provides energy for driving all weather systems. Radiation reaches the earth from the sun in about eight minutes and can be specified by either wavelength or frequency. The electromagnetic spectrum includes all the different wavelengths of radiation. Sunlight is composed of ultraviolet, visible, and infrared radiation, with 95% of the radiation having wavelengths between 0.25 and 2.5 μm. Radiation from the earth is in the infrared part of the spectrum, with 95% of the radiation having wavelengths between 2.5 and 25 μm.

When sunlight strikes any substance, it is reflected, absorbed, or transmitted. Seasonal and diurnal temperature variations at the earth's surface lag behind the maximum receipt of energy since the earth continues to warm as long as more radiation is received than is emitted.

The heat budget for the earth-atmosphere system can be evaluated by considering what happens to solar and terrestrial radiation. The atmosphere absorbs about 20% of the solar radiation going through it, and the albedo of the earth-atmosphere system is about 30%. This leaves 50% of the solar radiation to be absorbed at the earth's surface. This energy is used to warm the earth—with subsequent loss by infrared radiation—to evaporate water, or to supply energy for convection in the atmosphere. There are indications that long-term inbalances occur in the global heat budget, causing the temperature to change over long periods of time.

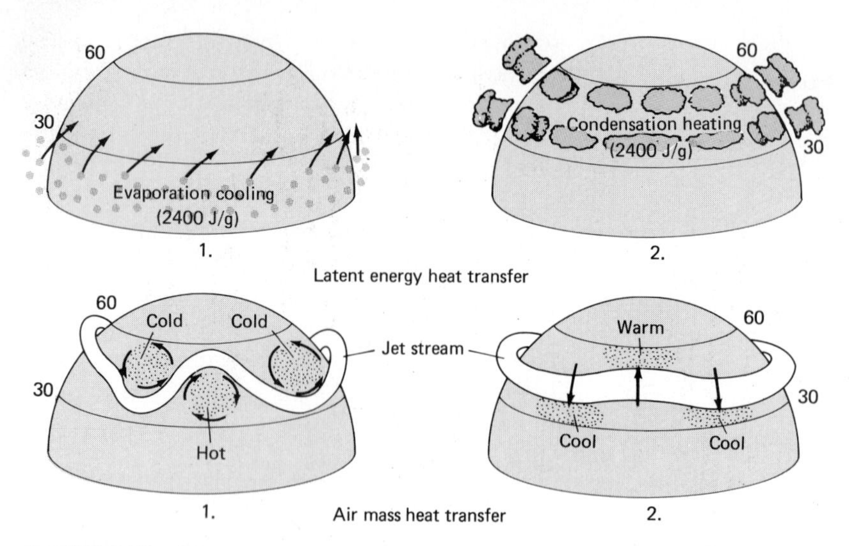

FIGURE 3-22 Heat exchanges occur as latent energy and warm air are transferred northward. The evaporation of water in tropical regions and the condensation of moisture over midlatitudes transfers large amounts of heat. Also, cold air masses are brought southward while warm air masses are carried northward by the meandering nature of the jet stream.

Although a direct relationship exists between solar heating and some atmospheric circulations, many other weather systems are indirectly related to solar heating through the jet stream. The jet stream is important in heat exchange as well as in driving smaller storm systems.

Large quantities of heat are transferred across midlatitudes as the jet stream meanders between the boundary separating cold polar air from warm equatorial air. Heat is transported northward both by the transfer of latent energy and the displacement of warm air masses northward and cold air masses southward. Additional heat is redistributed by ocean currents, which carry warm water northward in some locations and cold water southward in other locations.

STUDY AIDS

1. Explain why the earth intercepts the same amount of sunlight as would be intercepted if it were a flat disk of the same diameter placed perpendicular to the sun's rays.

2. Explain why polarized sunglasses are effective.

3. Select two different parts of the electromagnetic spectrum and compare the characteristics of energy at these wavelengths.

4. Compare solar and terrestrial radiation.

5. Calculate the wavelength of maximum emission for an object that has a temperature of 320 K.

6. Explain the atmospheric window.

7. List and describe the gains and losses of energy by the atmosphere.

8. Discuss the heat budget of a desert surface.

9. Explain how the jet stream transports large quantities of heat across midlatitudes.

TERMINOLOGY EXERCISE

Use the glossary to reinforce your understanding of any of the following terms used in Chapter 3 that are unclear to you.

- Solar constant
- Radiation
- Electromagnetic spectrum
- Visible light
- Infrared radiation
- Ultraviolet radiation
- Blackbody emission
- Albedo
- Solstice
- Equinox
- Atmospheric window
- Absorption bands
- Greenhouse effect
- Heat capacity
- Long wave

THOUGHT QUESTIONS

1. Do you think the weather proverb "Red sky in morning, sailors take warning" has any basis in fact? Explain.

2. Can you explain why dew forms on objects at the ground instead of falling from the air?

3. What do you think is the main heat transfer process away from the surface of the moon?

4. If a sheet of black metal and a black cloth were both placed on top of a snow cover, which of these would cause the snow to melt faster? Explain your answer.

5. If an unexpected cloud cover develops after the minimum temperature has been forecast for a night, how would you expect this to affect the accuracy of the forecast? Explain your answer.

6. What kinds of relationships can you describe between topography and the removal of snow cover on a sunny day?

SPECIAL TOPIC 4

DUST DEVILS

You may have noticed small vortices picking up dust on a warm summer day. These dust devils or whirlwinds are especially common in hot deserts (see Fig. S4-1). They are formed as heated air rises in "thermals" due to convection. Beneath the rising air current, a vortex may form as surface air flows inward toward the region that has a reduced air pressure. Dust devils are related to heating at the ground and therefore have something in common with hurricanes. However, they are very small atmospheric vortices that are typically too weak to cause damage to people or their property.

Measurements in dust devils have shown that the inner core contains a downdraft, and that the direction of rotation is cyclonic for some dust devils and anticyclonic for others. Their size is so small that no preference seems to exist for the direction of rotation, in contrast to larger atmospheric vortices which all rotate cyclonically in the northern hemisphere. The only other exception to this physical principle is the formation of extremely rare anticyclonic tornadoes. We can conclude that the earth's rotation affects the direction of rotation of most tornadoes and all larger atmospheric vortices, but does not influence the direction of rotation of atmospheric vortices smaller than tornadoes.

FIGURE S4-1 A dust devil in action a few miles south of Phoenix, Arizona. (Courtesy of Sherwood B. Idso.)

CHAPTER FOUR
HORIZONTAL WINDS

WIND SPEED AND DIRECTION

Winds blow because of differences in pressure. Pressure gradients may develop on a global scale or on a local scale because of differences in heating. On the global scale the resulting atmospheric motion consists of the jet stream flow and movement of large air masses. On the local scale a smaller quantity of air is involved.

The surface wind speed and direction can be quite variable over a short period of time (Fig. 4-1). The wind speed also varies between day and night at any one location. Surface winds are lower at night and have less turbulence because there is no solar heating. Wind speed increases during the daytime, with large fluctuations developing as the earth's surface warms unevenly. These fluctuations are expressed as gustiness of the wind during the daylight hours.

Wind directions change as the pressure patterns change. At midlatitude locations in the central and eastern United States, shown in Figure 4-2, the average wind direction during the summer months is from the south. Wind directions from the east or west are much rarer. Winter winds are from the north more frequently, but a high percentage of the winds are still from the south; again the smallest frequency of winds is from the east or west. On an annual basis, winds are most frequently from the south.

Wind directions in the western United States are more frequently from the west and northwest. In mountain regions the wind directions are influenced by the topography and this is shown by the variety of predominant winds at the ground.

In order to understand wind direction or speed, specific cases must be considered in relationship to the pressure distribution. The surface synoptic map in Figure 4-3 shows that the winds at different locations are related to the pressure distribution. The wind is blowing cyclonically and slightly in toward the center of the low-pressure area, and anticyclonically and slightly away from the high-pressure area. Southwest winds occur to the south of a low-pressure center while northeast winds blow to the north of a low-pressure center. At any given time, the pressure pattern governs the wind direction as well as the wind velocity.

Prior to the arrival of a low-pressure system that is traveling to the north of a given location, the wind will be blowing from the south (Fig. 4-4). As the low-pressure center moves directly north of the location, the winds may shift more to a westerly direction. It is more likely, however, that the wind will shift suddenly to the northwest, as the cold front accompanying the low-pressure center passes the location. As the low moves farther eastward, the winds shift to a more northerly direction.

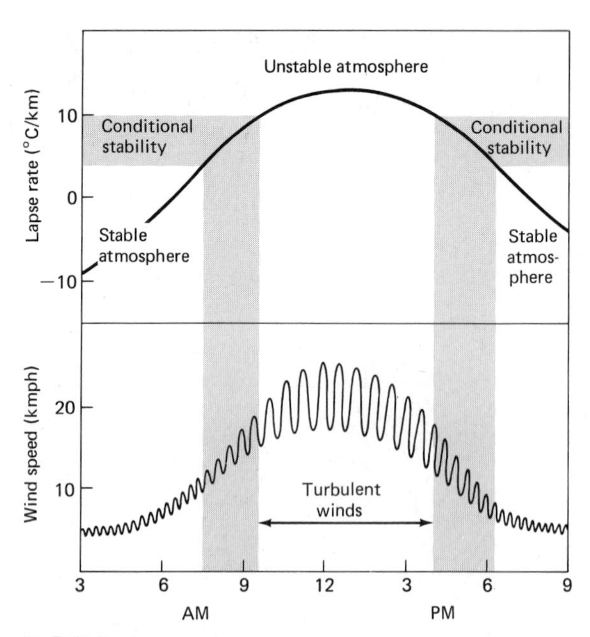

FIGURE 4-1 Typical diurnal wind speed variations and related atmospheric stability conditions.

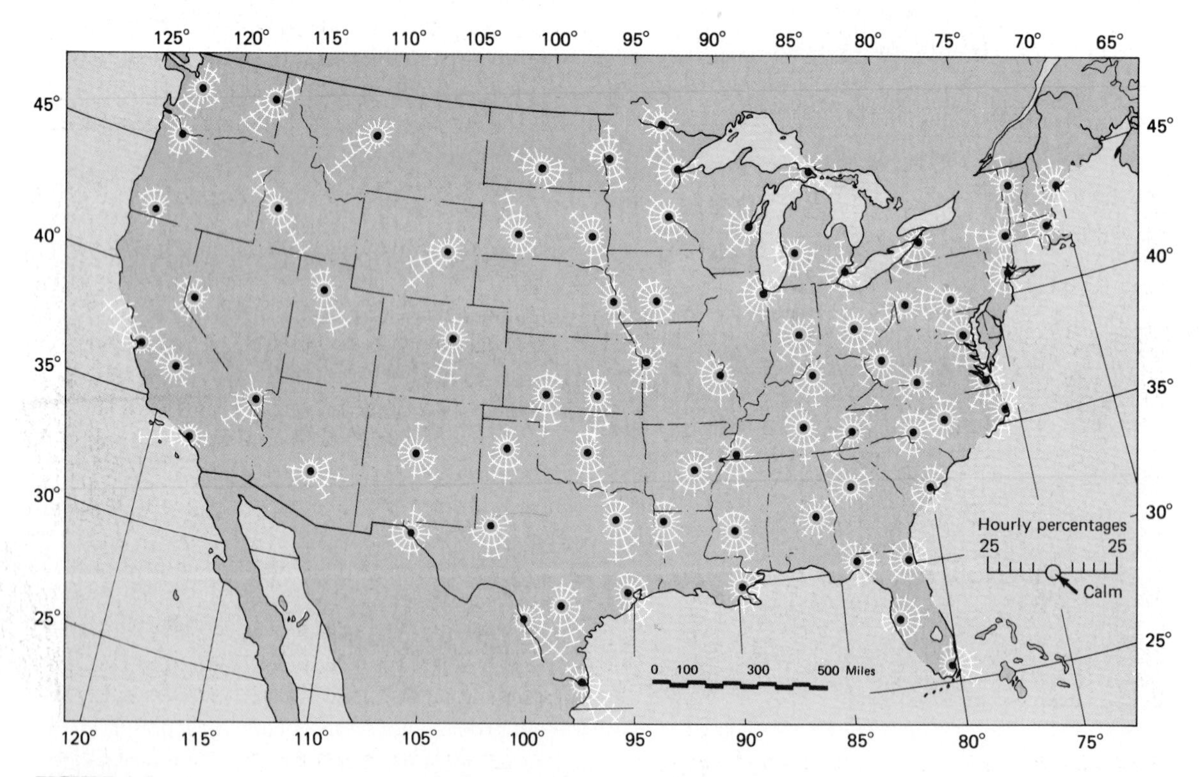

FIGURE 4-2 Average annual percentage frequencies of wind direction. Each location shows the percentage of time that the wind blew from various directions. (Environmental Data and Information Services, NOAA.)

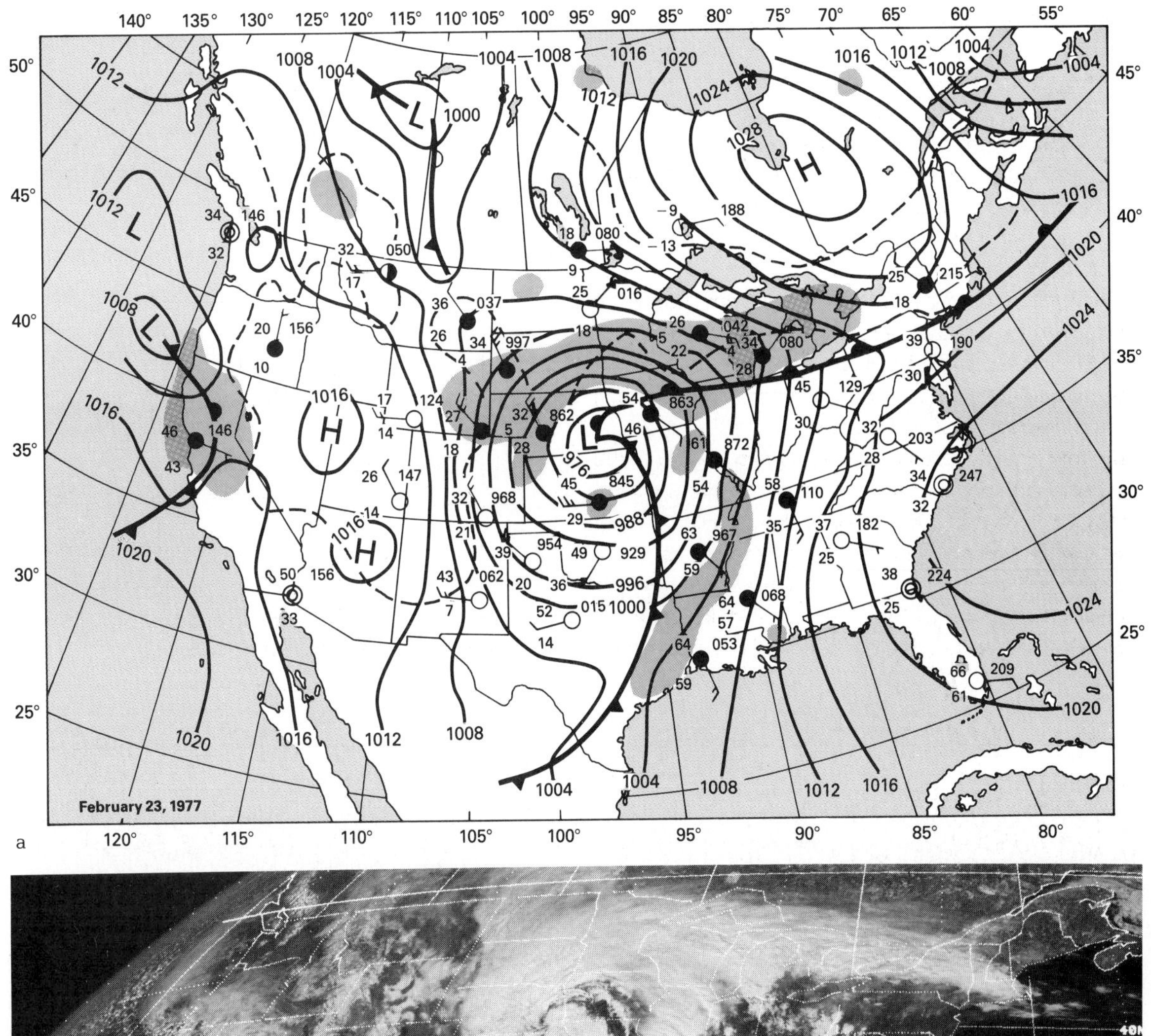

FIGURE 4-3 (a) Synoptic weather map for February 23, 1977, showing a large midlatitude cyclone in the central United States, with winds cyclonic and slightly toward the center of lower pressure. An area of high pressure is over the western United States with smaller pressure differences and lighter winds. (b) Satellite photograph of this storm.

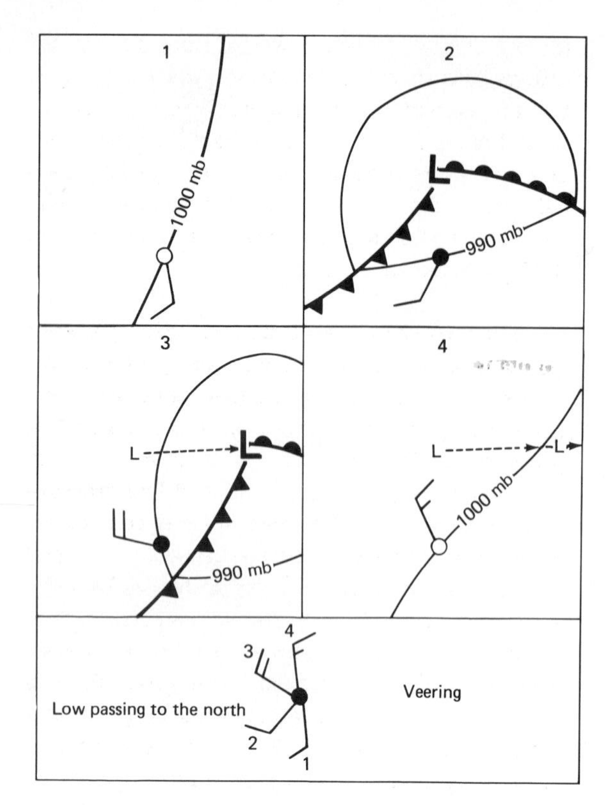

FIGURE 4-4 The change in wind direction as a low-pressure center (L) passes to the north of a particular location. In this case the winds veer with time.

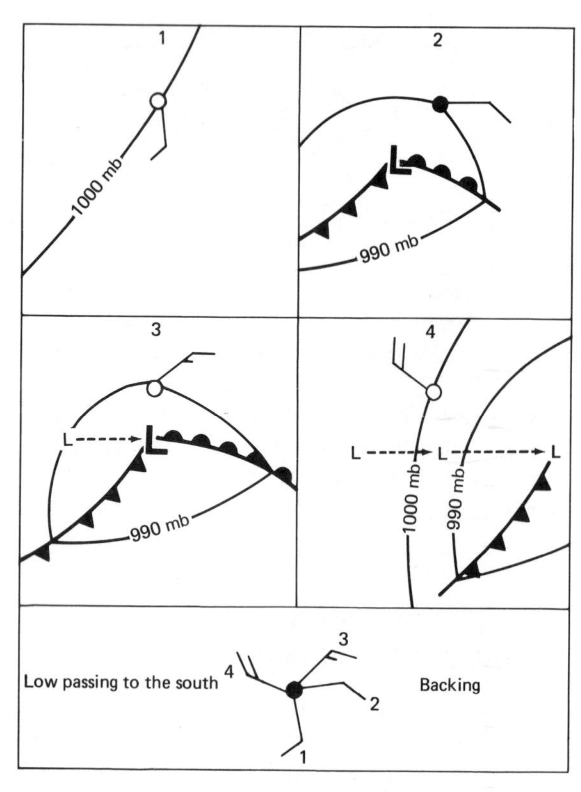

FIGURE 4-5 Wind direction changes as a low-pressure center passes to the south of a particular location. When this happens, the winds back with time.

It is a general rule that a wind shift from the southwest to the west or northwest indicates that a low-pressure center has moved past to the north of that location. When the winds shift in such a way (clockwise direction) they are said to **veer** with time.

If a low-pressure system passes to the south of a location, the wind shifts in the opposite direction (Fig. 4-5). Prior to the arrival of a storm center that is to the west, the wind direction will be from the south. As the center moves to the south of the location the wind will shift to an easterly direction, then to the northeast, and eventually to the northwest. The change will therefore be from the south to the east and then to the northwest. Such a wind change in a counterclockwise direction is called **backing** with time.

The weather at a particular location is influenced by the position of the low-pressure center. The area of greatest snowfall is generally located to the north of the low-pressure center. This is also the area that has east winds. Because of this, east or northeast winds can be used as an indication of snow if the temperature is slightly below freezing and the moisture content of the air is high.

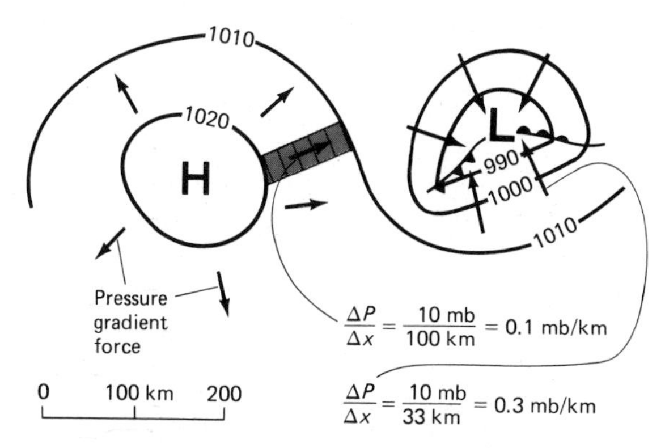

FIGURE 4-6 The pressure gradient force is related to the change in pressure (ΔP) per unit distance (Δx) and is always directed from higher toward lower pressure.

FORCES THAT CAUSE HORIZONTAL WINDS

Atmospheric motion in a horizontal direction is caused by a limited number of forces and accelerations. The force that starts the action is the pressure gradient force, which arises from differences in the horizontal distribution of atmospheric pressure. After air is in motion, it is affected by acceleration developed from the earth's rotation (Coriolis acceleration), centrifugal acceleration, and frictional deceleration in the lower atmosphere.

Newton's second law states that forces applied must equal the amount of acceleration multiplied by the amount of mass. If a given force is applied to a particular mass the acceleration (change in velocity with time) of the mass is determined. These physical principles apply to the atmosphere as well.

Accelerations in the atmosphere are the result of the sum of all the forces acting on any unit mass of the atmosphere. If the forces acting on the atmosphere are known, the resulting acceleration is known. Since Newton's second law states that accelerations are equal to forces per unit mass, the Coriolis acceleration, centrifugal acceleration, and frictional deceleration are commonly treated as forces. This is correct, however, only if force per unit mass is specified. Since the acceleration of air that arises from the changing north-south and east-west directions on our rotating earth is easier to visualize than an imaginary force causing such an acceleration, the winds of the atmosphere will be explained by using accelerations, with reference to the forces that cause them.

Pressure Gradient Acceleration The **pressure gradient force** is determined by the pressure pattern at any given time. A gradient of pressure is a change in pressure per unit distance. Consider a pressure distribution, as shown in Figure 4-6, with closed isobars around the high- and low-pressure centers. The isobars are frequently spaced closer together around a low-pressure center than around a high-pressure center. If the distance from the 1020-mb to the 1010-mb pressure is 100 km, then the pressure gradient will be 0.1 mb/km, directed from the higher toward the lower pressure. The pressure gradient is more than twice as great near the low-pressure center. The air would accelerate along the pressure gradient toward the lower pressure if this were the only force acting on the air mass.

Routine pressure measurements over land surfaces give the pressure pattern at any particular time. An isobaric analysis shows the position of high- and low-pressure centers and the gradients between them. Quantitative information on the pressure gradient is therefore available. Accelerations that arise from the pressure gradient force are always directed from higher to lower pressure, with magnitudes determined by the change in pressure per unit distance. Thus accelerations and resulting wind velocities are related to the pressure gradient, as observed on any surface weather map. Measured wind directions in the atmosphere are not directly from high to low pressure because more than one factor affects the movement of the air at any time.

Coriolis Acceleration The second major factor in determining atmospheric motion is the **Coriolis acceleration,** which arises because of the rotation of

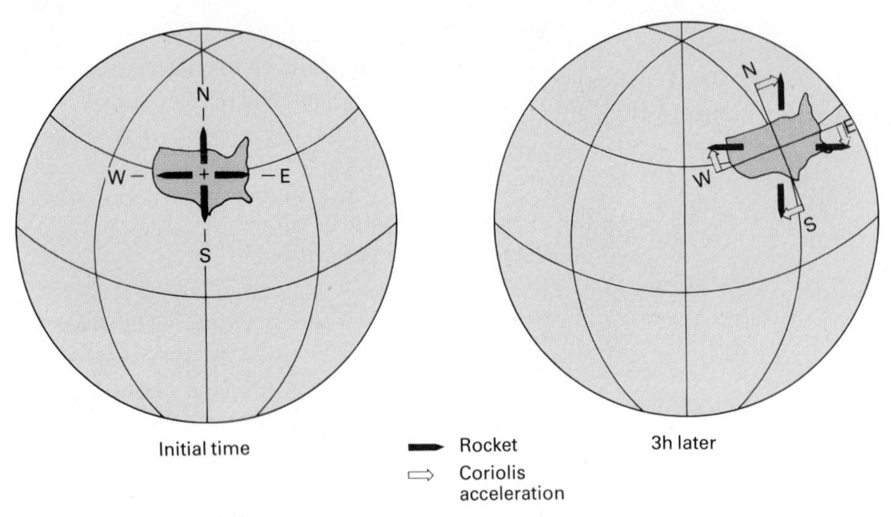

FIGURE 4-7 Coriolis acceleration develops because of the earth's rotation. Consider four rockets fired from the central United States. Three hours later the earth has rotated enough to change the north-south and east-west directions significantly in relation to a fixed point in space. This produces a real acceleration to the right of the path of motion, in the northern hemisphere, as an observer looks from the rotating earth at the departing rockets.

the earth. Any object moving in the atmosphere, or space, not located over the equator, travels along a curved path, as observed from the rotating earth. Consider rockets that are fired northward and southward, as in Figure 4-7. In only three hours the earth has completed one-eighth of an entire rotation. Accompanying the rotation is a change in the northerly direction, if observed from a fixed point in space, or from the rockets, which are no longer attached to the earth. The earth has, in effect, twisted the north-south and east-west directions beneath the rockets. Since we measure directions from fixed points on earth, the rockets have a real acceleration to the right of their path of motion when compared with the shifting north-south line and viewed as they move away from the observer. As indicated in Figure 4-7 this acceleration—the Coriolis acceleration—is greater toward the poles and less toward the equator.

East-west motion is also accompanied by an acceleration to the right of the path of motion because the earth is rotating. Although the rockets, for example, could have straight paths, as observed from a fixed point in space, their paths are curved to an observer on the rotating earth (Fig. 4-7). The earth actually moves "out from under" the rockets toward the left in the northern hemisphere. Because of the Coriolis acceleration, all wind currents in the northern hemisphere are accompanied by deflection to the right of their path of motion.

The earth's rotation affects all wind currents in the southern hemisphere in the same way, except for the change in perspective with respect to the earth's rotation (see Fig. 4-8). All air currents in the southern hemisphere are deflected to the left of their path of motion, instead of to the right, as in the northern hemisphere.

The Coriolis acceleration also has magnitude that depends on the latitude, the velocity of the moving mass, and the speed of the earth's rotation. Since the latitude and speed of the earth's rotation are constant at a particular location, the magnitude there is determined directly by the velocity of the moving mass. This relationship exists because the twisting of the earth-oriented directions is accentuated as masses (rockets in Figs. 4-7 and 4-8, for

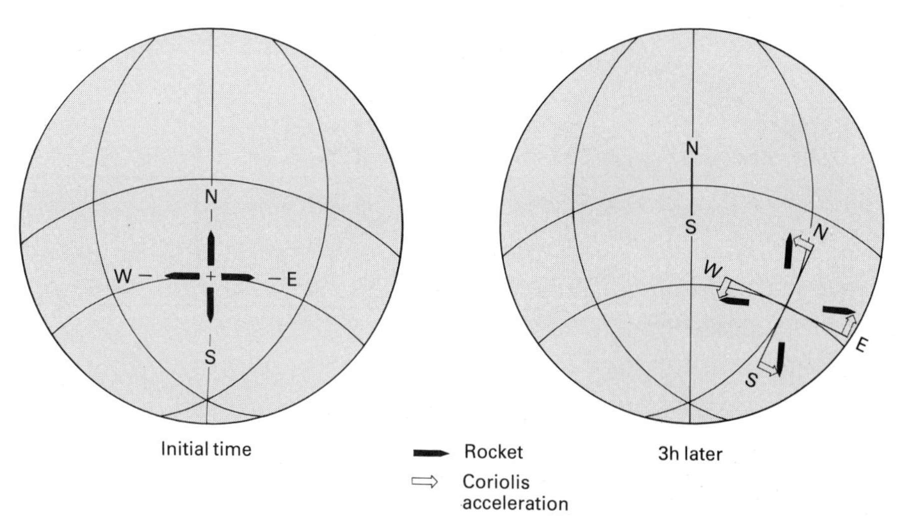

FIGURE 4-8 Coriolis acceleration is to the left of the path of motion in the southern hemisphere because of the earth's rotation. It is zero at the equator because the primary directions do not change there as the earth rotates. The Coriolis acceleration increases with distance away from the equator since the maximum change in orientation of the primary directions, fixed on the rotating earth, occurs at the poles.

example) move farther away from the observer. Thus, a moving mass will experience a greater Coriolis acceleration to the right in the northern hemisphere if the velocity is greater (Fig. 4-9). An object with an initial velocity of 30 km/h at 40° N latitude will have a Coriolis acceleration sufficient to produce a velocity of 8 km/h to the right of its original path of motion after only one hour. The acceleration to the right is even greater for more northerly latitudes in the northern hemisphere and is at a maximum at the poles and zero at the equator.

Centrifugal Acceleration Air currents seldom follow a straight path for any great distance. As the air flows around high- and low-pressure centers, it is influenced by **centrifugal acceleration.** This acceleration is directed outward from both high- and low-pressure centers. It is therefore in the same direction as the pressure gradient acceleration around high-pressure centers, but is in the opposite direction around low-pressure centers.

Centrifugal acceleration is much more important for circulations smaller than the midlatitude cyclone. The strong winds in a hurricane and tornado are sufficiently curved for the centrifugal acceleration to overshadow the Coriolis acceleration, leaving the pressure gradient and centrifugal accelerations as the major factors. If the low pressure in a strong tornado were suddenly eliminated there would be no pressure gradient acceleration to hold the surrounding winds in a circular path, and the centrifugal acceleration would cause a parcel of air to travel in a straight path. When a balance exists between these two opposing accelerations with curved flow, the winds are called **cyclostrophic.**

Frictional Deceleration Another factor that affects atmospheric motion is friction. Friction can exert an influence only after the air is in motion. Frictional drag acts in a direction opposite to the path of motion and can cause only deceleration (**frictional deceleration**) (Fig. 4-10). However, a reduction in velocity is accompanied by a reduced Coriolis acceleration, so the direction of the wind is also affected by friction. Frictional effects are limited to the lower kilometer above the topography,

up to 1 km above surface.

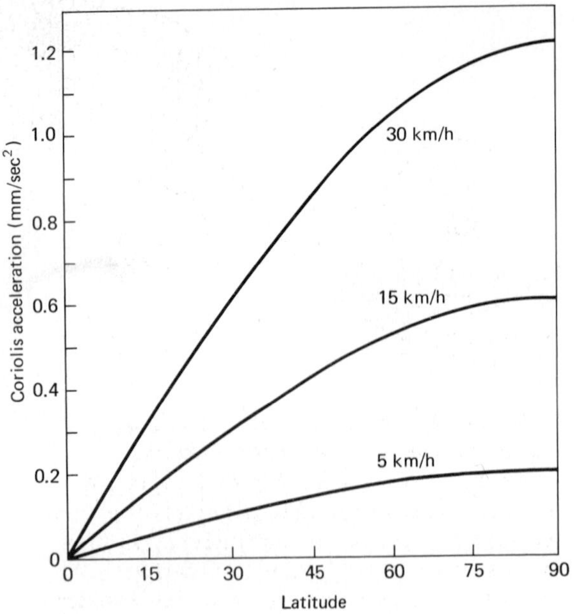

FIGURE 4-9 Magnitude of the Coriolis acceleration depends on the latitude and the velocity of the moving object. The acceleration is greater near the poles and is greater for higher velocities.

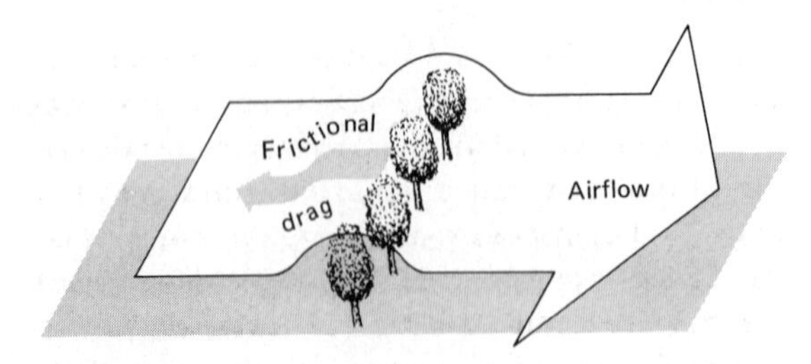

FIGURE 4-10 Frictional drag occurs in the lower atmosphere as air flows over any obstruction. This drag decreases the speed of the winds.

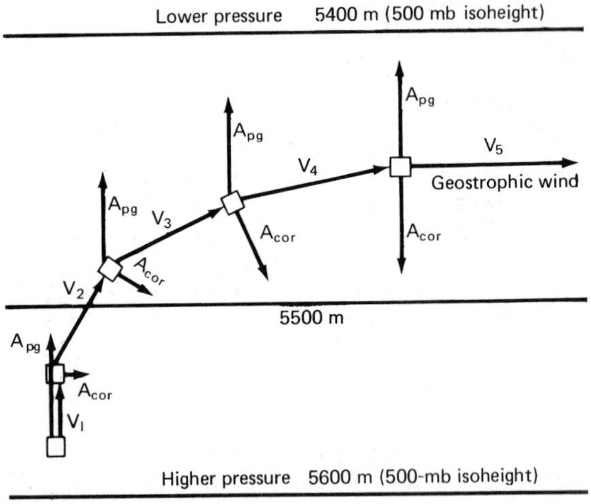

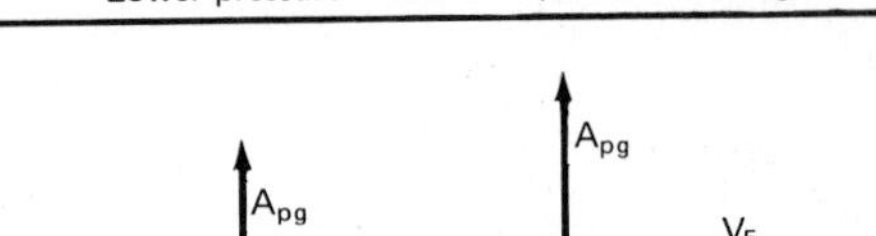

FIGURE 4-11 Geostrophic wind develops as the air begins to move because of a pressure gradient. As the velocity (V) increases, the Coriolis acceleration (A_{cor}) increases until it balances and opposes the pressure gradient acceleration (A_{pg}).

where the presence of trees, buildings, hills, and valleys exerts a frictional drag on the winds blowing over them. The wind velocity in the lower kilometer of the atmosphere is determined by the sum of these factors: acceleration from the pressure gradient, Coriolis acceleration, centrifugal acceleration, and frictional deceleration.

GEOSTROPHIC WIND

The winds above 1 km are affected only by the pressure gradient acceleration and Coriolis acceleration if the isobars are straight, since observations show that frictional effects can be ignored. It has also been observed that the total accelerations (changes in wind speed) are small after the air is in motion. The **geostrophic wind** is the name given to the resulting straight wind obtained by assuming that no frictional effects occur. The geostrophic wind represents a balance between the pressure gradient acceleration and Coriolis acceleration.

The development of the geostrophic wind can be illustrated by starting with a given pressure distribution and no initial wind speed (Fig. 4-11). As-

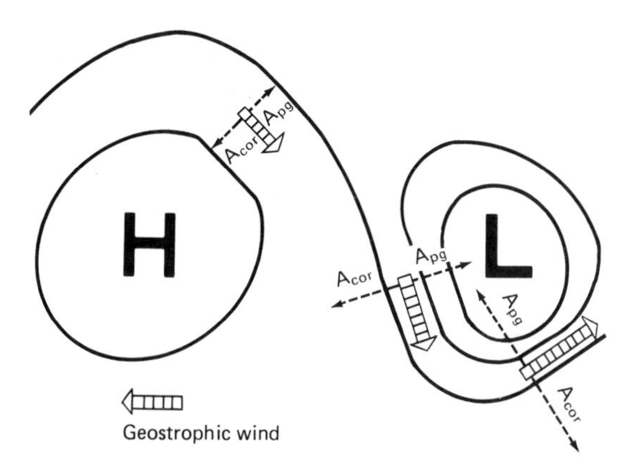

FIGURE 4-12 The geostrophic wind is a straight wind that represents a balance between the pressure gradient acceleration (A_{pg}) and the Coriolis acceleration (A_{cor}). It is stronger where the isobars are closer together and weaker where they are farther apart because of the influence of the pressure gradient.

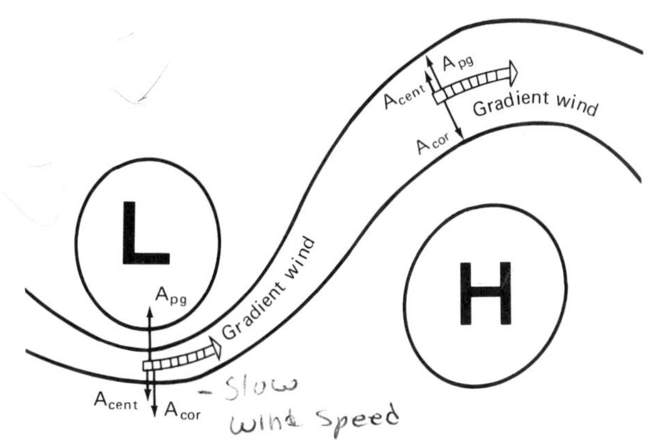

FIGURE 4-13 The gradient wind is a curved cyclonic flow around low-pressure areas and anticyclonic flow around high-pressure areas. It represents a balance between the pressure gradient acceleration (A_{pg}) and the combined effects of the Coriolis acceleration (A_{cor}) and centrifugal acceleration (A_{cent}) around the low-pressure center. The combined pressure gradient acceleration and centrifugal acceleration balance the Coriolis acceleration around high-pressure areas. If the transition from geostrophic to gradient flow is not rapid enough, mass is transported away from both high- and low-pressure centers.

sume the height of the 500-mb pressure is 5600 m at one location and 5400 m at another. The pressure gradient will be perpendicular to the 5600-m **isoheight** (line of constant height) directed toward the lower 5400-m isoheight. Since the height (and hence pressure) gradient force will be the only force acting under these specified conditions, the air will start to move directly toward the lower height (or pressure). As soon as the air starts to move, it will acquire a Coriolis acceleration and deflect slightly to the right in the northern hemisphere. The air parcel will continue to gain speed as long as the pressure gradient acceleration remains greater than the Coriolis acceleration. Since the Coriolis acceleration increases as the wind speed increases, the winds will soon reach an equilibrium speed that will allow the Coriolis acceleration to exactly match and oppose the pressure gradient acceleration. The resulting geostrophic wind (Fig. 4-12) will flow along the height contours (also lines of constant pressure) with the pressure gradient acceleration perpendicular and directed to the left of the flow direction toward lower pressure and the Coriolis acceleration directed perpendicular and to the right of the flow direction.

GRADIENT WIND

Since upper airflow patterns are usually curved, the influence of centrifugal acceleration must also be considered. As straight geostrophic winds approach a trough, the cyclonic curvature causes a centrifugal acceleration in the same direction as the Coriolis acceleration (Fig. 4-13). If the wind speed is reduced immediately, the result is a smaller Coriolis acceleration, with the sum of the centrifugal and Coriolis accelerations balancing the pressure gradient acceleration. The curved flow resulting from this balance is called the **gradient wind.** As the air approaches a ridge, the centrifugal acceleration is in the same direction as the pressure gradient acceleration. Thus, the air parcel must increase its speed, and hence the Coriolis acceleration, to achieve a balance.

A common feature of upper air pressure patterns is the closer packing of isobars around a low-than around a high-pressure center. This closer packing of isobars occurs as the winds approaching a trough fail to decrease in speed soon enough to achieve gradient flow. Thus, the combined Coriolis and centrifugal accelerations are greater than the pressure gradient acceleration, and some air moves outward from the low-pressure center in opposition to the pressure gradient acceleration. This outflow of air intensifies the low-pressure center and causes the isobars to be spaced closer together.

Airflow around a high-pressure center is also outward from the high-pressure center if the wind speed does not increase fast enough for the Coriolis acceleration to balance the opposing pressure gradient and centrifugal accelerations. But outflow of air away from the high-pressure center has the effect of weakening it, since the pressure on earth is related to the mass of air above that location. Thus, the isobars spread apart around the high- and are closer together around the low-pressure areas because of these effects.

The development of the gradient wind explains why the wind always blows counterclockwise or cyclonically around a low-pressure center in the northern hemisphere and clockwise (but also cyclonically) around a low-pressure center in the southern hemisphere. Because of the earth's rotation the wind is always deflected to the right of its path of motion in the northern hemisphere (left in the southern hemisphere), allowing it to blow only cyclonically around a low-pressure area.

The wind velocity at the point of equilibrium between the pressure gradient and Coriolis accelerations is determined by the magnitude of the pressure gradient. The magnitude of the pressure gradient is apparent from observation of the isobars on any weather map. If the isobars are close together, the pressure is changing faster with distance and the acceleration due to the pressure gradient is greater. Thus, a large pressure gradient demands that the resulting geostrophic or gradient wind be greater.

The observed winds above 1 km in the atmosphere usually flow along the isobars (height contours) indicating that a balance exists between the pressure gradient and the Coriolis acceleration (Fig. 4-14). The winds do not blow from high to low pressure. If they did, the low-pressure areas would be filled rapidly and could not exist for several days, as is observed in the atmosphere. The airflow cyclonically around the low-pressure center allows the cyclone to persist with less air above the center, thereby maintaining the lower pressure at the surface.

The geostrophic and gradient winds are useful concepts in understanding the major wind systems in the atmosphere.

FRICTION-LAYER WIND

Near the surface below 1 km, frictional effects must be considered. Frictional drag arises from any surface feature that interferes with airflow. If total acceleration changes of the wind are insignificant, after equilibrium of the forces and accelerations is achieved, the accelerations from the pressure gradient must be balanced by the sum of the Coriolis acceleration and frictional drag for straight winds. **Friction-layer wind** can be illustrated by starting with the geostrophic wind (Fig. 4-15). The air in a geostrophic wind flows along the isobars with a balance between the pressure gradient and Coriolis accelerations. If surface friction is introduced, the wind velocity decreases. As the wind velocity decreases, the Coriolis acceleration decreases, resulting in flow toward the lower pressure because

FIGURE 4-14 The surface and 500-mb synoptic weather charts for May 6, 1975. Notice that the surface winds are across the isobars, slightly toward the lower pressure and cyclonic around it. The wind velocities at the surface are higher around low-pressure centers because the isobars are closer together and the pressure gradient is greater. Winds are frequently calm or very light at the surface near centers of high pressure because of pressure gradient forces. At 500 mb the winds are along the lines of constant pressure, with the magnitude of the wind also related to the spacing of the isobars, since this is determined by the pressure gradient acceleration.

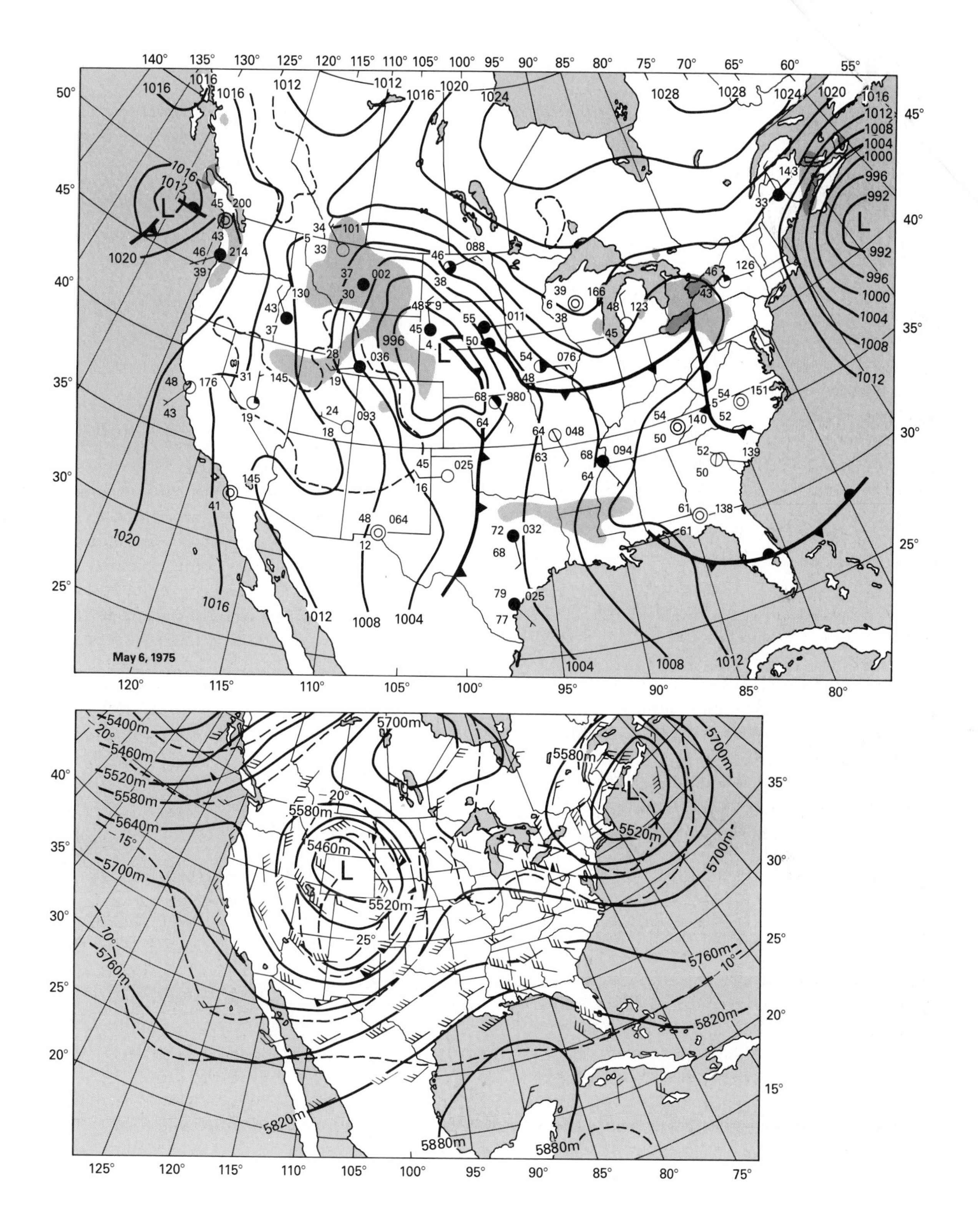
May 6, 1975
5400m
5460m
5520m
5580m
5640m
5700m
5760m
5820m
5880m
5460m
5520m
5580m
5700m
5760m
5820m
5880m

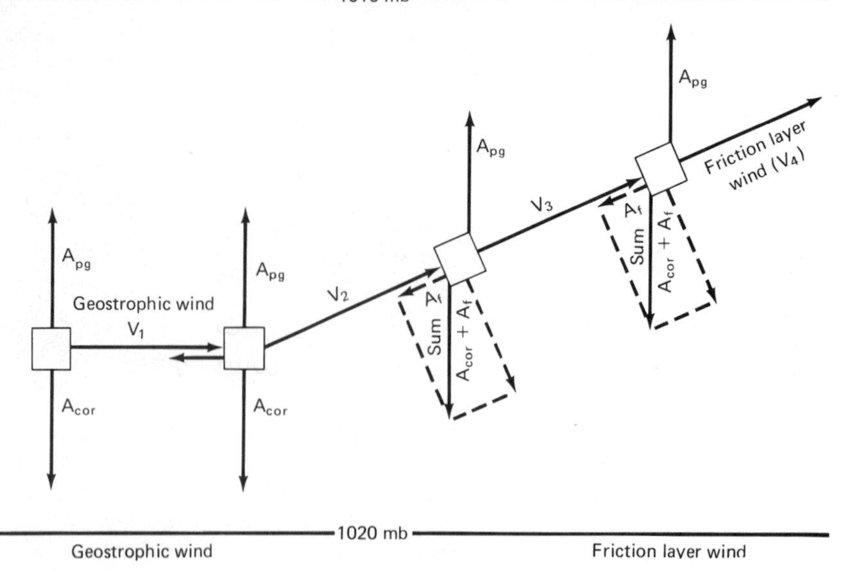

FIGURE 4-15 Within the friction layer, winds result from a balance between the pressure gradient acceleration (A_{pg}) and the sum of the Coriolis acceleration (A_{cor}) and frictional deceleration (A_f). This causes the wind to flow across the isobars, slightly toward lower pressure.

the pressure gradient acceleration is now greater. The resulting airflow reaches a balance where acceleration from the pressure gradient is matched by the combined Coriolis acceleration and frictional deceleration. Friction-layer wind is therefore neither parallel nor perpendicular to the isobars, but is between, with the exact direction determined by the amount of friction exerted by the surface.

OBSERVED SURFACE AND UPPER-LEVEL WINDS

Surface weather maps (see Fig. 4-3) show that the airflow is across the isobars toward lower pressure, with an angle of about 45° for land surfaces and about 15° over the smoother ocean surfaces. Therefore, frontal cyclones have a component of flow near the surface toward the lower pressure. Unless the cyclone has compensating outflow in the upper atmosphere, it cannot last for very long. Because of the cause and effect relationship between the pressure and the wind direction and speed, the surface and upper air pressure patterns can be used to obtain information on the wind flow patterns at that level. A comparison of a surface weather map and a 500-mb map (Fig. 4-14) shows that the air flows cyclonically around the low-pressure centers in both cases. The wind speed is greater where the isobars are closer together in both cases, but the wind direction is parallel to the isobars (or height contours) at 500 mb while it flows slightly toward the lower pressure at the surface.

It is possible for the wind direction at the surface to be opposite to the wind direction in the upper atmosphere. A **cold core high-pressure system** (anticyclone with colder air near its center) at the surface can be associated with a low pressure at 500 mb because the air is more compressed near the surface above the high-pressure area (Fig. 4-16). If the total mass of air is greater above a location, the pressure at the surface is higher, but if most of this mass is contained below the 500-mb

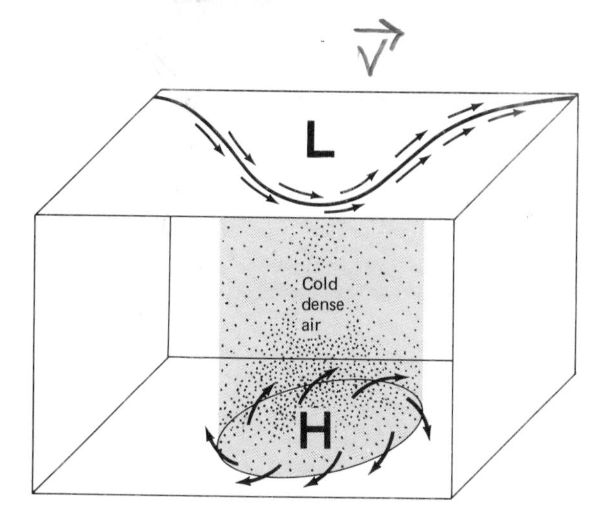

FIGURE 4-16 The pressure distribution can change with height depending on the temperature structure. If the air is very cold and dense, high-pressure areas can exist at the surface, while a pressure reversal in the upper atmosphere allows a low-pressure center to be located over the high surface pressure.

FIGURE 4-17 This photograph shows the development of the lake breeze over Chicago during the day because of the warmer land surface. Note the stream of smoke coming inland from sources located on the shore.

level, the pressure there may be lower. This is possible if less air is above the 500-mb level at this location than in surrounding areas. When this pattern occurs, the winds at each level blow in the direction prescribed by the pressure pattern, but the wind direction at the surface is opposite that at 500 mb. Cold air outbreaks frequently occur in midlatitudes during the winter accompanied by this type of pressure and wind distribution.

LOCAL WIND SYSTEMS

Land and Sea Breezes Heating and cooling cycles that arise from solar heating of the atmosphere create several localized wind systems that are too small to be affected by the Coriolis acceleration. The onshore and offshore breezes near bodies of water are a good example. Reservoirs of water cause the temperature of the air above to be quite different from the air temperature over the land. During the daytime, land surfaces heat faster than the water and reach higher temperatures because of the different thermal characteristics of land and water.

Warmer air above the land is more buoyant because it is less dense than air over the adjacent cooler water. This leads to rising air currents over the land, with descending denser air over the water. A localized pressure gradient is thus developed that initiates a circulation that brings air from over the water across the land during the daytime, causing onshore breezes (Fig 4-17). The climate of Chicago is influenced by the presence of Lake Michigan and the resulting lake breeze. The temperature can drop 10°C or more during the daytime as the lake breeze develops.

Such lake or **sea breezes** can be observed when the large-scale pressure gradients are weak with corresponding light winds. The temperature field (horizontal distribution of temperature), and resulting pressure gradient, reverses during the night as the water retains its heat and the land cools rapidly. This results in warmer, less dense air above the water and colder air above the land, causing the local winds to shift from the land to the water and creating offshore or land breezes.

Mountain and Valley Breezes In mountainous terrain, **mountain breezes** develop at night in

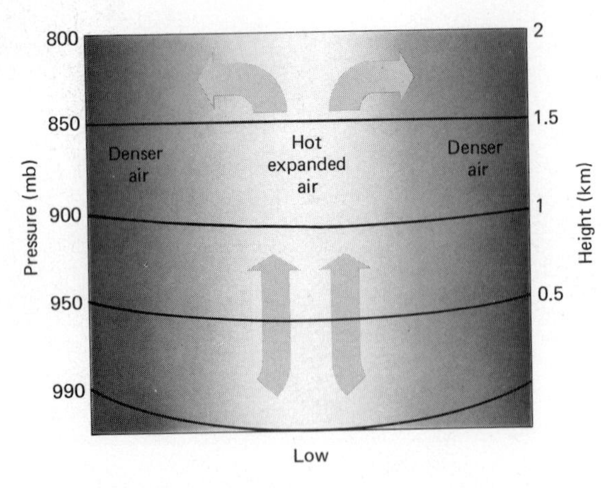

FIGURE 4-18 The thermal low develops as a mass of air is heated by solar radiation and expands and rises in the atmosphere, where, for example, an increase in pressure is experienced at the 800-mb surface. If this pressure increase occurs, the air moves outward from the higher pressure leaving less air above the surface, which results in the development of a lower surface pressure.

response to infrared radiation cooling. This cooling causes valley locations to be much colder at night, as the cold heavy air drains toward lower elevations. During the daytime the air above the terrain warms from solar heating, and valley breezes rise up the mountain slopes. Mountain winds may be concentrated by the narrowing terrain in some areas and may acquire speeds as high as 150 km/h. These winds are stronger when the large pressure patterns from traveling high- and low-pressure systems provide the proper pressure gradient for accelerated drainage winds down the mountain slopes.

Drainage Winds A smaller amount of solar radiation in comparison to terrestrial radiation losses may cause strong winds in cold regions of the world. As air moves across the snowfields of Greenland, Norway, and Alaska, it is cooled and becomes much denser, resulting in an accelerated drainage wind, which is a flow of denser cold air down the slopes into valleys below. These cold drainage winds are called **katabatic winds.** They are frequently quite light but may attain higher velocities if the general pressure distribution contributes to the airflow in a similar direction. Katabatic winds are given special names in some localities where they are particularly severe, such as southward from the Austrian Alps where they are called bora and along the French coast of the Mediterranean Sea where they are called the mistral.

RADIATION-CONTROLLED PRESSURE FIELDS

Thermal Lows In dry regions of the world where cloud cover is sparse, the abundant solar heating may produce lower atmospheric pressure at the ground. This pressure field is called a **thermal low** since it is directly related to surface heating. Average summer pressure patterns over the United States show a low pressure in the Arizona–New Mexico region. The development of the thermal low can be illustrated by considering two adjacent volumes of air. If they are initially at the same temperature, the pressure at the surface will be the same at both locations provided that the same amount of air is above in each case. If more solar heating occurs above one of the locations, the air will expand and be forced upward and eventually out into surrounding areas (Fig. 4-18). Since less air will then be located above the area of solar heating, the surface pressure will be lower than those in surrounding surface locations.

When viewed from the side, the isobars through a thermal low are arranged with decreasing pressure upward and increasing pressure outward in a horizontal direction at the surface. Since the warm air is more expanded above the center of the thermal low the distance between isobars is greater. Therefore the horizontal pressure differences have equalized at a height above 1 km. For this reason, the thermally induced low pressure does not show up on the 500-mb chart (about 5.5 km) and may not exist on the 850-mb chart (about 1.5 km).

Monsoons The onset of **monsoon** rains in India is related to the development of a thermal low-pressure area. India is located southward from Siberia, a large land mass that gets very cold during the wintertime, thus cooling the air above it to very low temperatures. This cold, dense air forms a large high-pressure system over Siberia with associated anticyclonic winds around it. These winds bring air with small amounts of rainfall to India from the cold, dry continental regions during the winter. During the summertime the airflow patterns reverse, as solar heating warms the land and produces a thermal low over India. The resulting cyclonic flow of air around the low-pressure center brings air from the ocean over the land, causing heavy rainfall. This pattern produces the wet monsoon season as the moisture is released from the saturated air. Other factors also contribute to the onset of the monsoon; the jet stream, for example, moves over the Himalaya mountain range, and the **intertropical convergence zone** (the location where the tropical trade winds from the northern and southern hemispheres converge) moves over India, sustaining the monsoon circulation.

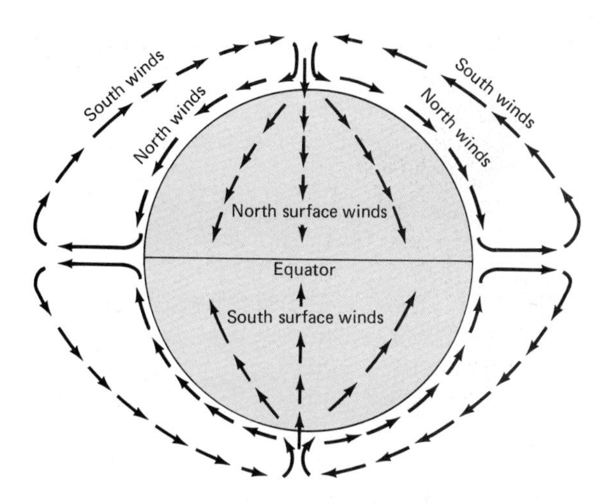

FIGURE 4-19 A simple thermal circulation would result if the earth were not rotating and had uniform surface material, with heating at the equator and cooling at the poles.

GENERAL ATMOSPHERIC CIRCULATION

The laws governing the flow of air are useful in promoting an understanding of some of the features of the general circulation of the atmosphere. In order to consider the global atmospheric circulation, it is helpful to make some simplifying assumptions. These assumptions are (1) the earth is not rotating; (2) the earth is composed of uniform material; (3) hotter, less dense air exists at the equator and colder, denser air at the poles.

Simple Thermal Circulation The three simplifying assumptions made in the last section lead to a model of simple thermal circulation. If the earth were not rotating, the atmospheric circulation would be much simpler because the Coriolis acceleration would be eliminated. The assumption of uniform surface materials eliminates the complications introduced by the contrasting heating rates of oceans and continents. With these assumptions the warmer tropical air near the equator would rise and flow northward in the upper atmosphere and the cold dense air from the poles would sink and flow toward the equator, giving a direct **thermal circulation** (Fig. 4-19). The surface winds would blow from the north in the northern hemisphere and from the south in the southern hemisphere.

Theoretical Three-Cell Model If the first assumption is eliminated and the earth were rotating with a uniform surface material, the resulting circulation would be composed of three cells (Fig. 4-20). The upper atmospheric air flowing northward from the equatorial regions could not travel directly to the North Pole. The Coriolis acceleration would cause it to be deflected to the right, making it a westerly wind. An excess mass of air would then accumulate in the upper atmosphere at about 30° N latitude. This would cause a band of high pressure to develop, with subsiding air at this latitude. This would complete one of the three circulation cells located between 0° and 30° latitude. A second circulation cell would develop between 90° (the poles) and 60° latitude as the air near the poles sinks and, near the surface, flows southward in the northern hemisphere (northward in the southern hemisphere) until it encounters air from the opposite direction at about 60° N latitude. The third cell would be located in midlatitudes between the other two cells.

The surface winds in the northern hemisphere would be from the northeast from 0° to 30° N latitude because of the southward flow from the high pressure at 30° and the deflection to the right by the Coriolis acceleration. The air moving northward from the band of high pressure at 30° would be affected by the Coriolis acceleration and become southwesterly winds from 30° to 60° N latitude. Near the north pole the north winds would be deflected to the right and become a band of northeasterly winds. These surface winds would be mirrored in the southern hemisphere giving a convergence zone at the equator.

In the upper atmosphere the air moving northward from the equator would be affected by the Coriolis acceleration, which would transform it into westerly winds, thus creating a jet stream. Two bands of westerly winds in the upper atmosphere would occur: one at about 30° and the other at about 60° N latitude. The northern jet stream would be the **polar jet stream** and the southern one the **subtropical jet stream.**

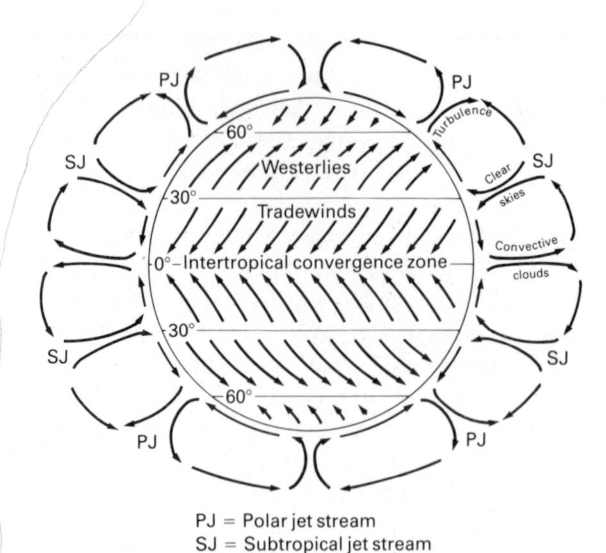

FIGURE 4-20 A three-cell circulation would occur if the earth had uniform surface material, with heating at the equator and cooling at the poles. The surface winds include tropical trade winds, midlatitude westerlies, and polar easterlies.

Realism of the Three-Cell Model Even though uniform surface material has been assumed for the earth's surface, the resulting three-cell circulation has some similarities to the observed wind systems. The northeasterly **trade winds** from the equator to about 30° N latitude, the intertropical convergence zone near the equator, and the southeasterly trade winds in the southern hemisphere are parts of the observed wind system. These winds are so consistent that they were used by early sailing vessels in crossing the ocean. If the ships veered too far north of the equator in the northern hemisphere they became stalled in the calm winds at about 30° N latitude and had to dump their cargo. It has been said that this occurred frequently to ships that were sailing to the New World loaded with horses, and the ocean in this region eventually had so many dead horses that it became known as the horse latitudes.

The polar cell also has similarities to the observed winds, but the central cell of the three-cell circulation model is not as realistic. The presence of continents complicates the atmospheric circulation so that belts of pressure at 0°, 30°, and 60° do not exist in midlatitudes.

Even though the three-cell model is not totally realistic, names have been given to two cells after those who first suggested their occurrence. The equatorial cell is a **Hadley cell** and the middle cell is a **Ferrel cell.** The Ferrel cell is in the region that is frequently described as the **prevailing westerlies.** However, as was shown previously by typical wind directions for midlatitudes, the surface winds are not predominately from the west except along the west coast. However, upper atmospheric winds are frequently from the west, causing thunderstorms and midlatitude cyclones to travel from the west or southwest. This is the only sense in which this region has "prevailing westerlies."

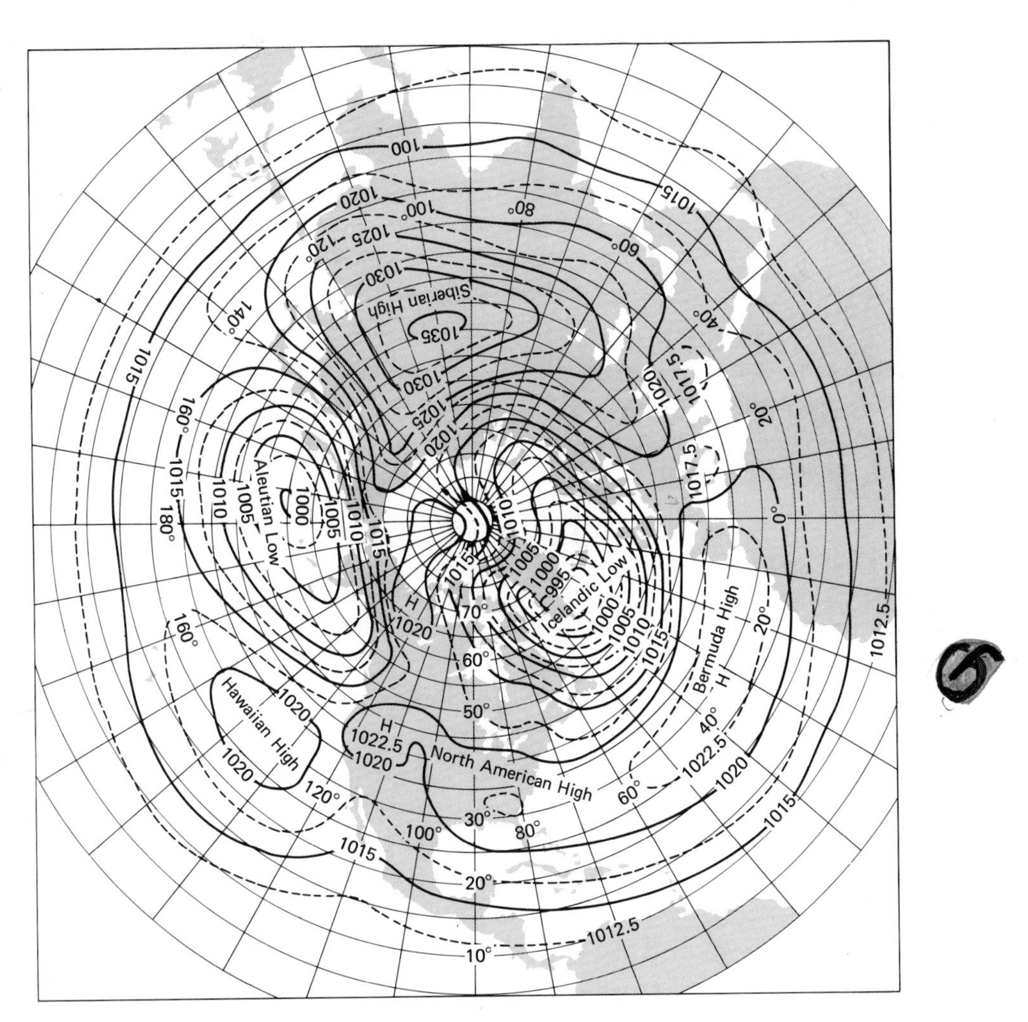

FIGURE 4-21 The distribution of average sea-level pressure in millibars over the northern hemisphere in January. (From National Weather Service, NOAA.)

AVERAGE SURFACE PRESSURE AND WIND

In order to understand the general circulation of the atmosphere, the average pressure patterns must be considered. The average surface pressure patterns in January show that some areas have much higher pressure than others (Fig. 4-21). Higher pressure exists at about 30° N latitude over both the Atlantic and Pacific Oceans while lower pressure occurs at about 60° N latitude over the Atlantic and Pacific Oceans. Higher pressure exists over the continents during January, with no major bands of pressure such as were suggested by the three-cell model. In the southern hemisphere there are fewer continents and a greater tendency toward a band of lower pressure at 60° S latitude. Table 4-1 summarizes these patterns for the northern hemisphere.

During July the average pressure distribution shows large high-pressure areas over both the Atlantic and Pacific oceans at about 35° N latitude with some indication of lower pressure over the oceans at about 60° N latitude (Fig. 4-22). The continents have lower pressure over them during the summer.

These surface pressure patterns during the summer and winter determine the wind directions and other weather features. The general circulation near the surface is related to the average pressure patterns. The surface circulation in the northern hemisphere is much more complicated than the three-cell circulation model indicates, as shown by the average pressure distribution and resulting winds. The surface atmospheric circulation at any one time is even more complicated than indicated by the average pressure distribution if the occurrence of traveling high- and low-pressure systems in midlatitudes is considered. Although average pressure patterns and simplified models are helpful in understanding some of the general features of the atmospheric circulation, it must be remembered that the actual circulation at any time is determined by the pressure gradients at that time. Traveling midlatitude cyclones are a distinguishing feature of a significant part of the world and are therefore an important part of the general atmospheric circulation.

TABLE 4-1

Location and strength of semipermanent high- and low-pressure cells

Ocean locations	Position	Winter	Summer
Aleutian low (Pacific)	60°–65°	Strong	Weaker
Icelandic low (Atlantic)	60°–65°	Strong	Weaker
Hawaiian high	30°–35°	Weaker	Stronger
Bermuda high	30°–35°	Weaker	Stronger
Continental locations			
Siberian high		Strong	Absent
North American high		Strong	Absent
SW U.S. thermal low		Absent	Strong
India thermal low		Absent	Strong

JET STREAMS

A general relationship exists between solar radiation and larger atmospheric circulations such as jet streams.

Surplus energy is absorbed by the earth-atmosphere system in tropical areas. More than 4000 trillion watts must be transferred northward across midlatitudes to maintain normal temperatures in the tropical and polar regions. Jet streams with their associated smaller storm systems are an important part of the earth's heat exchange system because of their high wind speeds and momentum. Both the subtropical jet stream and polar jet stream are important features of the upper atmospheric circulation.

The subtropical jet stream, which is located at about 25° N latitude, is generated because of excess heating in tropical regions and the northward movement of air in the upper regions of the atmosphere. As air moves northward from the equator, its speed toward the east must increase if it retains the same speed that it acquired from the rotating earth at the equator. The rotational speed of the earth is 1669 km/h at the equator. Northward moving air from the equator would acquire very high speeds relative to a point on the earth if its energy were not dissipated in some way. Air moving southward from the North Pole would acquire a very high speed even more quickly because of the more rapid increase in the radius of rotation.

Northward-bound air from the equator would become a westerly wind of 224 km/h at 30° N latitude in the vicinity of the subtropical jet stream and 507 km/h at 50° N latitude in the region of the polar jet stream (Fig. 4-23). Although these are calculated values, they compare very favorably with the maximum measured wind speeds at these latitudes. The greatest recorded wind speed associated with the polar jet stream is about 500 km/h. The polar easterlies, developed from southward moving air from the North Pole, are also observed features of atmospheric circulation.

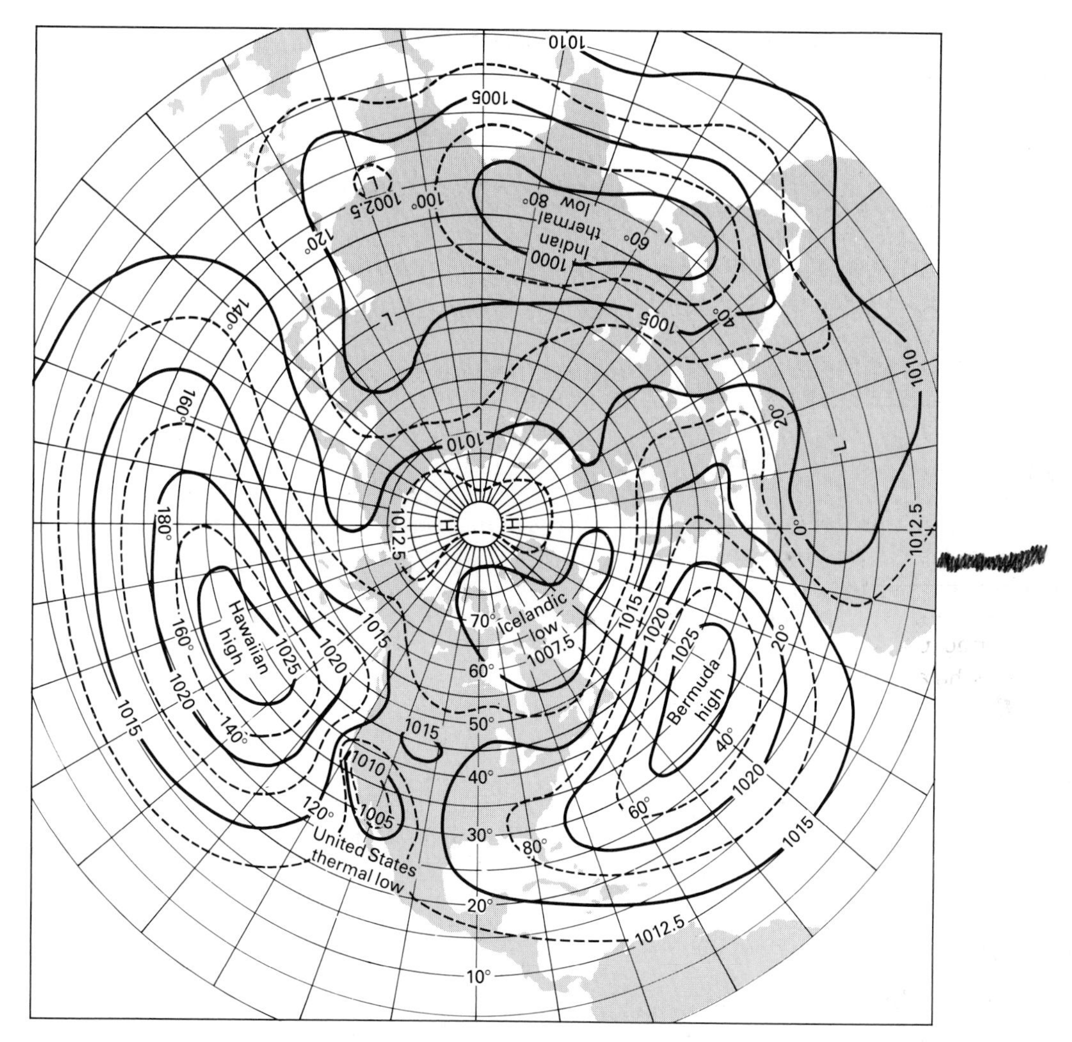

FIGURE 4-22 The distribution of average sea-level pressure in millibars over the northern hemisphere in July. (From National Weather Service, NOAA.)

Polar Jet Stream The most important feature of the upper atmospheric circulation is the polar jet stream. Like the subtropical jet stream, this jet stream derives its energy from the hemispherical distribution of solar radiation. After the jet stream is in motion, however, its momentum and energy are responsible for the generation and maintenance of many smaller atmospheric storm systems and circulations.

It can be demonstrated, by considering its seasonal location and velocity, that the polar jet stream is related to radiation at the surface. The polar jet stream is located at about 35° N latitude during the winter in the northern hemisphere (Fig. 4-24). Its velocity averages about 130 km/h during this season. During the summertime it moves northward so that its average location is closer to 50° N latitude and the velocity averages only about

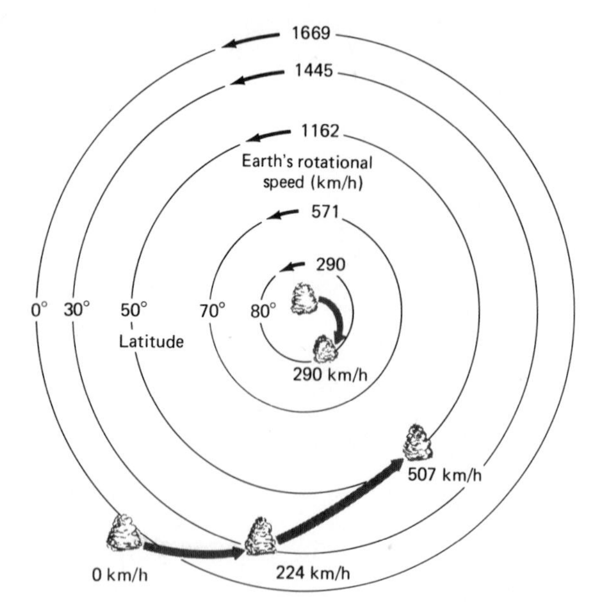

FIGURE 4-23 The velocity that air starting from the equator would acquire as it moved northward and the velocity that air from the North Pole would acquire as it moved southward, starting from 0 velocity. These calculated velocities are comparable to the actual measured maximum velocities.

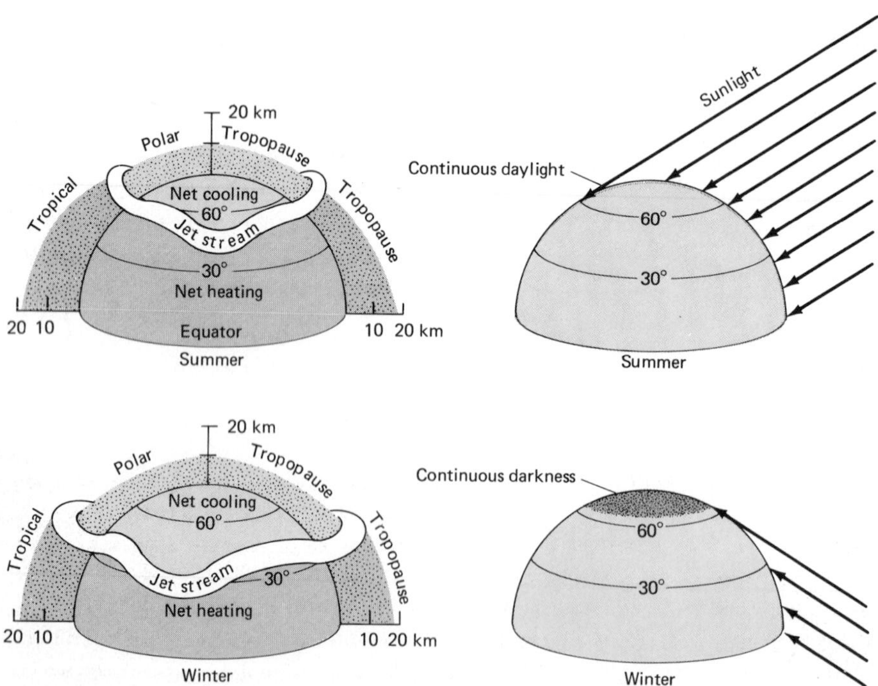

FIGURE 4-24 The seasonal migration and speed of the polar jet stream is related to solar radiation. During the summer the North Pole receives continuous energy so that the contrast in temperature between polar and equatorial air is much less. During the winter the North Pole has continuous darkness, creating a much greater contrast in temperature between polar and equatorial air. For this reason, the jet stream is displaced southward and acquires a much higher velocity.

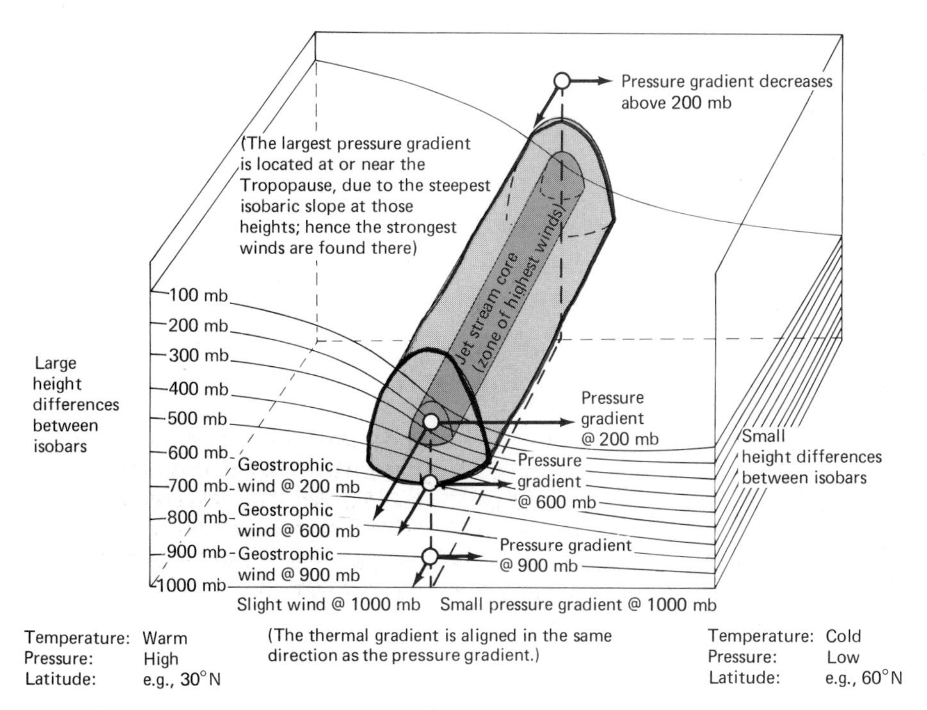

FIGURE 4-25 The pressure gradient determines the speed of the jet stream. The pressure gradient increases up to the 200-mb level and then decreases above this level in response to global differences in heating.

65 km/h. These seasonal differences are directly related to solar heating.

During the winter the North Pole is turned away from the sun so that the earth's surface there receives no energy, while the equatorial region receives surplus energy. Thus, a large contrast exists between the heating of the atmosphere in tropical regions and that in polar regions. This creates large differences in the density of the air in these locations. Density differences translate to pressure differences and resulting wind speeds, as indicated in Figure 4-25. Therefore, the greatest wind speeds in the polar jet stream occur during the winter, when the North Pole receives no energy and the contrast in the density of air at different latitudes is the greatest. For the same reason, the speed of the polar jet stream is less during the summer, when the North Pole receives some solar energy and the resulting contrast in density of the air is much less.

The average position of the polar jet stream is related to the seasonal change due to the tilt of the earth's axis with respect to the sun. It moves northward in the summer as the boundary of the heated warm air masses moves northward. For the same reason, it moves southward during the winter and maintains higher velocities because

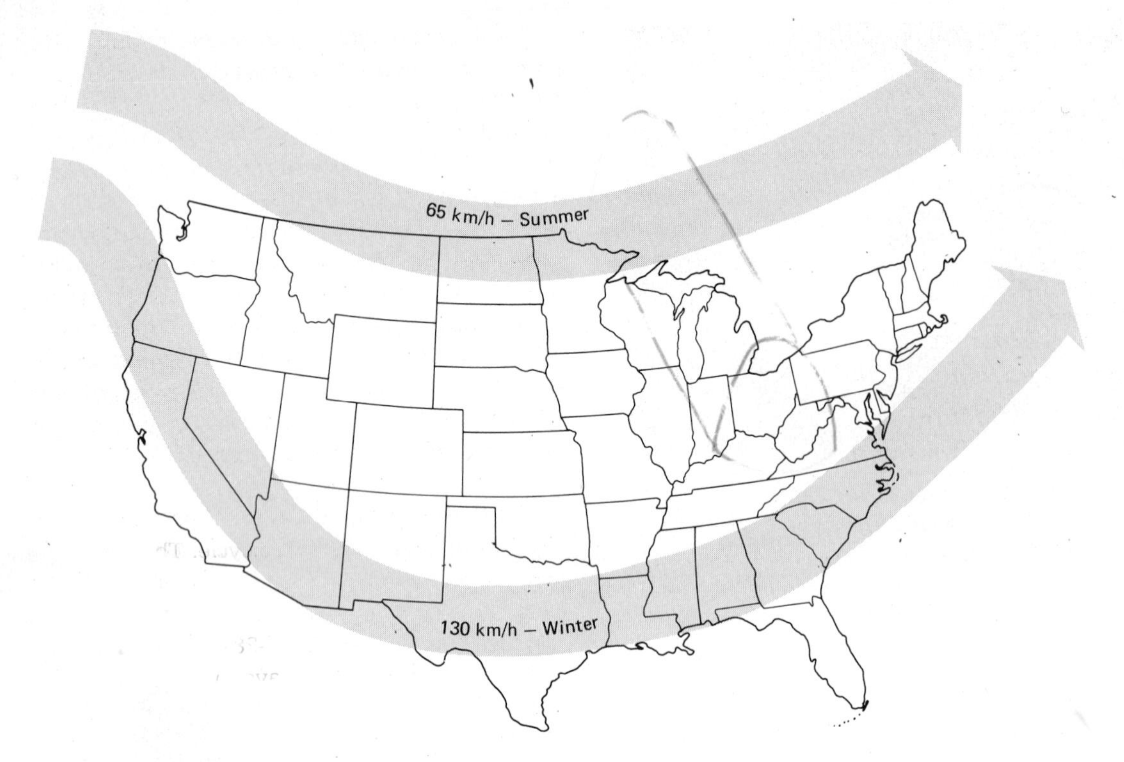

FIGURE 4-26 The average position of the polar jet stream changes with the seasons so that it is much farther south in the wintertime. The velocity of the polar jet stream also changes with seasons so that its velocity is almost twice as great in the wintertime as in the summertime.

the temperature contrast between air masses is greater. The position and speed of the polar jet stream then are related to solar energy distribution.

The position and the speed of the polar jet stream are very influential in determining surface weather of midlatitude locations (Fig. 4-26). The location of midlatitude cyclone development is determined by the jet stream. Tornadoes are more frequent in the spring than during the fall because of the transfer of energy from the jet stream and surrounding winds to a thunderstorm that produces a tornado. Since the speed of the polar jet stream is much greater in the spring following the period of darkness at the North Pole, it has more energy to transfer to thunderstorms and tornadoes as it crosses the midsection of the United States, where a large contrast between surface air masses is likely in the spring. In the fall the speed of the jet stream is not as great as it moves southward, where the contrast in surface air masses may be just as great as during the spring. Severe weather is not as likely during the fall months because of this characteristic of the polar jet stream.

Many other examples of the influence of the polar jet stream on surface weather could be cited. During 1976 it stayed farther northward than usual, resulting in record-breaking high temperatures in January and February. In the winter of 1977, exactly the opposite occurred, when the po-

FIGURE 4-27 The rotating dishpan experiment simulates the formation of long waves in the atmosphere. (Photograph courtesy of Dave Fultz, Hydrodynamics Laboratory, Dept. of the Geophysical Sciences, University of Chicago.)

lar jet stream was displaced far to the south of its normal position over the eastern United States. The result was the coldest winter in 170 years for most of the nation east of the Rocky Mountain states. Hot dry weather in the United States during the summer is associated with very stable anticyclonic curvature of the polar jet stream over the northern United States and Canada. An example of this also occurred in 1977 when an unusually dry fall and warm winter occurred for the southwestern United States and Alaska; the northward displaced polar jet stream was associated with moisture deficits of 25 to 50% and with temperatures 15 to 30°C warmer than normal.

If patterns of the polar jet stream could be accurately predicted, it would be possible to forecast the weather for a week or more in advance. It is only possible at present to indicate the relationship of the polar jet stream to solar heating. Figure 4-24 shows that the jet stream is related to the discontinuity between the tropical and polar tropopause, but little information is available that provides more quantitative relationships useful for forecasting future patterns of the polar jet stream.

Some experiments, such as the rotating dishpan, have been used to simulate hemispherical atmospheric circulation in order to gain more information on the long waves in the atmosphere (Fig. 4-27). When a pan (filled with water to simulate the atmosphere) is rotated at the proper speed, while being cooled near the center and heated near the outer edge, flow patterns are established that have similarities to airflow patterns of the jet stream. Such experiments have been conducted to better understand the atmosphere. Proper understanding is necessary before it will be possible to predict precisely the flow patterns and related surface weather several weeks in advance.

Jet Stream Cycles The polar jet stream frequently completes a specific cycle. The first stage of the cycle consists of air streams that stretch uniformly from west to east across midlatitudes of the northern hemisphere (Fig. 4-28). In the next stage small meanders or long waves develop in the jet stream. The presence of these long waves in the jet stream means that some locations have cyclonic curvature and other regions have anticyclonic curvature. The temperature field at the surface remains roughly aligned with the jet stream: In general, the warm air stays to the south and the cold air to the north. The jet stream determines this boundary to some degree since it influences the movement of air masses. However, the existence of the polar jet stream is related to differences in heating at different latitudes, as previously described. The jet stream is able to maintain such a dual role since the development of high speeds in the jet stream give it energy and momentum for moving air masses and driving lesser atmospheric circulations such as midlatitude cyclones and tornadoes.

As the polar jet stream passes through various stages with the cold air to the north and warm air to the south, the meanders eventually become so large that pools of cold air are isolated and forced southward by large cyclonic meanders (Fig. 4-29). At the same time pools of warm air are forced northward by large anticyclonic meanders. Eventually, the flow patterns become so curved that the

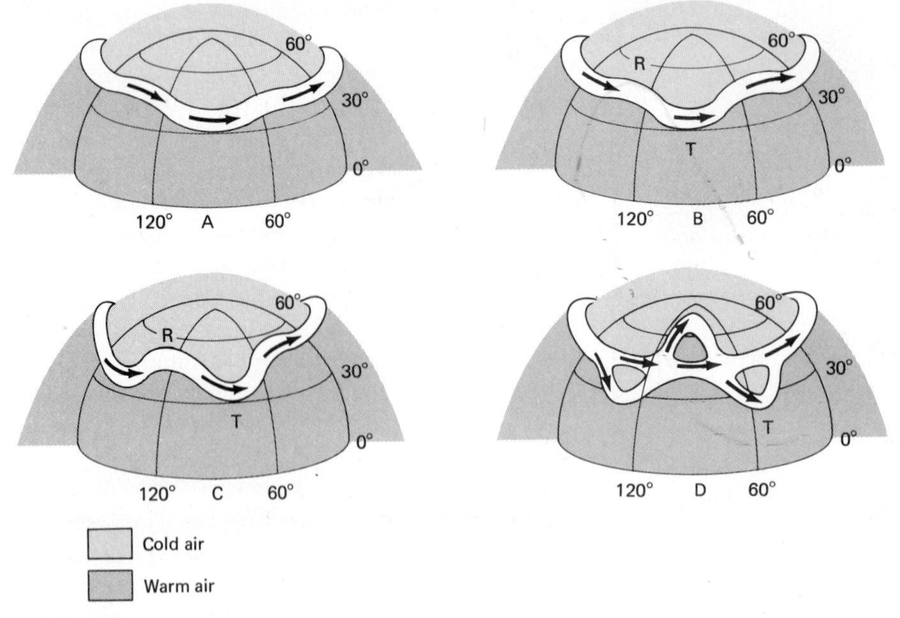

FIGURE 4-28 The polar jet stream frequently completes a cycle composed of increasing meanders that continue until the warmer air, which is normally located to the south of the jet stream, is cut off into pools north of the jet stream ridges (labeled R in diagram), while colder air from the north is left in pools farther southward. This latter occurs as a trough, labeled T in the diagram, and increases in intensity until the jet stream cuts off the meander that formed it.

FIGURE 4-29 As the general circulation of air at the jet stream develops into meanders, it affects large areas. In this satellite photograph, a ridge in the jet stream produces clear skies or scattered clouds throughout the western United States, while a southward meander of the jet stream creates a trough over Florida. A storm embedded within this trough brings overcast skies and rain to most of the northeastern United States. (Photograph courtesy of NOAA.)

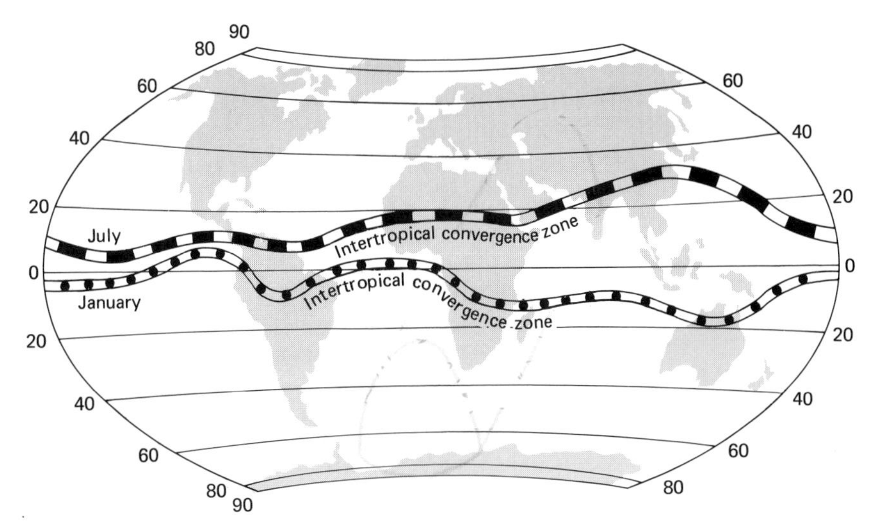

FIGURE 4-30 The intertropical convergence zone, representing the thermal equator, shifts its position much more in some parts of the world than in others. Such shifts bring rain to India in July and to Australia in January.

meanders are cut off by the jet stream, leaving pools of air with abnormal temperatures displaced northward and southward. The jet stream is then ready to start the cyclic process over again. The cycle may take a month or more to complete.

Large cyclonic bends in the jet stream are able to generate large midlatitude cyclones by supplying circulation and energy. Midlatitude cyclones, in turn, are able to generate smaller storms including thunderstorms and tornadoes. These storms are much smaller in size, with the average diameter of a tornado being less than 300 m. The thunderstorm that produced it may be 30 km across, while the midlatitude cyclone may be more than 800 km in diameter. The diameter of the cyclonic curvature of the jet stream is even larger. Thus, the polar jet stream is very influential in determining surface weather and climate.

INTERTROPICAL CONVERGENCE ZONE

One of the most influential aspects of tropical weather is the **intertropical convergence zone (ITCZ)**. Like the jet stream, the ITCZ is directly related to solar radiation. However, unlike the jet stream it is a surface feature of weather in the tropics. The ITCZ represents the **thermal equator** as the northeasterly trade winds of the northern hemisphere meet the southeasterly trade winds of the southern hemisphere. As the trade winds converge, the deep convective rain clouds of tropical regions are enhanced.

The movement of the intertropical convergence zone with the tilt of the earth's axis causes extreme differences in the weather of many tropical locations (Fig. 4-30). The monsoons that accompany the passage of the ITCZ are the primary source of rain in some locations, particularly in India and the semiarid Sahel of tropical Africa (the countries directly south of the Sahara). Monsoon weather in many parts of the world is limited to lands between the equator and the ITCZ. Its farthest northward extent varies from year to year and is thought to be a primary factor in the Sahelian drought of the early 1970s, since the ITCZ remained about 2° latitude farther southward across Africa. When the jet stream and its surrounding airflow moves farther southward during the summer, the northward movement of the ITCZ is thereby limited. This condition has been called the Sahelian effect by Reid Bryson, of the University of Wisconsin.

As can be seen in Figure 4-30 the north-south movement of the intertropical convergence zone is greatest between Asia and Australia. This movement brings monsoon rains to the northern tip of Australia. In the Pacific Ocean, movement is very limited, as it stays near the equator. The average position of the ITCZ is 5° N latitude displacing the thermal equator from the geographic equator by this amount.

SUMMARY

Atmospheric winds are driven by differences in heating on a global or local scale, although many of the atmospheric winds are related indirectly to heating. Wind speed and direction change as the pressure patterns change. Surface winds blow cyclonically around low-pressure centers and slightly in toward the lower pressure. Surface winds blow anticyclonically around an area of high pressure and slightly away from the center.

Horizontal winds are caused by a limited number of forces and accelerations. The sum of the pressure gradient acceleration, Coriolis acceleration, centrifugal acceleration, and frictional deceleration determines the total acceleration of a parcel of air. The acceleration from the pressure gradient is determined by the distribution of pressure patterns and is always directed from higher to lower pressure. The Coriolis acceleration results because of the earth's rotation. Rotation of the spherical earth causes an acceleration to the right or left at any location except the equator. Accelerations are to the right of the path of motion in the northern hemisphere and to the left in the southern hemisphere. The magnitude of the Coriolis acceleration is greater if the velocity of the moving mass is greater. Frictional deceleration occurs in the lower atmosphere as the wind speed is reduced because of flow over topographic features, trees, buildings, and other physical objects that exert a drag on the winds above.

Geostrophic winds occur as a balance is reached between the pressure gradient and Coriolis acceleration. These straight winds are frequently observed above 1 km in the atmosphere. Gradient winds include centrifugal accelerations to produce curved flow. The geostrophic and gradient winds flow along lines of constant pressure or height with very little or no crossflow. Friction-layer flow occurs in the lower kilometer because of the additional effect of frictional drag on atmospheric winds. This causes some flow of air across the isobars from higher to lower pressure at an angle, depending on the amount of friction.

The general circulation of the atmosphere can be simplified by making several assumptions, including the assumption of uniform surface material. This results in the development of a theoretical three-cell model for the general circulation. The equatorial cell (Hadley cell) corresponds very well with the observed northeasterly trade winds from the equator to 30° N latitude and southeasterly trade winds from the equator to 30° S latitude, with an intertropical convergence zone near the equator. The middle cell (Ferrel cell) is not as realistic. The presence of continents complicates the atmospheric circulation sufficiently to void the uniform surface material assumption. Large differences exist in the way that water and land heat as solar radiation is absorbed. The reflection, absorption, heat capacity, and conduction of heat are all very different for land surfaces and water. These differences cause the oceans to warm very slowly and retain heat for long periods of time in comparison to land surfaces.

The average pressure at the earth's surface can be used to understand the general circulation. High-pressure areas tend to occur over continents in the wintertime with low pressure during the summertime. High-pressure areas also occur over the oceans at about 30° N latitude with low pressure over oceans at about 60° N latitude. Although average pressure patterns are helpful in understanding the general atmospheric circulation, traveling midlatitude cyclones and anticyclones are a further complication in the general circulation.

In the upper atmosphere the polar jet stream is an important feature of the circulation. The jet stream derives its energy from the hemispherical distribution of solar radiation. After its speed is developed, the momentum of the jet stream helps to generate and drive lesser atmospheric circulations. The position of the polar jet stream is related to seasonal changes, with advances farther north in the summertime and south in the wintertime in the northern hemisphere. Greater amounts of precipitation and cloudiness are associated with those areas beneath cyclonic bends in the jet stream while hot dry weather in the summertime is associated with stable anticyclonic curvature of the jet stream. The jet stream frequently completes a specific cycle in which the meanders become larger and larger until eventually the flow around the earth becomes straighter leaving cold air pools isolated southward from the jet stream and pools of warm air isolated northward from the jet stream.

STUDY AIDS

1. Describe and compare some of the different characteristics of surface and upper atmospheric wind speed and direction.

2. Explain the difference between backing and veering winds, and describe their relationship to a low-pressure system.

3. Explain each of the forces that affect horizontal winds.

4. What is the direction and magnitude of the pressure gradient force if Denver, Colorado, reports 1030 mb of pressure and Kansas City, Missouri, reports 1010 mb of pressure?

5. Draw a diagram to illustrate the Coriolis acceleration for a westward-moving object.

6. Use Figure 4-9 to determine the magnitude of the Coriolis acceleration as the velocity of an air parcel changes from 0 to 30 km/h at a location 45° N latitude.

7. Is it possible for an air parcel of 15 km/h to experience the same change in magnitude of the Coriolis acceleration as in the previous study aid if its position changed from the equator to the North Pole?

8. Explain why winds always blow cyclonically around low-pressure areas in the northern hemisphere.

9. Explain the geostrophic wind.

10. Compare the geostrophic wind and friction-layer flow with the observed surface and upper air winds.

11. Discuss the realism of the three-cell general circulation model.

12. What effects do the earth's surface materials have on atmospheric pressure patterns?

13. How is the position and speed of the polar jet stream related to solar heating?

14. Explain jet stream cycles.

TERMINOLOGY EXERCISE

Use the glossary to reinforce your understanding of any of the following terms used in Chapter 4 that are unclear to you.

Veer
Backing
Newton's second law
Pressure gradient force
Coriolis acceleration
Centrifugal acceleration
Cyclostrophic wind
Frictional deceleration
Geostrophic wind
Isoheight
Gradient wind
Friction-layer wind
Cold core high-pressure system
Sea breeze
Mountain breeze
Katabatic winds
Thermal low
Monsoon
Intertropical convergence zone (ITCZ)
Thermal circulation
Polar jet stream
Subtropical jet stream
Trade winds
Hadley cell
Ferrel cell
Prevailing westerlies
Thermal equator

THOUGHT QUESTIONS

1. What effect does the Coriolis acceleration have on the shape of cloud patterns associated with traveling low-pressure systems in the northern and southern hemispheres?

2. Assume that a commercial jet travels from San Francisco to New York within the core of a typical winter jet stream. If the normal ground speed of the jet were 500 km/h, calculate the number of hours it would take for this flight. Under the same conditions, how long would the return flight take?

3. Would you expect certain regions of a jet stream to be more turbulent and cause greater difficulties for aircraft than other regions? Which ones?

4. Would you expect a relationship to exist between the average position of the polar jet stream and the position of the intertropical convergence zone? Explain your answer.

5. Can you speculate on ways of controlling and using the jet stream more effectively in weather applications?

SPECIAL TOPIC 5

EL NIÑO AND SOUTHERN OSCILLATION

Several decades ago a seesaw pattern of atmospheric pressure was discovered in the tropical Pacific and Indian oceans. This pressure pattern has become known as the Southern Oscillation. In the 1960s a connection was discovered between the Southern Oscillation and sea-surface anomalies. In 1969, J. Bjerknes suggested that sea-level and temperature fluctuations were connected with equatorial atmospheric circulation in a positive feedback manner. He also suggested that if the equatorial easterlies slackened, the warmer waters in the western equatorial Pacific would tend to surge eastward. Residents of Peru and Ecuador have traditionally called this unusually weak current of warm water *El Niño,* since it arrives near the Christmas season. El Niño and the Southern Oscillation are now viewed as striking examples of global-scale climate interactions. El Niño becomes much stronger about every ten years, and extensive ocean warming occurs, which kills or drives away the anchovies from Peru and Ecuador, having a drastic effect on the local fishing industries.

The effects of an extensive ocean warming are also global, as many climatic abnormalities are related. Flooding in Ecuador, Peru, Cuba, and the southern United States often occurs in strong El Niño years. Droughts in Australia, Indonesia, the Philippines, and Southern Africa have been traced to the effects of El Niño. One of the positive effects of El Niño, though, is extraordinarily few Atlantic tropical storms and hurricanes.

An indication of the beginning of El Niño is a southward displacement of the intertropical convergence zone. This movement of the ITCZ is associated with light winds, abnormally high sea-surface temperatures, and a deep thermocline in the southeastern equatorial Pacific Ocean. These changes occur early in the year, when the intensity of the southeast trade winds is at a seasonal minimum, when sea-surface temperatures in that region are already at a seasonal maximum, and when the seasonal migrations of the ITCZ take it to its lowest latitude. In its early stages El Niño is simply an amplification of the seasonal cycle.

El Niño differs from the yearly warming of ocean waters as the areas of warming begin to expand westward from the coasts of Peru and Ecuador. By October, high sea-surface temperatures are spread over the entire tropical Pacific Ocean. El Niño's mature phase is from November to January, 10 to 13 months after its onset.

The 1982–83 El Niño was particularly pronounced. Wide-spread warming first appeared in the equatorial Pacific Ocean in May 1982. In June, ocean-surface temperatures were 1° to 2° C above normal from the South American Coast to 170° E. In September, about three months after the collapse of the equatorial easterlies, rapid ocean warming continued in the eastern equatorial Pacific. The collapse of the easterlies also caused a rise in sea level in the eastern Pacific Ocean. By December 1982, the ocean temperatures were more than 4° C above normal. Such extremely high sea-surface temperatures persisted in the eastern equatorial Pacific Ocean through June 1983.

The weather of 1983 included some of the most unusual and severe events of this century. Many of these occurrences have been blamed on the unusually strong El Niño of 1982 and early 1983. A prolonged dry spell affected Hawaii December through March. Much-below-normal winter rainfall in Hawaii is common during El Niño years, since the jet stream is stronger and farther south than normal, placing a ridge over Hawaii.

El Niño events seem also to be related to the development of extremely low pressure in the Gulf of Alaska. In February 1983, the mean monthly sea-level pressure was the lowest of this century. This caused the polar jet stream and tracks of storms to shift hundreds of kilometers southward as they entered North America from the Pacific. This persistent trough yielded very high winds, excessive rains, and much property damage in California, Washington, and Oregon.

In Ecuador and northern Peru, extremely heavy rains caused widespread flooding. Drought was widespread in southern Africa, Sri Lanka, southern India, the Philippines, Indonesia, and Australia.

Some positive effects may also accompany El Niño. The winter of 1982–83 was one of the warmest in 25 years in the temperate north latitudes. This may have occurred because of a relationship between the warm ocean temperatures and unusually warm winter temperatures. Another possible positive effect of El Niño is the low number of hurricanes in the Atlantic. The 1982 Atlantic hurricane season was the quietest in 52 years. During El Niño, a stronger jet stream may shear off the top of any tropical storm forming in the Atlantic.

El Niño seems destined to play an important role in understanding and predicting climatic patterns of the future. Although all the interactions between ocean temperatures and weather events are not yet known, much evidence is accumulating on their direct links. Such links may prove to be important elements in the future development of better long-range forecasts, as well as keys to a better understanding of seemingly unrelated weather events.

CHAPTER FIVE

STABILITY AND VERTICAL MOTION

Pressure Gradient Force is like Saturated Vapor Pressure.

VERTICAL FORCES

Vertical motion in the atmosphere develops in response to the forces acting on the air, as specified by Newton's second law. However, the forces giving rise to vertical motion are different from those that cause horizontal winds. The force of gravity affects any volume of air, with a greater force on a volume that contains more mass. All the molecules within the atmosphere would be pulled down to the earth's surface if there were no counteracting upward force. The upward force is a vertical pressure gradient force that exists because of the pressure decrease with altitude (Fig. 5-1). When the vertical pressure gradient force balances the downward gravitational force, no new vertical air currents are initiated since the two vertical forces are in equilibrium.

Since the gravitational pull on the molecules that comprise the air and the resulting decrease in pressure with altitude is very constant, the major cause of imbalances in the two vertical forces is variation in the density of the air. Air density is changed if the air temperature changes, since warm air expands and becomes less dense. The measured **atmospheric lapse rate** specifies the temperature at various heights in the atmosphere. The measured lapse rate is, therefore, the means of identifying features of vertical motion in the atmosphere.

NORMAL ATMOSPHERIC LAPSE RATES

The atmospheric lapse rate determines the vertical motion and stability of the atmosphere. A **stable atmosphere** has few or no vertical air currents, while

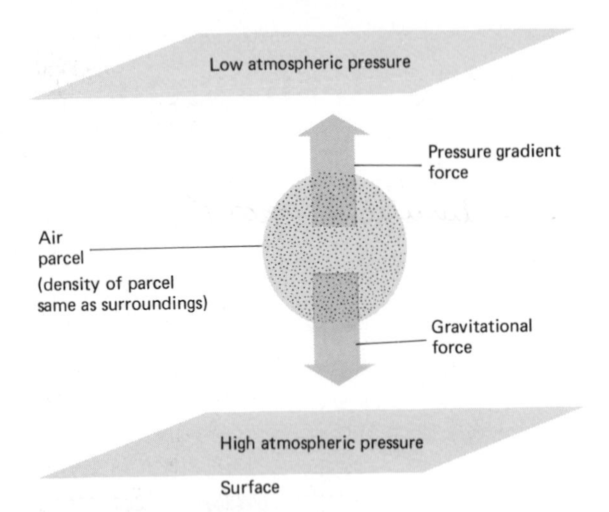

FIGURE 5-1 Vertical motion occurs in the atmosphere when the vertical forces—pressure gradient force and gravitational force—are unbalanced.

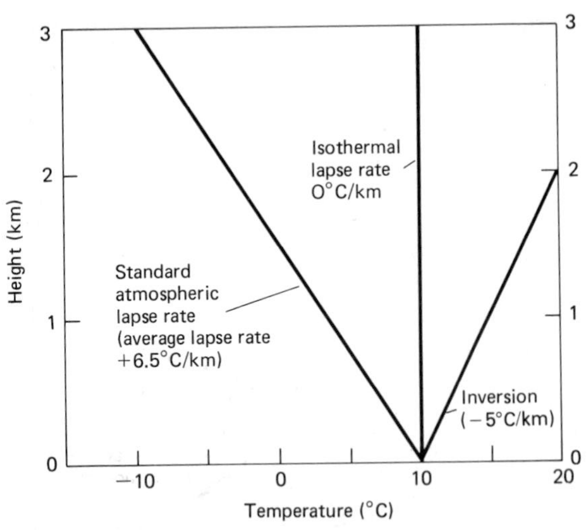

FIGURE 5-2 The standard lapse rate decreases with height at the rate of 6.5°C/km, while an isothermal lapse rate shows no change with height and an inversion has increasing temperature with height.

an unstable atmosphere has considerable vertical mixing from air currents. A bubble of air that becomes warmer than its surroundings is buoyant and begins or continues to rise through the surrounding air. Measurements of the vertical temperature profile or lapse rate are used to determine the stability of the atmosphere at particular times. Throughout the troposphere the temperature changes from an average global surface temperature of 15°C to −56°C at 11 km, giving an average lapse rate of 6.5°C/km. (A temperature *decrease* with height is defined as a *positive* lapse rate.) Temperature increases with height (inversions) are, mathematically, *negative* lapse rates. If the temperature is constant with height, an isothermal lapse rate exists (Fig. 5-2). For positive lapse rates, larger numbers correspond to faster decreases in temperature with increasing altitude.

Temperature profiles through the troposphere are important in so many atmospheric processes that measurements are made at almost 100 upper air stations located in the United States, as explained in Chapter 2. These stations send up radiosondes twice each day at 0000 and 1200 GMT. These measure the actual lapse rate through the troposphere for stability determinations and other purposes. The average of these measurements gives a lapse rate of 6.5°C/km. The lapse rate is most variable near the ground in the lowest kilometer. Heating during the day and cooling during the night cause the lapse rate to change (Fig. 5-3), with inversions common at night. The average diurnal temperature change in many parts of the United States is 15°C. This amount of variation in the surface temperature causes large changes in the lapse rate in the lower part of the atmosphere and this affects the stability of the lower atmosphere.

Measurements can be made in the lower atmosphere by a small, fast-response thermometer that shows the development of turbulence from an **unstable atmosphere.** Measurements taken every few minutes after sunrise on a sunny day would show a gradual warming in the lower atmosphere. The

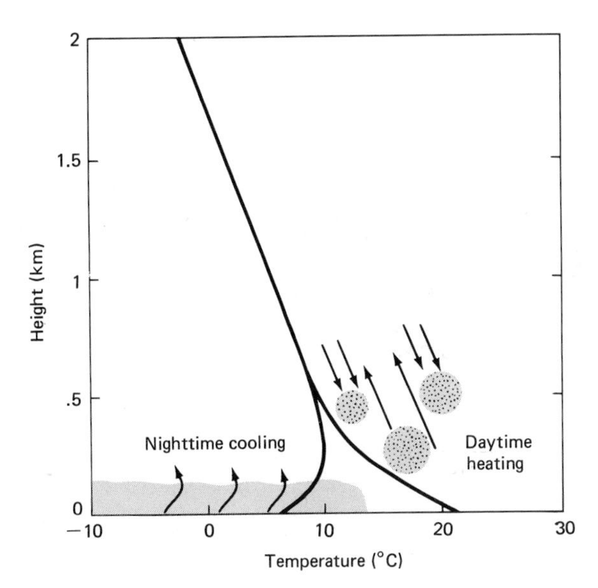

FIGURE 5-3 The lapse rate in the lower kilometer of the atmosphere varies between day and night, as solar radiation is absorbed during the daytime and terrestrial radiation is lost at night.

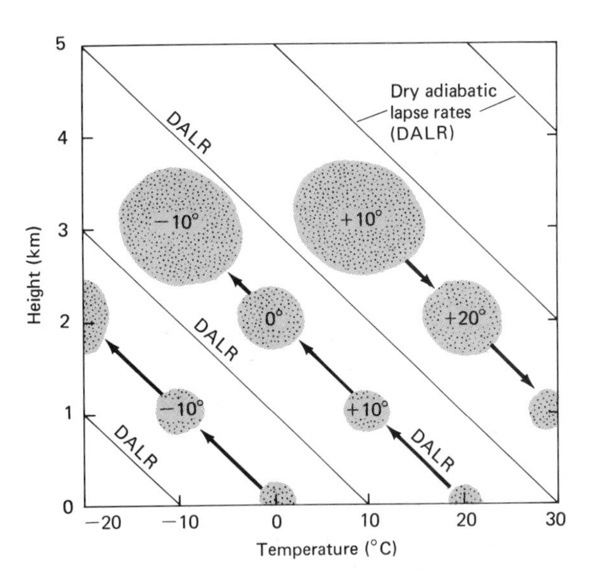

FIGURE 5-4 The dry adiabatic lapse rate represents a constant expansion cooling (or heating) rate of 10°C/km.

measured temperature would constantly increase until about 9 A.M. when large oscillations in the temperature would start to occur with the onset of thermal convection in the atmosphere. The turbulence corresponds to surface heating, which increases the lapse rate with the result that the atmosphere becomes unstable and develops convective air currents.

An unstable atmosphere also corresponds to increased gustiness of the wind (see Fig. 4-1). Wind measurements during the night show a laminar (smoother, less turbulent) airflow. After sunrise, the turbulence and vertical fluctuations increase markedly in association with surface heating.

DRY ADIABATIC LAPSE RATE

The stability of the atmosphere is related to the measured lapse rate since a greater lapse rate means hotter surface air in comparison to the air above. The warmer air close to the surface is more buoyant, leading to greater vertical currents in the atmosphere. If a rising bubble of air does not mix with the surrounding air, it cools at the expansion cooling rate of 10°C/km rather than at the measured external air temperatures, which may be called the **environmental lapse rate.** This constant expansion cooling rate inside the air bubble, called the **dry adiabatic lapse rate,** occurs as air is lifted in the atmosphere into lesser atmospheric pressure where the rising air expands (Fig. 5-4).

The expansion process is intrinsically a cooling process. This can be explained by the molecular activity or it can be demonstrated in several different ways. If a piston is used to apply pressure to compress air, its increase in temperature can be measured as the pressure is increased. Similarly, the amount of cooling can be measured as the pressure is decreased.

The dry adiabatic lapse rate can also be calculated by the appropriate equations that govern the behavior of gases. These equations are used in Appendix A to calculate the dry adiabatic lapse rate of

9.8°C/km. This lapse rate of approximately 10°C/km (5.5°F/1000 ft) therefore represents the amount of cooling that occurs as air expands or the rate of heating as it contracts. The dry adiabatic lapse rate is applicable only if the air is unsaturated, as indicated by the term *dry*. As long as the relative humidity remains less than 100%, rising or descending air follows the dry adiabatic lapse rate if it does not mix with surrounding air. *Adiabatic* is a term used to indicate that no heat additions or losses occur. It is assumed that no radiation is absorbed and no mixing develops with the surrounding air.

Expansion cooling has many practical applications. A refrigerator, for example, works because of expansion and compression of a gas. In one part of the refrigerator the gas is expanded, with the expansion of the gas and latent heat effects cooling this compartment, while the compressor in the lower part of the refrigerator compresses the gas thereby liberating heat.

Cooling or heating is an important effect of rising or descending air. If air rises 1 km in the atmosphere it cools 10°C, while descending air heats at 10°C/km. If the amount of vertical motion were as great as horizontal motion (it is in general much less), the expansion cooling and compression heating would play a dominant role in surface temperatures.

Air in the atmosphere that has a certain temperature and occupies enough volume that mixing at the edges can be ignored must cool at 10°C/km regardless of the temperature of the surrounding air. Rising unsaturated air, therefore, follows the dry adiabatic lapse rate rather than the lapse rate measured by radiosondes at any one time. The only exceptions are if radiation is absorbed by the air or if mixing significantly affects the temperature of the air at the edges of the rising bubble of air. If some of the environmental air is incorporated into the edges of the bubble of rising air the resulting temperature at the edges is an average of the environmental air temperature and the interior temperature of the bubble. This process occurs as thunderstorms develop. The rising bubble of unsaturated air may be large enough that the inside cools at the dry adiabatic lapse rate while some mixing with the environmental air occurs at the edges of the buoyant bubble of air. This mixing is called **entrainment.**

STABILITY OF UNSATURATED AIR

Parcel Method The stability of the atmosphere can be determined from measured lapse rates. The actual lapse rate is measured at specific times, 6:00 A.M. and 6:00 P.M. CST, at all the upper air weather stations. The measured temperature profile together with the dry adiabatic lapse rate gives information on the stability of the atmosphere.

The **parcel stability method** can be used to determine the stability of the atmosphere. Stability considerations with the parcel method consist of assuming a parcel of air is displaced upward when there is no compensating downward motion occurring close enough to the parcel to affect its temperature. This assumption is possible since a rising parcel of air that has been displaced upward is accompanied by compensating downward motions that are scattered over a much larger volume in the atmosphere. A second assumption is that the parcel does not mix with the surrounding air. This is possible if the parcel is large enough that the inside of the rising parcel of air experiences no mixing with the surrounding atmosphere.

With these two assumptions, the unsaturated atmosphere is stable if a displaced parcel of air returns to its original position. The atmosphere is unstable if the displaced parcel is accelerated away from its original position. Neutral stability occurs if a displaced parcel of air remains in equilibrium after displacement. Therefore, on the basis of parcel displacement the atmosphere is determined to have stable, unstable, or neutral stability.

The behavior of a parcel of air after displacement is determined by the relationship of the measured lapse rate to the dry adiabatic lapse rate. If the measured lapse rate is different from the dry adiabatic lapse rate, then a displaced parcel in the atmosphere follows the dry adiabatic lapse rate. This rate determines its temperature after dis-

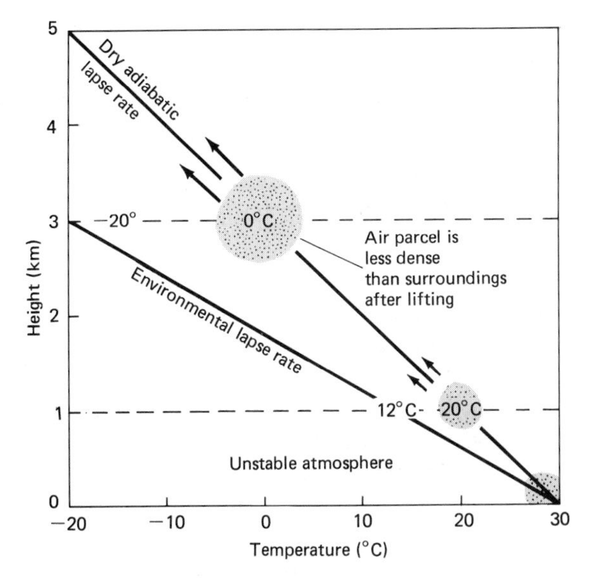

FIGURE 5-5 An unstable atmosphere exists if a displaced parcel of air accelerates away from its original position. This occurs if the environmental lapse rate is greater than the dry adiabatic lapse rate. The pull of gravity is less on the rising parcel than on surrounding air since its mass per unit volume is less.

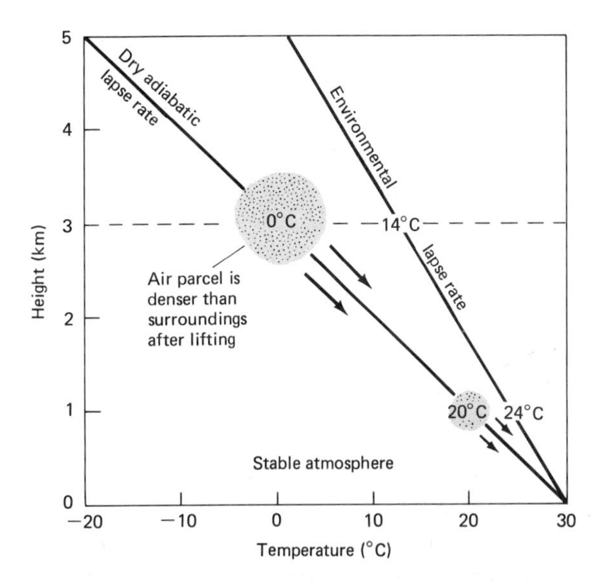

FIGURE 5-6 A stable atmosphere exists if a displaced parcel of air returns to its original position. This occurs when the environmental lapse rate is less than the dry adiabatic lapse rate. The pull of gravity is greater on the displaced parcel than on surrounding aii since its mass per unit volume is greater.

placement. If the measured lapse rate is greater than the dry adiabatic lapse rate, the lifted parcel of air will have an internal temperature higher than the environmental air (Fig. 5-5). If the temperature of the parcel is hotter, it will be less dense and more buoyant, resulting in acceleration away from its original position. This illustrates an unstable atmosphere since vertical currents are intensified.

The atmosphere is stable if the displaced parcel is accelerated back to its original position. This occurs when the measured lapse rate is less than the dry adiabatic lapse rate. As shown in Figure 5-6, when the measured lapse rate is less than the dry adiabatic lapse rate, a parcel displaced upward will cool at the dry adiabatic lapse rate of 10°C/km and arrive at a higher elevation with a colder temperature than its surroundings. If the temperature of the parcel is colder, it is denser and will fall back to its original position. This leads to stability of the atmosphere since vertical currents are eliminated. A stable atmosphere means that convection, turbulence, thunderstorm development, and mixing of pollutants are all less likely.

To further illustrate the concept of atmospheric stability, assume that a lapse rate of 15°C/km exists. With these conditions (Fig. 5-7), the temperature at 1 km in the atmosphere will be 0°C and at 3 km it will be −30°C. Assume also that a bubble of air large enough to avoid mixing of the bubble's center with the surrounding air starts to rise from the surface with the surface temperature of 15°C. Displacement may be initiated by flow over hills and valleys or even smaller surface features such as trees and buildings. The temperature within the displaced parcel will be 5°C at 1 km since it will cool at the dry adiabatic lapse rate. If the temperature inside the parcel at 1 km is 5°C and the surrounding air temperature is 0°C, the parcel is warmer and more buoyant. Therefore, it will rise further. At 2 km the temperature inside the parcel will be 10°C

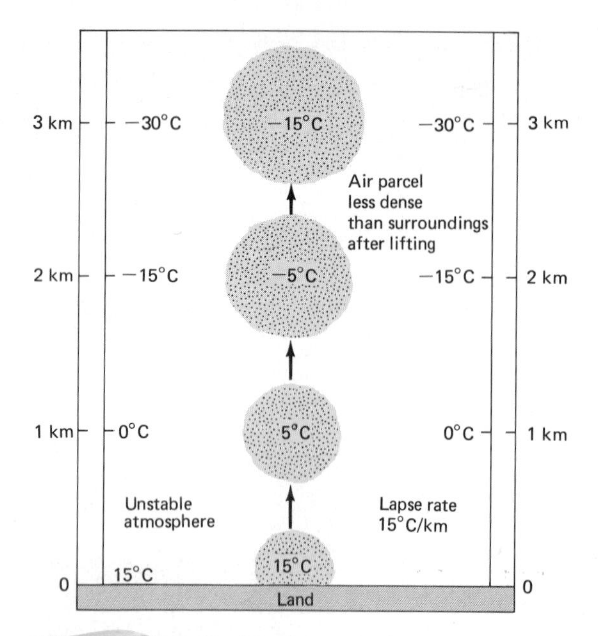

FIGURE 5-7 A displaced parcel of air will continue to rise as long as the temperature inside the parcel of air remains warmer than the surrounding atmosphere. This occurs with an unstable atmosphere.

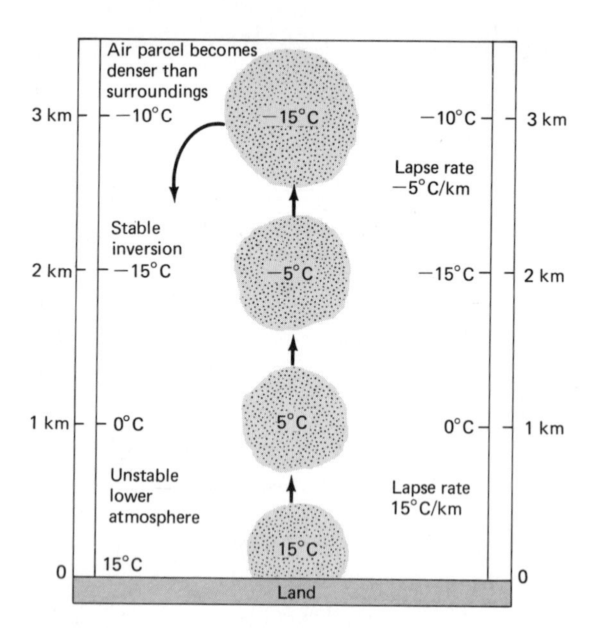

FIGURE 5-8 An inversion layer is a very stable atmospheric characteristic because it causes the atmosphere surrounding a rising parcel of air to become warmer than the inside of the air parcel, thus preventing it from gaining buoyancy.

colder, at a temperature of −5°C, and a 10°C temperature difference will exist in comparison with the surrounding air. It will be even more buoyant and will keep rising as long as the measured lapse rate is greater than the cooling rate of 10°C/km. This illustrates an unstable atmosphere with a measured lapse rate greater than the dry adiabatic lapse rate.

An inversion is very effective in stopping convective activity since it is a very stable layer. If the measured lapse rate used in the previous example is changed only slightly by adding an inversion from 2 to 3 km, its effect is illustrated in Figure 5-8. If the same measured lapse rate of 15°C/km exists below 2 km and an inversion of −5°C/km exists from 2 to 3 km, the measured temperature at 3 km will be 5°C warmer than at 2 km because of the negative lapse rate.

If it is assumed that a bubble of air rises from the surface as before, the temperature of the parcel and surroundings will be the same as in the previous example to a height of 2 km. If the parcel of air at 2 km with a temperature of −5°C, compared to −15°C in the surrounding air, is lifted through the inversion to 3 km it will cool, giving a temperature inside the parcel of 10°C cooler or −15°C. The temperature of the surrounding air will be 5°C warmer since the inversion layer increases in temperature with height. The environmental temperature will be −10°C compared to a temperature of −15°C inside the parcel. Thus, the parcel will be colder, making it denser than its surroundings. The only way it could have gotten to a height of 3 km is if it had enough inertia to rise past its equilibrium level because of its momentum. Otherwise, because of the shallow **upper air inversion,** it could

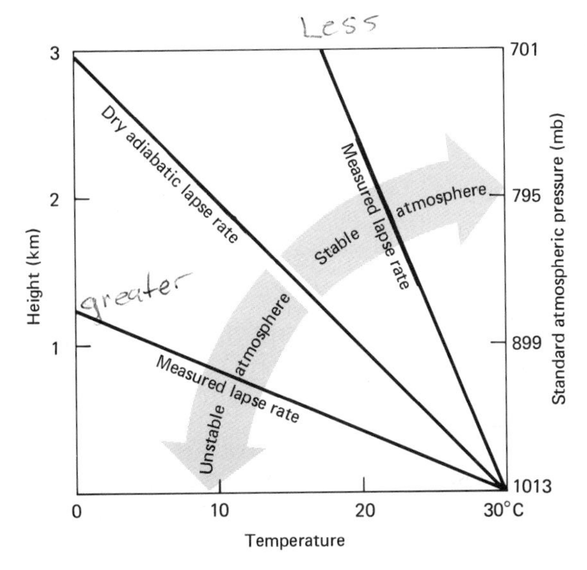

FIGURE 5-9 The stability of the atmosphere can be determined by comparing the measured lapse rate with the dry adiabatic lapse rate. If it is less than the dry adiabatic lapse rate, it is stable. If it is greater than the dry adiabatic lapse rate, it is unstable.

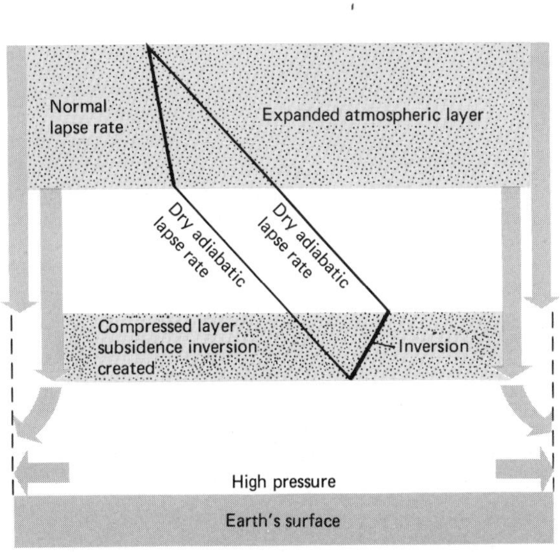

FIGURE 5-10 Subsiding air above a high-pressure system produces a more stable atmosphere since the top of a subsiding layer travels a greater distance than the bottom of a subsiding layer. This tends to create an upper air inversion.

never have risen to that height. This illustrates the stabilizing effect of an inversion in the atmosphere. Even a shallow inversion layer decreases convective activity and mixing in the atmosphere.

Lapse Rate Method The stability of the atmosphere can also be determined by bypassing the parcel method and simply comparing lapse rates. If the measured lapse rate is less than the dry adiabatic lapse rate, stable atmospheric conditions exist for unsaturated air (Fig. 5-9). If the measured lapse rate is greater than the dry adiabatic lapse rate, then unstable conditions occur. If the measured lapse rate through the lower troposphere is 10°C/km (dry adiabatic lapse rate), neutral stability of the atmosphere exists. Therefore the measurements can be used to compare with the dry adiabatic lapse rate to determine the stability of the atmosphere. Charts are available (**thermodynamic diagrams**) that have dry adiabatic lapse rates plotted for various temperatures. The radiosonde data (observed temperature profile) are then plotted on the same chart giving detailed information on those layers in the atmosphere that are stable and unstable.

High-pressure areas are nearly always stable and have fair weather associated with them. One of the reasons for this is the subsiding air associated with high-pressure areas. If air is subsiding in the atmosphere it heats adiabatically (Fig. 5-10). This tends to develop an upper air inversion because the top of an atmospheric layer descends farther than the bottom as it moves downward in the atmosphere and is compressed by higher atmospheric pressure. Therefore upper air inversions are common with high-pressure areas. This causes them to

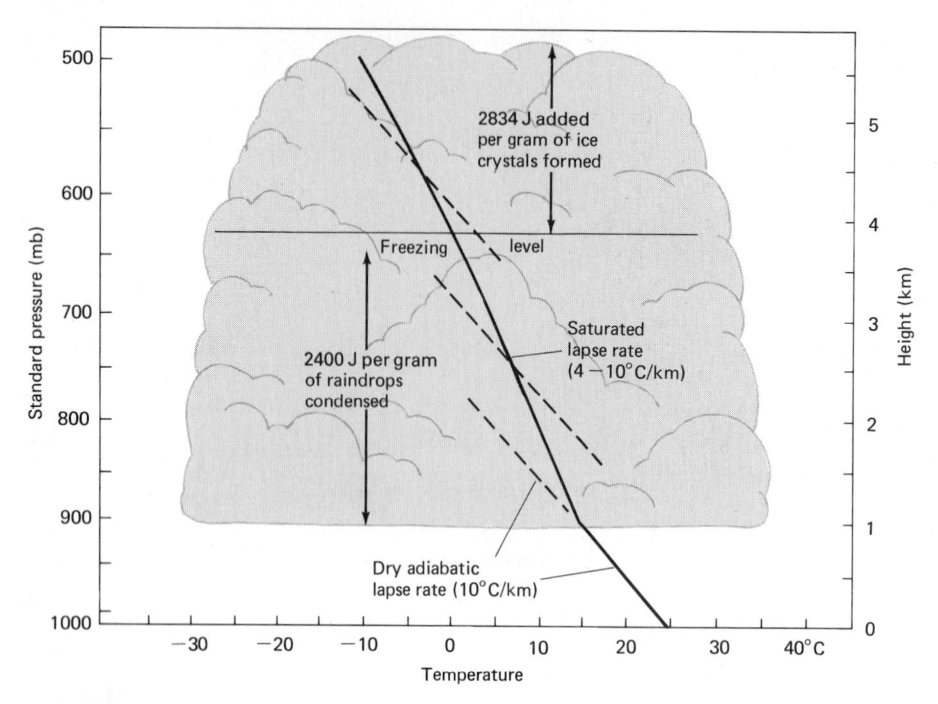

FIGURE 5-11 The saturated lapse rate is less than or equal to the dry adiabatic lapse rate because of the release of latent energy as moisture condenses or deposits in a cloud.

have not only fair weather but also the possibility of problems such as air pollution associated with them. A good example of this is the high-pressure area frequently located near Los Angeles that creates stable atmospheric conditions and decreases the mixing of pollutants from the surface with the air higher in the atmosphere.

SATURATED ADIABATIC LAPSE RATES

As air becomes saturated, another factor must be considered. The latent heat of condensation and the latent heat of deposition supply heat to rising saturated air. Changes in the state of water vapor are associated with energy absorption or release, as described in Chapter 1. As air becomes saturated and clouds form, heat is released (Fig. 5-11). When liquid water condenses as cloud droplets, the latent heat of condensation is released, averaging about 2400 J/g (575 cal/g) of water. As ice crystals or snowflakes form, the latent heat of deposition is released as vapor deposits directly to solid, bypassing the liquid state. In this case 2834 J (677 cal) are released into the atmosphere for every gram of snowflakes. Eighty calories (335 J) per gram of ice are released as liquid drops change to sleet or hailstones in the atmosphere. These energy transfers within the atmosphere decrease the cooling rate of rising saturated air.

Beneath the base of a thunderstorm the bubble of rising air cools at the dry adiabatic lapse rate, but within the thunderstorm the addition of latent heat must be considered, since this decreases the expansion cooling rate. Rising air within a cloud cools at the saturated lapse rate or **moist adiabatic lapse**

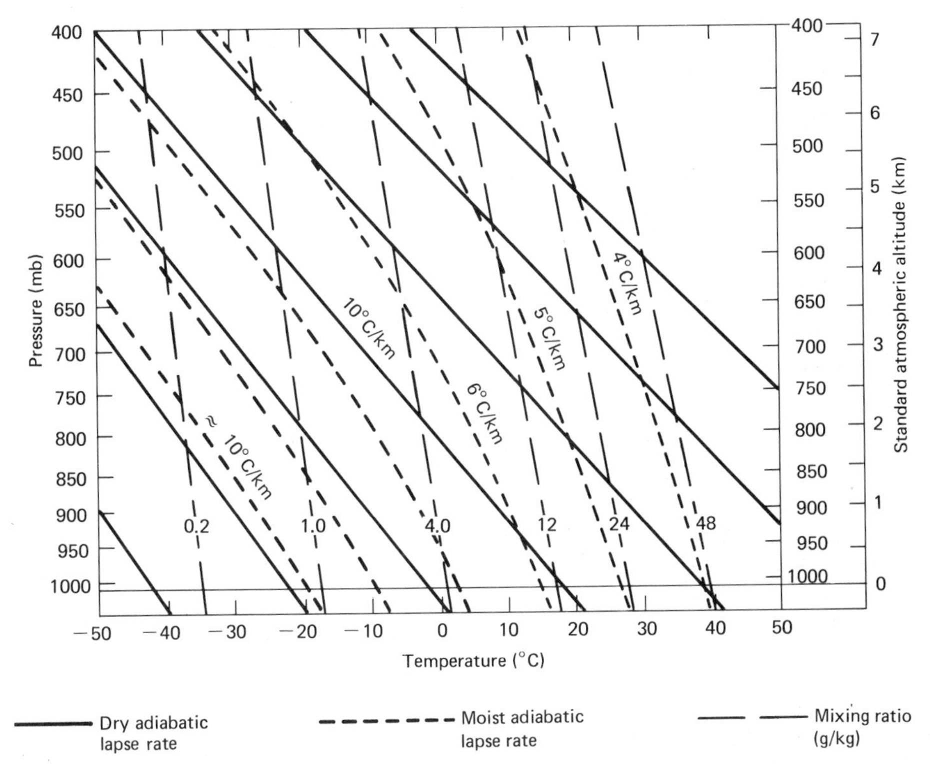

FIGURE 5-12 Thermodynamic diagrams are useful for plotting upper atmosphere temperature profiles since the measured data can be compared directly with the information on the thermodynamic diagram. This information consists of lines of dry adiabatic lapse rates, moist adiabatic lapse rates, and lines of constant mixing ratio.

rate. The saturated or moist adiabatic lapse rate includes the heat gained because of condensing water vapor and is always less than or equal to the dry adiabatic lapse rate. It varies with altitude and temperature, since warm air near the surface holds much more water vapor to condense than cold air near the tropopause. The moist adiabatic lapse rate varies from about 4°C/km at normal cloud base heights to the dry adiabatic lapse rate (10°C/km) near the tropopause, where nearly all the water has condensed out of the rising air. A typical value within the atmosphere is 7°C/km (4°F/1000 ft). The actual value of the saturated lapse rate is obtained from charts, with the rate plotted with height along with the dry adiabatic lapse rate and mixing ratio of the air. These are used by the National Weather Service for plotting radiosonde measurements. The most common type are the thermodynamic diagrams (Fig. 5-12). Since these charts already contain the moist and dry adiabatic lapse rates, the plotted measured lapse rate readily gives the stability of the atmosphere.

STABILITY OF SATURATED AIR

The stability of the atmosphere may depend on whether the air is saturated or not. If the air is unsaturated and the measured lapse rate is greater than the dry adiabatic lapse rate, the atmosphere is unstable. If the measured lapse rate is slightly less than the dry adiabatic lapse rate, then the atmosphere is stable if the air is unsaturated. However, if the air is saturated it may be unstable.

This same method is used for determining the stability of saturated air except that the measured lapse rate is compared with the saturated lapse rate rather than the dry adiabatic lapse rate. If the measured lapse rate is greater than the saturated lapse rate the air is unstable. If the measured lapse rate is less than the saturated lapse rate then the atmosphere is stable.

GENERAL STABILITY CATEGORIES

It is sometimes not possible to determine from the morning radiosonde data whether the air will be saturated or unsaturated during the rest of the day. General stability categories can be used to allow for either case. **Absolute stability** exists in the atmosphere when the measured lapse rate is less than the saturated lapse rate (Fig. 5-13). If the measured lapse rate is less than the saturated lapse rate, it is also less than the dry adiabatic lapse rate since the saturated lapse rate is always less than or equal to the dry rate. In this case the atmosphere has absolute stability since it is stable regardless of whether the air is saturated or unsaturated.

Conditional stability exists when the measured lapse rate is between the saturated lapse rate and dry adiabatic lapse rate. If the atmosphere is unsaturated, it is stable since it is less than the dry adiabatic lapse rate (Fig. 5-13). If the air becomes saturated from being pushed upward, it will be unstable since it has a greater lapse rate than the saturated lapse rate. Therefore, it is called conditional stability since it depends on whether the air is saturated or unsaturated.

The other stability category is **absolute instability.** Absolute instability exists if the measured lapse rate is greater than the dry adiabatic lapse rate (Fig. 5-13). Since the dry adiabatic lapse rate is the larger, a greater measured lapse rate indicates an unstable atmosphere for either saturated or unsaturated air.

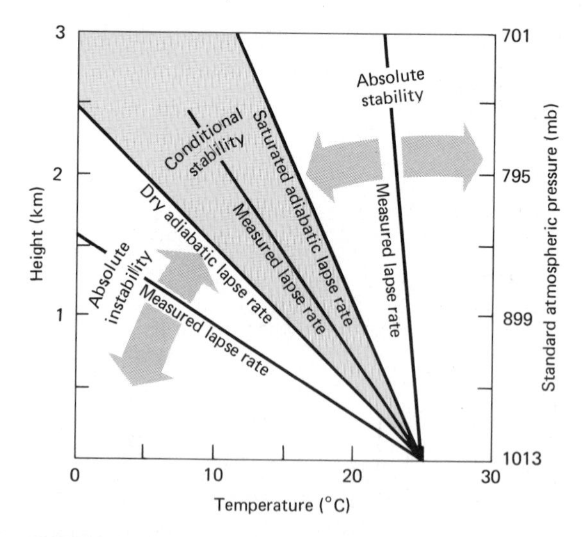

FIGURE 5-13 General stability categories are the following: absolute stability, which occurs when the measured lapse rate is less than the saturated adiabatic lapse rate; conditional stability, which occurs when the measured lapse rate is between the saturated and dry adiabatic lapse rate; and absolute instability, which occurs when the measured lapse rate is greater than the dry adiabatic lapse rate.

APPLICATIONS OF ATMOSPHERIC STABILITY

Various manifestations of the stability of the atmosphere occur. One of these is the appearance of the pollutants from smokestacks. If the lapse rate is very stable with an inversion, smoke plumes are dispersed very slowly (Fig. 5-14). The plume looks different for various lapse rates. A lapse rate greater than the dry adiabatic lapse rate (superadiabatic) causes pollutants to mix more rapidly, for example, and may result in a looping appearance of a plume. Fumigation and pollution episodes may result from an upper air inversion that traps pollutants in a surface atmospheric layer.

Stability of the atmosphere is also related to severe weather. The occurrence of more tornadoes between 4 and 6 P.M. than any other time is related to atmospheric stability. While the lapse rate at 6 A.M. may indicate a stable lapse rate, surface heating during the day increases the lapse rate in the

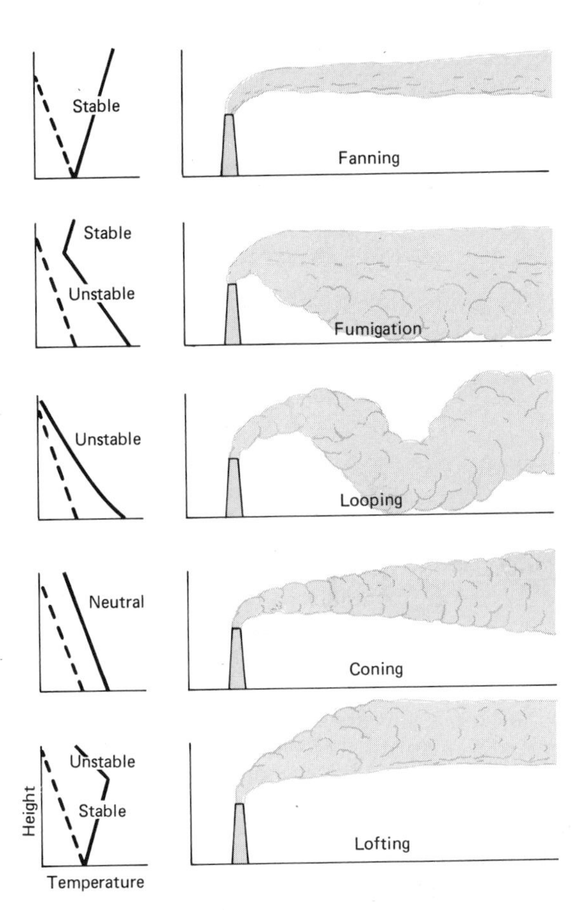

FIGURE 5-14 The appearance of plumes from smokestacks is related to atmospheric stability conditions. An unstable atmosphere causes more mixing in a looping plume while a very stable atmosphere may lead to fanning, fumigation, or lofting depending on where the stable atmospheric layer is located. Coning occurs when the atmospheric stability is neutral.

FIGURE 5-15 Tornadoes typically form in the late afternoon, when the ground and lower atmosphere have been heated by sunlight. (Photograph courtesy of NOAA.)

lower part of the atmosphere. This increase in the lapse rate may be just enough to cause the atmosphere to become unstable and develop large thunderstorms. Thus, because of the influence of solar heating on atmospheric stability, the maximum time for severe weather is between 4 and 6 P.M. (Fig. 5-15).

The stability of the atmosphere can be quantified by determining the **lifted index.** The lifted index is a measure of atmospheric stability that is used in tornado forecasting. The first step in calculating the lifted index is to find the average mixing ratio for the lowest 100 mb in the atmosphere. This can be obtained from the morning radiosonde data. Using the forecasted maximum temperature for the day, assume a parcel of air is lifted adiabatically to 500 mb (Fig. 5-16). It will rise first at the dry adiabatic lapse rate until saturation occurs at the **lifting condensation level** (height where the relative humidity reaches 100%). Above this level the parcel of air will cool at the saturated adiabatic lapse rate to 500 mb. The lifted index is then equal to the measured temperature at 500 mb minus the temperature of the lifted parcel of air in whole degrees. It is negative if the temperature of the lifted parcel of air is greater than the temperature of the environment as measured by the morning radiosonde. A negative lifted index indicates an unstable atmosphere, and a positive lifted index a stable atmosphere. Tornado activity is associated with

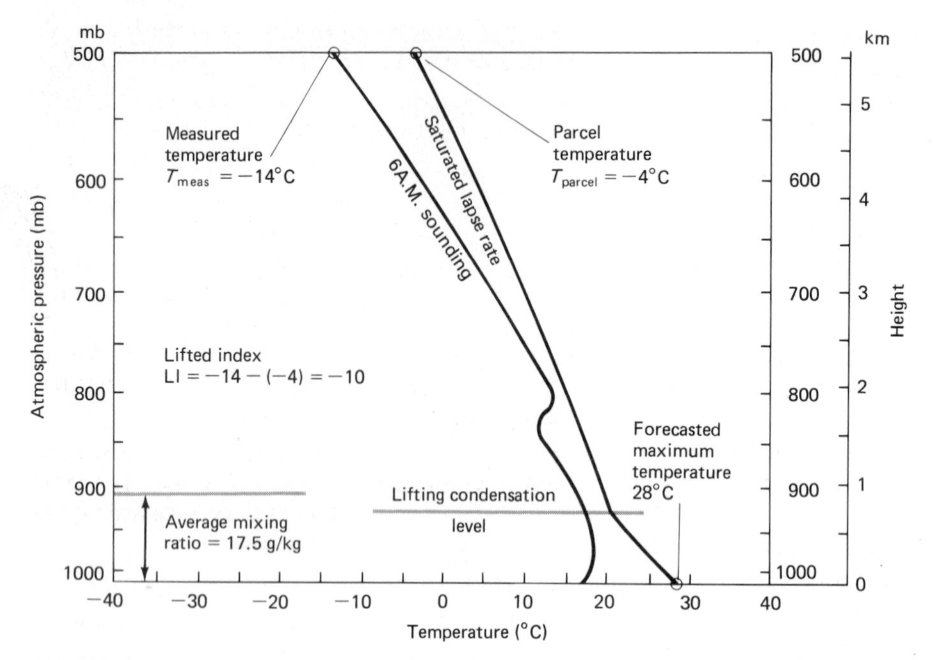

FIGURE 5-16 The lifted index is determined by assuming a parcel of air, having the forecasted maximum temperature and average mixing ratio of the lowest 100 mb, is lifted to the 500-mb level, where its temperature is then compared with the measured temperature at that height.

lifted indices more negative than −2. For example, the lifted index prior to a major tornado in Topeka, Kansas, in 1966 was −6.

SUMMARY

Vertical winds develop in the atmosphere if the vertical forces are unbalanced. The atmospheric temperature at various heights determines the amount of upward force and, therefore, the amount of vertical motion and stability of the atmosphere. The atmospheric lapse rate changes hourly as the earth's surface warms during the daytime producing greater lapse rates, and cools at night contributing to the development of surface inversions.

The dry adiabatic lapse rate is the calculated expansion cooling rate as air is lifted and the compression heating rate as air descends to lower elevations. The stability of unsaturated air can be determined by the parcel method. If a parcel of air is lifted in the atmosphere and it accelerates away from its original position, then an unstable atmosphere is indicated. If the displaced parcel returns to its original position, then a stable atmosphere is indicated. A neutral stability exists if a displaced parcel of air is still at equilibrium after displacement.

Atmospheric stability can also be determined by simply comparing lapse rates. If the measured lapse rate is less than the dry adiabatic lapse rate, a stable atmospheric condition exists. If the measured lapse rate is greater than the dry adiabatic lapse rate, an unstable atmospheric condition exists. Neutral stability exists if the measured lapse rate is 10°C/km.

After the air becomes saturated, latent heat additions must be considered in determining the expansion cooling rate. Latent heat additions because of condensation and deposition of water vapor in clouds contribute considerable heat to rising air. Therefore, the saturated adiabatic lapse rate is considerably less than the dry adiabatic lapse rate. If the air is saturated and water is condensing or depositing, the saturated adiabatic lapse rate varies from 4 to 10°C/km. The stability of saturated air is determined by comparing the measured lapse rate with the saturated lapse rate in a manner similar to the determination of the stability of unsaturated air.

General stability categories can be used if the stability or instability of the atmosphere is unknown. If the measured lapse rate is less than the saturated lapse rate, absolute stability of the atmosphere exists since the atmosphere is stable regardless of whether it is saturated or unsaturated. If a lapse rate is measured between the dry adiabatic lapse rate and the saturated adiabatic lapse rate, then conditional stability is indicated. Absolute instability occurs if the measured lapse rate is greater than the dry adiabatic lapse rate.

Atmospheric stability has many important applications. The dispersion of pollutants in the atmosphere is very much related to atmospheric stability conditions. Very little dispersion occurs in a stable atmosphere while a great deal of dispersion occurs with unstable atmospheric conditions. Severe weather is related to the occurrence of an unstable atmosphere. The atmosphere is most likely to be unstable during the late afternoon hours as solar heating occurs. This is the reason for the observed greater frequency of tornadoes in the late afternoon hours.

STUDY AIDS

1. Describe the vertical forces acting on an individual parcel of air. Relate this to the density and temperature of air.

2. Describe the diurnal variations in the lapse rate near the earth's surface.

3. Compare the stability of the atmosphere and the character of the winds.

4. Compare the operation of a refrigerator with the dry adiabatic lapse rate.

5. Explain why and when air follows the dry adiabatic lapse rate. Include the assumptions used in determining the dry adiabatic lapse rate.

6. Plot a measured lapse rate of 7°C/km and another measured lapse rate of 12°C/km and compare the temperature of a lifted parcel of air with its surroundings under these two different conditions as the parcel is lifted 1 km, 2 km, and 3 km and the parcel of air remains unsaturated.

7. Use the same two measured temperature profiles to compare a parcel of saturated air that is lifted 1 km, 2 km, and 3 km if the saturated lapse rate is 6°C/km through this layer.

8. How are thermodynamic diagrams used?

9. Describe why an inversion layer is so stable.

10. Explain the general stability categories: absolute stability, conditional stability, and absolute instability.

11. Discuss some of the applications of atmospheric stability.

TERMINOLOGY EXERCISE

Use the glossary to reinforce your understanding of any of the following terms used in Chapter 5 that are unclear to you.

Atmospheric lapse rate
Stable atmosphere
Unstable atmosphere
Environmental lapse rate
Dry adiabatic lapse rate
Entrainment
Parcel stability method
Upper air inversion
Thermodynamic diagrams
Moist adiabatic lapse rate
Absolute stability
Conditional stability
Absolute instability
Lifted index
Lifting condensation level

THOUGHT QUESTIONS

1. Explain how different colored soils and different materials at the earth's surface may affect the stability of the lower atmosphere.

2. How do you think the stability of different air masses would compare?

3. Explain how atmospheric temperature and moisture may cancel their stability effects.

4. Explain how you would expect the average height of cloud bases to vary in different parts of the United States.

5. If the time (hours of the day or a longer time period) of release of pollutants into the atmosphere could be chosen, what are some of the atmospheric conditions that should be selected to minimize pollutant concentrations?

SPECIAL TOPIC 6

SOARING

Soaring is a very rewarding sport for many people. As soon as the tow line releases the sailplane you must match your ability to remain airborne against the pull of gravity, which gently pulls you down. Time in the air can be increased significantly by applying a knowledge of meteorology. Lift may be provided by thermals, by airflow over hills or mountains, or by converging surface air. (Fig. S6-1).

Thermals form as a bubble of air becomes less dense than its surroundings. This happens as sunlight heats a plowed field, bare soil, rocky ground, streets, or rooftops of cities. Poor thermal generators are lakes, green fields or forests, and wet terrain. Visible indicators of the existence of a thermal are dust devils on the ground, developing cumulus clouds above, or soaring birds. The cumulus clouds that frequently cap a thermal may be blown downwind, thereby creating a line of clouds called a cloud street. Flying a glider under a cloud street is a very good way to complete a one-way cross-country flight because the same updrafts that create the clouds may also support a glider.

Circling birds such as hawks, vultures, eagles, gulls, and terns may provide a good indication of a large isolated thermal. Lift can be prolonged by circling the glider in an effort to remain within the updraft. Climbs to more than 1500 m (5000 ft) are possible within five minutes in a good thermal.

In mountainous terrain, lift is frequently present along the windward side of a mountain range. The glider pilot steers along the mountain slope and makes all turns away from the mountain. It's crucial to stay on the windward side, by never passing over the peaks, since lift is absent on the leeward side of the mountain just downwind from the peak. Farther away from the mountain peak, on the leeward side, lift can be found in standing lee waves.

More information on soaring can be obtained from the Soaring Society of America, P.O. Box 66071-A, Los Angeles, California 90066.

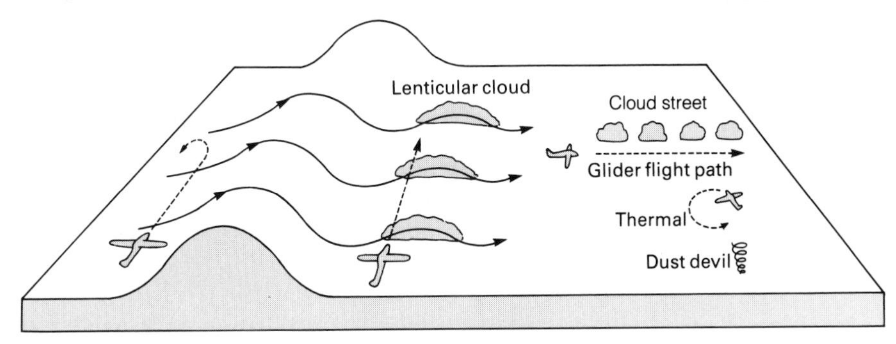

FIGURE S6-1 Regions of lift may be found near mountain chain, thermals, or cloud streets.

CHAPTER SIX
CLOUDS, PRECIPITATION, AND FOG

CLOUD OBSERVATIONS

Cloud observations are an important part of atmospheric study. Not only are many clouds quite beautiful, but the type and amount of cloud formation is an indication of the weather that can be expected at the earth's surface. Cloud information is obtained at all the synoptic weather stations and generally consists of amount of cloud cover; cloud movement; and cloud height, composition, and type.

The amount of local **cloud cover** is recorded as the number of tenths of visible sky covered by clouds. This is usually an estimate made by the meteorologist. The height of cloud base is very important over airports. Many airport weather stations have an automatic **ceilometer,** an instrument consisting of a projector, which sends a beam of light upward to be reflected from clouds, and a photocell, which detects the returned light. By measuring the angle to the lighted cloud base, the ceilometer measures the height of the cloud base.

The cloud base height or **lifting condensation level** can also be determined from surface temperature and dew point temperature measurements since the height that air must be raised to produce saturation depends on the temperature of the air at the surface and the amount of water vapor in it.

The cloud base height is greatest in drier regions of the southwestern United States (Fig. 6-1). The average cloud base varies from more than 2500 m in the desert region of southern Arizona to less than 350 m in northwestern Washington, while heights of about 700 m are typical in much of the central and eastern United States.

Clouds are composed of liquid water droplets with temperatures above freezing, liquid water

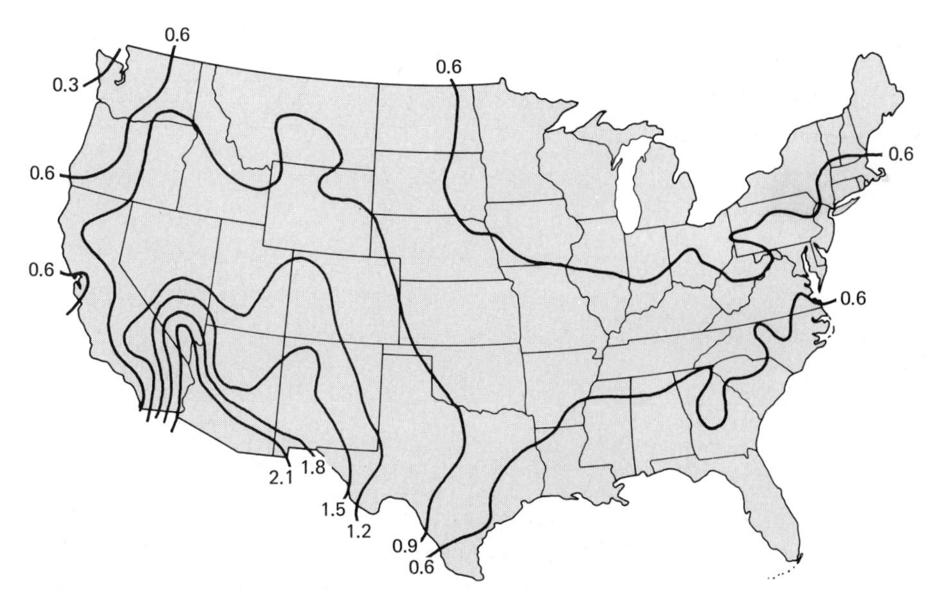

FIGURE 6-1 The mean annual lifting condensation level (in kilometers) in the United States shows large variations. Cloud base heights average more than 2 km in the southwestern desert region.

droplets with temperatures below freezing (supercooled), and solid particles. Supercooled cloud droplets can remain in the liquid state to temperatures as low as −40°C, but their normal occurrence in the atmosphere is from 0 to −25°C. Supercooled droplets condense from vapor to liquid at temperatures above freezing and are carried into colder regions of the cloud.

Solid water deposits, or changes from vapor to solid, in the atmosphere as ice crystals or snowflakes form at temperatures below freezing. Clouds that form in the upper troposphere are composed entirely of ice crystals and snowflakes, since the atmosphere is well below freezing (Fig. 6-2). Ice crystals grow in many different shapes around the basic form, which is hexagonal or six-sided (Fig. 6-3). The particular form of the crystal depends on its crystallization temperature and the temperature during further growth by **deposition,** that is, additional transfer of vapor to the solid form. If the atmospheric temperature is around freezing (0 to −3°C), then the ice crystal will form as a hexagonal plate of ice. If the temperature is colder (−8 to −25°C), the crystals again form as hexagonal plates. However, if the temperature is between −3 and −8°C, the crystals that grow are either prisms or needles. A prism is a six-sided structure that can be described as an elongated hexagonal plate or a pencil-shaped piece of ice; a needle is a very long, thin sliver of ice. Between −12 and −16°C, ice needles may grow into dendritic crystals, which are six-sided and have several elongated fingers of ice. If the crystallization and growth temperature is very cold—less than −25°C—prisms are the predominant crystalline type.

The *International Cloud Atlas,* published by the World Meteorological Organization, has established criteria for classifying clouds. Based on this system, clouds are classified into ten main groups, called **cloud genera,** which are mutually exclusive.

FIGURE 6-2 The typical ice crystal is hexagonal in shape because of the way water molecules fit together. Therefore, snowflakes are usually six sided, as growth occurs fastest at the six points. The beginning crystals near the centers of these snowflakes are hexagonal, but variations in growth produced a variety of shapes and number of points for these unusual snowflakes. (Photo courtesy of NOAA.)

CLOUD GENERA

The ten cloud genera (Fig. 6-4) are grouped according to height (high, middle, and low) and vertical development. The high clouds—cirrus, cirrocumulus, and cirrostratus—are composed of ice crystals since they form at heights from 6 to 14 km. The middle clouds—altocumulus and altostratus—form at heights from 2 to 7 km and are frequently composed of supercooled liquid water droplets mixed with ice crystals. The low clouds—nimbostratus, stratocumulus, and stratus—are composed of liquid water droplets and generally form below 2 km. Clouds with vertical growth, cumulus and cumulonimbus, ordinarily contain liquid water in low levels and ice crystals in the upper part of the cloud. The cumulus cloud is the smaller of the two, and may not extend far enough above the freezing level to contain ice crystals. In contrast, the anvil of a large cumulonimbus (the thunderhead) is usually above the freezing level, where the strong winds blow ice crystals away from the top of the thunder-

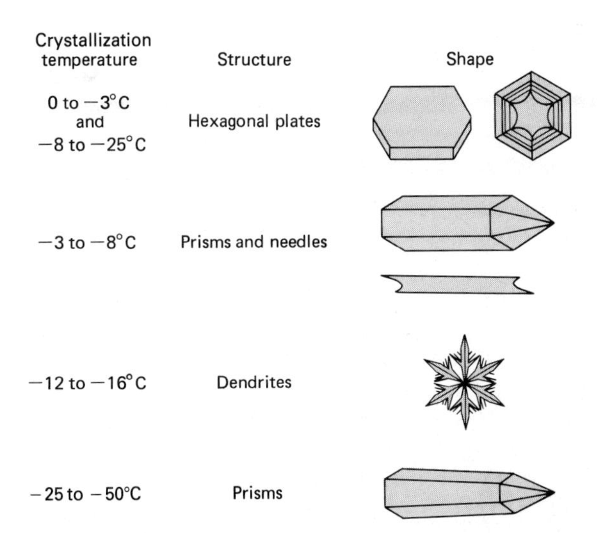

FIGURE 6-3 Various types of ice crystals and the temperatures at which they crystallize and grow.

storm. This is called cirrus blowoff and helps to form the anvil shape (like a blacksmith's anvil).

A particular type of cumulus cloud known as a lenticular cloud forms from airflow over rough terrain and is interesting since it shows the cloud formation process (Fig. 6-5). As waves form in the air streams, the regions of rising air form cumulus clouds while the areas with descending air are cloud free. Such clouds are frequently shaped like a lens and may be mistaken for flying saucers.

The type of cloud formation frequently indicates the kind of weather to be expected. For example,

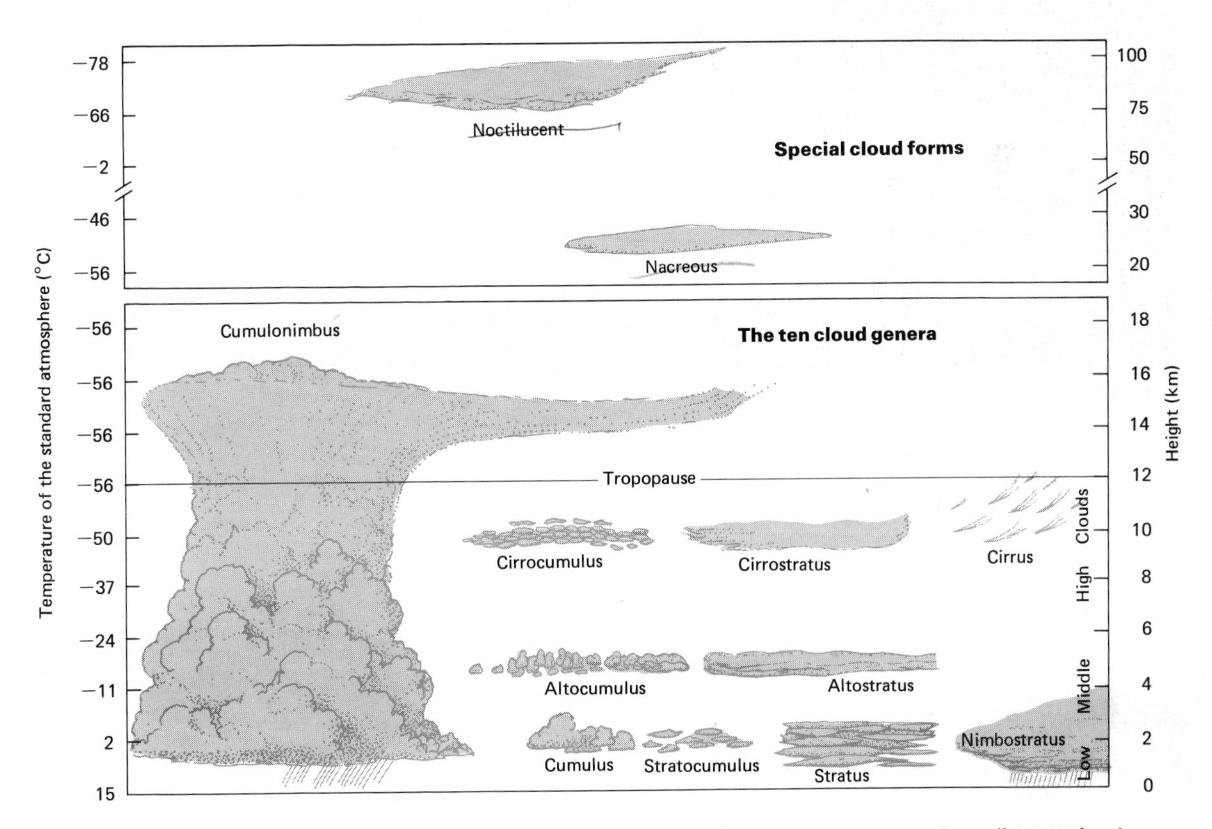

FIGURE 6-4 Eight of the cloud genera usually occur at specific heights and thus can be labeled high, middle, or low clouds. The other two cloud genera comprise clouds with vertical growth (cumulus and cumulonimbus). Nacreous and noctilucent clouds, shown at their common heights, are two special cloud types that may be seen after dark because of their height.

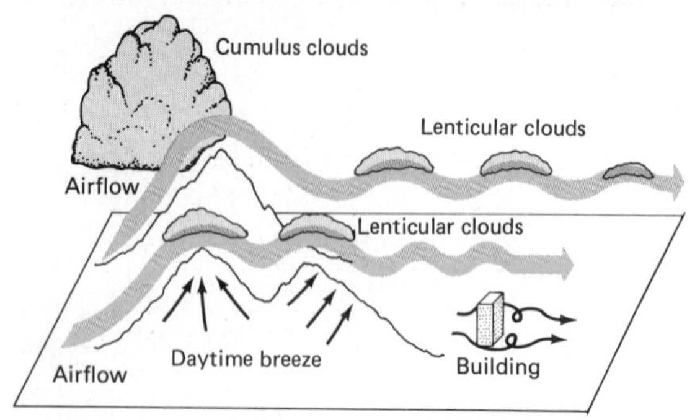

FIGURE 6-5 Topographic features may cause standing waves in the atmosphere and contribute to the development of lenticular clouds as the air rises within crests in the waves.

an approaching warm front that may provide extended rainfall in significant amounts is frequently preceded by as much as 1500 km by cirrus clouds. As the warm front approaches, the cloud type may change from cirrus to altostratus to stratus and then to nimbostratus. Cold fronts are frequently accompanied by cumulonimbus clouds. When cumulonimbus clouds form in a line they are called **squall lines.** These develop either along the cold front or in the warm air mass ahead of the front.

The symbols in Table 6-1 are used on synoptic charts to show the cloud type.

CLOUD FORMATION PROCESS

The ability of water to exist as invisible vapor in air is directly dependent on the temperature. As a rough guide, the amount of vapor that can exist in air is reduced by almost one-half for each 10°C temperature decrease. Thermodynamic diagrams (Fig. 5-12) or tables such as Table 6-2 are used to determine more accurately how much vapor can remain in the air at a particular temperature. As the table demonstrates, saturated vapor pressure is lower over water than over ice at the same temperature. This is because molecules can escape from liquid more easily than they can from ice. Similarly, saturation specific humidity is lower over water than over ice at the same temperature. As unsaturated air rises, it cools at the dry adiabatic lapse rate until it becomes saturated because of expansion cooling (Fig. 6-6). Further ascension of the air results in formation of cloud droplets around minute particles in the atmosphere with a cooling rate specified by the moist adiabatic lapse rate.

Several different natural lifting mechanisms cause atmospheric cooling. Simple **mechanical lifting** of air occurs as air flows over mountains, and is sometimes called orographic lifting. Such rising air expands as it becomes surrounded by less atmospheric pressure, and this results in cooling of the air at the dry adiabatic lapse rate that may eventually lead to saturation (Fig. 6-7). As a result of such expansion cooling, clouds frequently form on the windward (upwind) side of mountains.

The major atmospheric cloud-producing process is not mechanical lifting of air over mountains, but **dynamic lifting** of air into lower pressure with expansion cooling producing saturation. Dynamic lifting associated with weather fronts, cyclones, hurricanes, and thunderstorms causes cloud formation by expansion cooling (Fig. 6-8). Dynamic lifting occurs near weather fronts as warm moist air is lifted by a wedge of cold denser air causing the warm air to cool to saturation as its original volume expands.

The center of a midlatitude cyclone is an area of dynamic lifting as the air converges slightly toward the center of lower pressure and rises above it. As the air rises into lower pressure at higher altitudes, it cools at the dry adiabatic lapse rate until it becomes saturated and clouds are produced.

Hurricanes are also a source of dynamic lifting. As the air spiraling around the central eye of the storm rises with the help of latent heat additions, bands of clouds are produced around the eye of the hurricane.

TABLE 6-1

The ten cloud genera

Genus	Height[a] (km)	Composition	Symbol	Appearance from earth
		High clouds		
Cirrus (Ci)	6 to 14	Ice crystals		Detached, delicate fibrous structure in hairlike hooks or narrow bands. Sometimes called mare's tails.
Cirrocumulus (Cc)	6 to 14	Ice crystals		Thin white patches, sheet or layer of cloud. Often arranged in rows without dark bases.
Cirrostratus (Cs)	6 to 14	Ice crystals		A thin, transparent milky sheet with indefinite borders. May have fibrous or smooth structure. Causes a halo around the sun or moon.
		Middle clouds		
Altocumulus (Ac)	2 to 7	Ice crystals or water droplets (possibly supercooled)		White and gray patches, sheets or layers. May have dark bases. Often arranged in lines or rows.
Altostratus (As)	2 to 7	Ice crystals or water droplets (possibly supercooled)		More dense than Cs but partially translucent, gray sheet. May not show fibrous structure. Sunlight penetrates as through ground glass. Does not produce haloes.
		Low clouds		
Nimbostratus (Ns)	1 to 6	Chiefly water droplets		Gray or dark layer with no distinct cloud element. Thick enough to obscure the sun. Produces precipitation and may be obscured by lower stratus clouds.
Stratocumulus (Sc)	Below 2	Water droplets		Gray and white patches or layers that may occur as solitary clumps or solid overcast. May resemble puffs of cotton. When overcast, they produce an irregular pattern of light and dark patches larger than Ac.
Stratus (St)	Below 2	Water droplets		Soft, dull gray layer with uniform base or ragged patches. Resembles fog, but does not rest on ground.
		Clouds with vertical development		
Cumulus (Cu)	Vertical development from 1 to 5	Water droplets		Detached clouds with sharp outlines. Swelling domes or towers. Horizontal base that is usually dark; top often resembles a cauliflower.

(Continued on next page)

TABLE 6-1 *(Continued)*

Genus	Height[a] (km)	Composition	Symbol	Appearance from earth
		Clouds with vertical development		
Cumulonimbus (Cb)	Vertical development from 1 to 16	Water droplets and ice crystals		Heavy dense cloud with considerable vertical extent. Produces lightning, thunder, and precipitation. Top of cloud is high enough to be composed of a spreading cirrus plume which gives a characteristic anvil shape.

[a]Heights given are applicable to midlatitudes. For the corresponding heights in polar regions, multiply by 0.6; for tropical regions, multiply by 1.4.

TABLE 6-2

Saturation vapor pressure and saturation specific humidity for various temperatures at standard atmospheric pressure

Temperature (°C)	Saturation vapor pressure over water (mb)	Saturation vapor pressure over ice (mb)	Saturation specific humidity over water (g/kg)	Saturation specific humidity over ice (g/kg)
−20	1.25	1.03	0.78	0.64
−15	1.91	1.65	1.20	1.04
−10	2.86	2.59	1.79	1.62
−5	4.21	4.01	2.63	2.51
0	6.10	6.10	3.80	3.80
5	8.72		5.44	
10	12.28		7.67	
15	17.05		10.7	
20	23.37		14.7	
25	31.67		20.0	
30	42.42		26.9	
35	56.22		35.8	
40	73.75		47.3	
100	1013.25		622.0	

Thermal heating that initiates **convection** (heat transfer by vertical mixing) currents and thunderstorms is another source of dynamic lifting. Thermal heating occurs as the earth's surface and lower atmosphere heat from solar radiation and form buoyant bubbles of air in the warm air. Some of these bubbles of air rise and undergo sufficient expansion cooling to form clouds that develop further into large thunderstorms with strong updrafts and downdrafts. Such thunderstorms as well as much of the precipitation produced by updrafts within the thunderstorms result from expansion cooling.

Saturation of the air to produce clouds can result either from cooling of the air with no change in water vapor content or by an increase in the amount of water vapor. Since water vapor is added to the

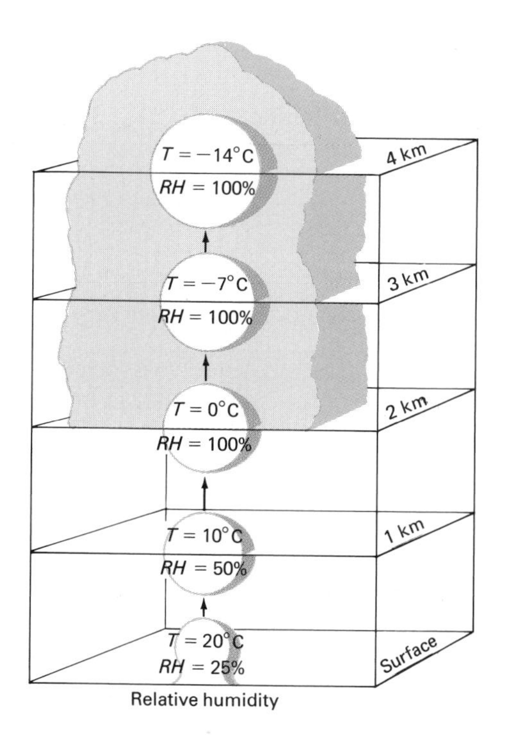

FIGURE 6-6 Rising parcels of air cool at the dry adiabatic lapse rate of 10°C/km until the air becomes saturated. Within a cloud, the cooling from expansion is partially counterbalanced by latent heat additions, as moisture condenses or deposits, decreasing the cooling rate to as little as 4°C/km. A typical value of 7°C/km is shown for the moist adiabatic lapse rate. Temperature (*T*) and relative humidity (*RH*) changes are shown for air rising from the surface to 4 km.

FIGURE 6-7 Expansion cooling occurs as air is lifted into lower atmospheric pressure, just as gas in the coils of a refrigerator cools by expansion. Conversely, compression causes heating.

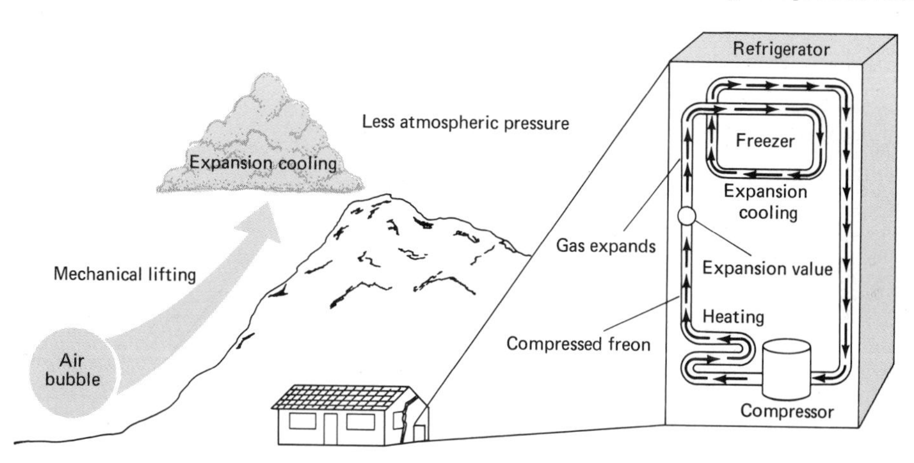

troposphere from the surface, it is more concentrated close to the surface and decreases in quantity with height, as was previously shown. Because the rate of addition of water vapor at the earth's surface is very slow compared to the rapid reduction in water-holding capacity of the air as it is cooling, the expansion cooling process is the most significant cloud formation process.

In mountainous areas of North America a relatively hot, dry wind, called the **chinook wind,** forms by a process that is the cloud formation process in reverse. Similar winds blow in the Alps, where they are called foehn winds. The word *chinook* or "snow eater" originated in Oregon where it was applied to a warm wind that came from the direction of the Chinook Indian Territory. Because of its

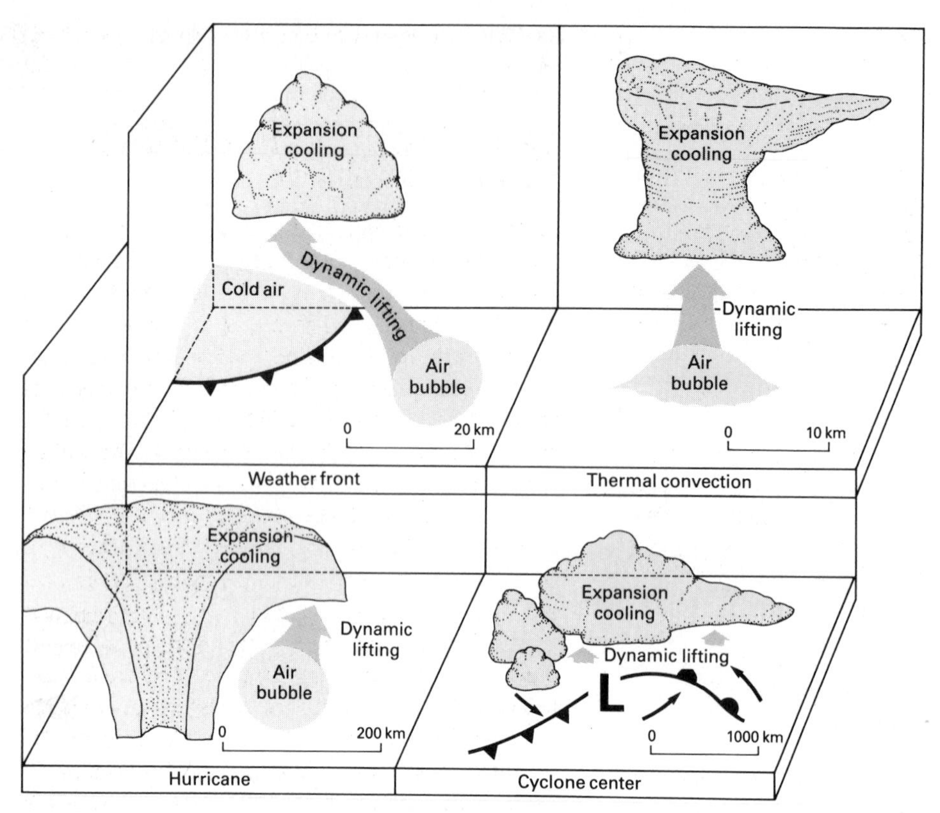

FIGURE 6-8 Various forms of dynamic lifting cause expansion cooling and cloud formation. These include lifting at weather fronts, and within thunderstorms, hurricanes, and frontal cyclones.

hot, dry character it can melt 25 cm or more of snow in a very short period of time.

Chinook winds may develop as air flows over a mountain range (Fig. 6-9). Air rising on the windward side cools at the dry adiabatic lapse rate until clouds form, releasing latent heat that decreases the cooling rate. On the leeward side, the descending air warms at a faster rate (dry adiabatic) than it cooled on the windward side of the mountain. Water vapor that condensed to form cloud droplets released 2400 J (580 cal) of energy for every gram of water condensed. This heat plus the compression heating on the leeward slopes of mountain ranges produces a very warm dry wind.

Chinook winds may also develop as air travels for considerable distances at higher elevations, as occurs over the Rocky Mountains. As the air moves hundreds of kilometers across such regions, it absorbs heat from the terrain below, especially during periods of fair weather. It may then descend into the central Great Plains, undergoing compression heating that results in a large temperature increase as well as a large humidity decrease. Conversely, with the appropriate pressure gradient, higher pressure to the east, air may descend from the east across the coastal ranges and Sierra Nevada into southern California. These so-called Santa Ana winds reach high speeds and increase the fire hazard because of their low humidity and high speeds.

Extreme temperature increases in short periods of time have been observed. On one occasion the

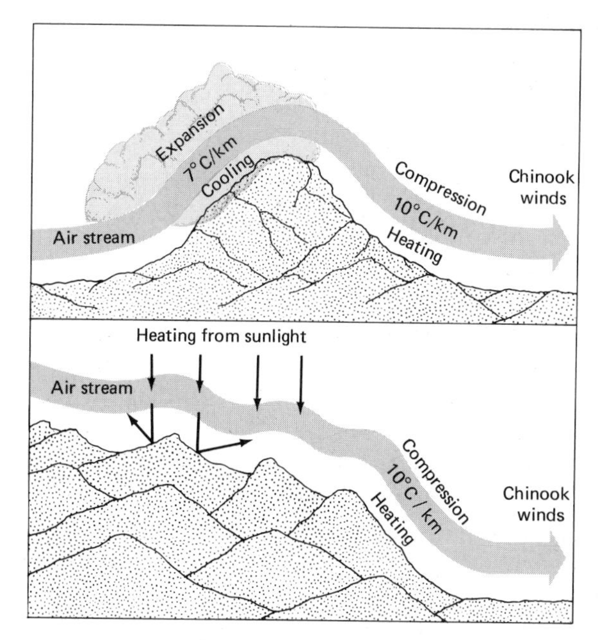

FIGURE 6-9 Chinook winds are caused by compression heating as air moves downward, encountering higher atmospheric pressure. Two different contributing factors are the release of latent heat as clouds form and radiation heating under clear skies.

temperature increased from −10°C to 6°C in only ten minutes at Rapid City, South Dakota. On January 22, 1943, at Spearfish, South Dakota, the temperature increased from −20°C to 7°C in only two minutes.

These unusual winds, which are most common after periods of cold weather, snowstorms, and blizzards during the spring and winter months, are allegedly to blame for symptoms of ill health in some persons. Some experience headache, nausea, sleeplessness, lethargy, and irritability, and increases in pulse rate and drops in blood pressure have been reported. These winds were blamed for frequent gunfights during the Gold Rush days. If the effects of the chinook are drastically felt, it is usually for less than a day. Often, these effects can hardly be noticed and blend in with the prevailing winds.

GROWTH OF CLOUD DROPLETS AND ICE CRYSTALS

In order to understand the formation of precipitation within clouds we must first consider the condensation and deposition of moisture in more detail. Air that is lifted by any of the various mechanisms cools by expansion until it may eventually become saturated. As water condenses or deposits on small particles a cloud is formed. The particles involved in the condensation process are called **condensation nuclei;** those in the deposition process are called **deposition nuclei** (also, ice or freezing nuclei). There are, in general, plenty of condensation nuclei in the atmosphere; these consist of dust, salt, and various other solid particles that are small enough to allow tiny liquid water droplets to condense around them. If there is a deficiency of condensation nuclei, supersaturation occurs, with relative humidities greater than 100%. The greatest relative humidities that are commonly observed are about 101%. This means that as the air reaches saturation a sufficient number of condensation nuclei are available to allow cloud droplets to form immediately.

Variations in the number of solid particles in the atmosphere exist in different geographical locations. The atmosphere over oceans contains fewer particles, but these are primarily salt particles, which are very effective condensation nuclei since they are **hygroscopic,** which means they readily absorb moisture in the same way that table salt absorbs moisture from a humid atmosphere. There seems to be little or no deficiency of nuclei for the condensation of liquid water.

Ice crystals require a different kind of nuclei—deposition nuclei—for their formation. The most effective are hexagonally shaped, since this is the basic shape of ice crystals. Selected clay and smoke particles may serve as deposition nuclei. Deficiencies of deposition nuclei for ice crystal formation, however, do occur in the atmosphere. In fact, the existence of supercooled liquid water in clouds

Skip

with a lack of sufficient nuclei is the basis of modern cloud seeding operations; such operations supply deposition nuclei for ice crystal formation artificially.

Solute Effect The **solute effect** is an important growth mechanism for small cloud droplets. If salt or other soluble particles serve as condensation nuclei for liquid water, they dissolve and form a solution. The vapor pressure of a solution is less than that of pure water. The vapor pressure is related to the evaporation of molecules from water. For example, if a closed container is half filled with water, evaporation of some of the water molecules will add water vapor to the space above (Fig. 6-10). The molecules of water vapor striking the walls and top of the container exert a certain pressure (force per unit area). The amount of pressure exerted by the molecules is the vapor pressure. When an equilibrium state is reached, with an equal number of molecules returning to the liquid state as are vaporized, the air above is saturated. The corresponding vapor pressure is the saturation vapor pressure. It is uniquely determined by the temperature, just as the saturation specific humidity is determined by the temperature. A comparison of the saturation vapor pressure and saturation specific humidity is shown in Table 6-2 for various temperatures.

The vapor pressure of a pure water droplet is greater than for a droplet containing a solution since it is easier for water molecules to evaporate from water than from a solution. If a solution containing salt is evaporated, the salt molecules do not vaporize but remain in the container. The salt molecules physically interfere with the escape of water molecules from a solution, thereby reducing the vapor pressure of the solution. Therefore, a small water droplet that has formed a solution by dissolving its condensation nucleus is more concentrated than a larger droplet.

More concentrated smaller droplets have a smaller vapor pressure with a smaller evaporation rate than larger droplets because of the solute effect (Fig. 6-11). Therefore, small droplets may grow at the expense of large ones, or they may grow in an environment that is marginal for raindrop growth. For example, if the relative humidity within a cloud were 99%, large drops would evaporate while smaller droplets would be able to grow larger by absorbing vapor from the surrounding environment.

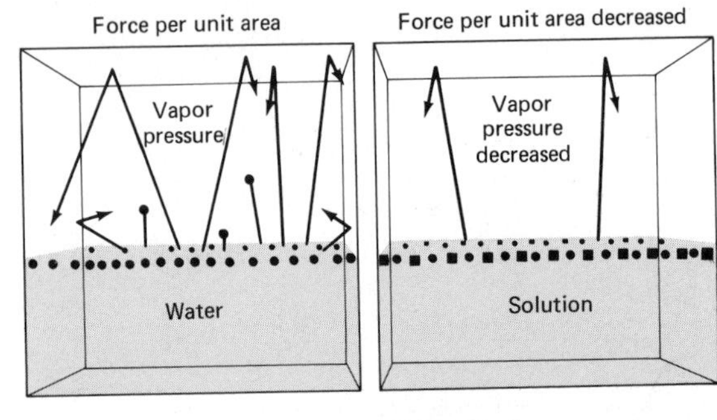

FIGURE 6-10 The vapor pressure of a solution composed of water and salt, for example, is less than the vapor pressure of pure water. This fact has various ramifications for precipitation processes.

Curvature Effect The solute effect on the droplet vapor pressure is partially counterbalanced by the **curvature effect.** A water molecule is composed of two hydrogen and one oxygen atom arranged in such a way that one end is positively charged and the other is negatively charged, making it a polar molecule. Water molecules arrange themselves in a particular way since the positive charges attract negative charges and repel other positive charges (Fig. 6-12). Such intermolecular forces cause water to be held together as droplets in the atmosphere. The charges are weak so that beyond a size of about 0.5 cm large raindrops can no longer be held together by these intermolecular forces and the raindrop splits into several smaller drops. On a flat water surface, each molecule is attracted by its

Skip

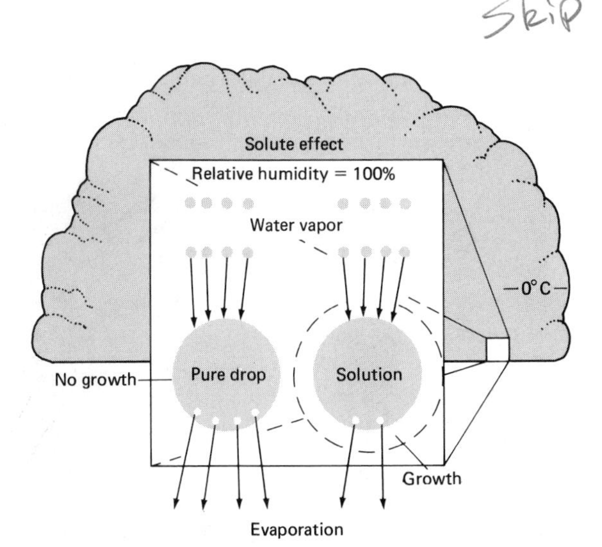

FIGURE 6-11 If a small cloud droplet condenses around a salt particle and dissolves the salt, forming a solution, it can grow in the same environment where a pure drop would experience no growth.

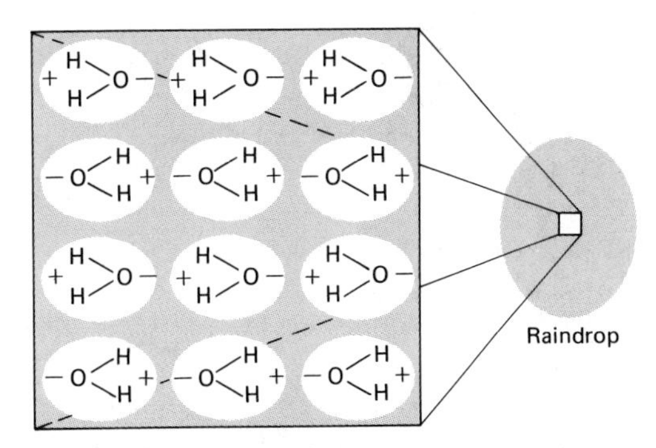

FIGURE 6-12 Water is a polar molecule, which means that it has weak charges that give rise to intermolecular forces that arrange the individual molecules in specific patterns. These charges are of sufficient magnitude to cause small droplets to be pulled together into spheres. Larger drops are held together until they reach a size of about 0.5 cm.

neighbors, but if the water surface is curved fewer neighbors exist for any particular surface molecule. With fewer molecules holding it in place, it is easier for the molecule to escape from a small curved droplet than from a larger one. The vapor pressure is greater for small droplets therefore, with a resulting larger evaporation rate (Fig. 6-13). If this were the only effect, small droplets in clouds would evaporate faster than larger droplets. The curvature effect on vapor pressure and the resulting evaporation rate of cloud droplets is opposite to that of the solute effect.

Combined Solute and Curvature Effects Together, the solute and curvature effects determine the growth or evaporation of small droplets, which are the most concentrated solutions since they have greater curved surfaces than larger drops. It is the combination of these two factors that is important in the growth of cloud droplets. If the relative humidity were 100% in a cloud, and if a flat water surface with no impurities were present, it would be in equilibrium with the saturated air. If

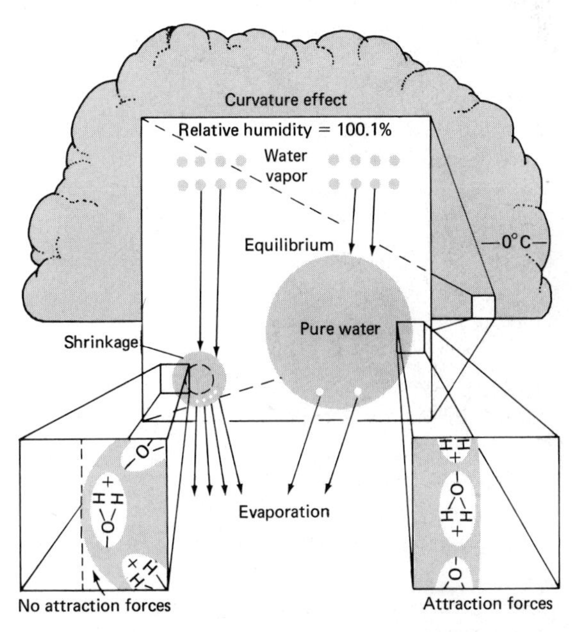

FIGURE 6-13 Because a small droplet is more curved than a larger water droplet, each surface molecule of water has fewer neighbors with electrical charges to hold it within the droplet. Therefore, evaporation is greater—small droplets may evaporate because of the curvature effect when larger droplets may be in equilibrium with the environment.

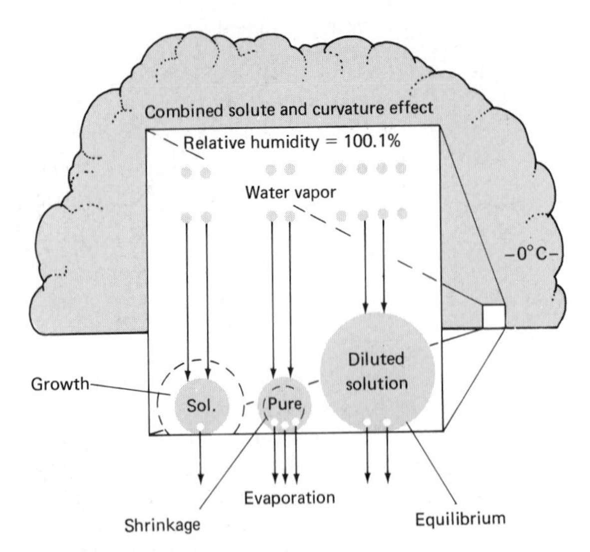

FIGURE 6-14 Combined solute and curvature effects determine which tiny cloud droplets may grow into raindrops, since a small droplet that dissolves its condensation nuclei to form a solution also has a curved surface. If the small droplet is more concentrated, the solute effect may allow it to grow into a large cloud droplet in the same environment where a small pure drop would evaporate. The solute effect is most pronounced in very tiny concentrated droplets and becomes less important as the droplet grows and becomes more diluted. It is therefore most important in the nucleation region of the cloud.

water condenses on a salt nucleus, it may grow in volume until its size results in an equilibrium relative humidity of perhaps 99%, or if it is more concentrated, it may even grow when the relative humidity is 97%. If the curvature effect is considered alone, a relative humidity for the surrounding air of at least 101% would be required for small droplets to grow. The combination of the solute and curvature effects results in a maximum cloud droplet size, since the amount of curvature and concentration of the solution both decrease as the cloud droplet grows (Fig. 6-14). If the size of the condensation nucleus is large enough to produce a concentrated solution, then the cloud droplet can grow large enough that it may eventually become a raindrop through additional processes to be described. Therefore, the solute effect is an important part of the raindrop growth process even though it is partially counterbalanced by the curvature effect. The solute effect is of primary importance in the nucleation region of clouds since it is negligible for droplets with diameters that exceed 20 μm.

Since every cloud that forms does not produce rain, another transition must occur between the condensation of cloud droplets and the formation of large raindrops that reach the surface. Rainfall requires the development of drops large enough to have significant fall velocities, in addition to withstanding evaporation losses between the cloud and the ground.

GROWTH OF RAINDROPS

Warm Clouds The principal mechanism for making raindrops from cloud drops in warm clouds is the **coalescence** process. The extremely varied sizes and types of condensation nuclei in the atmosphere produce through the solute effect cloud droplets of widely differing diameters. Direct observations have shown that cloud droplets of many different sizes exist in a typical cloud. Each droplet has a fall velocity that depends on its particular size. For example, if the diameter of the droplet is only 0.001 mm, the terminal fall velocity will be only 0.00004 m/sec, but if the diameter is that of a large raindrop, 5 mm, the fall velocity will be 8.9 m/sec. For an ordinary cloud droplet with a diameter of 1 mm, the fall velocity will be 0.076 m/sec. The size of the droplets, therefore, greatly affects the fall rates of cloud droplets and raindrops.

As various droplets fall through a cloud at different rates, collisions occur between droplets. Large droplets, for example, collect smaller ones as they fall, producing the coalescence or accretion process (Fig. 6-15). That is, small droplets coalesce with large drops as the leading surface of the large drop strikes them. In addition, some wake capture occurs; that is, small droplets collect on the back side of large drops. This latter effect happens for the same reason that a station wagon going down a dusty road collects dirt on the back window. Air is

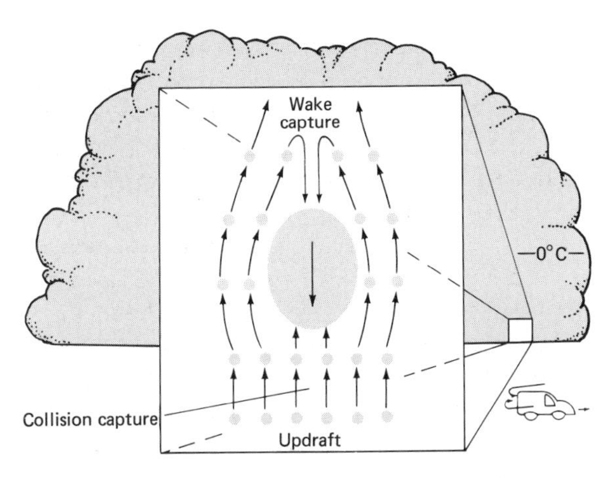

FIGURE 6-15 Raindrops may grow by coalescence with smaller cloud droplets through collision capture on their leading edge or wake capture as small droplets are pulled into the backside of the raindrop because of airflow patterns around an obstacle.

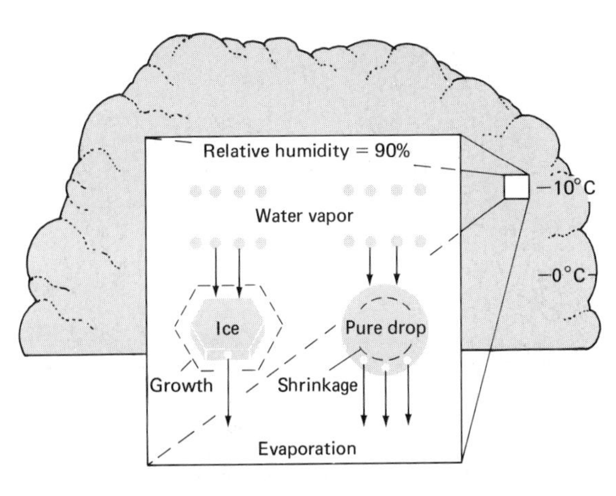

FIGURE 6-16 Above the freezing level, ice crystals and liquid water droplets frequently coexist. The evaporation rate of ice crystals is less than that of pure droplets, with the result that ice crystals may be able to grow in environments where the relative humidity is 90% or less for pure water droplets.

pulled around and impacts the back of the vehicle because of the nature of airflow around an obstacle. Vortices or circular air currents are produced by the sides of the vehicle and these force the air into the back window. As large droplets fall, small droplets are pulled into their back side with sufficient velocity to produce coalescence. Wake capture and capture from direct collisions form the coalescence process, a very important final step in the production of precipitation from warm clouds.

Cold Clouds Much of the world's precipitation originates in clouds that develop near the freezing level and extend upward into colder air. This cold environment produces a different raindrop formation process due to the coexistence of ice crystals and liquid water droplets. Both ice crystals and liquid water droplets are present in clouds with temperatures from −5 to −30°C. The vapor pressure of liquid water droplets and ice crystals is quite different even though they may have the same temperatures (Table 6-2). Since the ice crystals are solids, the individual molecules are held more rigidly and possess less kinetic energy than molecules in liquid droplets. Because of their greater mobility, as indicated by a greater vapor pressure, it is much easier for molecules to escape from the liquid water droplets (Fig. 6-16). The difference in vapor pressure between ice and water at the same temperature reaches a maximum at −12°C, where the difference is 0.3 mb. This difference in vapor pressure seems small, but it is quite important in allowing ice crystals to grow in an environment where liquid droplets would evaporate, since ice crystals are in a saturated environment if the relative humidity is only 85% for liquid drops. Ice crystals, then, can grow by collecting water vapor at the expense of liquid water droplets, as the liquid water evaporates into an environment that is already saturated for ice crystals. This is possible because of the reduced evaporation rate of ice crystals and their small vapor pressure.

This raindrop formation process may be called the **Bergeron-Findeisen precipitation process**, after the scientists who suggested it in the 1930s. First, the process assumes the coexistence of ice crystals and supercooled liquid water. The ice crystals with their lower vapor pressure and evaporation rate

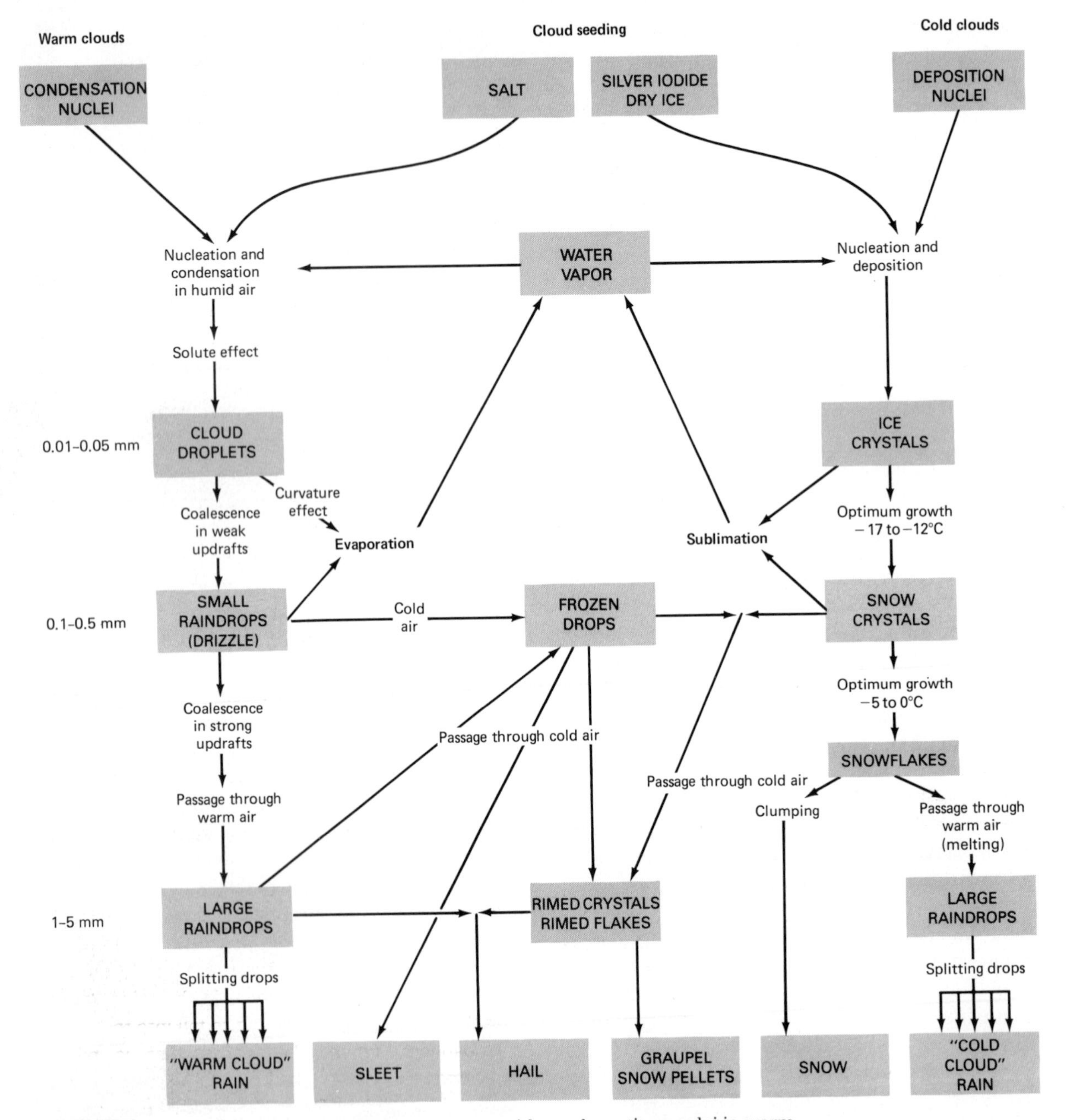

FIGURE 6-17 The precipitation process may start with condensation nuclei in warm clouds, resulting in growth by the solute effect and further growth by coalescence until large raindrops split into many smaller droplets and continue to grow by coalescence. In cold clouds, deposition nuclei are used in the growth of ice crystals. These may grow larger where they can melt in a lower, warmer atmosphere or arrive at the surface in various precipitation forms.

then grow at the expense of liquid water droplets until they are large enough to fall into the lower atmosphere, where they encounter warmer air. The large ice crystals then melt and form raindrops, with further growth through coalescence. This is probably the major precipitation process in midlatitudes. In tropical regions the coalescence process is most important since many tropical clouds do not extend above the freezing level.

Cascade Effects We all know that major downpours occur regularly in many locations. Therefore, the raindrop formation processes must work rapidly. Several of the precipitation formation processes, in addition to another effect, operate together to produce heavy rainfall. Initially, coalescence may occur on the larger ice crystals, with further growth by deposition as water vapor collects on the crystals. The ice crystals may encounter supercooled liquid droplets which freeze upon impact. This process is called **riming** and may fuse several ice crystals and snowflakes together. These may melt in the lower atmosphere to become raindrops (Fig. 6-17). **Cascade effects** may then operate, the first stage of which may be compared to dumping a bucket of water off a twentieth-story balcony. The water breaks up into drops before it hits the street. Coalescence produces larger and larger raindrops, but the intermolecular forces holding the drops together are exceeded at a size of about 5 mm, and the drops break apart into several smaller raindrops. These experience further growth by coalescence as they fall through another dense cloud. This process may be repeated several times to produce a downpour of raindrops at the surface.

FORMS OF PRECIPITATION

For the farmer who needs water to grow crops and for the rural or suburban commuter who needs highways clear of snow and ice, the type and amount of precipitation is frequently critical. The most common type of precipitation is rain, since the temperature in the lower atmosphere is generally above freezing in most parts of the world. The intensity of rainfall at the surface is quite variable and ranges from a record 30 cm/h to a very gentle mist that is hardly able to fall from the atmosphere. The amount of rainfall varies from more than 350 cm per year in the northwestern United States to less than 10 cm per year in the southwestern United States (Fig. 6-18).

Freezing rain may develop under the right temperature conditions. Freezing rain can quickly produce a glaze over streets, trees, and utility wires, resulting in traffic accidents, power blackouts, and various other problems (Fig. 6-19). In order for freezing rain to occur, the ground temperature must be 0°C or colder, with warmer temperatures in the atmosphere above. To provide the warmer atmosphere above, where water vapor can condense into liquid, while the cold surface causes the water to freeze, a surface temperature inversion is necessary (Fig. 6-20). Such surface inversions may develop from a variety of causes, but typically they occur near the leading edge of cold air from the north as it pushes southward.

Ice pellets (sleet) are transparent or translucent bits of ice less than 5 mm in diameter. These pellets require a slightly different atmospheric condition for formation: Raindrops form in the atmosphere where the temperature is above freezing and then fall into a lower layer of air with temperatures below freezing; here, the raindrops freeze, becoming small ice pellets. Ordinarily, to provide temperatures above freezing in the atmosphere above a colder atmospheric layer, an upper air inversion is required (Fig. 6-21). Raindrops can then form as vapor condenses in the warm layer. As they become large enough to fall into the lower atmosphere, where the temperature is below freezing, the raindrops freeze. The appropriate synoptic conditions for ice pellet formation may be set up by the passing of a low-pressure system. The thin surface layer of cold air traveling southward behind a low-pressure system lifts the warmer air currents from the south to produce cold surface temperatures and an upper air inversion that allows the formation of ice pellets.

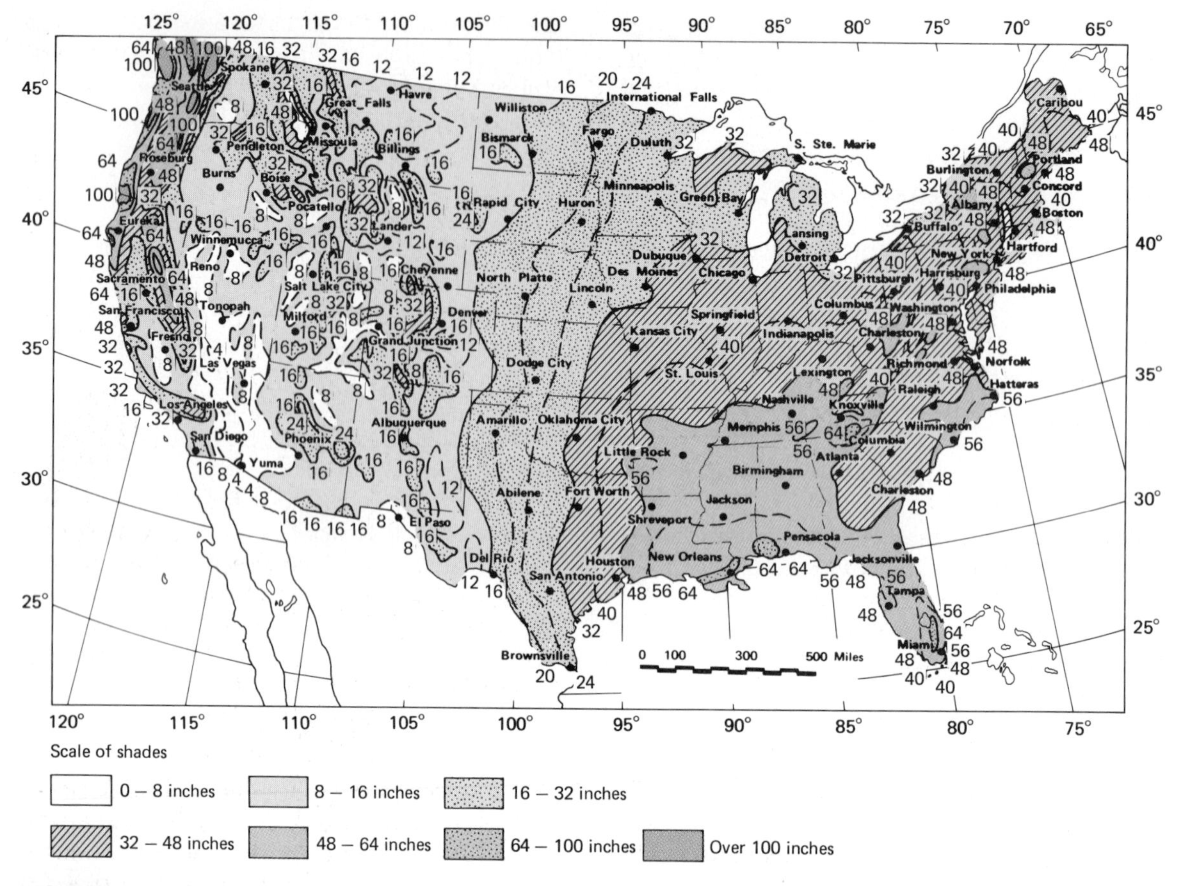

FIGURE 6-18 Distribution of rainfall throughout the continental United States. (Environmental Data and Information Services, NOAA.)

FIGURE 6-19 Ice storms are a very destructive form of precipitation. Utility wires, TV antennas, and trees are particularly vulnerable to these storms. (Photo courtesy of NOAA.)

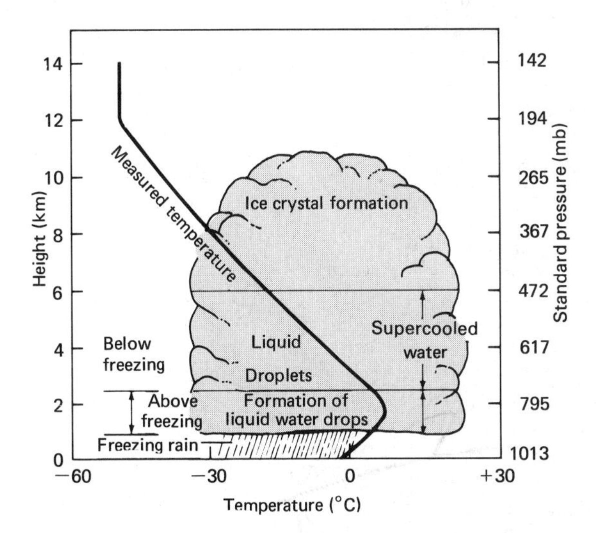

FIGURE 6-20 Atmospheric temperature profile required for the formation of freezing rain includes a surface inversion to provide a layer of air with temperatures above freezing for forming liquid raindrops.

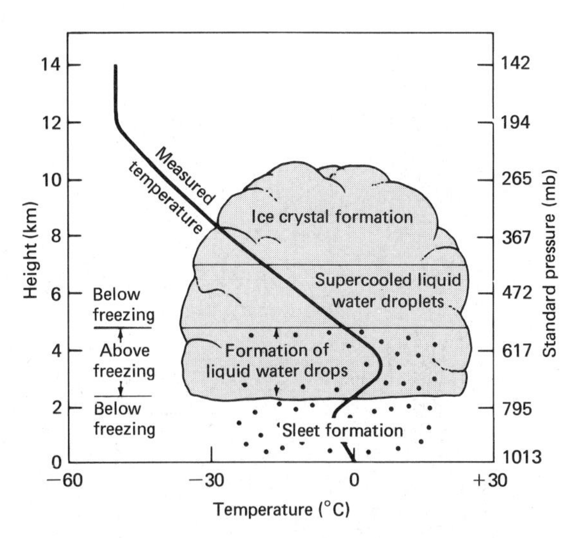

FIGURE 6-21 Atmospheric temperature profile required for the formation of ice pellets (sleet) includes an upper air inversion to provide a warm layer, where liquid raindrops can form. The cold layer below is required to freeze the raindrops into ice pellets.

Snow frequently falls in midlatitudes during the winter as low-pressure systems move through an area with accompanying temperatures below freezing. These systems may produce a sequence of weather, with the type of precipitation changing from rain to freezing rain, sleet, and then to snow, as the temperature drops to freezing and then below freezing. Snow can develop only from water vapor that deposits directly as a solid, bypassing the liquid state. A snowflake forms first as a very tiny six-sided hexagonal crystal. The crystal then grows fastest at the six points because the points are exposed to water vapor coming from more directions than locations along the sides. This accentuates the points and develops a six-sided snowflake.

Snow is most common in winter just north of the center of low pressure. As the warm moist air travels around the center of lowest pressure, it overrides colder air located north of the low and is cooled to its saturation temperature, producing rainfall and snow (Fig. 6-22). Snow generally occurs with east winds, since the winds at locations north of a midlatitude cyclone are from the east. A forecasting technique for locating the southern boundary of the snow line is to note the position of the 0°C isotherm on the 850-mb map since this has been observed to correspond with the snow line at the surface.

Snow pellets are white, spherical grains of ice 2–5 mm in diameter. They can be distinguished from packed snowflakes since snow pellets are firm enough to bounce when they hit the ground. Snow pellets, or **graupel,** grow as supercooled droplets freeze on ice crystals. They may fall for a brief period as the precipitation changes from sleet to snow.

Hail is a destructive form of precipitation that will be considered in more detail in a later chapter. Very briefly, hail is distinctly larger than ice pellets (5–190 mm in diameter) and forms by a much different process. The large updrafts and downdrafts in mature cumulonimbus clouds provide the mechanism for hail formation. Hailstones normally have

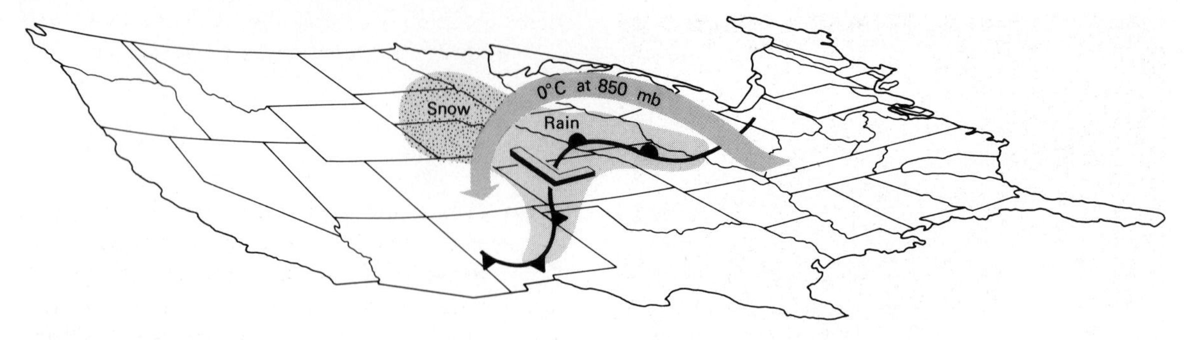

FIGURE 6-22 Atmospheric conditions that produce heavy snow include warm air that overrides a warm front to become east winds north of the low-pressure center. If the temperature at 850 mb is near 0°C or slightly below snow is indicated.

concentric shells of ice alternating between those with a milky appearance and those that are clear. The milky shells, containing bubbles and partially melted snowflakes, correspond to a period of rapid freezing, while the clear shells develop as the liquid water freezes much more slowly.

UNUSUAL PRECIPITATION

Among nature's flukes are a variety of colored forms of precipitation. Rain may be colored as a result of its condensation nuclei or from washout of particulate matter in the air. If the condensation nuclei are not conventional dust or salt, the precipitation may well take on the coloration of the nuclei. Another possibility for colored rainfall exists when a large quantity of material is suspended in polluted air to be washed out by rain falling through it. This may cause the rainfall to take on the color of the foreign material being washed out. On April 9, 1970, bright red rain fell on Thessalonika, Greece. A low-pressure system with strong winds had pulled in a huge load of clay minerals from the Sahara and lifted them well into the troposphere, where the upper-level winds carried the minerals into a shower system in Greece. Rainfall then washed out the particulate matter, giving Thessalonika a very unusual red rain.

Precipitation may also be colored due to more local weather events. A thunderstorm may sweep up particulate matter with its strong surface winds, and some of the particulates may travel through the storm's updraft-downdraft system. On June 6, 1959, a thunderstorm dumped a torrent of greenish-yellow rain on Dunstable, Massachusetts. The thunderstorm had apparently sucked up a gigantic mass of pollen and deposited it on the town. Visibility during the freak rain was cut to less than 10 m.

Reddish-yellow precipitation occurs every few years in Kansas and Missouri, as thunderstorms pick up colored dust from Oklahoma, New Mexico, and southern Colorado and carry it along as they travel northeastward. It is particularly noticeable if the showers are light, leaving dirty cars, windows, and so on after the storm passes.

Various colors of snow have also been observed. On March 9, 1972, Saharan particulate matter was

FIGURE 6-23 Snow rollers give an unusual appearance to a snow field as the wind combines with thawing temperatures to create natural snowballs or rolls resembling rolls of carpet. These snow rollers formed near Marshall, Missouri, on February 18, 1977, following the passage of an intense cyclone that brought heavy snow. (Photograph by George R. Clemens, courtesy of Marty Kugel.)

swept into the precipitation sector of a low-pressure system. Because of copper minerals in the dust, a blue snow fell from this storm in the French Alps.

Snow can also change color after it falls. This frequently happens in the Russian taiga, the Caucasus, and Tien Shan ranges on the Russian border. The snow cover there turns a variety of colors during the spring as microorganisms in the snow concentrate minerals such as iron. Colonies made up of algae, fungi, and bacteria grow, and their pigmentation colors the snow cover. Such "snow blooms" are triggered by warm frontal passages in late spring.

Also, snow cover distortions can occur that are quite unusual. The best known is probably the **snow roller,** a formation resembling a rolled-up carpet (Fig. 6-23). Snow rollers form when the temperature is just above freezing, producing a small amount of melting of the snow. High winds then blow the snow into rollers just as a child rolls the snow into a snowball to make a snowman. The most common locations for the formation of snow rollers are rooftops, since the slope of the roof helps the wind overcome the force of gravity, and on hills and river banks for the same reason.

FOG

If the air near the ground is cooled sufficiently, it becomes saturated and a fog develops. By definition, a **fog** exists if visibility is reduced to 1 km (5/8 mi) or less, although the National Weather Service commonly reports visibility reductions before they become this limited.

Fogs can be classified according to the process that causes the air to cool, as shown in Figure 6-24.

Radiation fog, often called ground fog, is generated as the earth's surface cools by loss of radiation to space at night. The fog layer may be quite shallow, for example, just up to the headlights of an automobile so that the top of the fog layer is visible but very little is visible beneath.

Upslope fog is generated by airflow over topographic barriers. As the air is forced to rise upward where the atmospheric pressure is less, it is cooled by expansion and produces a fog on the windward slopes of hills or mountains. It is not uncommon for upslope fog to develop over large sections of the high plains, including Colorado and Kansas, when the pressure pattern produces east winds.

Advection fog may be generated by winds that contrast in temperature with the earth's surface. Warm air advection can produce fog through contact cooling with a cold surface, while cold air advection over warm water or warm, moist land surfaces results in fog formation as water evaporates into the cold air. This is sometimes called steam fog.

Frontal fog is produced as weather fronts, especially warm fronts, pass through an area. Precipitation falling into the colder air ahead of the warm front may evaporate enough water to cause the formation of small droplets as fog near the ground.

The typical fog produced by one of these processes is composed of liquid water droplets. However, these processes may produce an ice fog in arctic air as water deposits on nuclei in the atmosphere, creating minute ice crystals. Such ice fog

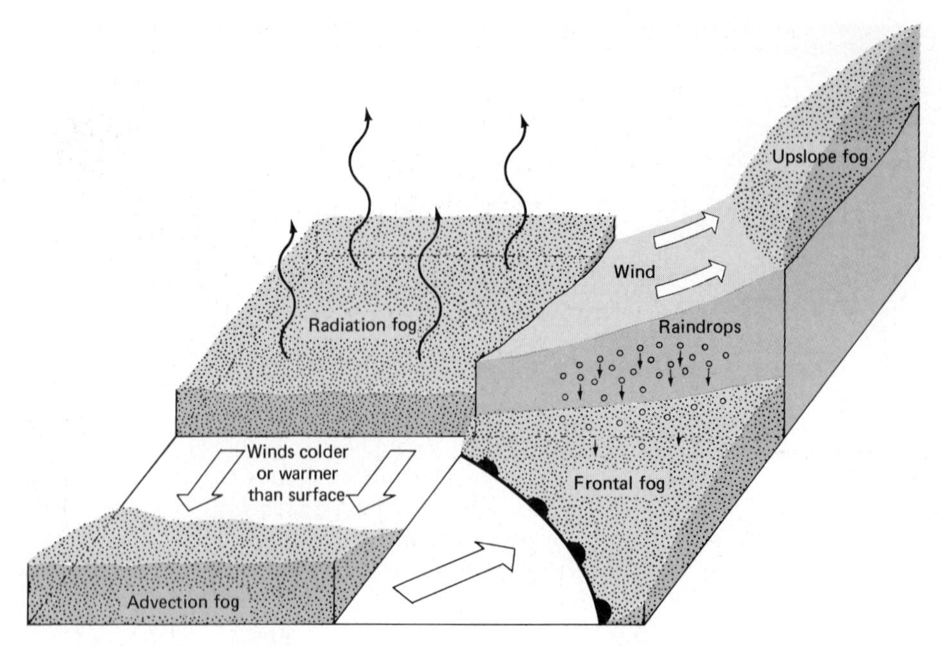

FIGURE 6-24 Four different types of fog classified according to formation factors are: radiation fog, upslope fog, advection fog, and frontal fog. Fogs form as radiation cools the lower atmosphere, as air rises over topographic barriers, as contrasts exist between the air temperature and the earth's surface, or as fronts cause saturated surface air.

reduces visibility and may even be harmful to a person's respiratory system, causing breathing difficulties if the ice crystals grow large.

The mean annual number of days with heavy fog (visibility less than 0.5 km) is shown in Figure 6-25. Fogs are most common along coastal areas, where they form by the advection process. They are especially frequent in spring and fall as the contrast between land and water temperatures is most pronounced. The effects of fog are felt on ships that are forced to travel through them, as well as on permanent residents of coastal areas. Other undesirable effects of fog include the loss of millions of dollars each year by aviation and numerous deaths and damages resulting from automobile accidents.

SUMMARY

Clouds form as air rises and cools by expansion. The dry adiabatic lapse rate of 10°C/km gives the expansion cooling for unsaturated air, while the moist adiabatic lapse rate of 4 to 10°C/km is the expansion cooling rate for saturated air. Expansion cooling occurs as air is lifted by mechanical or dynamic means including orographic lifting and the dynamic lifting caused by hurricanes, frontal cyclones, and convective thunderstorms. Chinook winds develop from the cloud formation process in reverse.

The type of weather to be expected can frequently be anticipated from observations of the

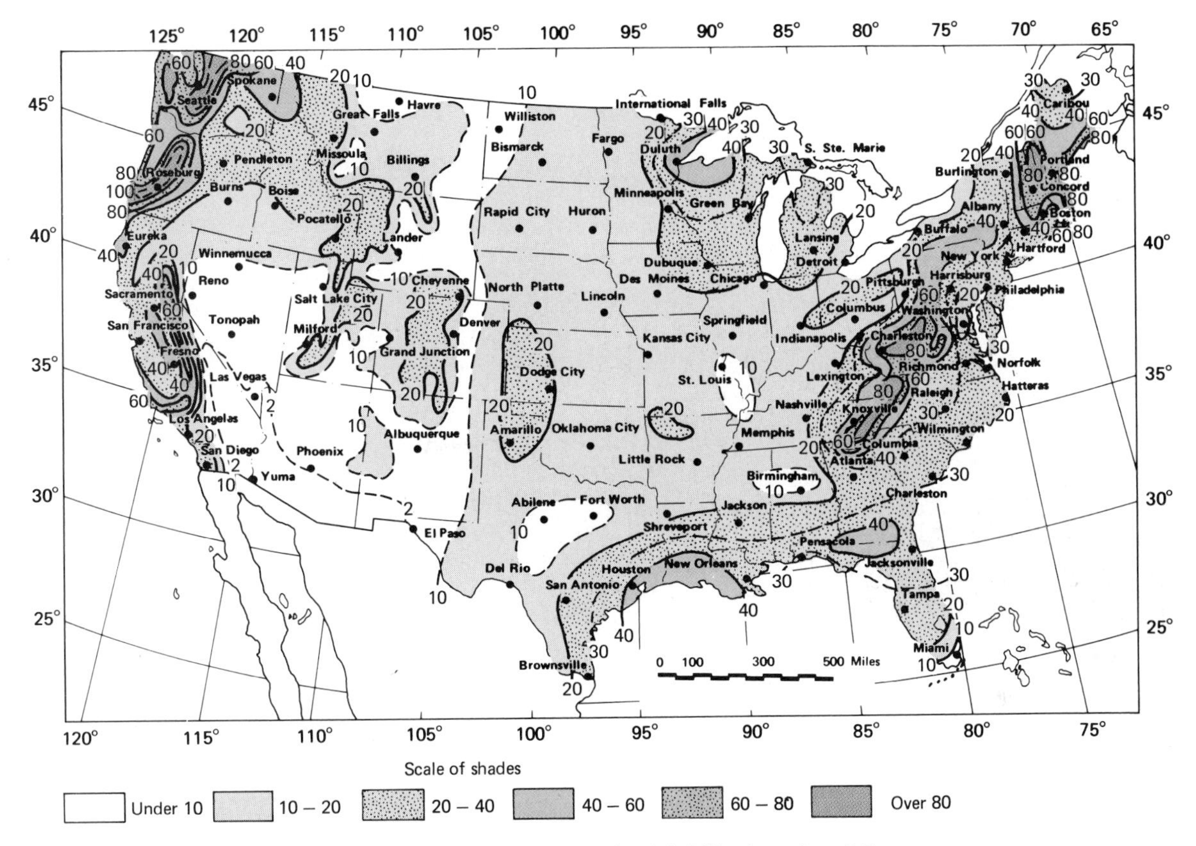

FIGURE 6-25 Mean annual number of days with heavy fog (visibility less than 0.5 km). (Environmental Data and Information Service, NOAA.)

particular type of clouds. In addition to cloud type, meteorologists frequently gather information on the amount of local cloud cover, the direction of cloud movement, and the height of the cloud base.

Liquid water droplets condense in clouds around condensation nuclei. Condensation nuclei may be dust particles, salt particles, or other solid materials. Ice crystals require a different kind of nuclei for deposition. Deposition nuclei are not as plentiful in the atmosphere, thus giving a basis for modern cloud seeding operations.

Not all clouds produce rainfall. Therefore the growth mechanism of raindrops is important in determining how much rainfall reaches the surface. The solute effect helps small droplets grow larger because of a reduced vapor pressure. The solute effect on the vapor pressure and evaporation of small droplets is partially counterbalanced by the curvature effect since small droplets are more curved than larger ones. Because cloud droplets of various size exist in a cloud, the coalescence process is an important growth mechanism in warm clouds. When ice crystals and liquid water droplets exist in cold clouds, the ice crystals will grow more rapidly because of a lower vapor pressure and less evaporation. These various raindrop formation processes together with cascade effects are important in midlatitudes. In tropics the coalescence process and cascade effects are most important, as many clouds do not extend above the freezing level.

Common forms of precipitation are rainfall and snowfall, with other possible forms that include freezing rain, ice pellets, and hail. Unusual colors of rain and snow are caused by dust, pollen, and minerals in the atmosphere.

Fog forms from several different cooling mechanisms and is a significant problem to aviation, shipping, and automobile travel.

STUDY AIDS

1. Keep a record of the clouds that you observe every day for a two-week period.

2. Take ten cloud photographs and record the type of weather observed. Use the photographs to classify the clouds.

3. Calculate the height of the cloud base if the measured temperature is 15°C and the dew point temperature is 10°C.

4. Explain the solute effect.

5. Is the curvature effect greater for small or large droplets? Explain.

6. Which raindrop growth mechanisms operate in a cloud above the freezing level?

7. Explain the cascade effect.

8. Explain the atmospheric conditions for the formation of sleet and freezing rain.

9. Explain the formation of colored rain and snow.

10. Where and when would you expect to find snow rollers?

TERMINOLOGY EXERCISE

Use the glossary to reinforce your understanding of any of the following terms used in Chapter 6 that are unclear to you.

Cloud cover
Ceilometer
Lifting condensation level
Deposition
Cloud genera
Squall line
Mechanical lifting
Dynamic lifting
Convection
Chinook wind
Condensation nuclei
Deposition nuclei
Hygroscopic
Solute effect
Curvature effect
Coalescence
Bergeron-Findeisen precipitation process
Riming
Cascade effects
Freezing rain
Ice pellets
Snow
Snow pellets
Graupel
Hail
Snow roller
Fog

THOUGHT QUESTIONS

1. Explain the conditions under which a small cloud droplet will evaporate even though the relative humidity is 100%.

2. Describe some of the differences that may contribute to the growth or reduction in size of liquid water droplets and ice crystals in the same relative humidity environment within a cloud.

3. Do you think clouds that form over the oceans or over land are most efficient in producing rainfall? Give reasons for your answer.

4. Based on your knowledge of precipitation processes, describe and explain some of the atmospheric factors that would be required to produce a heavy snowfall.

5. Compare expected rainfall amounts for two different clouds, one with a much lower cloud base than the other.

SPECIAL TOPIC 7

ACID RAIN

Excessive pollutants in the atmosphere may change the acidity of rainfall. Concern for this problem has led to extensive investigations by industry, government, and university scientists. Some people in southeastern Canada and the northeastern United States are concerned about the levels of sulfur dioxide and other pollutants from Midwestern utilities and factories that burn coal. The major wind systems of the atmosphere carry pollutants toward the east or northeast, depending on the looping nature of the jet stream over the United States (Fig. S7-1). When pollutants are incorporated into the raindrop formation process or are washed out of the atmosphere, they produce acid rain, which can destroy lakes, kill aquatic life, and damage buildings. Some people also contend that greater acidity of lakes may produce elevated concentrations of toxic metals in the food chain, which could represent a potential threat to human health.

Research indicates that natural sources of sulfur dioxide are not as important in contributing to acid rains as emissions from midwestern power plants and factories. Solutions to the problem could involve burning only high-quality coal, removing the sulfur from coal before it is burned, or removing it during the burning process.

Many of the issues concerning acid rain will certainly continue to be evaluated to determine precise relationships between pollution sources, atmospheric processes prior to deposition as acid rain, and effects of acid rain after it reaches the surface.

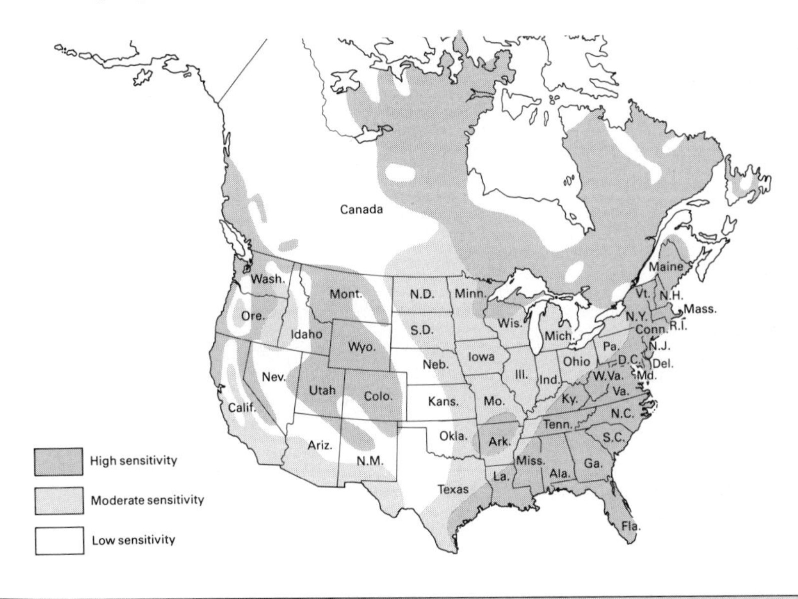

FIGURE S7-1 Certain areas in the United States and Canada are more sensitive to acid rain than others. (From G. Tyler Miller, *Living in the Environment,* 3d ed. Copyright © 1982, Wadsworth, Inc.)

SPECIAL TOPIC 8

FROSTS AT ABOVE-FREEZING TEMPERATURES

Ice does not form unless the temperature is at or below the freezing temperature, but you might be surprised to see frost on the ground when the air temperature is reported as 4°C or 40°F. Are the thermometers wrong or is there another explanation? Variations in thermometers do occur, as do variations in air temperature over short distances at the earth's surface—but even greater variations are typical with increasing height above the surface. During the night the temperature normally increases with height since cooling of the ground occurs rapidly as radiation is lost. This in turn cools the air near the surface.

Reported air temperature measurements are obtained within a standard instrument shelter. The thermometer is exposed to air at a height of almost 2 m above the ground. The nighttime inversion may be so well developed that the temperature on the ground can be at the freezing level while the temperature at instrument height is several degrees warmer. Thus, frosts can easily occur when the reported minimum temperature never reached the freezing level during the night.

PART THREE
ATMOSPHERIC STORMS

CHAPTER SEVEN

FRONTAL CYCLONES

EARLY CONTRIBUTIONS

Prior to the early 1900s, the nature of traveling low-pressure systems was not well understood. Weather forecasting techniques ignored the rapid changes associated with different air masses across weather fronts, although it was known that lower pressure as measured by a barometer brought clouds and precipitation. The first general recognition of the importance of weather fronts and their association with centers of low pressure was made by Norwegian scientists during World War I. In 1918, J. Bjerknes published an article describing the structure of frontal cyclones (Fig. 7-1). Bjerknes's theory, which became known as the polar front theory, made a significant contribution toward advancing our information on the structure of frontal cyclones.

The cyclone model proposed by Bjerknes had significant applications in forecasting, as meteorologists became aware of the importance of warm and cold fronts. The stages in the life cycle of a frontal cyclone shown in Figure 7-1 can be frequently observed on weather maps. However, information on the wind above the surface was not available in 1918, and its contribution to the structure of frontal cyclones has been recognized only recently.

ATMOSPHERIC ENVIRONMENT FOR FRONTAL CYCLONES

As we have already seen, the global distribution of heat is important in driving wind systems. The winds, in turn, keep the tropics from overheating and prevent polar regions from being quite so cold. The origin of the jet stream is related to excess

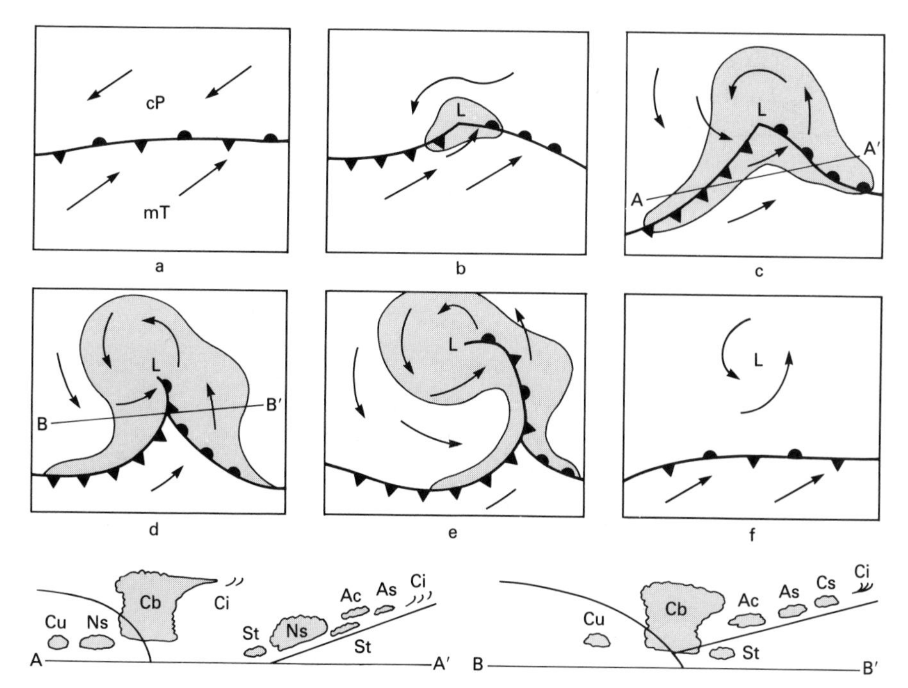

FIGURE 7-1 Stages in the life cycle of a midlatitude cyclone, as proposed by J. Bjerknes. Diagrams a–f show a view from above, as on weather charts; the other two views are side views of cross sections through frontal cyclones.

heating in tropical regions and net cooling in polar regions, but after the jet stream has developed speed and momentum it influences the movement and mixing of air masses. Of the several jet streams that may exist at any one time, the polar jet is most important in redistributing heat and generating storms since it occurs in midlatitudes where the contrast in air masses is greatest.

Frontal cyclone activity is related to the polar jet stream. Large swirls in the atmosphere feed lesser swirls as energy is transferred through scales of motion ranging from the jet stream to local turbulence and molecular activity (Fig. 7-2). As the jet stream circles the earth, it develops meanders and completes cycles with the result that troughs with cyclonic curvature and ridges with anticyclonic curvature frequently occur. Jet stream troughs are regions of frontal cyclone development (**cyclogenesis**); within them, atmospheric swirls are generated with a much smaller radius of circulation than that of the jet stream (Fig. 7-3).

The speed of the jet stream is greatest during the winter season as it migrates southward in the northern hemisphere, and this intensifies frontal cyclones. The region of cyclonic activity follows the jet stream as it shifts its position between 25° and 55° N latitude.

FRONTAL CYCLONES

Frontal cyclones (also called midlatitude and wave cyclones) are large traveling atmospheric vortices with centers of low atmospheric pressure. An intense frontal cyclone may have a surface pressure of 970 mb. Several closed isobars normally surround the center of lowest pressure even if the pressure is only slightly lower than average (Fig. 7-4). Surface winds blow cyclonically around the low-pressure center and slightly toward the center. A cold front normally extends southwestward from the center of the cyclone with a warm front extending eastward. Most midlatitude cyclones have a cold front associated with them but many do not have a warm front since the contrast in temperature and humidity between the air located toward the southeast and east of the low-pressure center is frequently insignificant.

Precipitation is associated with the rising air located over the center of the midlatitude cyclone and over weather fronts. Air is forced to rise near the center because of the convergence of air toward the lower pressure; air rises along a front as the warm air is forced upward by the colder, denser air.

Frontal cyclone formation frequently occurs beneath the trough of the jet stream. The dynamics of the jet stream flowing over mountain ranges favor the formation of a trough on the leeward side. The Rocky Mountains, therefore, contribute to trough development in the jet stream that initiates midlatitude cyclone formation in the central United States. This process makes the weather more difficult to forecast there than in many other locations in the United States since the movement of a midlatitude cyclone is easier to forecast than its development.

The midlatitude cyclone is seldom stationary. The direction and speed of movement are determined primarily by the winds in the upper atmosphere at the jet stream level. One simple means of forecasting the future position of one type of midlatitude cyclone is to project the storm movement at 50% of the 500 mb winds. Thus, rawinsonde measurements of the upper winds are useful in calculating the anticipated future speed and direction of a particular storm.

Another estimate of future movement is given by the winds behind the cold front. If the winds behind the cold front are 50 km/h the front can be projected across the earth's surface at this speed. A midlatitude cyclone that has moved 1200 km in one day can be projected to move 1200 km the next day. Thus, persistence or the past history of movement of a midlatitude cyclone is a third technique for projecting the location of a midlatitude cyclone at some future time.

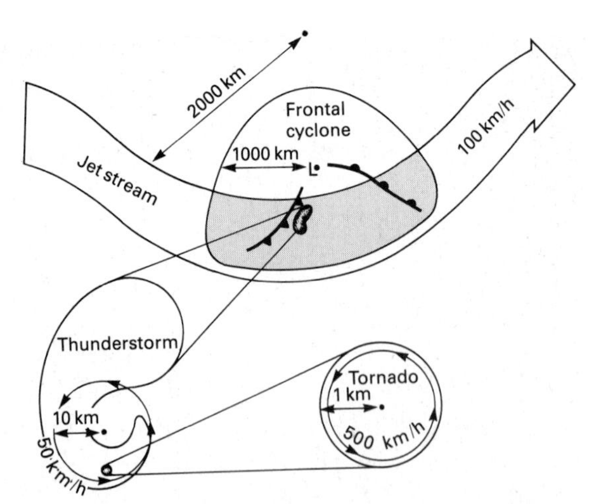

FIGURE 7-2 Energy is transferred within weather systems through various scales of motion in the atmosphere, ranging from that of the jet stream (perhaps 2000 km) to that of a tornado. (From J.R. Eagleman, *Severe and Unusual Weather,* [New York: Van Nostrand Reinhold, 1983].)

These simple estimates may be used to supplement more accurate techniques such as numerical weather prediction performed at the National Meteorological Center, as discussed in Chapter 2.

AIRFLOW IN FRONTAL CYCLONES

Airflow in a frontal cyclone (Fig. 7-5) consists of rising air above the center of low pressure, with surface winds converging as they rotate cyclonically toward the center of low pressure. Horizontal convergence of the air exists near the surface, while in the upper atmosphere outflow of air exists because of horizontal divergence. Thus the structure above

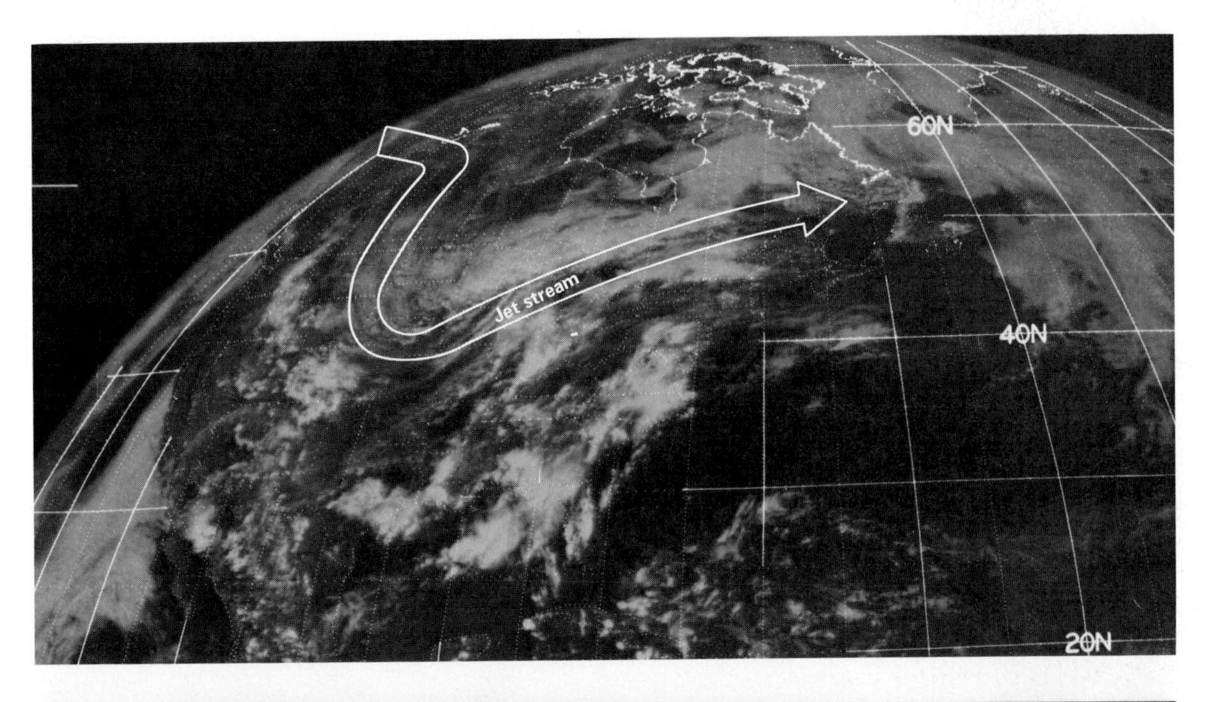

FIGURE 7-3 The weak cyclonic circulation indicated by the cloud patterns on August 2, 1974 (top) was intensified by the following day as energy from the jet stream was transferred to the smaller cyclonic storm. Smaller thunderstorms within the frontal cyclone represent another step in the energy transfer process. (Photos courtesy of NOAA.)

FIGURE 7-4 A typical frontal cyclone consists of a low-pressure center with a warm front extending east or southeastward and a cold front extending south or southwestward. In this case, for June 12, 1975, at 1700 GMT, the low-pressure system also contained an occluded front that extended eastward from the low-pressure center of 996 mb. Cloud cover and precipitation typically occur in the warm air mass ahead of the cold front, in advance of the warm front, and near the center of the frontal cyclone. The jet stream typically flows above or south of the center of low pressure, as in this case, and frequently causes an intrusion of drier surface air south of the low-pressure center. This produces clearing beneath the jet stream southward from the low-pressure center until the jet stream intersects the frontal system. (Photo courtesy of NOAA.)

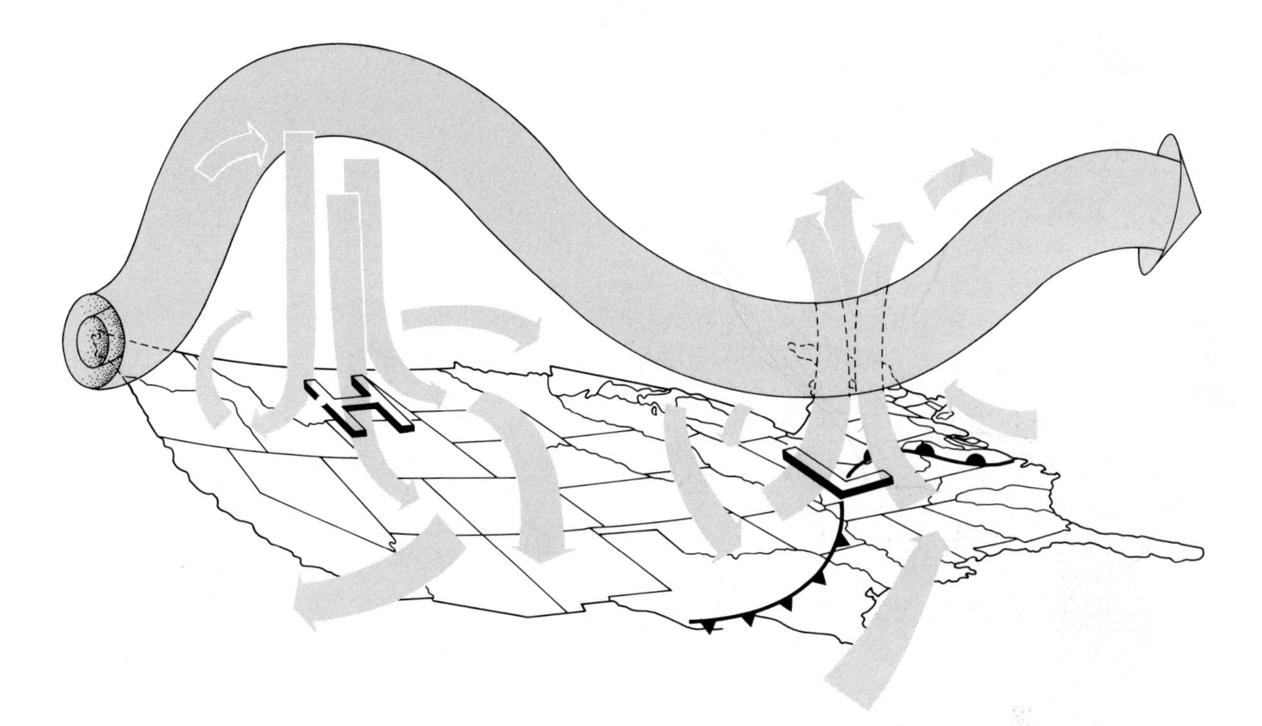

FIGURE 7-5 The three-dimensional structure of a frontal cyclone consists of cyclonic or counterclockwise airflow at the surface around the center of low pressure with some convergence toward the center. Above the low pressure the air rises in response to the horizontal divergence and outflow of air that typically occurs near the jet stream. Anticyclones have the opposite structure, with subsiding air above the high pressure, horizontal convergence in the upper atmosphere, and horizontal divergence at the surface.

a low-pressure system consists of rising air with horizontal convergence at the surface and divergence in the upper atmosphere. Airflow patterns in an **anticyclone** (high-pressure system) are just the opposite, consisting of subsiding air above the high pressure, diverging air at the surface, and horizontally converging air in the upper atmosphere.

THE JET STREAM AND CYCLOGENESIS

Early explanations of the development of frontal cyclones emphasized opposing air masses and surface features of frontal cyclones. However, we must include the winds above the surface to gain a more complete understanding of these atmospheric storms.

Frontal cyclone development is related to jet stream features. In fact, the low-pressure center at the surface is generated because a smaller amount of air is above that particular location. The jet stream is primarily responsible for the outflow of air in the upper atmosphere, which results in less weight of air above that particular location and gives a lower surface pressure. As a center of low pressure develops, the air starts to flow toward it to fill the partial vacuum. The Coriolis acceleration causes the air to travel to the right of its path of motion in the northern hemisphere, and this produces

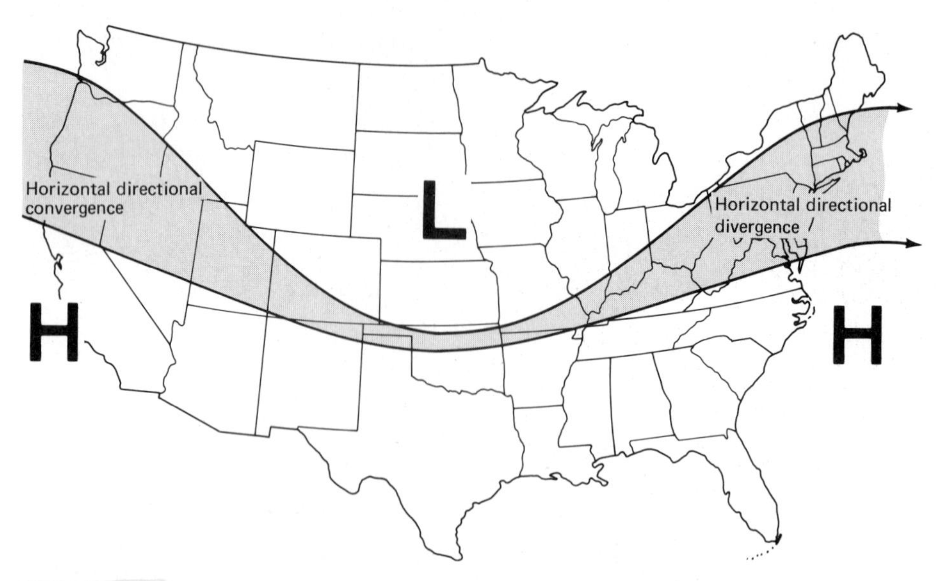

FIGURE 7-6 Horizontal directional divergence caused by diffluence of the air typically occurs downstream from a trough in the jet stream where the trough arises from flow around a lower atmospheric pressure. Horizontal directional convergence caused by confluence typically occurs from an upper air ridge to an upper air trough. Upper air diffluence is one of the contributing factors in frontal cyclone generation.

cyclonic airflow around the center of low pressure, thereby generating a midlatitude cyclone.

Upper Air Divergence An important factor in the generation of a midlatitude cyclone is the amount of **upper air divergence.** Horizontal divergence in the upper atmosphere can arise either from directional divergence or velocity divergence. Directional divergence arises if the winds are spreading outward, as in Fig. 7-6. If the winds are diverging this represents a net outflow of air, and if it occurs over a very large area it leaves less air above the underlying surface, leading to the development of low pressure at the surface.

Speed divergence can produce the same result. Speed divergence occurs in air streams behind areas of maximum winds (Fig. 7-7). If the downstream winds were 200 km/h, for example, but the upstream winds were only 150 km/h, the area between would experience speed divergence. More air would be moving out of this particular locality than was moving into it, creating divergence.

Particular regions in the atmosphere are more likely to develop divergence than other regions. Directional divergence is most common from the upper air trough to the ridge since the isobars are typically packed closer together around low-pressure systems than around high-pressure systems.

Cyclonic Vorticity Upper air divergence is only one of several related factors that are needed to generate a midlatitude cyclone. **Cyclonic vorticity** is also important. The vorticity of the atmosphere is related to the tendency of air to rotate. Vorticity can be illustrated by considering a large disk

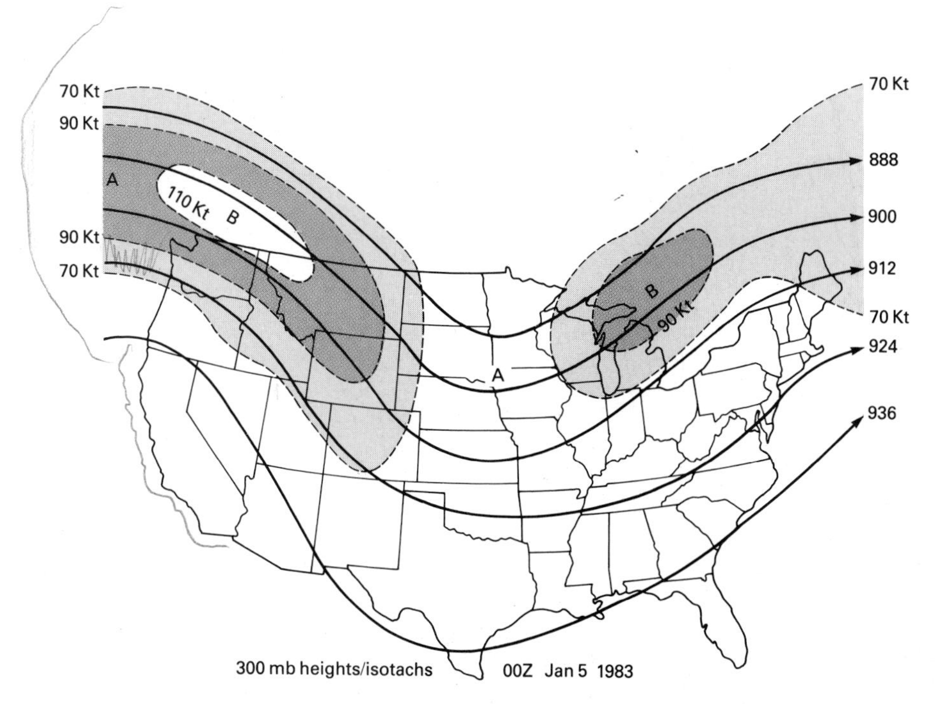

FIGURE 7-7 The jet stream frequently contains regions of maximum speed (labeled B). Regions of speed divergence are located upstream from these maximums (from A to B in the diagram). Such regions of speed divergence move along the jet stream and are one of the contributing factors in cyclogenesis. In the diagram, Kt = knots (unit of wind speed).

placed in the atmosphere with the air flowing around it (Fig. 7-8). If such a disk were placed in a region of cyclonic curvature in the jet stream, the air would exert a sufficiently greater force on the south side of the disk than on the north side to cause it to rotate cyclonically. Similarly, a large mass of air within the cyclonic curvature of the jet stream begins rotating cyclonically because of the forces acting on it. Cyclonic vorticity in the trough of the jet stream is an important factor in the formation of midlatitude cyclones.

Wind Shear Another factor in generating midlatitude cyclones is **wind shear,** which technically is another aspect of vorticity. Wind shear exists in the atmosphere whenever a difference in velocity occurs between two different layers of air. For vertical wind shear, this may be a layer of air moving from the south overlain by a layer of air moving from the west. Wind shear also frequently exists in a horizontal direction. Such wind shear, important in the generation of midlatitude cyclones, occurs just north of the jet stream. A horizontal slice through the jet stream (Fig. 7-9) would show the highest wind speeds in the central core, with decreasing wind speeds northward and southward. If a large disk were placed just north of a straight jet stream, it would rotate cyclonically because of the wind shear since the disk would be affected more by the high wind speeds on the south side than by the lower wind speeds on the north. The opposite would occur just south of the jet stream core, where anticyclonic wind shear exists. Since the region just north of the jet stream core contains cyclonic wind shear, it contributes to the generation of cyclonic rotation in midlatitude cyclones.

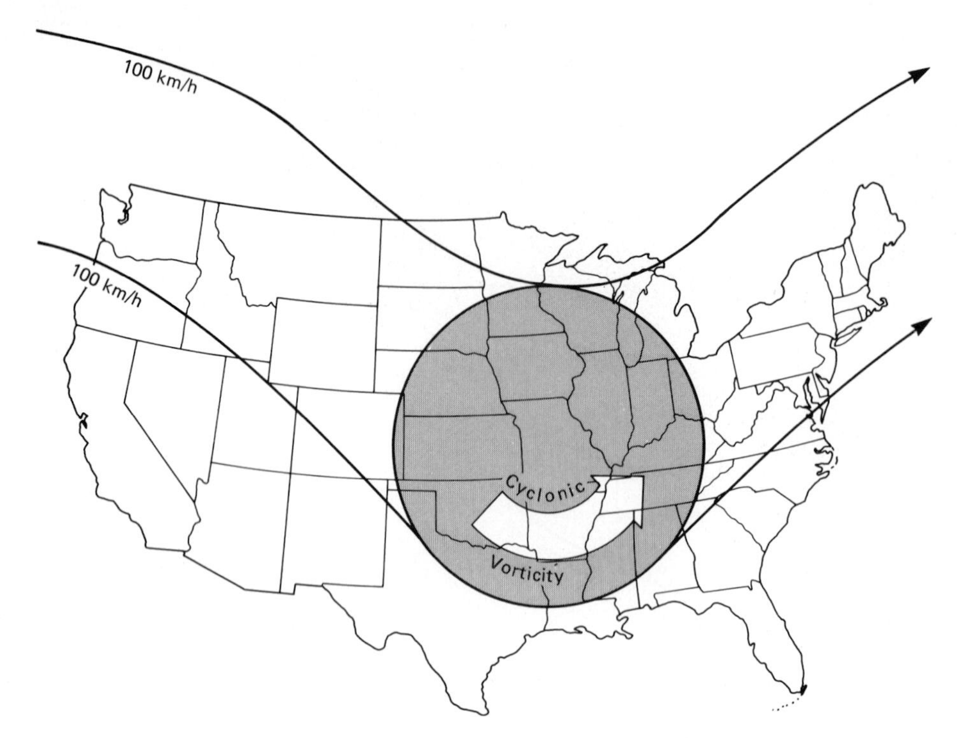

FIGURE 7-8 Cyclonic vorticity within the trough of the jet stream would initiate cyclonic rotation of a large disk or a large circular air mass placed within the jet stream. This rotation would occur because the cyclonic winds of the jet stream would exert a greater force on the southern side of the disk than on the northern side. Cyclonic vorticity is one of the factors contributing to cyclogenesis.

DEVELOPMENT AND LIFE CYCLE OF FRONTAL CYCLONES

Several factors work together in cyclogenesis. Upper atmospheric factors cause the surface pressure to decrease, and this produces surface winds that converge toward the lower pressure.

Surface convergence then contributes to the formation of cyclonic rotation at the surface around the low-pressure center. As surface air converges (Fig. 7-10), the Coriolis acceleration deflects the moving air to the right of its path of motion. Air that was converging toward the center of the low-pressure area thus acquires cyclonic rotation because of Coriolis acceleration. Therefore, horizontal convergence of the surface air initiates cyclonic rotation near the ground.

Surface convergence caused by upper air divergence, cyclonic vorticity, and wind shear initiates and intensifies cyclonic rotation. These various cyclogenesis factors occur in certain preferred locations in the atmosphere in the vicinity of the jet stream. Frontal cyclones develop in particular locations determined by the position of the jet stream.

As troughs and ridges form in the jet stream, the isobars are packed closer together around upper air low-pressure systems. Thus, a region of directional divergence normally exists from the upper air trough to the ridge, giving support for a surface low-pressure system. The generation region for most frontal cyclones is the trough of the jet stream (Fig. 7-11) since upper air divergence, cyclonic vorticity, and cyclonic wind shear cause lower surface

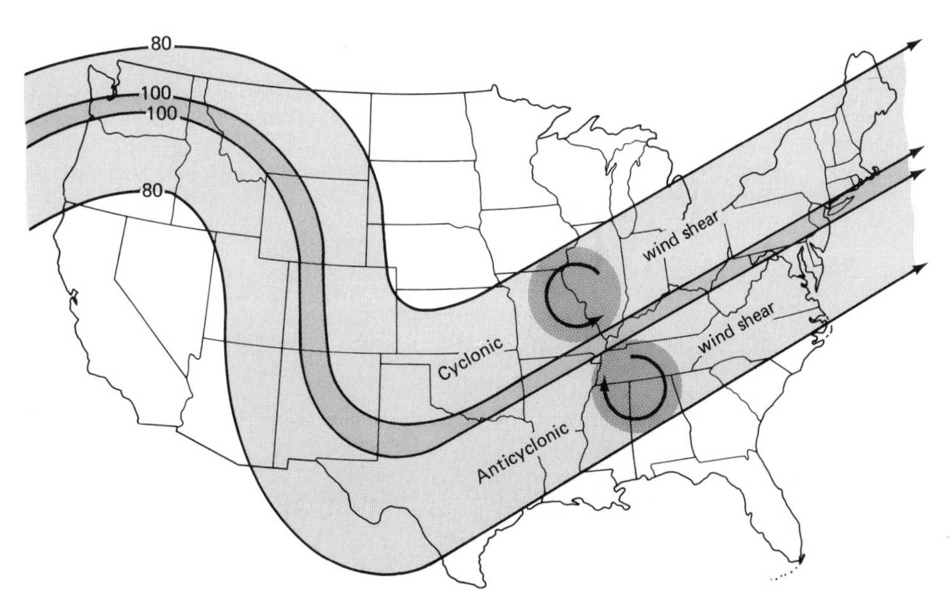

FIGURE 7-9 Cyclonic wind shear occurs north of the jet stream because of the decreasing wind speed toward the north. Anticyclonic wind shear occurs south of the jet stream as the jet stream wind speeds decrease southward. Cyclonic wind shear north of the jet stream is one of the contributing factors in cyclogenesis.

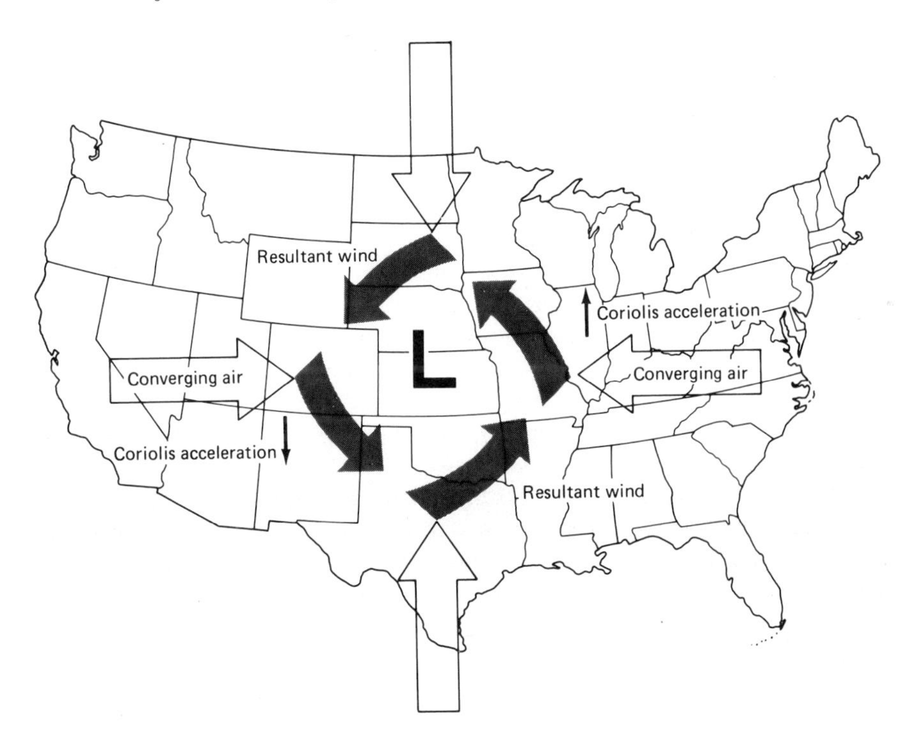

FIGURE 7-10 Cyclonic rotation is intensified near the ground by air converging toward the center of low pressure as it is accelerated to the right of its path of motion due to the Coriolis effect. Because of this effect even a small area of ascending air above the low-pressure center causes cyclonic rotation at the ground to cover a much larger area.

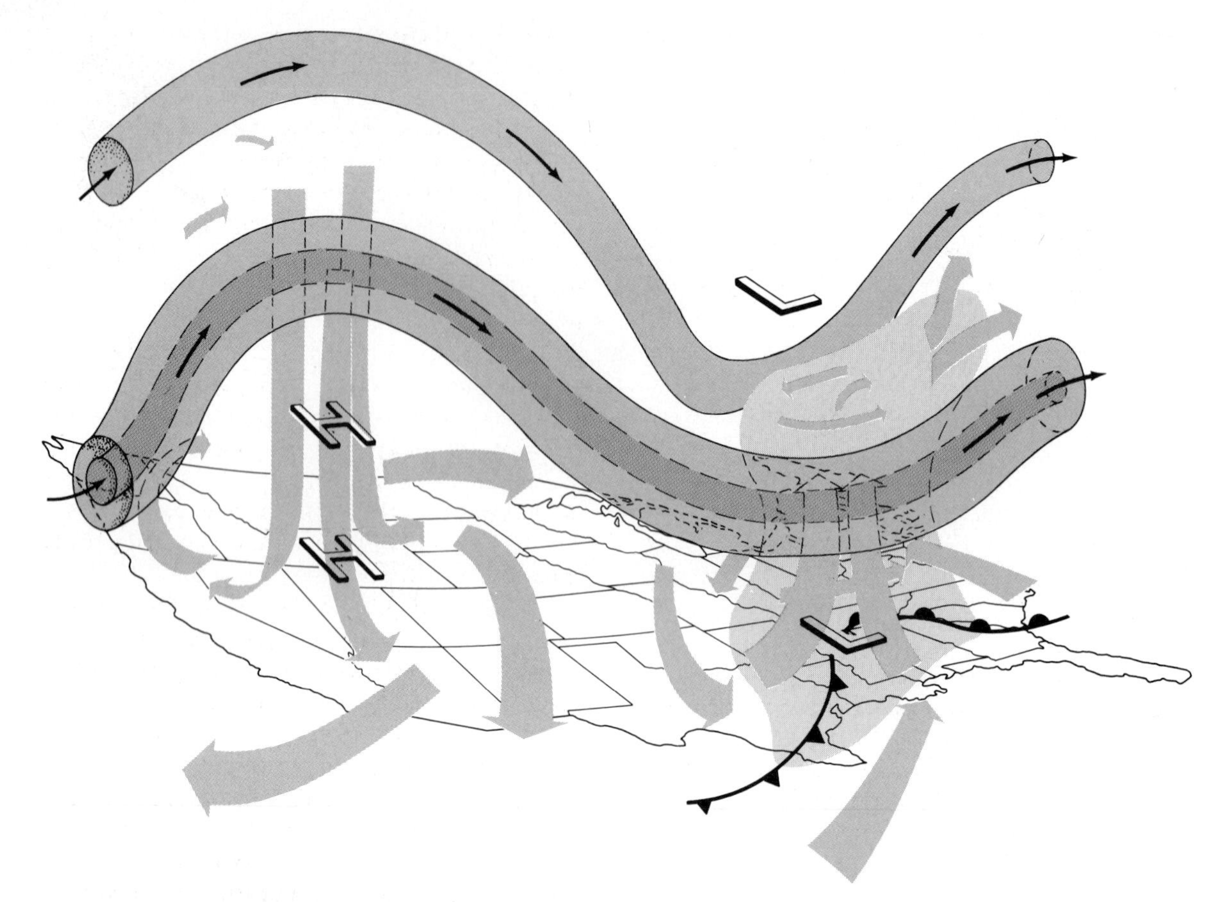

FIGURE 7-11 Upper air divergence, cyclonic vorticity, and wind shear cause frontal cyclones to develop beneath the trough of a jet stream. Upper air support for the cyclone causes it to complete its life cycle as it travels from the trough to the ridge of the jet stream.

pressure and cyclonic rotation. The frontal cyclone typically completes its life cycle underneath the region from the upper air trough to the ridge (Fig. 7-12). The frontal cyclone can survive for several days in the region from the trough to the ridge since the horizontal divergence provides enough outflow of air in upper levels to replace the surface air that converges inward to fill the cyclone. Thus **upper air support** is provided for a midlatitude cyclone in the region from the trough to the ridge.

The region in the upper atmosphere from the ridge downwind to the trough generally has horizontal convergence. The structure of anticyclones is more compatible with upper air horizontal convergence (see Fig. 7-5). Therefore anticyclones would be more likely to be located in the region from the upper air ridge to the trough since upper air support for anticyclones occurs there. A midlatitude cyclone that moves into the region from an upper air ridge to a trough would experience filling of air both at the surface and in the upper atmosphere, a situation that would lead to the elimination of the lower pressure and the midlatitude cyclone.

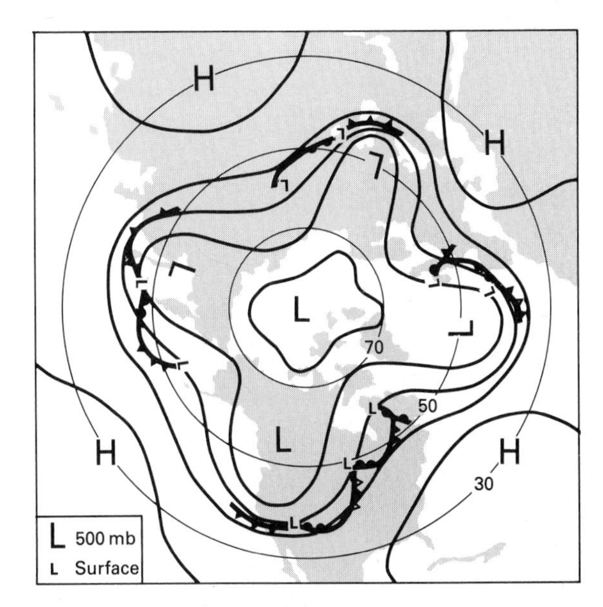

FIGURE 7-12 Schematic circumpolar chart showing a four-wave upper airflow pattern. The solid lines represent contours of 500-mb pressure while the surface fronts show a series of midlatitude cyclones representing four different cyclone families. In each case the oldest cyclone originated near the trough of the jet steam and is in the decaying stage as it approaches the ridge in the 500-mb flow patterns.

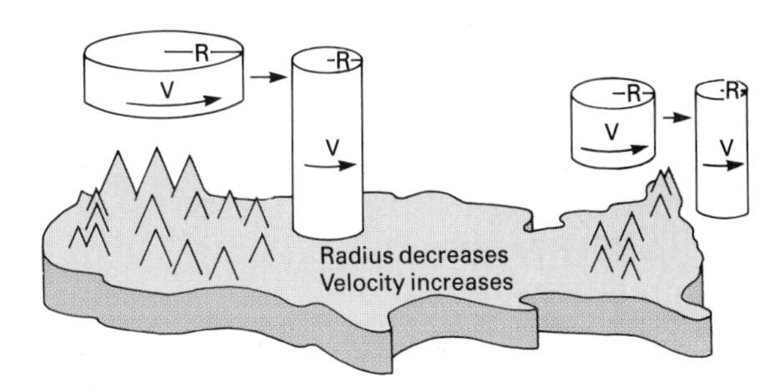

FIGURE 7-13 Topography also affects frontal cyclone development. As an air mass descends over the Rocky Mountains into the Great Plains, or over the Appalachian Mountains into the eastern coastal region, it stretches vertically. Horizontal circulation increases significantly in the process, in the same way that a skater who pulls in his or her arms during a twirl concentrates body mass and increases rotational speed. Therefore, cyclogenesis or intensification of existing storms occurs more often on the leeward side of major mountain ranges.

Because of the relationship between the upper atmosphere and storm systems midlatitude cyclones with associated fronts, cloudy skies, and precipitation occur and move through regions determined by the jet stream flow patterns, while anticyclones persist beneath other regions of the jet stream. Since these patterns may persist for weeks or months at a time, general weather conditions for extended periods of time are directly related to jet stream patterns.

The geographical location of cyclogenesis is determined, in some cases, by the topography. As air descends downwind from a mountain range the horizontal circulation is increased from vertical stretching, as shown in Figure 7-13. This increase, combined with normal jet stream contributions, causes greater cyclogenesis downwind from major mountain ranges.

The life cycle of a midlatitude cyclone is illustrated in Figure 7-14. The satellite photographs show the characteristic comma shape of the cloud shield corresponding to the cold front and center of low pressure. As the midlatitude cyclone matures and moves eastward, its shape changes as the jet stream carries cold air around its south side, causing **occlusion** and later decay of the storm. Occlusion is the process whereby the cold front extending from the center of a frontal cyclone travels faster than the warm front and overtakes it as in Figure 7-1(d) and 7-1(e). Therefore, occlusion occurs in the mature stage of a frontal cyclone.

TYPES OF MIDLATITUDE CYCLONES

Trough Cyclones Three basic types of midlatitude cyclones can be identified by observing the weather patterns for any extended period of time. One of the most common types of midlatitude cyclones may be called a **trough cyclone** since it is generated in a trough of the jet stream (Fig. 7-15). The trough is the upper atmospheric region with cyclonic curvature that also provides cyclonic vorticity. A region of cyclonic wind shear just north of the jet stream is also likely. In addition, the region

a. June 7, 1974
1902 GMT

b. June 8, 1974
1956 GMT

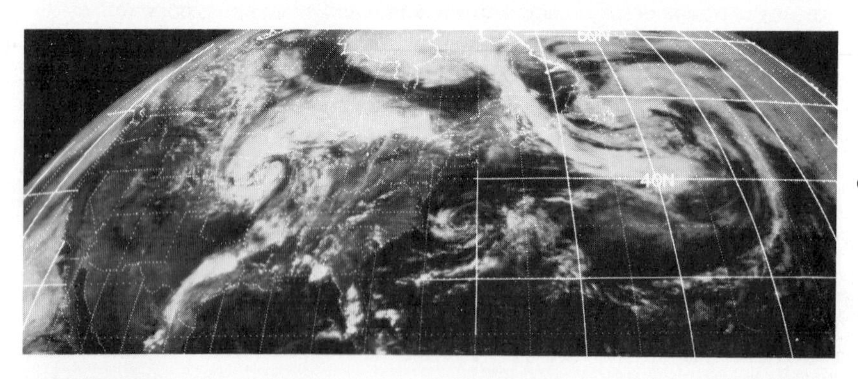

c. June 9, 1974
1855 GMT

d. June 10, 1974
1909 GMT

just downstream from the trough normally has greatest directional divergence. These contribute to the formation of the trough cyclone.

The trough cyclone develops in the jet stream trough and completes its life cycle from the trough to the ridge of the upper air pattern. The life cycle of the trough cyclone may include the occlusion process. The occlusion process consists of the cold front overtaking the warm front, thus producing an occluded front as the midlatitude cyclone reaches the ridge, where it soon dissipates. The trough cyclone typically travels at about 50% of the 500-mb winds. The direction of movement is also determined by the upper winds. The path of the jet stream can be used to project the path of the trough cyclone since the latter almost always travels along the flow of the former.

Long-Wave Cyclones The second type of midlatitude cyclone is a **long-wave cyclone** (Fig. 7-16). A long-wave surface cyclone is different from a trough cyclone, since the low-pressure area at the surface is connected with the low-pressure center in the upper atmosphere. The center of the long-wave cyclone may be displaced slightly southward from the center of the upper air low pressure but it is still north of the trough of the jet stream and is a part of the low-pressure center that exists in the upper atmosphere. This type of midlatitude cyclone frequently develops in March, with high wind velocities that are a part of the airflow patterns in the upper atmosphere. The pressure is also lower at the surface than for the trough cyclone thus contributing to packed isobars and higher velocity surface winds. Long-wave surface cyclones typically move very slowly. Their rate of movement is the same as the displacement speed of the whole meander in the jet stream since the surface low pressure is related to the upper air low-pressure area that forms the trough within the upper air circulation.

Long-wave cyclones may develop in the winter months as well as in the spring. When they occur in winter, blizzard conditions accompany them because of the high winds, snowfall, and resulting drifts. The isobars may be more circular around this type of cyclone and more packed, resulting in winds as high as 80 km/h. The major snow area is normally just north of the center of low pressure, with smaller amounts behind the cold front. The high winds cause problems with snowdrifts, blowing snow, and chilling temperatures. Snow rollers such as those shown in Figure 6-23 may also be produced. This type of cyclone may complete a life cycle that includes the occlusion process.

Short-Wave Cyclones The third type of midlatitude cyclone is a **short-wave cyclone.** A short-wave cyclone is the most difficult to forecast. It originates as a small disturbance in the major flow patterns in the jet stream (Fig. 7-17). If a small wave develops on the upper airflow patterns, it may be

FIGURE 7-14 Several different cyclonic storms are shown in this series of satellite photographs. A large persistent storm can be seen at 50° latitude over the Atlantic Ocean. The life cycle of a frontal cyclone of shorter life span is shown over the United States as it originated on June 7 as a diffuse region of clouds centered over Colorado. On June 8 it was beginning to develop into the characteristic comma shape, as Colorado, Nebraska, Missouri, and parts of Arkansas were covered by clouds. Single thunderstorms that developed hail and tornadoes can be seen over Oklahoma and southern Kansas. On June 9 the center of the frontal cyclone was near the Nebraska–Missouri border, with the cloud pattern showing the cyclonic flow around the low pressure. By June 10, 1974, the low-pressure center had moved over the Great Lakes and the storm was beginning to lose some of its intensity. (Photos courtesy of NOAA.)

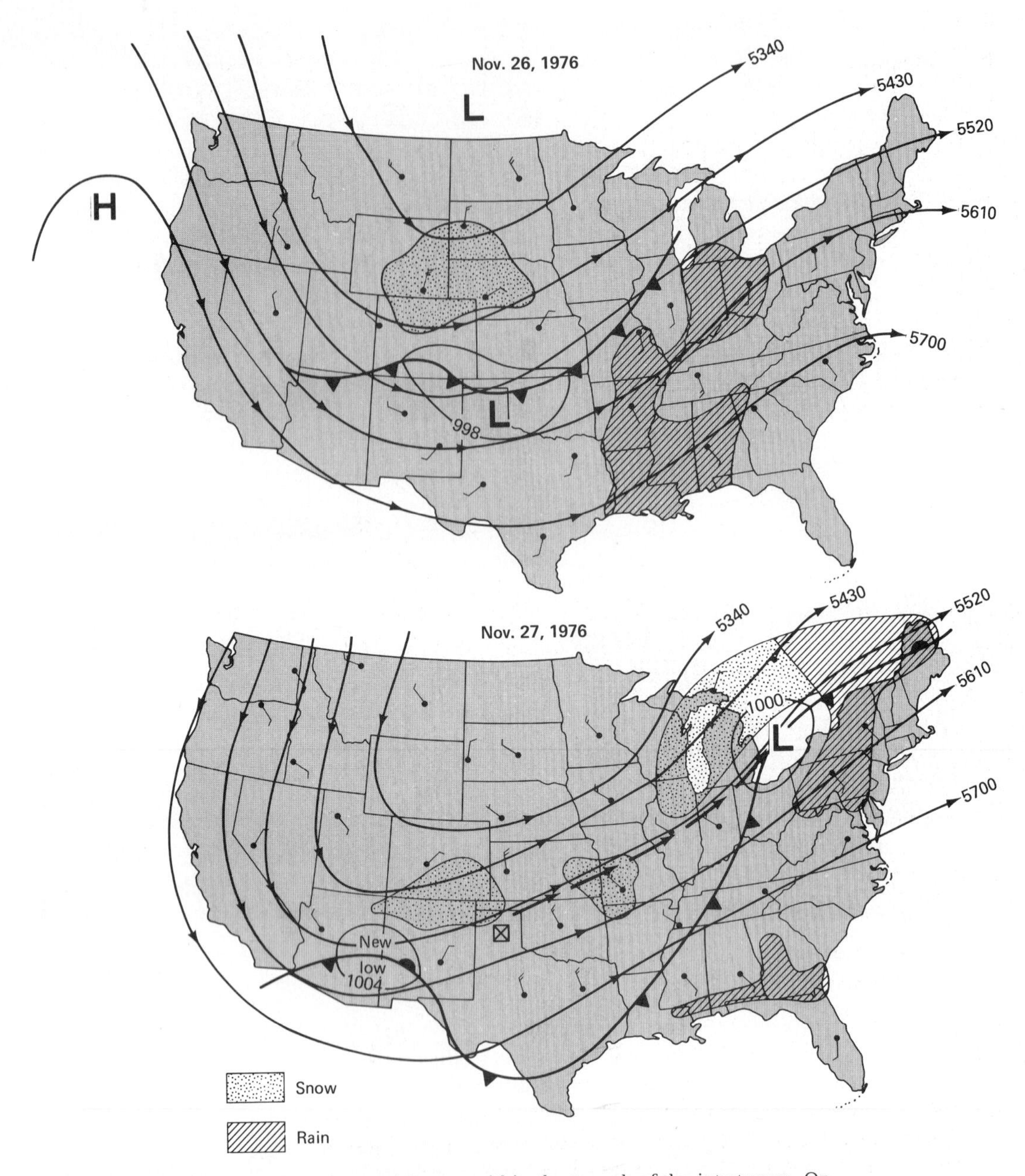

FIGURE 7-15 The trough cyclone originates within the trough of the jet stream. On November 26, 1976, a surface low pressure center was located over the Oklahoma Panhandle. Twenty-four hours later the cyclone had moved to the Great Lakes and a new cyclone was developing in the trough. The speed of movement of the first cyclone was 80 km/h as it traveled about 1900 km in 24 hours.

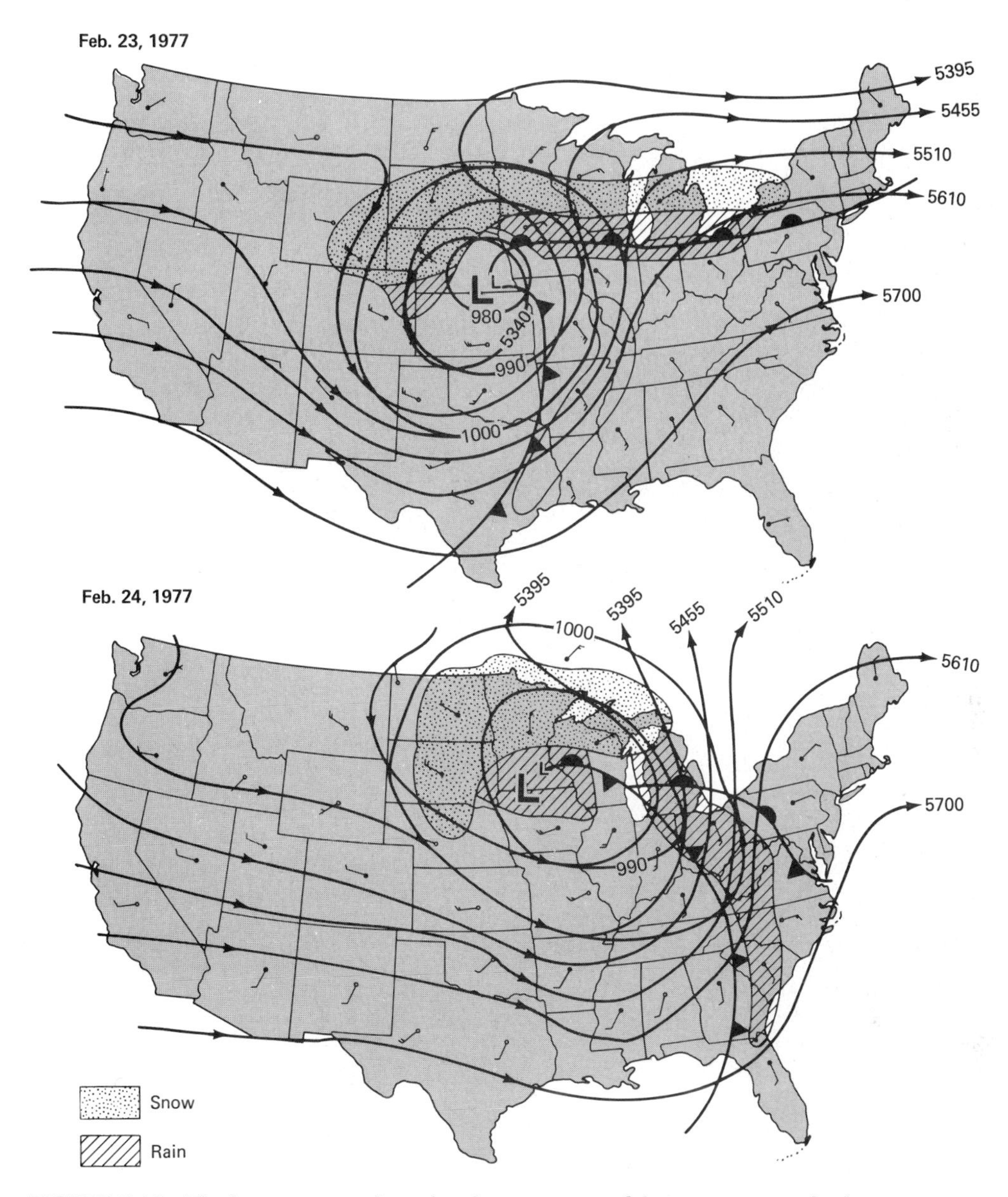

FIGURE 7-16 The long-wave cyclone develops as a part of the upper atmospheric low pressure. This connection causes the long-wave cyclone to have a lower surface pressure with closer packed isobars and greater wind speeds near the ground than other types of frontal cyclones. The long-wave cyclone travels at the same speed as the upper atmospheric low-pressure area, as shown here for February 23 and 24, 1977. This long-wave cyclone moved only about 600 km in 24 hours, giving it a rate of travel of only 25 km/h.

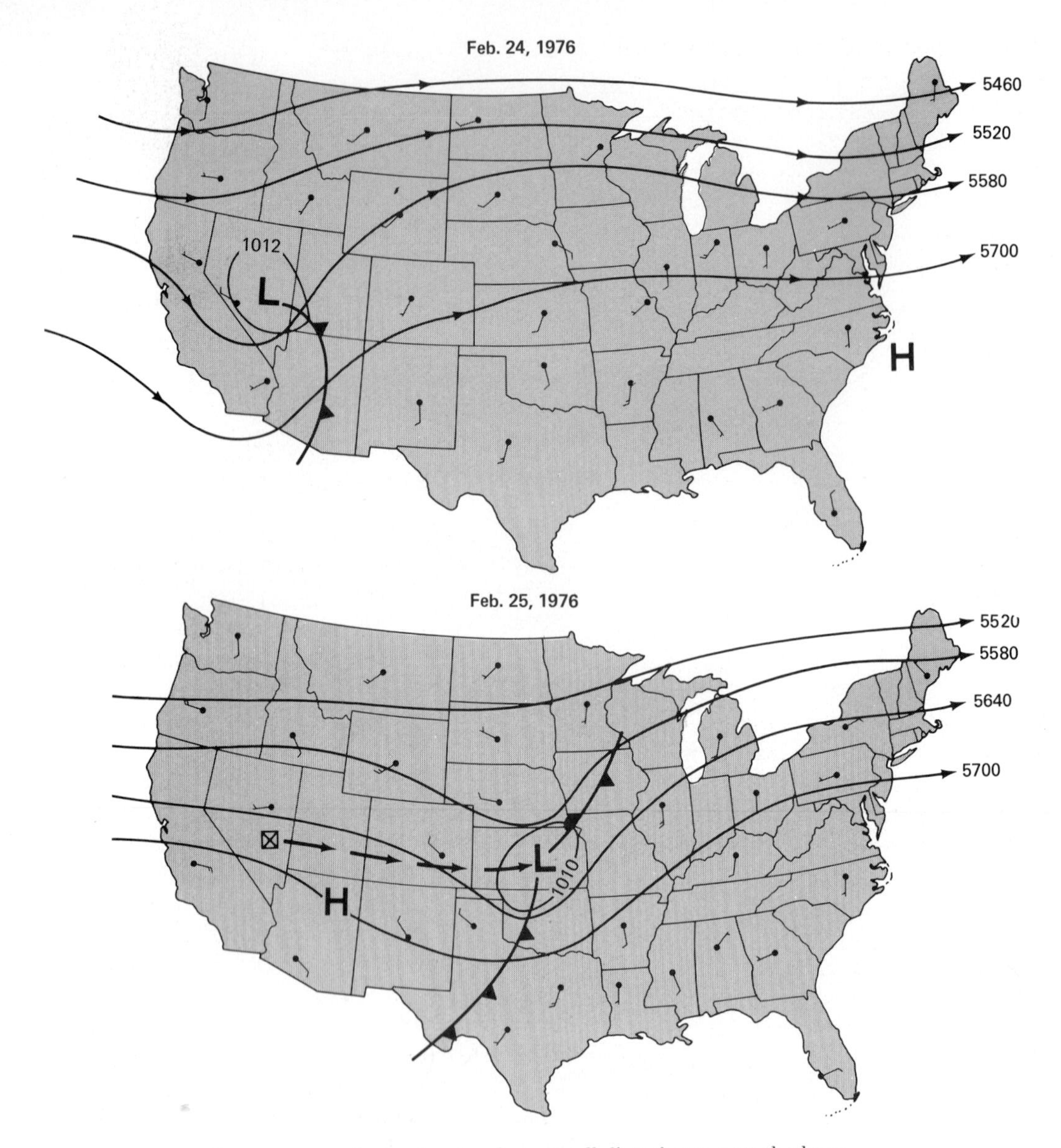

FIGURE 7-17 Short-wave cyclones develop from small disturbances on the long-wave upper atmospheric airflow patterns. On February 24, 1976, an upper air short-wave cyclone was centered over Nevada with a weak low pressure at the surface. Twenty-four hours later the center of the surface low was over Kansas as the disturbance in the upper atmosphere moved eastward. The short-wave cyclone moved about 1200 km in 24 hours, giving it a speed of travel of 50 km/h.

sufficient to cause precipitation beneath it without the presence of a surface low-pressure system. Stronger disturbances generate corresponding surface low-pressure areas beneath the wave in the upper airflow patterns. The short-wave cyclone travels much faster than the long-wave cyclone because the disturbance on the jet stream can move along with the air stream although its forward velocity is not as great as the air flowing through it. The difficulty in forecasting any changes in the weather associated with them arises because of their speed of movement and the weak surface low associated with them.

OCCURRENCE OF MIDLATITUDE CYCLONES

The type of midlatitude cyclone that is most likely to develop depends on the season. Long-wave cyclones are most common during the spring and winter and are very rare in the United States during the summer. Their peak occurrence is in the month of March, but they frequently develop during the winter months as well. The other two cyclone types occur during all seasons, with the short-wave cyclone predominating during the summer and the trough cyclone slightly more likely during the fall and winter.

The location of cyclone development is directly related to the jet stream and 500-mb airflow. The different cyclone types tend to travel along slightly different paths, as indicated in Figure 7-18. Short-wave cyclone paths are shorter and frequently go from west to east. Both the long-wave and trough cyclones frequently have long paths, with the direction of trough cyclones having a slightly greater northward component than that of the long-wave cyclone.

The typical lifetime of the short-wave cyclone is two days, the typical lifetime for the trough cyclone is four days, and that of the long-wave cyclone is five days. The short-wave and trough cyclones travel faster than the long-wave cyclone. Typical movement of trough and short-wave cyclones is 50 km/h in winter and 40 km/h in summer. Both of these occasionally travel at speeds of 100 km/h or more, however. The long-wave cyclone is the slowest traveler, with a typical speed of 30 km/h in spring and winter and 15 km/h in the fall.

The wind speed at the surface is greatest in a long-wave cyclone, with typical sustained velocities of 40 km/h. Wind gusts of 100 km/h are not unusual in a long-wave cyclone. In comparison, the trough cyclone frequently has surface winds of 30 km/h, and the winds in a short-wave cyclone are more frequently 20 km/h.

TORNADO-PRODUCING MIDLATITUDE CYCLONES

Severe thunderstorms and tornadoes in the central and eastern United States are characteristically associated with a particular type of midlatitude cyclone. Figure 7-19 shows the typical synoptic patterns associated with tornadoes. The most severe tornado-producing midlatitude cyclone is a trough type that has a low central pressure and is displaced slightly farther northward than the ordinary trough cyclone. This allows it to take on some of the characteristics of both the long-wave and trough cyclones. The strongest stream of upper atmospheric winds is across the cold front, causing rapid movement of the cooler air into the territory previously dominated by warm moist air. The surface low-pressure center is nearer the upper atmospheric low pressure, thus allowing some of the strong circulation from the upper atmosphere to be brought to the surface. This arrangement of surface air masses and upper atmospheric winds, together with several other factors that are important in tornado formation set the stage for the explosive development of very large thunderstorms that are most conducive to the generation of tornadoes on the ground.

BLIZZARDS AND STORM SAFETY

According to National Oceanic and Atmospheric Administration statistics, more than one hundred persons are killed each year by winter weather.

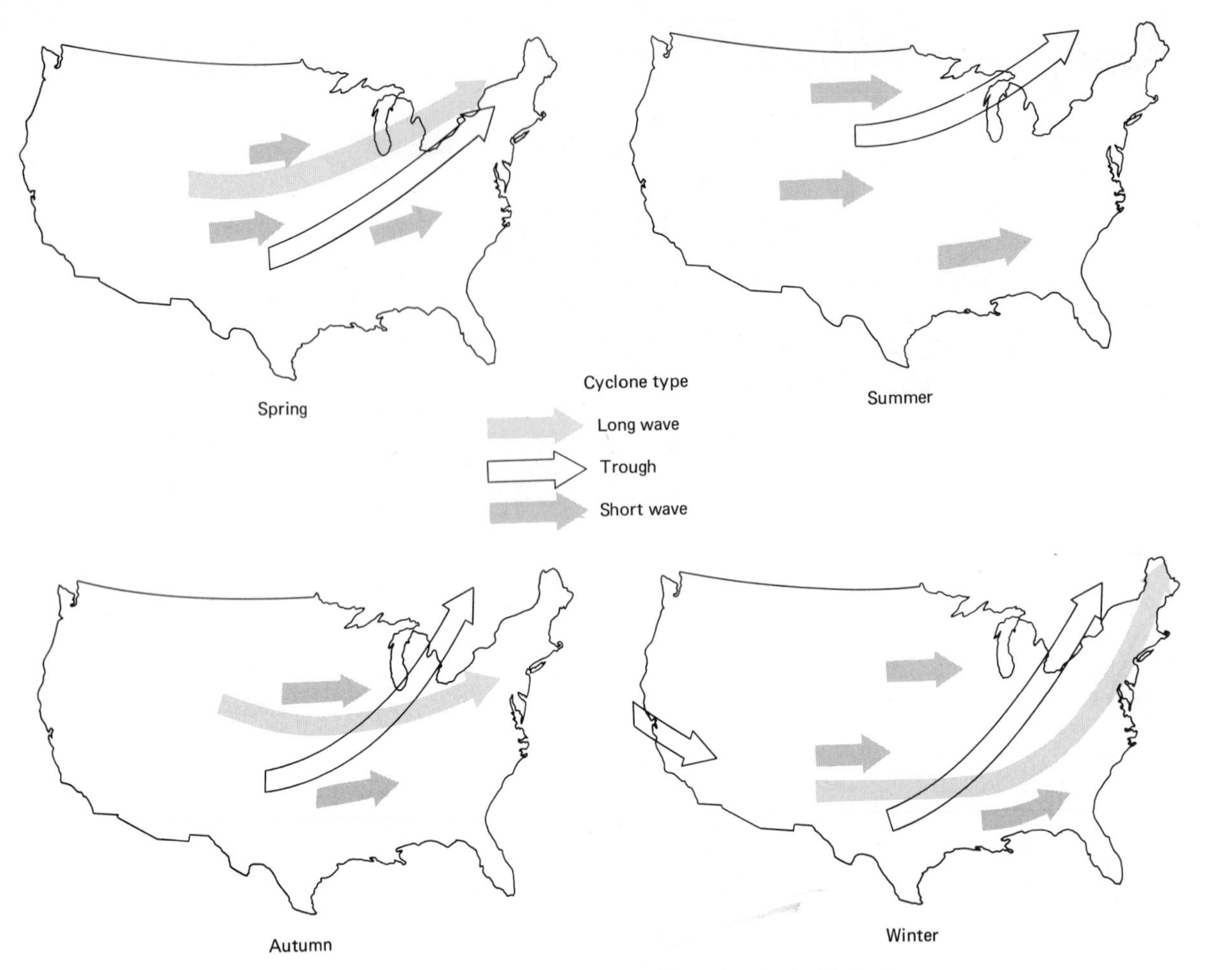

FIGURE 7-18 Frontal cyclone activity varies with season and location in the United States. Short-wave cyclones tend to travel for shorter distances and are more widely dispersed in each of the seasons. Long-wave cyclones are rare during the summertime and generally follow a more easterly path than trough cyclones in the other seasons.

About one-third of these die of heart attacks induced by overexertion in heavy snow. Another one-third die in storm-related accidents, mostly in cars. One-tenth freeze to death, and the remainder die from a variety of mishaps including falls, home fires, and carbon monoxide poisoning. Most of these deaths could be prevented if winter weather warnings were heeded and a few common sense safety rules were followed.

The National Weather Service uses the alerting words *watch* and *warning* for winter storms, just as it does for hurricanes, tornadoes, floods, and other natural hazards. A **watch** alerts people that a storm has formed and is approaching. The **warning** means that a storm is imminent and immediate action should be taken to protect life and property. When a **blizzard** is forecast, people can expect winds to be at least 60 km/h and accompanied by

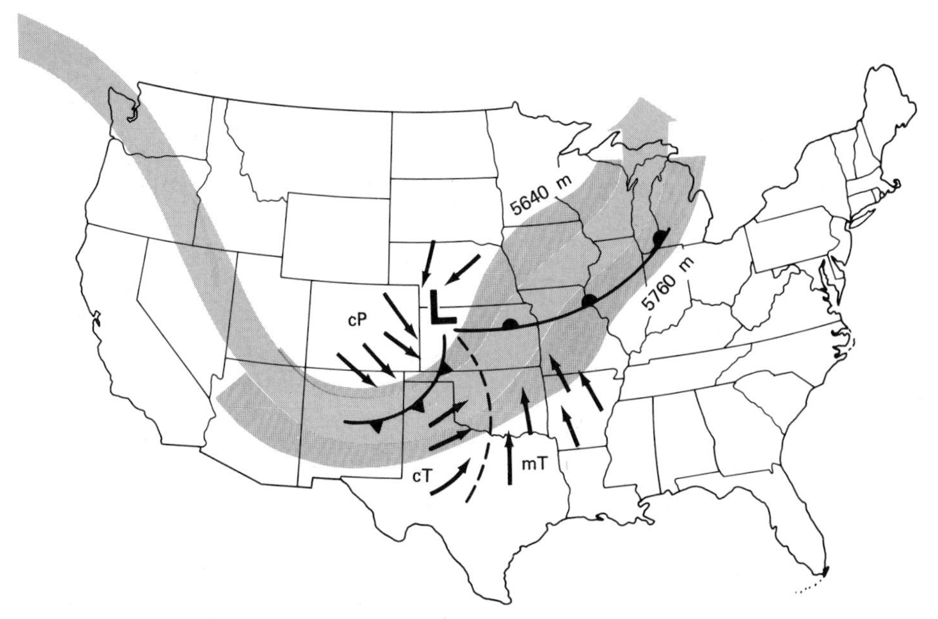

FIGURE 7-19 Typical tornado-producing synoptic weather patterns are shown here for 1500 CST, June 8, 1974, corresponding to a tornado outbreak in Kansas and Oklahoma. The darker band shows the position of the jet stream, and the wider band shows the 500-mb heights. Tornado activity occurred along the dry line separating the maritime tropical air mass from the continental tropical air mass and beneath the jet stream and upper air divergence. (See Figure 7-14 for ATS satellite photographs corresponding to this date.) The tornado-producing cyclone is frequently a hybrid between a long-wave and a trough cyclone, taking on some of the characteristics of each. It travels faster then the long-wave cyclone and is more intense than the trough cyclone.

freezing temperatures and considerable falling or blowing snow. Severe blizzard warnings are issued if the winds are expected to be at least 70 km/h, accompanied by densely falling or blowing snow and temperatures of −12°C or lower.

If snow is forecast without a qualifying adjective such as *occasional or intermittent*, then snow will probably fall for several hours without let up. If the forecast includes **heavy snow,** 10 cm or more is expected in a 12-hour period or 15 cm or more in 24 hours. This amount may vary slightly for different regions of the United States. Blowing and **drifting snow** forecasts ordinarily accompany blizzards. A forecast of drifting snow indicates that strong winds will pile the snow in traffic-impeding drifts. Only a few centimeters of snowfall during high winds can be blown into drifts high enough to stall traffic and close roads. In addition to public watches and warnings, the National Weather Service issues **travelers' advisories** when blizzards are expected to make travel difficult.

Blizzards occur as long-wave cyclones move across the country during the wintertime. Since the surface low-pressure center is connected with the upper atmospheric low-pressure center, the pressure is lower at the surface, and surface winds are much greater. The intense long-wave cyclone with its slow speed of travel gives the blizzard its devastating impact.

When planning for extended travel during winter months, automobiles should be checked in advance to make sure they are in proper operating order. The winter traveler should be prepared for the worst. It is advisable to keep a safety kit in the automobile. Among things to include are tire chains, snow shovel, sack of sand, flashlight, flares, extra gasoline, fire extinguisher, booster cables, external heater, first aid kit, compass, and blankets or sleeping bags.

Travelers should check the latest weather information before beginning a journey. If a blizzard traps you on the road, try to stay in the car. This is where rescue units are likely to reach you the soonest. If you are in deep snow don't try to push the car out or risk overexertion by frantic shoveling. Don't try to walk through a blizzard. Getting lost outside with very cold temperatures and blowing snow means almost certain death. While waiting for help keep the passenger compartment ventilated by opening a window slightly. Run the motor and heater sparingly. Carbon monoxide can build up and is deadly. Try not to remain motionless for long periods of time, and don't allow all occupants of the car to sleep at once. At night turn on the dome light so that workers can spot you easily.

In the home it is advisable to be as self-sufficient as possible. Assume that there will be no electricity, no central heating, no deliveries, no way of getting groceries, and no way of getting out for a day or two. Plans should be made in advance for battery powered equipment such as flashlights and radios. It is also advisable to have a supplementary heating system and a stock of extra food, including some that requires no cooking. It is also advisable to be conscious of potential fire hazards that may arise with prolonged use of stoves or space heaters not designed for continual operation.

The tragic death toll from the freezing temperatures and snow accompanying blizzards can be eliminated or minimized by proper planning and awareness of the potential hazard of these winter storms.

SUMMARY

Frontal cyclones develop in particular regions of the atmosphere beneath troughs in the jet stream. They are the largest traveling vortices common to the earth's atmosphere and usually have cold fronts and warm fronts associated with them. Cold fronts normally travel faster than warm fronts, producing cumulonimbus clouds and more violent weather. Stationary fronts and occluded fronts may also be associated with midlatitude cyclones.

Frontal cyclones are generated by the jet stream through specific effects related to upper air directional or speed divergence, cyclonic vorticity, upper atmospheric wind shear, and surface convergence. These midlatitude cyclone formation factors occur predominantly in the trough of the jet stream, initiating midlatitude cyclone development there. A midlatitude cyclone normally completes its life cycle from the trough to the ridge in the jet stream since upper air support is provided in this region. An anticyclone thrives in the region from an upper atmospheric ridge to a trough since upper air support is provided for it in this region.

Three basic types of midlatitude cyclones can be identified, including trough, long-wave, and short-wave cyclones. Trough cyclones occur within the trough of the jet stream. Long-wave cyclones occur north of the trough in the jet stream and beneath the upper atmospheric low-pressure center. Short-wave cyclones are related to the development of short-waves superimposed on the jet stream flow patterns. The most intense midlatitude cyclone is the long-wave cyclone. It typically has the longest lifetime, contains the highest surface winds and the lowest atmospheric pressure. This type of midlatitude cyclone occurs during the winter and spring and is responsible for blizzards in the wintertime and high winds during the month of March. Trough cyclones are the most common type of midlatitude cyclones and are responsible for considerable precipitation in the United States.

Tornado-producing cyclones are frequently hybrids between trough and long-wave cyclones. The center of lowest pressure at the surface is located beneath the region of maximum cyclonic wind shear in the jet stream, where it can take on some of the characteristics of both the long-wave and trough cyclones.

Blizzards are produced by long-wave cyclones that occur during the wintertime. Because blizzards are responsible for numerous deaths each winter, caution should be taken when driving or doing outdoor activities when blizzards are forecast.

STUDY AIDS

1. Describe the atmospheric environment for cyclogenesis.

2. Explain some of the differences in weather associated with cold fronts compared to warm fronts and relate these to the different characteristics of each.

3. Describe the structure of cyclones and anticyclones.

4. List and explain at least three features of the jet stream that contribute to cyclogenesis.

5. Describe a surface feature that contributes to cyclogenesis.

6. Explain why midlatitude cyclones frequently develop beneath the trough of the jet stream.

7. Describe the three basic types of midlatitude cyclones.

8. Compare the upper atmosphere formation factors for the trough and long-wave cyclone.

9. Compare the seasonal occurrence of the different kinds of midlatitude cyclones.

10. Describe the tornado-producing midlatitude cyclone.

11. Describe the blizzard-producing midlatitude cyclone.

12. List some important precautions to take during a blizzard warning.

TERMINOLOGY EXERCISE

Use the glossary to reinforce your understanding of any of the following terms used in Chapter 7 that are unclear to you.

Cyclogenesis
Frontal cyclones
Anticyclone
Upper air divergence
Cyclonic vorticity
Wind shear
Upper air support
Occlusion
Trough cyclone
Long-wave cyclone
Short-wave cyclone
Watch
Warning
Blizzard
Heavy snow
Drifting snow
Travelers' advisory

THOUGHT QUESTIONS

1. If blizzard conditions are occurring in Iowa, what kind of weather would you expect in southern Missouri? Explain your answer.

2. Many frontal cyclones do not have a warm front associated with them. Can you explain why they may develop in this way?

3. Is it possible for cyclonic storms to cross the equator? Explain your answer.

4. What major differences would you expect between frontal cyclones in the northern and southern hemispheres?

5. Explain how you might use a classification of the different types of frontal cyclones to make better weather forecasts.

SPECIAL TOPIC 9

SOLAR HEATING

The use of sunlight for heating homes has many attractions. Among them are no pollution, no cost except for a means of harnessing the energy, and no depletion of the supply of energy. Solar collectors or solar cells can be purchased, but passive solar heating can also be designed into a house. This is accomplished by making use of knowledge of the angle of the sun's rays for your particular latitude.

You can calculate the angle of the sun above the horizon at noon by the following method:

1. Calculate the difference between your latitude and the latitude when the sun is directly overhead.

2. Subtract this difference from 90°.

Suppose you live at $40^1/_2$° N latitude. On June 21, step 1 gives you $40^1/_2 - 23^1/_2 = 17°$ and step 2 gives you $90° - 17° = 73°$. Repeating for December 21, step 1 gives $40^1/_2° - (-23^1/_2°) = 64°$, and step 2 gives $90 - 64 = 26°$. You can use this information to position solar collectors at right angles to the winter sun by designing the slope of the roof at this angle or by placing the solar collectors at this angle even though the roof may have a different angle.

Passive solar heating can be designed into a house by use of south windows and the proper overhang for your latitude. Figure S9-1 shows how you would choose the right overhang for your particular conditions. The overhang can be made large enough to prevent summer sunlight from striking directly through south windows, while permitting winter sunlight to shine through to help warm the house.

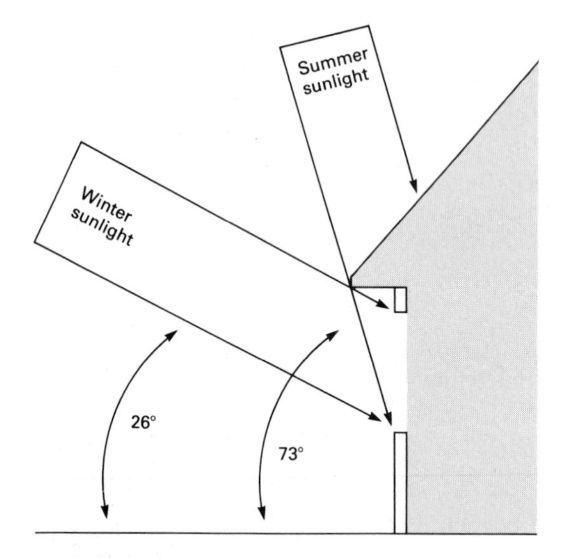

FIGURE S9-1 In designing a house for passive solar heating, the length of the roof overhang (eaves) can be designed to allow sunlight to shine through southern windows during the winter but not during summer, as shown in this illustration.

CHAPTER EIGHT
HURRICANES AND TROPICAL CYCLONES

TROPICAL WEATHER

The tropical region, bounded by the Tropic of Cancer and the Tropic of Capricorn, includes all the places on earth where the sun shines directly overhead. The temperature of the tropics does not change much from season to season because the direct rays from the sun produce considerable heat even when other regions of the earth are experiencing their cool season.

Weather in the tropics is dominated by convective processes and disturbances in the trade winds. Excess solar heat warms the atmosphere and evaporates water to feed the convective process. Therefore, rain showers and thunderstorms are regular features of tropical weather.

Disturbances in the trade winds, called **easterly waves** (Fig. 8-1), are also centers of cloudiness and precipitation. Streamlines, or paths of air parcels, that depict airflow in the tropics are more useful than sea-level pressure maps since pressure variations are normally small. An easterly wave in the tradewinds moves slowly westward. The weather it brings is influenced by the surface convergence of air on its eastern side and the divergence of air on its westward side. Convective showers accompany the surface convergence.

The region of the trade winds near 20° latitude frequently has sinking air associated with subtropical high-pressure centers. This creates the **trade wind inversion.** A well-developed inversion inhibits thunderstorm and hurricane development.

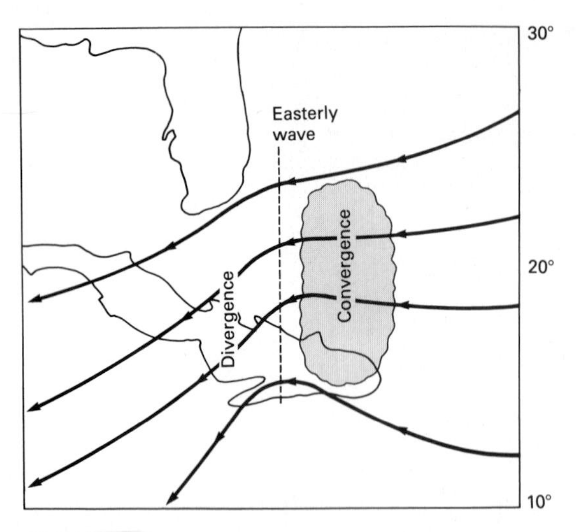

FIGURE 8-1 Waves in the tropical trade winds create regions of convergence. Clouds and showers are associated with such regions.

TROPICAL STORMS

Warm tropical waters develop their own unique storm systems. Intense tropical storms are known by various names. They are called **typhoons** near China and Japan, **hurricanes** in the United States, and cyclones near India in the Indian Ocean. All of these are similar tropical cyclones that have a very different structure from the midlatitude cyclones described in Chapter 7, although a tropical cyclone can be modified into a midlatitude cyclone as it moves out of tropical waters and into midlatitudes. A good example of this occurred in 1961 as the storm that had been Hurricane Carla passed through the central United States as a modified hurricane or strong midlatitude cyclone.

DEVELOPMENT OF A HURRICANE

A hurricane originates as a **tropical disturbance** with no strong winds or closed isobars around an area of low pressure containing cloudiness and some precipitation (Fig. 8-2). Many such disturbances exist at any given time within easterly waves in the tropical trade winds. It is not known why one of these occasionally develops into a large hurricane while others do not.

A disturbance that is to develop into a hurricane passes into the next stage, called a **tropical depression.** A tropical depression has at least one closed isobar accompanying a drop in pressure in the center of the storm (Fig. 8-2). The winds have increased but are still less than 60 km/h. In the next stage (**tropical storm**), the winds have increased to speeds between 60 and 120 km/h. Distinct rotation exists around a central low pressure that has several closed isobars around it. The next stage is the hurricane stage. Pronounced rotation has developed around the central core, with winds greater than 120 km/h. A tremendous atmospheric vortex has been generated that is smaller in size than the midlatitude cyclone, but much more intense.

CHARACTERISTICS OF HURRICANES

Hurricanes have several distinguishing features. They occur primarily during certain times of the year, with more hurricanes developing in September than in any other month in the northern hemisphere. One reason for this timing is that they form only over ocean areas where the water temperature is greater than 27°C. This requirement of very warm oceans restricts the area where hurricanes form that strike the United States. A small area (Fig. 8-3) has ocean temperatures greater than 27°C in the months of August and September. During the month of February, no area in the Gulf of Mexico or the southern Atlantic has temperatures averaging over 27°C. There is, however, an area north of South America where the average temperature is 27°C. This temperature is almost warm enough for a hurricane to develop and occasionally this does happen in January or February.

Hurricanes have no fronts associated with them as do midlatitude cyclones, but the isobars and winds are circular around the center of low pressure. Mature hurricanes normally have an **eye.** The eye of a hurricane may be cloud free or have an

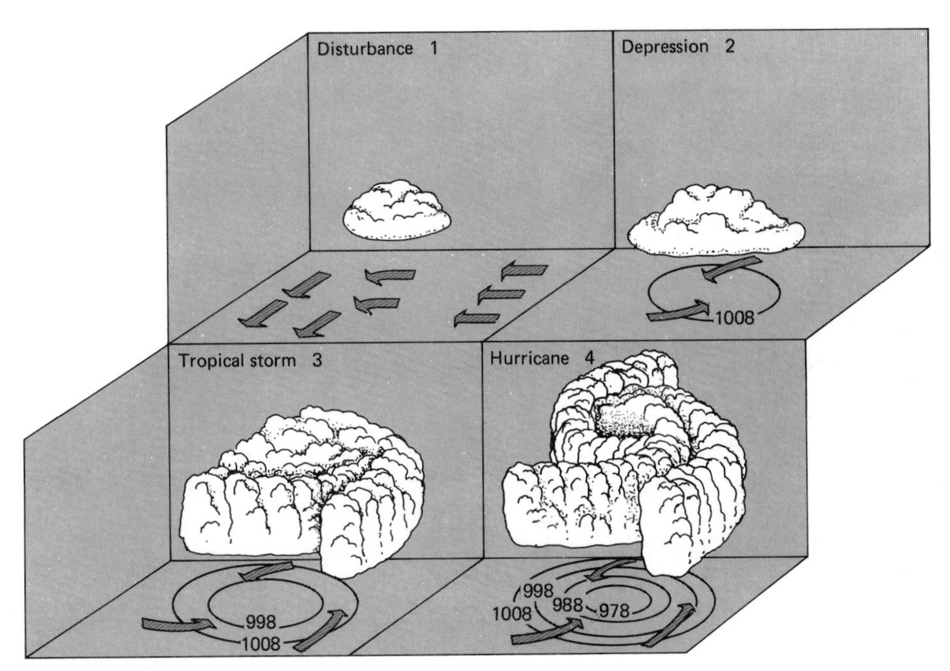

FIGURE 8-2 Various stages of hurricane development. First, tropical disturbances become tropical depressions when the surface pressure begins to fall. The developing storm becomes a tropical storm as several closed isobars are present at the surface, and winds and cloud patterns are circular around the center of low pressure. In the mature or hurricane stage the eye of the hurricane becomes well developed, corresponding to the lowest atmospheric pressure near the center of the storm. Rain is in bands surrounding the eye of the hurricane. The hurricane stage is followed by dissipation of the storm.

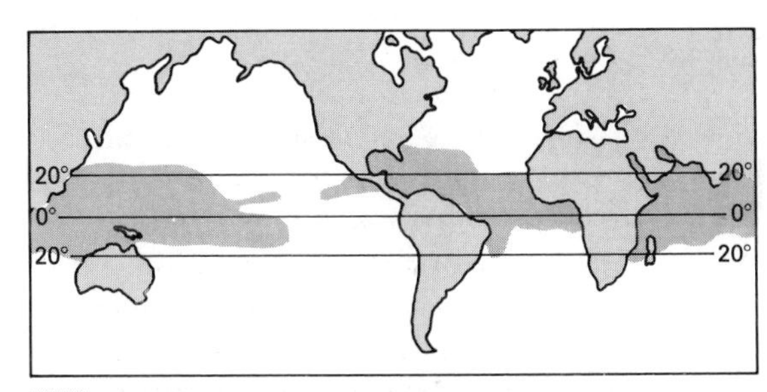

FIGURE 8-3 Tropical cyclones are generated in locations where the sea surface temperature is greater than 27°C. Maximum water temperatures normally occur in September in the northern hemisphere and March in the southern hemisphere, and thus these are the times of most frequent hurricane formation.

overcast of clouds that produce much less rain than surrounding bands of clouds. The eye of the hurricane may be 30 km in diameter and develop as the winds increase and become circular around the central core of low pressure.

Hurricanes are smaller than midlatitude cyclones. A typical hurricane may be 500 km in diameter while a midlatitude cyclone may cover almost 2000 km. Hurricane winds, however, are much more damaging than those of midlatitude cyclones because of their high speeds. One of the strongest hurricanes on record was Hurricane Camille in 1969 which had wind speeds over 320 km/h. The typical hurricane has winds of 160 km/h; only the strongest reach 300 km/h. Strong hurricanes are very damaging since winds with speeds greater than 160 km/h can take roofs off buildings and damage other objects. Winds as great as 200 km/h are very destructive.

A surprising characteristic of hurricanes is that they do not develop at the equator. Since warm ocean temperatures are required for hurricane development, it would seem that hurricanes should form near the equator, but they seldom occur within 5° or 500 km of it. The reason for this is related to the Coriolis acceleration due to the earth's rotation. Because of the earth's spinning at the equator, the only acceleration at that point is an outward or vertical acceleration. Since the Coriolis acceleration is the amount of horizontal acceleration, it is zero at the equator and reaches a maximum at the North and South Poles. The potential hurricane needs this rotational help to initiate spiraling air, and thus develop the storm's structure.

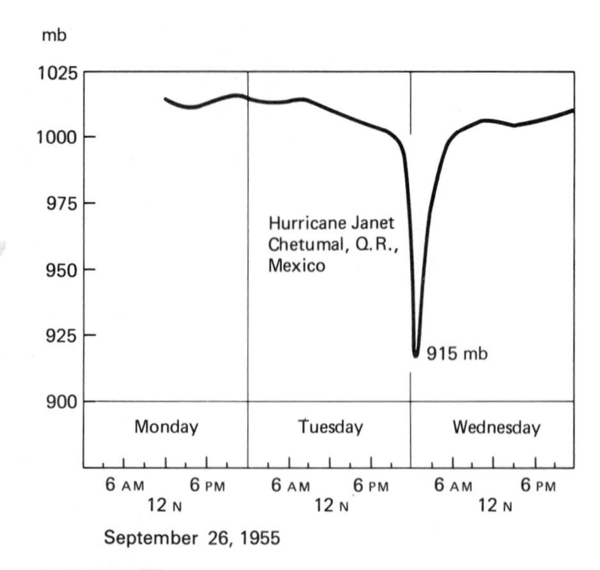

FIGURE 8-4 The center of a hurricane is accompanied by rapidly decreasing barometric pressure, as was measured in Hurricane Janet as it passed over Mexico in 1955. This hurricane produced near record low-pressure readings, as shown in this barograph trace. (From G. E. Dunn and B. I. Miller, *Atlantic Hurricanes* [Baton Rouge: Louisiana State University Press, 1964].)

STRUCTURE OF HURRICANES

The center of the hurricane has the lowest pressure. In the United States, one of the lowest pressures ever measured was in 1900 during the severe hurricane that struck Galveston, Texas, killing several thousand people. The lowest recorded pressure during the storm was 936 mb. The world record low pressure is 888 mb recorded as Hurricane Ida passed over the Philippine Islands in 1958. The eye of the hurricane has the lowest pressure, as shown by considering isobars in side view (Fig. 8-4). In plan view, the isobars are circular around the eye of the hurricane.

The temperature structure of a hurricane is such that the eye of the hurricane is warmer by several degrees than the surrounding regions (Fig. 8-5). The warmer temperatures in the eye of the hurricane extend from the surface to the top of the hurricane. If isotherms are drawn through the eye, they bulge upward in the central region of the hurricane. Warmer eye temperatures are due to the compression heating associated with subsiding air in the center of the hurricane.

This central structure of the hurricane is quite different from a midlatitude cyclone. The main driving force for the hurricane is the very warm humid air from over the ocean. The air flows into the base of the hurricane and spirals around the **eye wall.** Tremendous energy is supplied to the hurricane updrafts as cloud droplets condense since 580 cal are released for each cubic centimeter of condensed water; the rainbands are an integral part of the large vortex. This is much different from tornadoes, where the driving force is located above the storm, or from midlatitude cyclones, where the diverging air at the jet stream level provides the mechanism for a large central updraft over the low-pressure center. Because the hurricane acquires its energy from warm humid surface air, it loses its intensity very rapidly if it passes over land.

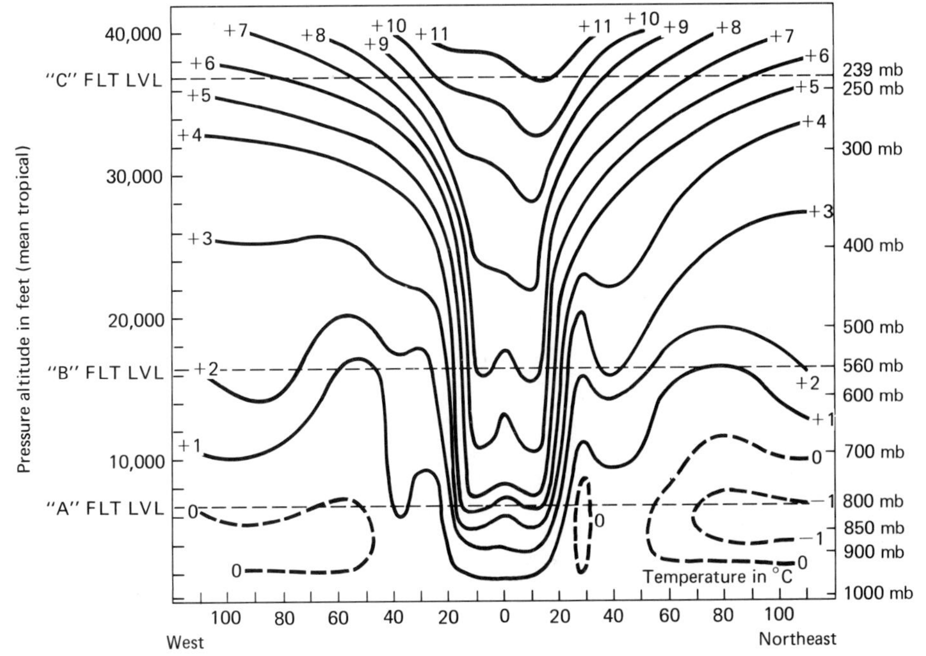

FIGURE 8-5 Warmer atmospheric temperatures are produced in the eye of a hurricane by the subsiding air as it heats adiabatically. These measurements from Hurricane Cleo in 1958 show a vertical cross section of temperature, with temperatures several degrees warmer in the eye near the surface and more than 10°C warmer in the eye of the hurricane in the upper atmosphere. (From B. I. Miller, *Science,* vol. 157, September 22, 1967, copyright 1967 by the American Association for the Advancement of Science.)

Since the hurricane is surface generated, with spiraling winds around the center of low pressure, and it is driven by warm air currents, the core consists of subsiding air that is only replacing the air thrown outward by centrifugal force. Thus, the hurricane has a very different structure from the midlatitude cyclone or the tornado.

The winds within a hurricane reach their maximum velocity just outside the eye of the hurricane. For example, in Hurricane Daisy in 1958 (Fig. 8-6), the maximum winds were about 20 km from the center of the hurricane. The winds at higher levels in the hurricane were less than at the surface. These measurements showed winds at 11 km of 130 km/h; at 6 km the wind speed was 185 km/h; and at 4 km the wind speed was 280 km/h. In general, the highest winds are located in a narrow band around the eye of the hurricane, with fairly calm winds within the eye. The surface winds decrease in velocity outward from the band of most intense winds and the wind speeds decrease above the surface.

MATURE AND DECAY STAGES

A typical hurricane reaches its peak intensity in about six days. In this mature, or hurricane, stage the wind and rainbands become organized and spiral inward toward the eye of the hurricane

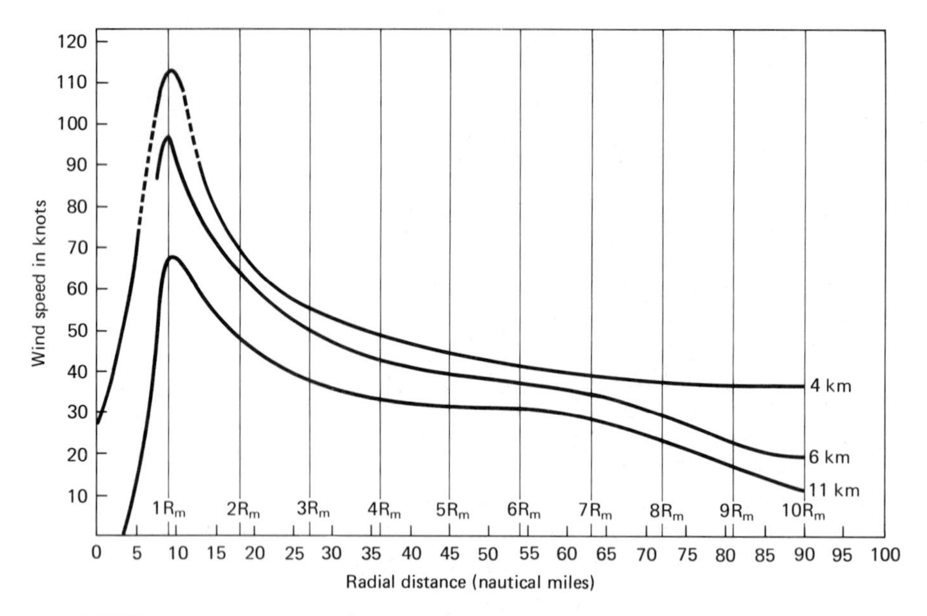

FIGURE 8-6 The wind profile through Hurricane Daisy in 1958 based on measurement by aircraft shows that the greatest wind speed occurred close to the earth's surface and in a narrow band around the eye of the hurricane. The wind speeds at a height of 11 km had a similar profile outward from the eye of the hurricane although the wind speeds were lighter at all distances. (From B. I. Miller, *Science,* vol. 157, September 22, 1967, copyright 1967 by the American Association for the Advancement of Science.)

(Fig. 8-7). The pressure has decreased at the center of the storm and the winds have formed a large atmospheric vortex. The eye develops as the air subsides in the center of the storm to replace the void that would otherwise form in the center of the vortex. In the hurricane stage, a lower pressure exists in the upper atmosphere above the storm center as well as at the surface. The low pressure in the upper atmosphere may have a ring of higher pressure around it due to the mass of air that is transported upward through the eye wall, with centrifugal force throwing it outward from the lower pressure at the center. At the surface, the lowest pressure is at the center and the isobars are circular around it, with the pressure increasing outward. The mature hurricane has rising air outside the center of the eye and rainbands caused by the rising air. Subsiding air is present in the center of the hurricane; spiraling air around it forms the vortex storm.

The last stage in the life cycle of the hurricane is the decaying stage. This starts as the hurricane moves northward out of the region where the ocean temperatures are warm, or as it crosses land where the humid, hot winds from over the ocean that feed it are missing. The typical hurricane lasts for several days and has an average life span of nine days. A hurricane decreases in intensity quite rapidly after **landfall;** its power for destruction is cut in half by the time it has traveled only 240 km inland.

OCCURRENCE OF HURRICANES

Autumn is the hurricane season in the northern hemisphere. Statistics on the number of hurricanes show that more hurricanes occur in September

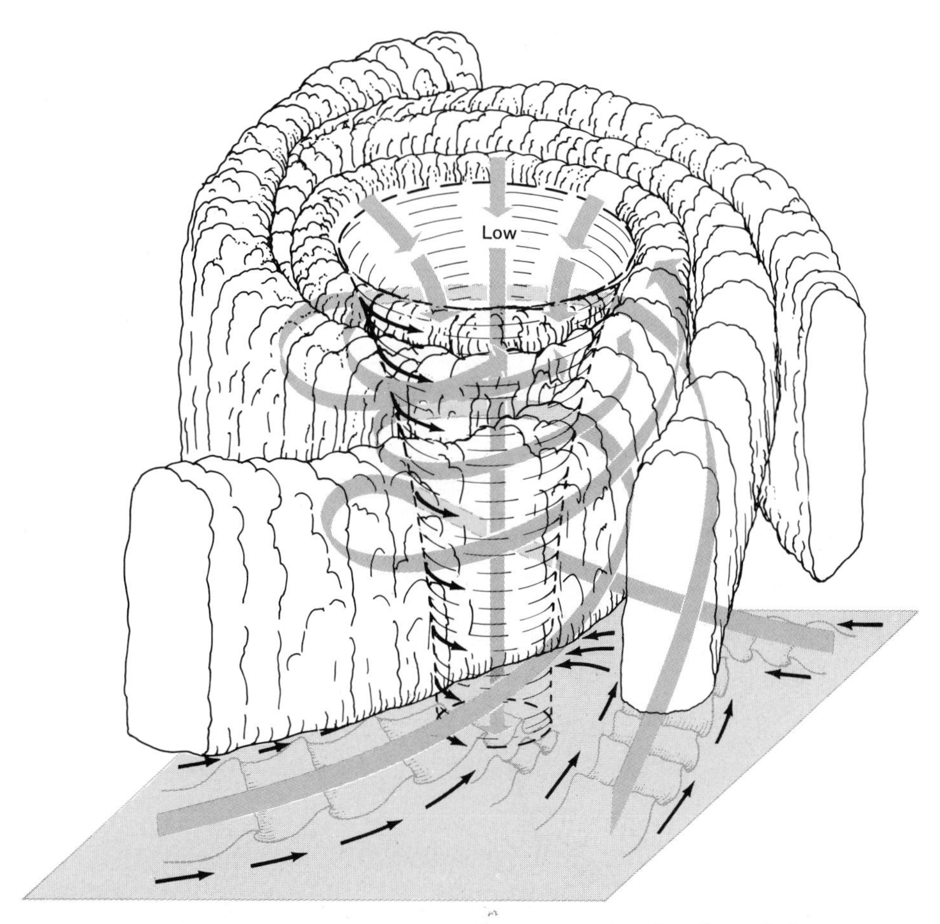

FIGURE 8-7 Airflow patterns in a mature hurricane consist of surface winds spiraling inward toward the eye of the hurricane, with rising air currents in bands of thunderstorms around the eye. The core of the vortex contains subsiding air that is replacing air thrown outward by centrifugal force. The hurricane is the only surface-generated atmospheric vortex that encompasses the entire atmospheric layer. Other atmospheric storms, besides the dust devil, have different driving mechanisms, such as the jet stream for the frontal cyclone and the thunderstorm for the tornado.

(Fig. 8-8) than in other months. A considerable number also occur in August and October, but very few occur in the winter months. The first hurricanes of the season are frequently the strongest storms. Hurricanes Agnes, Betsy, and Camille are good examples. An average of two hurricanes per year strike the United States (Fig. 8-9). This has varied from zero to five hurricanes per year since 1930.

Because of the requirement of warm ocean temperature hurricane formation is most common in the northern hemisphere between 10° and 30° N latitude. The region of hurricane generation for the United States is in the Atlantic Ocean and the Gulf of Mexico in the band from 10° to 30° N latitude (Fig. 8-10). The generation region shifts slightly month by month as the band of maximum heating shifts with the seasons.

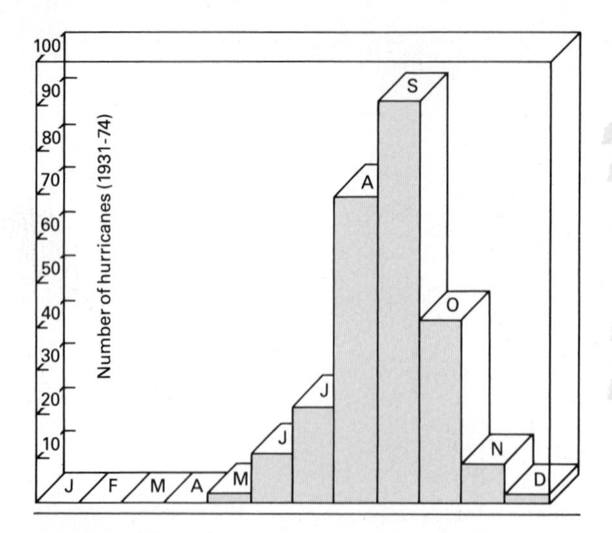

FIGURE 8-8 The monthly distribution of hurricane occurrence shows they are most common in the month of September. (Based on data for the North Atlantic for the years 1931 to 1974 as given in *Climatology Data, National Summary.*)

MOVEMENT OF HURRICANES

Typical paths of hurricanes show movement first toward the northwest, as they are guided by the trade winds and the average circulation between the surface and the tropical upper atmosphere. As hurricanes reach more northerly latitudes and move into the jet stream region, they are eventually influenced more by upper atmospheric winds from the west. At this point, they start to curve toward the northeast, giving a typical path of a hurricane (Fig. 8-11).

The path of a hurricane is quite important in coastal areas. Consider Hurricane Donna in 1960. It started in the southern part of the north Atlantic Ocean (Fig. 8-12). Nine days after reaching hurricane intensity, it narrowly missed Cuba and a day later struck Florida from the Gulf side. After crossing Florida and regaining intensity in the Atlantic, the hurricane hit the east coast of the United States. Crossing Florida is frequently enough to

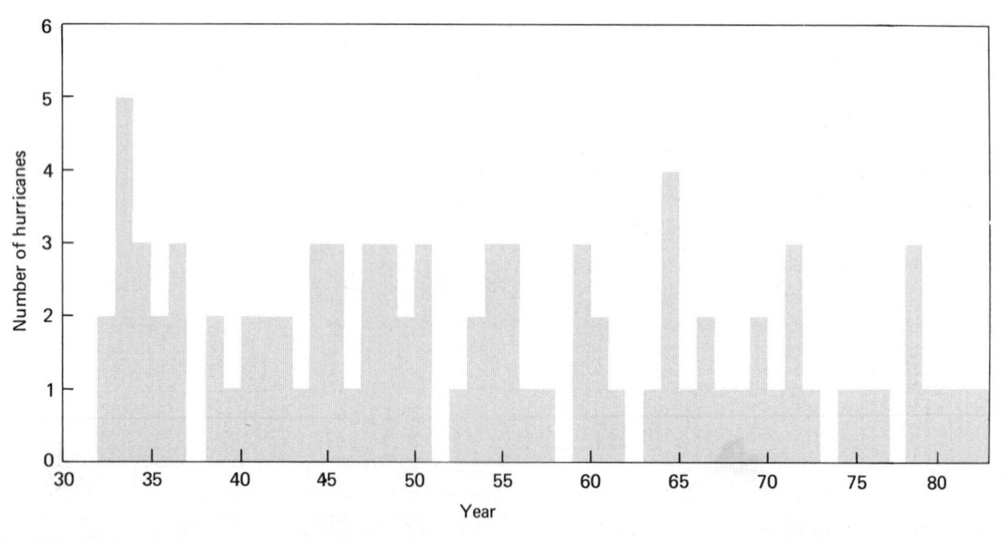

FIGURE 8-9 The number of hurricanes striking the United States has varied from a high of five in 1933 to none striking during several years. On average, almost two hurricanes per year hit the United States. (Data from *Climatology Data, National Summary.*)

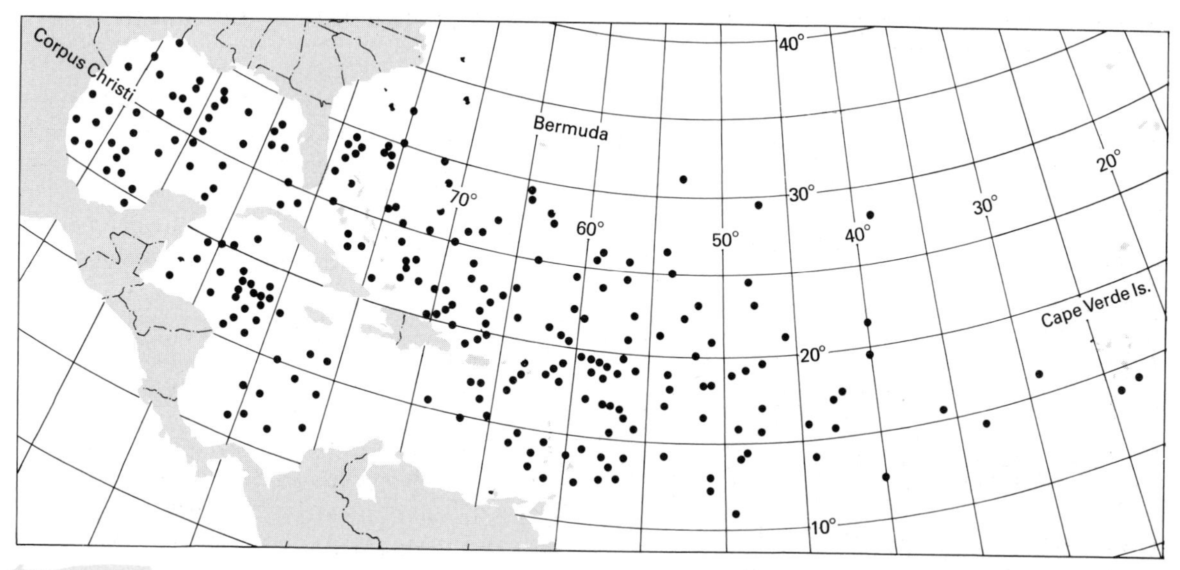

FIGURE 8-10 The location where tropical storms reached hurricane intensity is shown, based on all storms from 1901 to 1957. (U.S. Department of Commerce, *Hurricane Forecasting,* Washington, D.C., 1959.)

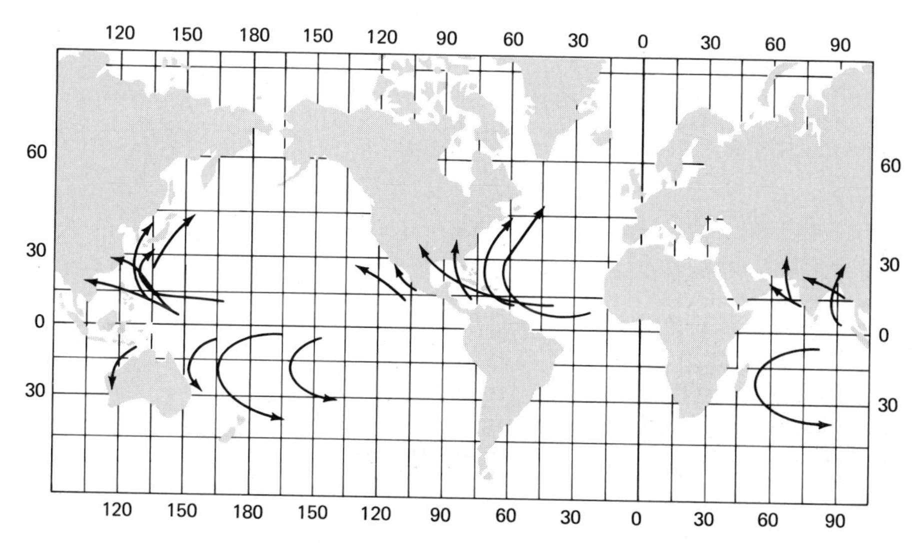

FIGURE 8-11 The typical hurricane originating in the northern hemisphere moves toward the west with bending northward and then northeastward as it moves into more northerly latitudes. Thus, the Gulf and east coast of the United States experience a greater number of hurricanes. (From NOAA.)

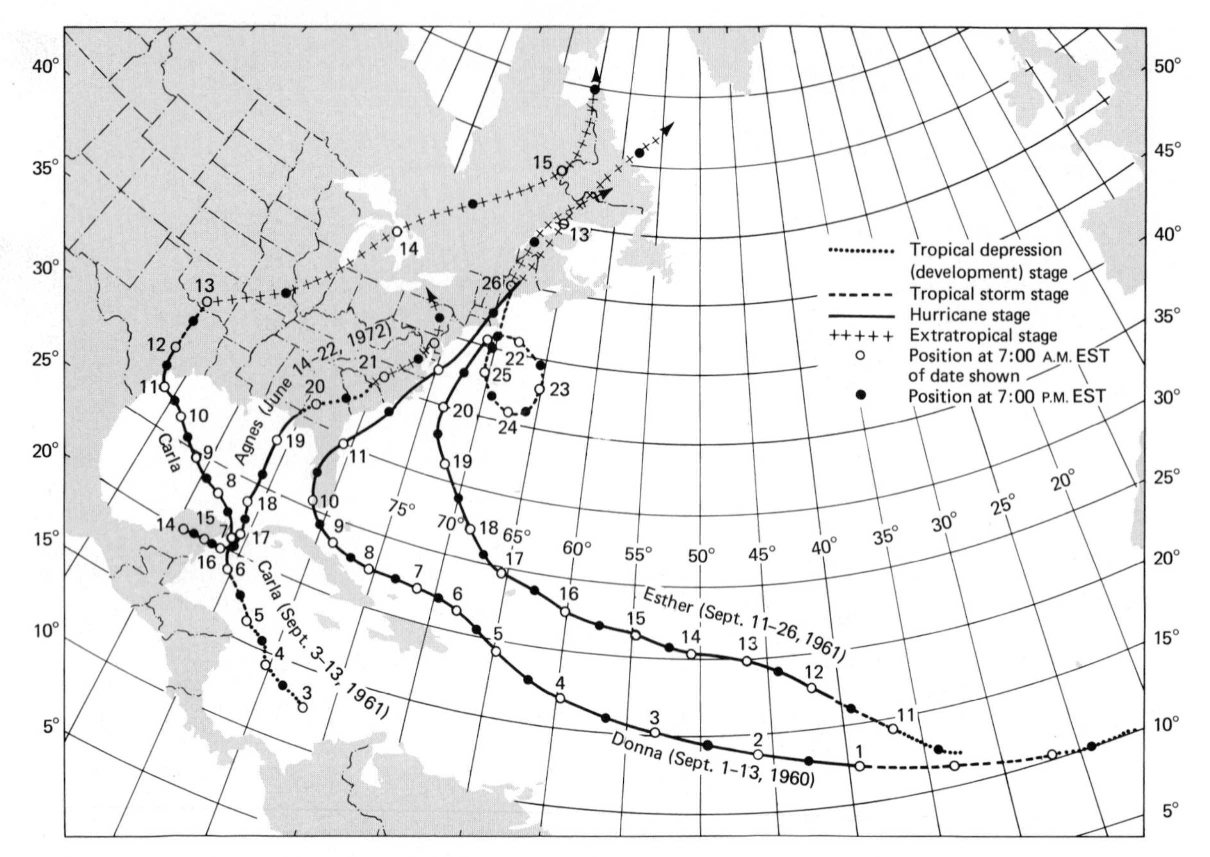

FIGURE 8-12 The paths of hurricanes Agnes, Carla, Donna, and Esther. Hurricane Donna was particularly devastating because of its path along the coastline. Positions of the hurricanes are shown at 7 A.M. EST of the date shown. (From NOAA.)

cause a hurricane to decay, but after Hurricane Donna crossed Florida and moved into the Atlantic Ocean it regained hurricane strength. As it traveled northeastward along the coast, it hit small sections of land, going back out over the water to reintensify and then striking more of the coastal area. Hurricane Donna was, thus, a particularly devastating hurricane.

Hurricane Carla in 1961 was also an intense hurricane and was generated in a similar location. When this storm was 11 days old, it struck Texas as a very destructive hurricane. One day later it was traveling across Texas and had decreased to the tropical storm stage. By the time it reached southern Oklahoma, it had been modified into an intense midlatitude cyclone. This required a complete change in the internal structure of the storm as the downdraft above the core of lowest pressure became an updraft with associated precipitation.

On September 11, 1961, the position of Hurricane Carla was just south of the coast of southern Texas (Fig. 8-13). The central pressure was 960 mb and the isobars were very circular around the center of this low pressure. By September 13, the center of the low was in southern Oklahoma and it was no longer a hurricane. This modification was typical of a hurricane that strikes land. Landfall is followed rapidly by a reduction in wind velocity and

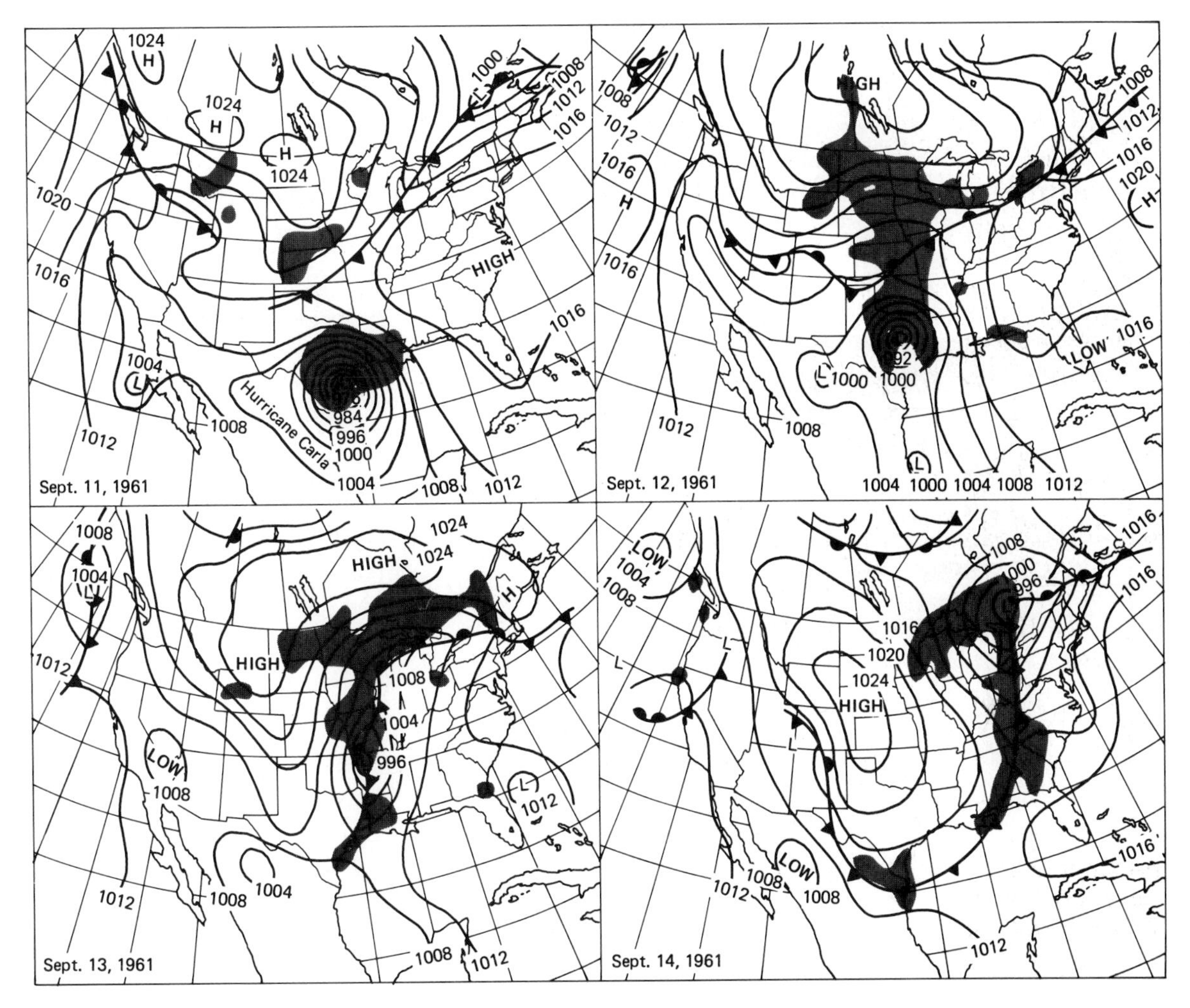

FIGURE 8-13 The position of Hurricane Carla is shown on September 11, 1961, just prior to landfall. By September 14 this storm had been modified into a midlatitude cyclone complete with frontal system.

an increase in the pressure, resulting in a much weaker storm that decays or transforms into a different type of storm if appropriate atmospheric conditions exist. The next day, September 14 (Fig. 8-13), the center of the low pressure was in Minnesota, where the storm had overtaken a frontal system. A cold front extended from the center of the low pressure giving the storm all the characteristics of a midlatitude cyclone.

Heavy precipitation occurred throughout the central United States from this storm. Kansas City, for example, received 30 cm of rainfall from the storm in comparison to the 50 cm of rainfall the Texas coast received as the hurricane passed over it. It took Hurricane Carla only three days to cross the United States after changing into a midlatitude cyclone. This tropical storm was atypical, however, since most tropical storms remain tropical storms throughout their lifetime, with completely different structures from midlatitude cyclones and decay after landfall.

The surface winds of a moving hurricane have different velocities on each side of the storm. For a hurricane moving toward the northwest, for example, the highest wind speeds occur to the right of its path of motion (Fig. 8-14). This results from a combination of the rotating winds and the translation speed of the storm.

If a symmetrical hurricane were stationary and had 130 km/h winds, the wind speeds would be equal on all sides of the eye. But if the hurricane were moving at a speed of 30 km/h, then the winds on the right-hand side of the path of motion would be composed of the wind rotating around the center of the hurricane plus the speed that it traveled across the ocean. Therefore, on the right-hand side it would have 160 km/h winds, while on the left-hand side the speed of translation (30 km/h) would be subtracted, resulting in only 100 km/h winds.

The difference between a wind of 100 and 160 km/h is substantial in terms of its destructive power. Navigators of ships at sea have long used this knowledge of differing velocities to avoid the highest velocity winds when caught in a storm. Today more information is available for the navigators since hurricanes can now be detected in satellite photographs as they develop (Fig. 8-15). In earlier years most of these storms in the open ocean were undetected until high winds or ocean swells—the first indications of a hurricane—were observed. Rules based on the direction the wind was coming from were used by ship navigators to set a heading that would take the ship out of the maximum winds known to occur to the right of the center of the path of the storm.

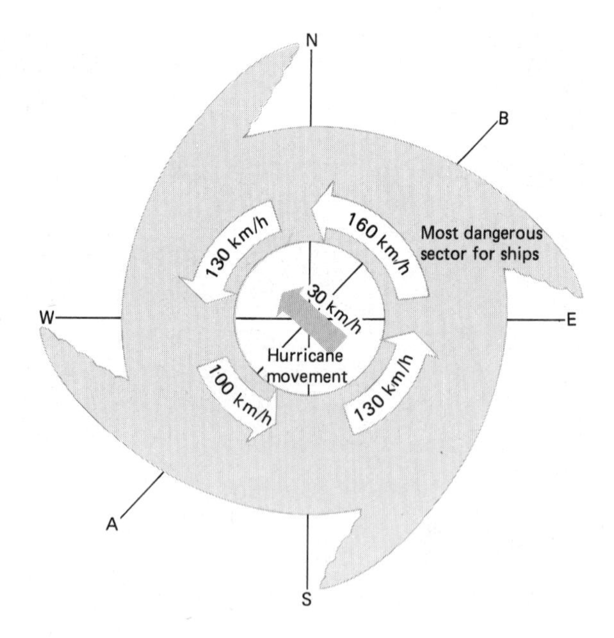

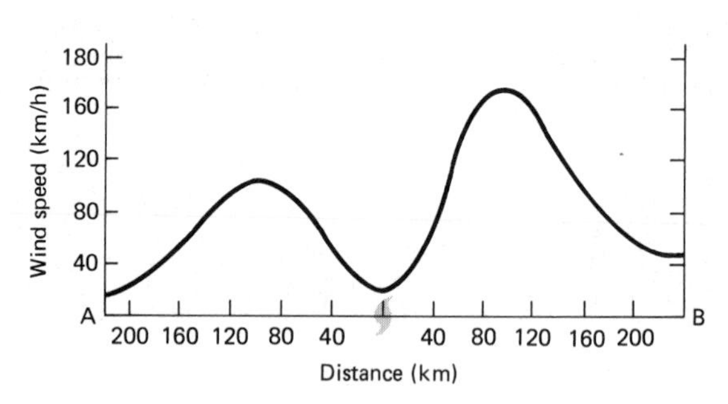

FIGURE 8-14 Wind speeds around the eye of a hurricane are influenced by the forward motion of the storm. A forward motion of 30 km/h toward the northwest would add to the winds in the northeast quadrant of the storm but would subtract from the winds in the southwest quadrant. This causes the right-hand side of the hurricane to be more dangerous and destructive, as indicated in the wind speeds actually measured in hurricanes.

DESTRUCTION FROM HURRICANES

Strong hurricanes are very destructive. Therefore coastal weather stations use radar to detect the rainbands of an approaching hurricane (Fig. 8-16). Flooding frequently occurs from hurricanes, for example. The record amount of flood damage occurred from Hurricane Agnes in 1972. More than 50 cm of rainfall associated with the passage of a hurricane is not uncommon (Table 8-1). The runoff systems in many cities cannot handle this much precipitation because of the gentle topography of many of the coastal areas where hurricanes occur (Fig. 8-17). Houston, Texas, for example, was

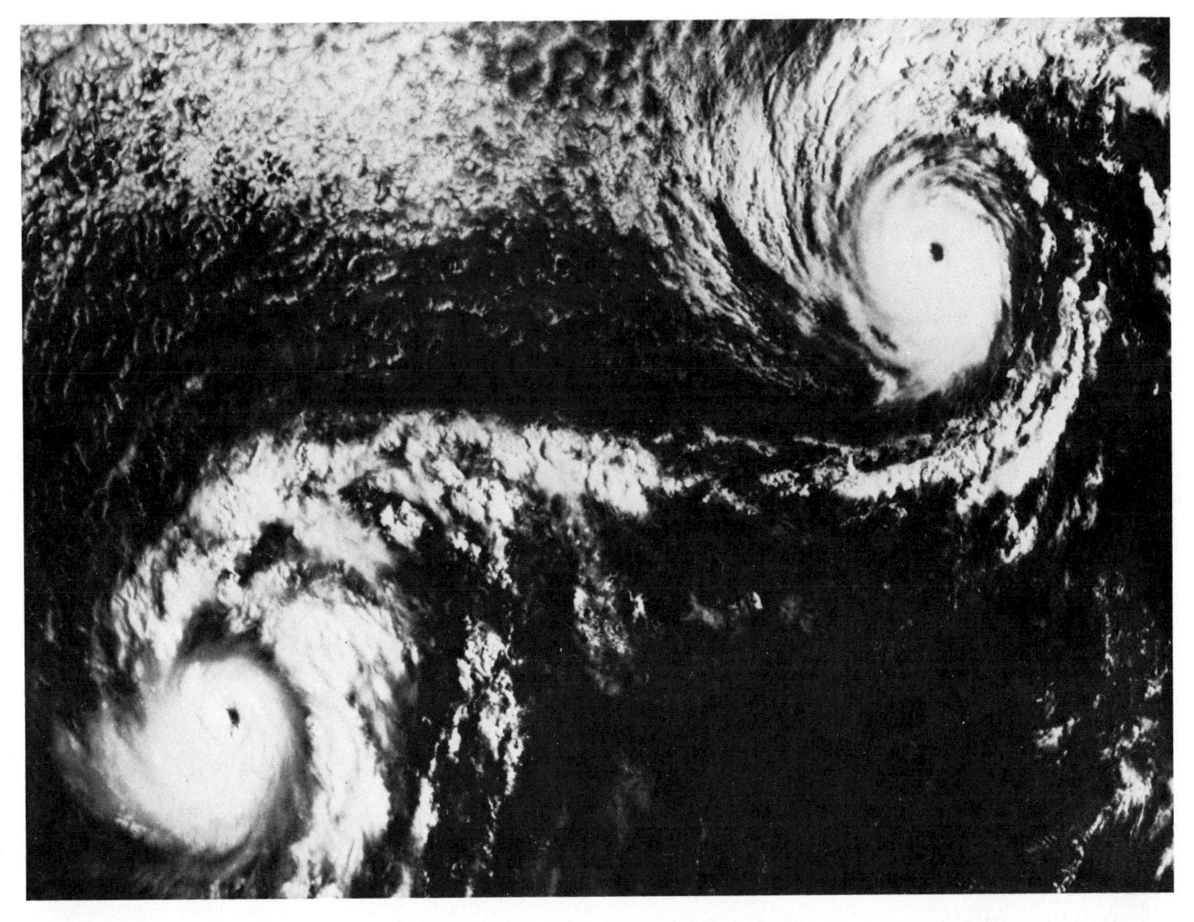

FIGURE 8-15 This NOAA-3 satellite photograph shows Hurricane Ione (left) and Hurricane Kirsten on August 24, 1974. The eyes of the storms are evident, as is the general circulation of the cloud patterns around the storms. (Photo courtesy of National Hurricane Center, NOAA.)

flooded on June 10, 1975, as 15 cm of rainfall occurred in a few hours during and prior to the early morning traffic. All the low areas on the highway from Galveston to Houston were covered with water, as were the low areas in Houston. At first, small cars were stalled or forced to park along the street; then larger automobiles were forced to do so. Automobiles were abandoned and left stranded on major Houston streets. Flooding associated with the heavy rainfall accompanying a hurricane can obviously be serious. If 15 cm of rainfall can create the problems Houston faced during that storm, a 50-cm rainfall can be devastating.

In addition to flooding and high winds associated with hurricanes, the ocean waves and **swells** are also a source of destructive power of a hurricane (Fig. 8-18). One rule of thumb for the magnitude of fully developed ocean waves associated with winds is that their height in meters is equal to 10% of the wind velocity in kilometers per hour. Thus a wind of 150 km/h could generate 15-m waves.

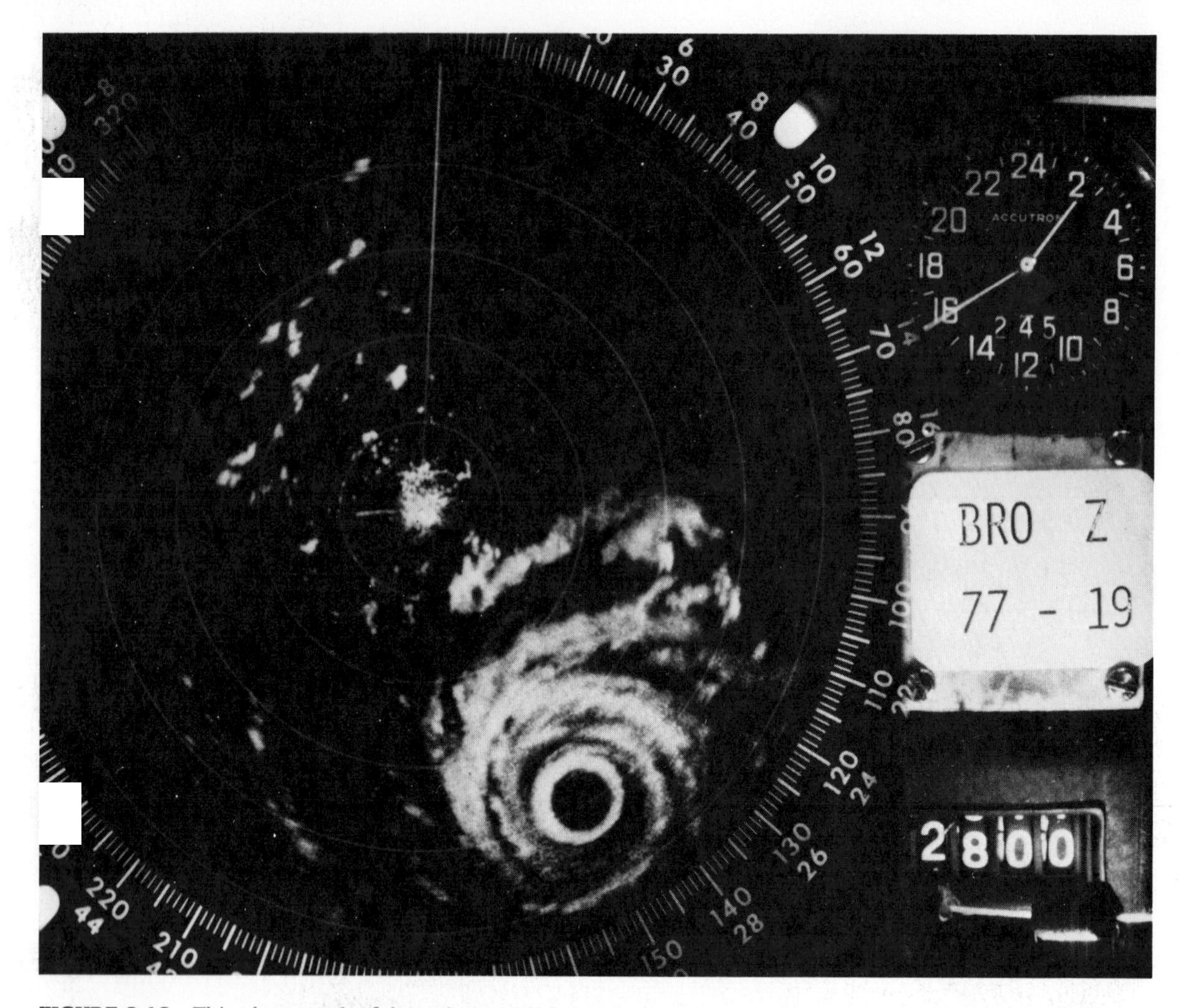

FIGURE 8-16 This photograph of the radar scope shows the eye and spiraling cloud bands of Hurricane Anita on September 2, 1977, as seen from Brownsville, Texas. Such radar systems located along the coastline are very useful in tracking approaching hurricanes. (Photo courtesy of Miles B. Lawrence, National Hurricane Center, Miami, Florida.)

Although the increase in ocean level associated with hurricanes is not as great in the Gulf of Mexico as in the Bay of Bengal, the ocean level may rise 4 m or more as a strong hurricane passes. The storm in Galveston that killed 6000 people in 1900 increased the ocean level by 4.4 m. This remained the United States record until 1969, when Hurricane Camille increased the ocean level 7 m at Pass Christian, Mississippi. In low-elevation coastal areas, destruction from hurricanes occurs because of rising ocean levels as well as because of the tremendous amounts of rainfall and the strong winds.

The generation of tornadoes is also a very damaging element of many hurricanes. Tornadoes occur most often in the leading half of a hurricane and are associated with large thunderstorms within the

TABLE 8-1

North Atlantic hurricane rainfall amounts and maximum recorded winds

Rainfall		
24-hour amounts (cm)	**Date**	**Location**
98	Sept. 5–6, 1950 (Easy)	Yankeetown, Fla.
97	Sept. 9–10, 1921[a]	Thrall, Tex.
81	Sept. 25, 1929	Rock Sound, Bahamas
77	Nov. 11, 1909	Silver Hill, Jamaica
76	Oct. 21, 1941	Trenton, Fla.
69	Aug. 19–20, 1969 (Camille)	Massies Mill, Va.
66	Sept. 16, 1936	Broome, Va.
61	Sept. 1, 1940	Ewan, N.J.
60	Aug. 8, 1940	Miller Island, La.
59	Oct. 9–10, 1924	New Smyrna, Fla.
Wind		
Wind speed (km/h)	**Hurricane and date**	**Observation site**
282	Janet, Sept, 27, 1955	Chetumal, Mex.
275	Camille, Aug. 17, 1969	Main Pass Block 92, La.
261	Oct. 18, 1944	Havana, Cuba
248	Sept. 17, 1947	Hillsboro Lighthouse, Fla.
245	Carla, Sept. 11, 1961	Port Lavaca, Tex.
230	Sept. 13, 1928	San Juan, P.R.

[a]Hurricanes designated only by dates occurred before it became customary to name them.

FIGURE 8-17 Hurricane Agnes caused extensive flooding along the Susquehanna River in Pennsylvania. A section of the bridge has been washed away, as shown here. (Photo courtesy of U.S. Coast Guard.)

spiraling rainbands. The strong damaging winds of the hurricane frequently cover the smaller tornado paths making it almost impossible to distinguish the damaging effects of each. Complete records of tornadoes associated with hurricanes are, therefore, not available, but it is known that hurricanes may produce large numbers of tornadoes. The record number of observed tornadoes was produced by Hurricane Beulah in September 1967, when it struck the Texas coast and produced 115 tornadoes.

HURRICANE TRACK PREDICTION METHODS

The agency responsible for predicting hurricanes in the United States is the National Hurricane Center located in Miami, Florida. Meteorologists there track and predict the future location of all hurricanes that affect the United States. No techniques are available for predicting the development of hurricanes—only their future position after they are observed on satellite photographs. The movement of most hurricanes is not very complicated and can be predicted very well by simply extrapolating their previous path of travel. In fact, the movement of seven out of ten hurricanes can be accurately predicted by simply noting their past travel history and projecting that into the future. The other three out of ten hurricanes, however, are much more difficult to understand.

The National Hurricane Center uses four different methods to predict hurricane movement. Only one of these is purely dynamic and is based on equations governing the condition and motion of the atmosphere. The other methods are statistical or combinations of statistical and climatological approaches. About 90% of the variation of a storm's movement is normally explained by the climatology plus persistence approach. In one method, statistics from the past history of the movement of all other hurricanes can be used to calculate the probability of a storm being in a certain position at a later time (Fig. 8-19). For a typical hurricane, there is a 45% probability that it will be in a small area 24 hours later and a 90% probability that it will be in a much larger area. The area of future location can be predicted more accurately for shorter times; the probability is described by cones, as in Figure 8-19.

FIGURE 8-18 The Richelieu Apartments on U.S. Highway 90 in Pass Christian, Mississippi, before and after Hurricane Camille, August 17, 1969. Twenty-five people decided to ignore the storm warnings and hold a hurricane party in this apartment building. Only two of them survived the storm, and they were severely injured as winds greater than 200 km/h and a storm surge of 7 m struck. (Photos by Chauncey T. Hinman.)

Another method is similar except that, instead of considering the climatology of all hurricanes, only those past hurricanes that have been located near the position of the current hurricane are used to determine the statistics.

A third forecasting method for the hurricane path consists of using these same parameters, climatology and persistence, in combination with cir-

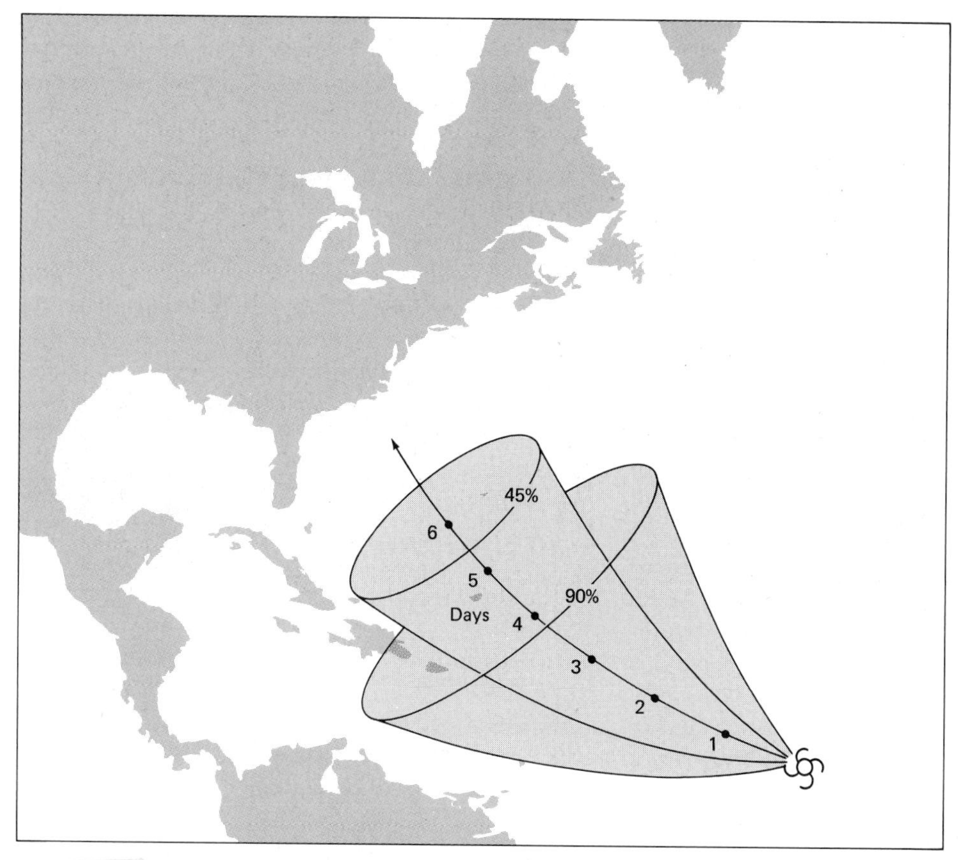

FIGURE 8-19 The typical path of hurricanes is toward the northeast, as the trade winds and other atmospheric winds affect the movement of the storm. A smaller, 45% probability cone describes the future position of the storm, while a much larger cone is required to describe the 90% probability of its future position.

culation features in the atmosphere. The specific atmospheric circulation may be important in the future motion of a hurricane since it has been observed, for example, that large midlatitude cyclones and anticyclones may cause a hurricane to take a route that avoids them.

The fourth method is the only one that is strictly dynamic. This procedure is based on equations and laws governing the structure and movement of the atmosphere. The equations are fed into a computer in order to calculate the mean winds from the surface to 18 km and thus to analyze the circulation of the winds and the projected center of the low pressure corresponding to the center of the hurricane.

No single one of these different methods consistently gives the best result when used to predict the movement of hurricanes. Landfall can be predicted to within about 180 km 24 hours in advance. However, a miss of 180 km frequently makes a tremendous difference in determining whether a given city will be subjected to the damaging winds and flooding associated with a severe hurricane.

IDENTIFICATION FROM SATELLITES

Satellites are very helpful in documenting the development and movement of hurricanes. In fact, today most hurricanes are first identified from satellite photographs. Once they are located, they can be studied further by the National Hurricane Center as specially equipped airplanes are flown through them to measure their pressure, temperature, and winds. The use of **remote sensing** methods for hurricane study are much less expensive and dangerous, however. Therefore, techniques have been investigated for using satellite photographs to gain information on the intensity of a particular hurricane. This is accomplished by comparing photographs of the current hurricane with similar features of past hurricanes that are related to the strength of the storm. For example, the **banding features** (characteristics of the rainbands) can be classified by their shape around the eye of the hurricane. A weak hurricane has a narrow band of clouds converging around and toward the eye, while a very intense hurricane has a wider band of clouds.

The appearance of the eye of the hurricane is also useful. If the hurricane does not have a well-defined eye, or clear area in the center of the storm, then the dense overcast at the hurricane's center gives some information. A central dense overcast that is very smooth and round in shape represents a much more intense hurricane than an irregularly shaped central core. A third factor that is used is the eye itself, if visible. If the cloud bands around it are broad, a large eye indicates that the hurricane is more intense.

These various factors have been put together into a classification system and a catalog of types of hurricanes has been generated (Fig. 8-20). Information on the banding features and characteristics of the eye of a new hurricane is extracted from satellite photographs, and that information is then compared to previously developed standards to arrive at a value on a scale of T1 to T8, where T8 represents the most intense hurricanes. This scale is further related to the expected pressure and wind speed associated with the hurricane. Thus particular satellite features are used to determine the expected highest wind speed and lowest pressure (Table 8-2). If the satellite features result in a low number on the scale, the storm is expected to have low wind speeds and higher pressure in the eye of the hurricane.

This method represents an interesting attempt to use remote sensing techniques to gain some quantitative information on hurricanes. The derived pressure and winds may not be entirely precise, but the method represents the beginning of new approaches that can be refined and perfected in the future to allow more detailed information to be gained from satellites.

HURRICANE SAFETY

Hurricanes are a major natural hazard along the Gulf and east coasts, partly because of human reactions to them. Hurricanes kill an average of 61 persons per year in the United States (based on data since 1936). However, a single storm in 1900 killed 6000 people in Galveston, Texas (Fig. 8-21). Hurricanes have a wide range of intensity. Weak hurricanes may strike a particular state for many years and people become accustomed to riding them out. In fact, hurricane parties are a common occurrence as a hurricane passes. This is great for weaker storms, but if the same behavior is carried over to a strong hurricane, it is an entirely different story.

General hurricane safety rules are suggested by the National Weather Service. A hurricane *watch* means a hurricane may threaten an area within 24 hours, while a hurricane *warning* means a hurricane is expected to strike the area within 24 hours. It is advisable to enter each hurricane season prepared. Check your supply of tools, boards, batteries, nonperishable foods, and other equipment you will need if a hurricane strikes your home. When your area is covered by a hurricane watch, continue normal activities, but stay tuned to radio or television for National Weather Service advisories.

When your area receives a hurricane warning, monitor the storm's position continuously through

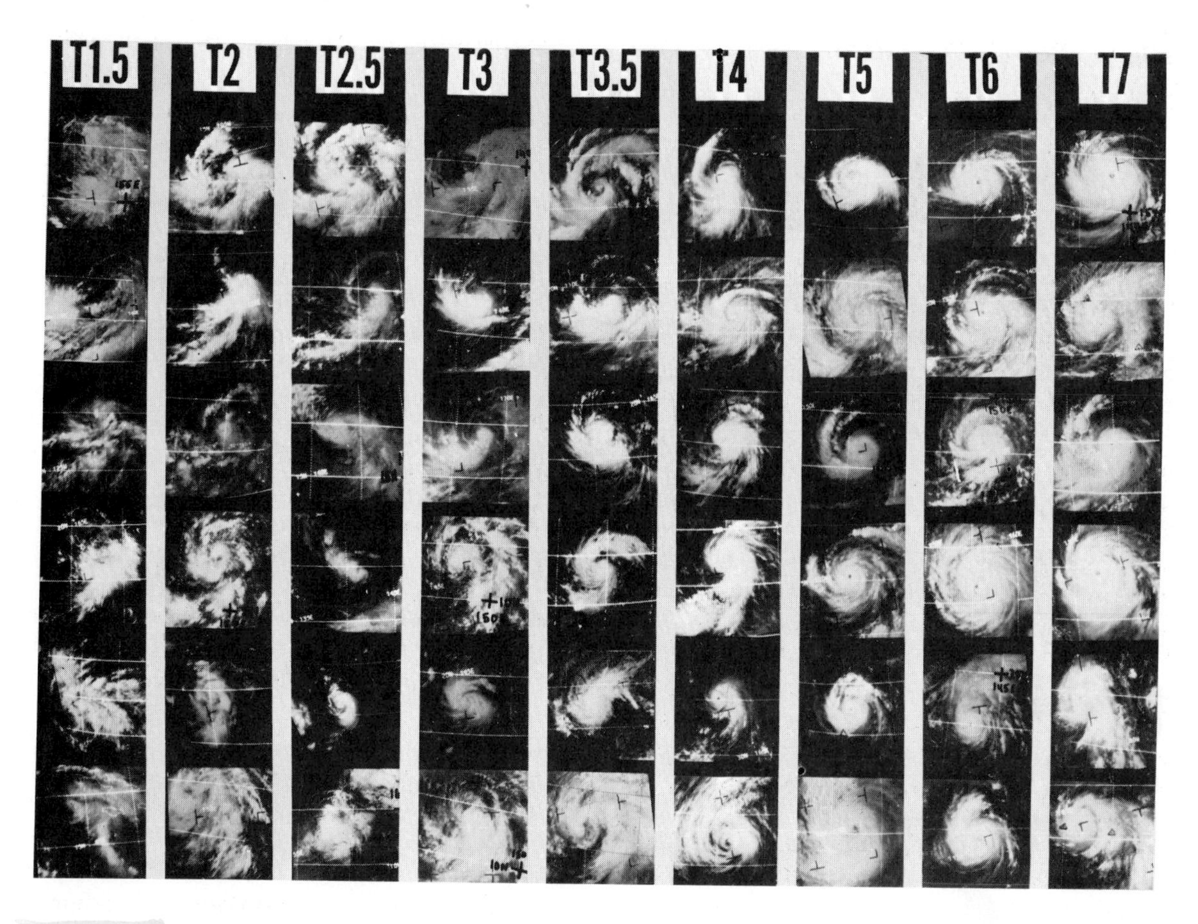

FIGURE 8-20 A catalog of satellite photographs of hurricanes can be compared against future satellite photographs of hurricanes to gain information on the winds and pressure within a storm. Numbers are assigned from T1 to T8 based on the banding features of cloud patterns and nature of the eye. (From Dvorak, U.S. Dept. of Commerce, NOAA TM NESS 36, 1973.)

Weather Service advisories with battery-powered equipment since a portable radio may become your only link with the outside world during the hurricane. Emergency cooking facilities and flashlights are essential if utilities are interrupted. Your automobile should be fully fueled. The windows of the house should be boarded up or protected with storm shutters. A supply of drinking water should be stored, since the city water supply may become contaminated or diminished by hurricane floods.

Leave low-lying areas when advised to do so. If you live in a mobile home, leave it for more substantial shelter. Mobile homes are extremely vulnerable to high winds. If your home is sturdy and at a safe elevation, remain indoors during the hurricane. Because hurricanes often cause severe flooding as they move inland, stay away from rivers and streams.

Tornadoes are often spawned by hurricanes and are among their lethal effects. When a hurricane

TABLE 8-2

The empirical relationship between the T number and maximum wind speed (MWS) and minimum sea-level pressure (MSLP)

T (Number)	MWS (km/h)	MSLP (Atlantic) (mb)	MSLP (NW Pacific) (mb)
1	46	1013	1007
1.5	46	1012	1006
2	56	1009	1003
2.5	65	1005	999
3	83	1000	994
3.5	102	994	988
4	120	987	981
4.5	143	979	973
5	167	970	964
5.5	189	960	954
6	213	948	942
6.5	235	935	929
7	259	921	915
7.5	287	906	900
8	315	890	884

Source: Vernon F. Dvorak, "A Technique for the Analysis and Forecasting of Tropical Cyclone Intensities from Satellite Pictures," NOAA TM NESS 36, 1973.

approaches, listen to radio or television for tornado warnings and seek shelter in the lowest and sturdiest part of the house in a location opposite from the approach direction of the tornado.

SUMMARY

Intense tropical storms occur over the oceans in many parts of the world. These are called by various names such as hurricanes, typhoons, and cyclones. Those affecting the United States are called hurricanes. Hurricanes originate over warm tropical oceans and pass through various stages of development. They originate as a tropical disturbance and become a tropical depression as one or more closed isobars develop in the beginning storm. They become tropical storms as the winds increase to speeds above 60 km/h, and they reach the hurricane stage as the winds exceed 120 km/h.

Hurricanes are smaller in size than midlatitude cyclones and have a different internal structure. They are characterized by descending air in the eye of the hurricane above the low surface pressure, with rising air currents in the cloudy rainbands surrounding the eye wall. Energy is provided for driving this large atmospheric vortex as water vapor condenses and releases its latent heat within the spiraling cloud layers around the eye wall. The band of strongest winds is just outside the eye of the hurricane and may contain wind speeds of 320 km/h, as in Hurricane Camille. The surface winds decrease quite rapidly outward from this band of intense wind, and wind speeds also decrease above the surface.

The average life of hurricanes is nine days and they occur most frequently in the autumn season. An average of two hurricanes per year strikes the United States. Hurricanes normally decay quite

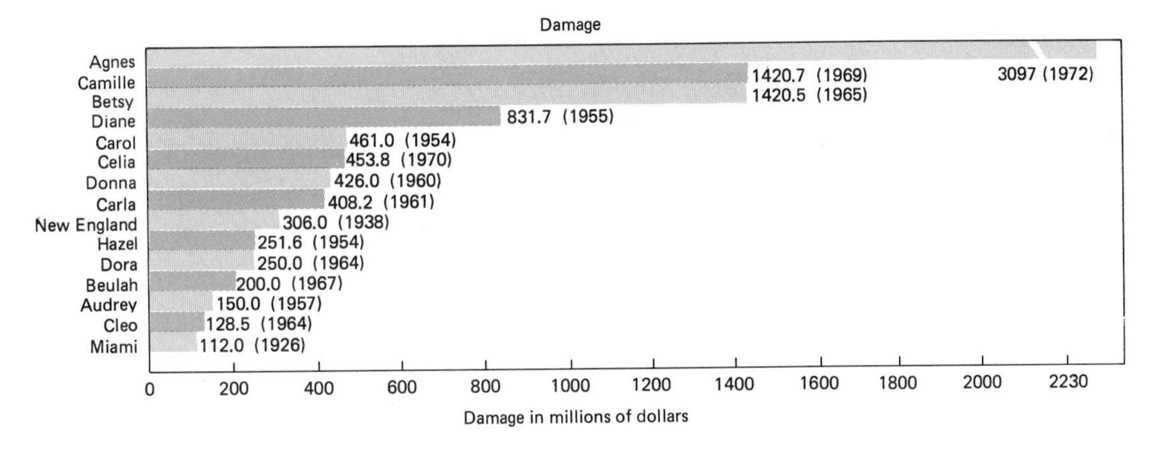

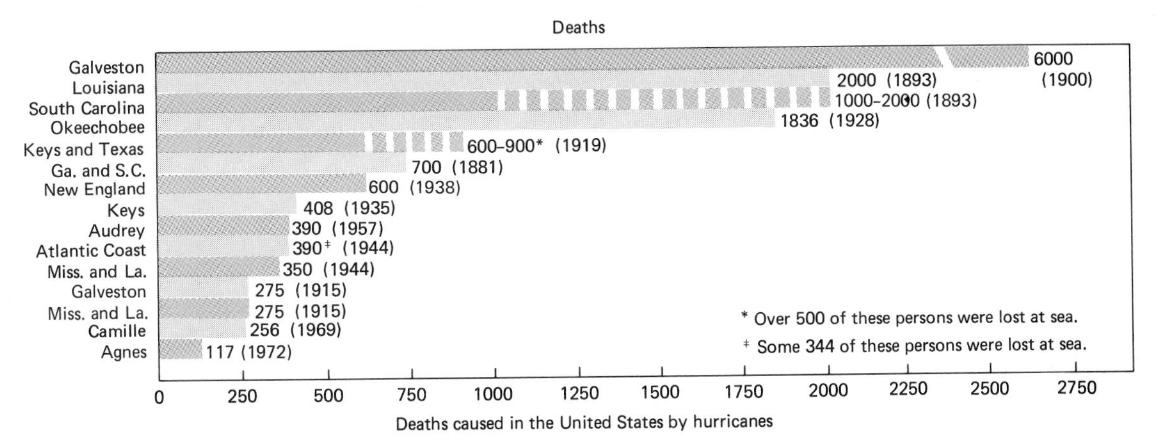

FIGURE 8-21 Damage and deaths caused by the major hurricanes in the United States. Hurricane Agnes tops the list for amount of damage, while the Galveston storm in 1900 caused the greatest number of deaths. (After *Weatherwise,* February 1971.)

rapidly as they strike land; however, it is possible for them to be modified into midlatitude cyclones. Destruction from hurricanes occurs because of strong winds and flooding from heavy precipitation and rising ocean levels. The ocean level rose 7 m as Hurricane Camille passed over Mississippi in 1969.

The National Hurricane Center tracks and predicts the movement of hurricanes after they are identified from satellite photographs or other means. Statistical and climatological forecasting techniques, in addition to a dynamic model, are used for predicting the future hurricane path. Satellites are used not only for identifying hurricanes but also to gain some quantitative information from the storms. The appearance of the eye of the hurricane and the nature of the rainbands surrounding the eye as observed on satellite photographs are compared with satellite photographs of past hurricanes in order to estimate the wind speed and pressure within the storm.

It is important to know that hurricanes occur with a wide range in intensity. The winds in many hurricanes are not strong enough to destroy buildings and cause other types of major damage. The stronger hurricanes, however, have winds strong enough to destroy buildings and cause major property damage. Sixty-one people, on the average, are killed each year by hurricanes in the United States. It is therefore important to be aware of the general characteristics of hurricanes and of safety precautions for residents and visitors to the Gulf and east coast states during the fall of the year.

STUDY AIDS

1. Describe the environment for hurricane development.

2. Make a list of some of the characteristics of hurricanes that are different from midlatitude cyclones.

3. Explain the warmer temperature and descending air in the eye of a hurricane.

4. Describe the stages in the development of a hurricane.

5. Describe and explain the typical movement of hurricanes.

6. Describe the changes in Hurricane Carla after landfall.

7. Why are greater wind speeds associated with the side of a hurricane to the right of its path of motion?

8. Explain the destructive components of the hurricane.

9. List and explain the different hurricane track prediction methods used by the National Hurricane Center.

10. Explain how satellites can be used to gain detailed estimates of wind and pressure in the storm.

11. List some hurricane safety rules that you consider to be important.

TERMINOLOGY EXERCISE

Use the glossary to reinforce your understanding of any of the following terms used in Chapter 8 that are unclear to you.

Easterly wave
Trade wind inversion
Typhoon
Hurricane
Tropical disturbance
Tropical depression
Tropical storm
Eye
Eye wall
Landfall
Swells
Remote sensing
Banding features

THOUGHT QUESTIONS

1. Explain how the wind directions and precipitation amounts would change as a hurricane passed over you, traveling toward the north.

2. Explain why the eastern coasts of some continents such as South America experience fewer hurricanes than others.

3. Explain the general appearance of a satellite photograph of a very strong hurricane.

4. How do you expect the amount of damage from hurricanes to change in the future? Can you offer suggestions for changing this?

5. Would you expect the shape of the coastline to affect the threat of flooding associated with an approaching hurricane? Explain your answer.

CHAPTER NINE
SEVERE THUNDERSTORMS

ORDINARY THUNDERSTORMS

It is not unusual for the weather to change rapidly from clear sunny skies to dark clouds with lightning and thunder. Such a change accompanies the development of thunderstorms. Thunderstorms grow when the atmosphere becomes sufficiently unstable because of localized surface heating or other causes. A rising bubble of air that becomes buoyant is cooled by adiabatic expansion until the air eventually reaches saturation and causes a cumulus cloud to form. If the atmosphere is unstable, the cumulus cloud grows vertically into a thunderstorm (cumulonimbus cloud).

Summer thunderstorms may be called **air mass thunderstorms** or **convective thunderstorms** since they frequently occur in warm air masses as the earth is heated from sunlight.

Detailed studies of air mass thunderstorms during the 1940s showed typical life cycles (Fig. 9-1). The initial cumulus stage begins as humid air within the warm air mass rises and cools to form single or clusters of puffy white cumulus clouds. Rising bubbles of warm air become visible and are seen as cumulus clouds because water vapor condenses as the rising air cools. The changing nature of these clouds is evident to anyone who has watched such clouds for only a few minutes. Some clouds are evaporated by the drier surrounding air and others build to the next stage, the mature thunderstorm.

The mature thunderstorm grows vertically to heights of 20 km in extreme cases and typically grows to 10 km, or more than 30,000 ft. As rain falls through the leading portion of a mature thunderstorm it creates downdrafts in this part of the storm while other regions of the thunderstorm contain updrafts. The thunderstorm is most intense

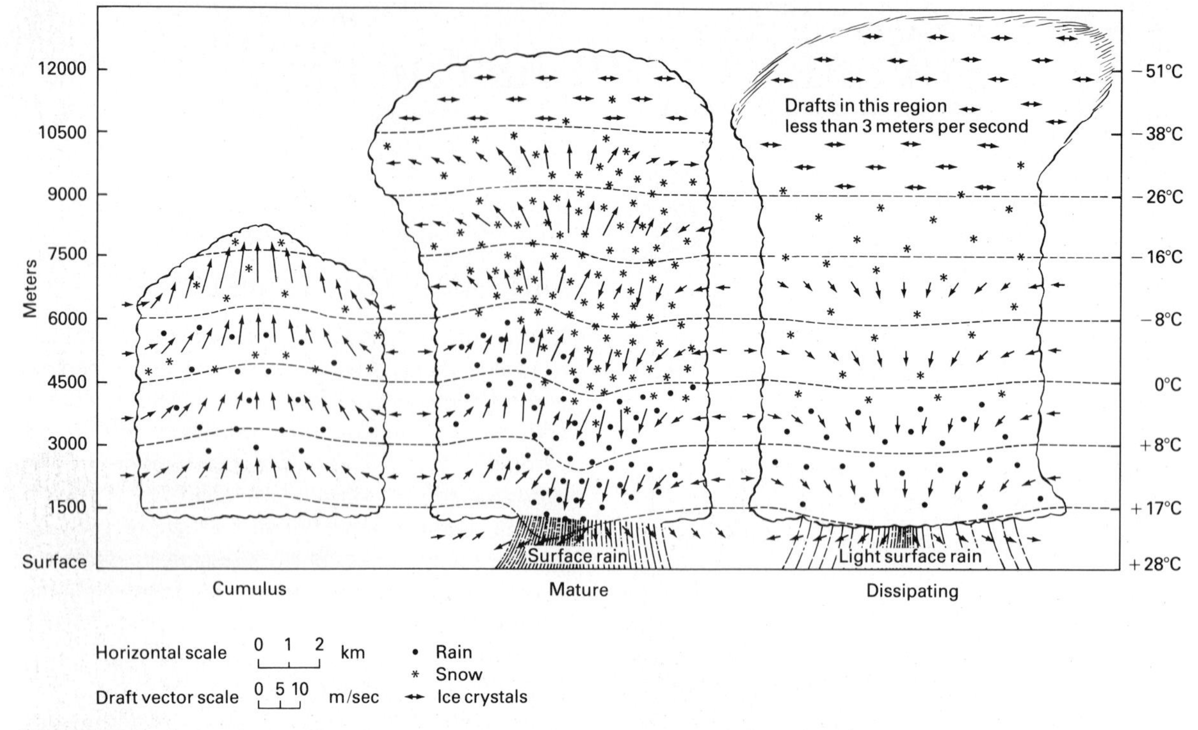

FIGURE 9-1 Stages in the life cycle of an ordinary thunderstorm. (From H. Byers and R. Braham, *The Thunderstorm,* U.S. Government Printing Office, 1949.)

during the mature stage and may develop the familiar anvil shape as the top of the storm grows into strong westerly winds near the tropopause. The mature air mass thunderstorm contains rain, thunder, and lightning, and produces wind gusts at the ground.

After about 30 minutes the mature thunderstorm typically begins to decrease in intensity and enters the dissipating stage. Air currents within the thunderstorm are converted to downdrafts. This shuts off the supply of warm moist air from the lower atmosphere and kills the storm. The typical air mass thunderstorm completes its life cycle in about an hour.

Air mass thunderstorms are more frequent in late afternoon hours because of the influence of surface heating in producing convection currents in the atmosphere. Thunderstorms may also form along cold fronts and as air flows over topographic barriers. Such forced lifting of warm air may produce thunderstorms with life cycles similar to that just described.

DEVELOPMENT OF SEVERE THUNDERSTORMS

Most thunderstorms are ordinary convective thunderstorms, but occasionally they grow larger than average and develop a different internal structure, becoming **severe thunderstorms**. These usually penetrate the tropopause and may grow to 20 km. Figure 9-2 shows the development of a severe thunderstorm on May 4, 1977.

Severe thunderstorms are defined as thunderstorms with frequent lightning accompanied by locally **damaging winds** or hail that is 2 cm in diameter or larger. Damaging winds are defined as sustained or gusty surface winds of 97 km/h or

FIGURE 9-2 The development of a severe thunderstorm on May 4, 1977, containing rain, hail, and a tornado south of Lawrence, Kansas. This thunderstorm grew rapidly through the troposphere spreading out into a characteristic anvil shape as it penetrated the tropopause. The view is toward the south showing the northern and western edges of the thunderstorm. (Photos by author.)

more. The various specific atmospheric factors that contribute to severe thunderstorm development will be considered in more detail.

Midlatitude Cyclone and Cold Front Severe thunderstorms are usually associated with cold fronts accompanying midlatitude cyclones. A cold front moving through a region forces warm air to rise just as surface heating causes ascending air. In addition, converging surface winds occur in various parts of a midlatitude cyclone (Fig. 9-3). The winds south of a midlatitude cyclone are ordinarily from the southwest while the winds to the west of the cyclone center in the region behind the cold front come from the northwest. This results in converging air along the cold front forcing the air above the front to rise. Vertical accelerations are important in contributing to thunderstorm development. A line of developing thunderstorms is not always located at the cold front but frequently occurs 150 km or more ahead of the advancing cold front. There are several reasons for this. Sometimes a boundary exists there between cT (continental tropical) air from the desert Southwest and mT (maritime tropical)

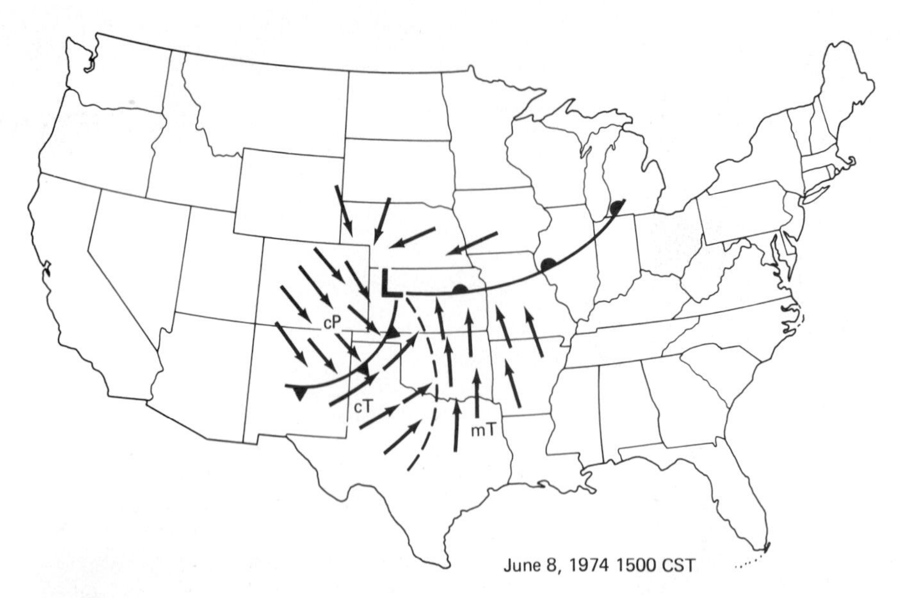

FIGURE 9-3 Surface convergence of air contributes to the development of severe thunderstorms. This typical synoptic map for severe weather is for June 8, 1974, at 1500 CST. Two areas of surface convergence occur, along the cold front and along the dry line separating continental tropical and maritime tropical air masses. Severe weather frequently occurs along the boundary between the continental tropical and maritime tropical air masses as surface convergence of air contributes to their development.

air from the Gulf of Mexico. If such an air mass boundary does exist it is called a **dry line** and represents an instability zone since the warm dry air is heavier than the warm humid air. Thunderstorms frequently develop along the dry line rather than along the cold front. (See the satellite photographs in Fig. 9-4.)

Another process that occurs ahead of the cold front and contributes to thunderstorm development there rather than near the front is similar to cloud formation on the leeward side of mountains. Air flowing over the mountain oscillates up and down after passing over the mountain. There are also indications that waves develop prior to flow over the mountains. Similarly, air flowing over a cold front may develop a wave in advance of the front (Fig. 9-5). Such waves may develop clouds in the ascending part of the wave pattern. The air ahead of the cold front may be so unstable because of its heat and water content that such a **prefrontal wave** may trigger the development of a line of severe thunderstorms.

Unstable Atmosphere Severe thunderstorm activity is more likely if the atmosphere is unstable, since an unstable atmosphere has a greater tendency for vertical currents to develop. Severe thunderstorms will form if absolute instability exists in a humid atmosphere. In this case the measured lapse rate of temperature is greater than either the dry adiabatic or the saturated lapse rates. The atmosphere is then unstable for either saturated or unsaturated air.

Thunderstorms usually form in a conditionally unstable atmosphere when the measured lapse rate is greater than the saturated lapse rate and

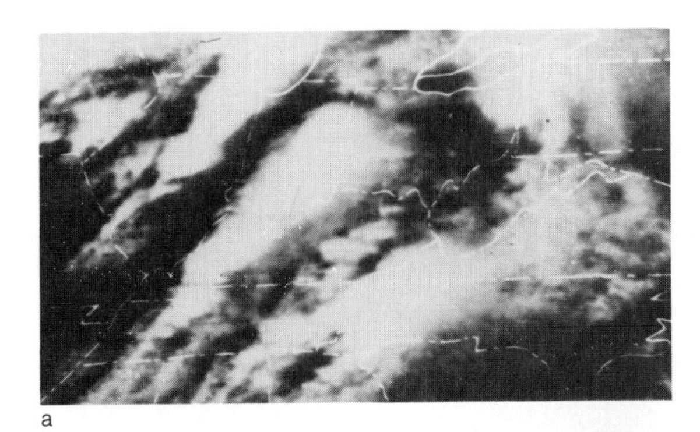
a

b

FIGURE 9-4 Advancing lines of thunderstorms (squall lines) can be seen in these satellite photographs. (a) Three separate squall lines extend from southwest to northeast in this photograph taken at 14:48 CDT on April 3, 1974. (b) As these squall lines moved eastward to the locations shown in this photograph, taken about 2 hours later (16:50 CDT), numerous tornadoes were spawned. (NOAA photos courtesy of National Environmental Satellite Center.)

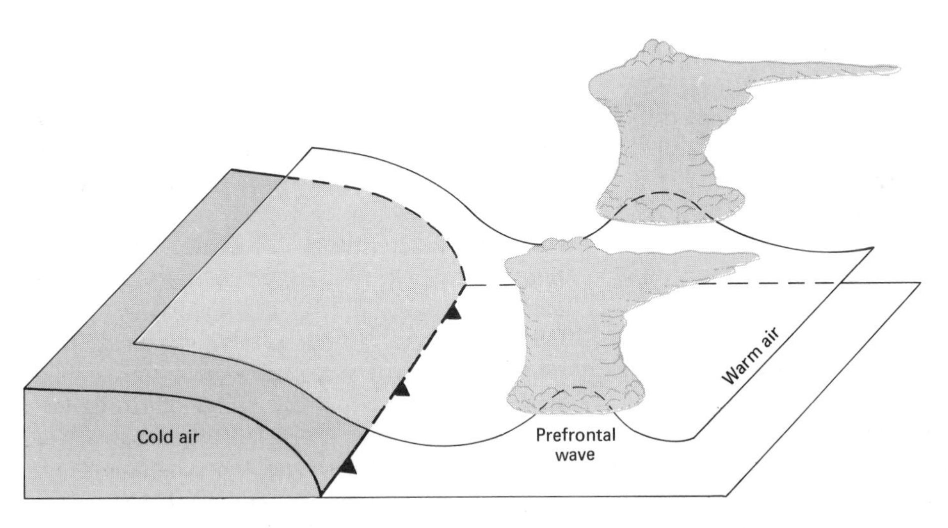

FIGURE 9-5 Prefrontal waves contribute to the development of severe thunderstorms. Such waves may spread outward from the advancing cold air, generating squall lines within the warm air mass. These waves travel in much the same way that waves move across a water surface.

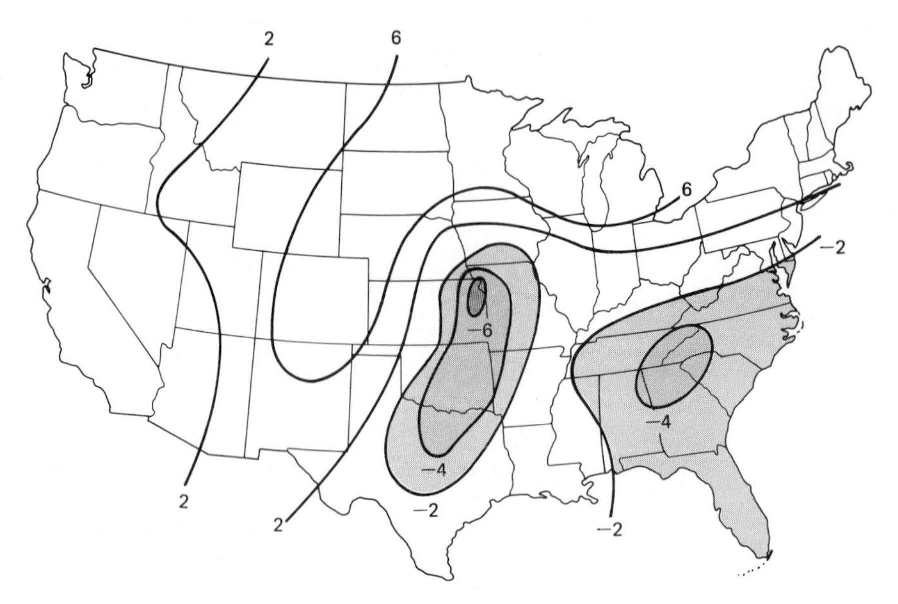

FIGURE 9-6 Atmospheric stability on June 8, 1966, at 7:00 P.M. CDT. The lifted index values shown are most negative in the central United States as tornadoes occurred in Topeka, Kansas, and eastern Oklahoma. The lifted index is a means of placing a number on the stability of the atmosphere, with more negative values indicating a greater chance of severe weather development.

less than the dry adiabatic lapse rate. With this atmospheric condition, slight increases in heating of the lower atmosphere, or rising air from any cause, may result in saturation of the air, leading to unstable conditions.

The lifted index is a useful measure of the stability of the atmosphere when severe thunderstorm development is possible. As previously explained, the lifted index is determined by considering the average mixing ratio of the lower atmosphere and assuming that a parcel of air having the forecasted maximum temperature for the day is lifted without mixing with the surrounding air. The resulting temperature of this parcel at 500 mb (5.5 km) is compared with the measured temperature there. The lifted index is the difference in temperature after the air parcel temperature is subtracted from the measured temperature at 500 mb. A negative lifted index indicates an unstable atmosphere since the temperature of the lifted parcel is warmer, making it buoyant. The more negative the number the more unstable the atmosphere becomes.

The lifted index becomes more negative with additional heating of the lower atmosphere or cooling of the upper atmosphere. Lifted index values are shown in Figure 9-6 for June 8, 1966, and correspond to a Topeka tornado. Eastern Kansas had a lifted index of -6, and negative values also occurred in Oklahoma. The two tornadoes on this day occurred in Topeka and central Oklahoma.

The importance of surface heating in creating an unstable atmosphere is shown in Figure 9-7. If the measured lapse rate is less than the dry adiabatic lapse rate in the lower atmosphere, and greater than the saturated lapse rate higher in the atmosphere, then surface heating may be very important. As the surface temperature increases, heat is transferred to the lower atmosphere causing the measured lapse rate to become greater than the

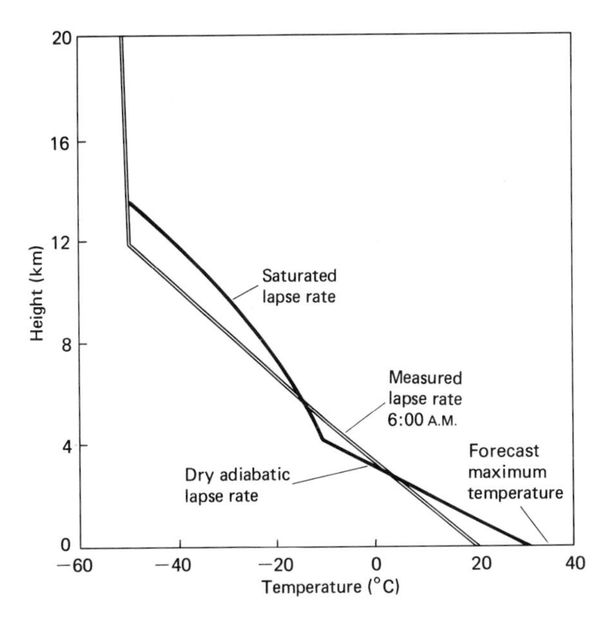

FIGURE 9-7 Surface heating is frequently a factor in severe thunderstorm development. The measured lapse rate shown for 6:00 A.M. becomes greater near the surface as the lower atmosphere warms. As it approaches the maximum temperature for the day, the lapse rate reaches its greatest value and is most likely to become unstable, allowing vertical motion and the development of severe thunderstorms. If surface heating is sufficient to cause the measured lapse rate to be greater than the dry adiabatic lapse rate in the lowest 4 km, the atmosphere becomes unstable to greater heights, as rising bubbles of air follow the saturated lapse rate above this level.

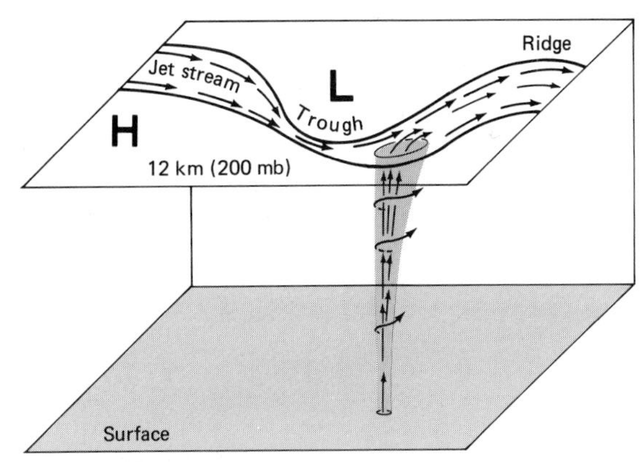

FIGURE 9-8 The jet stream contributes to severe thunderstorms by providing a favorable environment for updrafts beneath outflow regions that correspond to diffluence in the jet stream. The high wind speeds and cyclonic curvature of the jet stream also impart energy to severe thunderstorms.

dry adiabatic lapse rate. When this occurs the atmosphere is unstable and allows vertical growth in clouds and thunderstorm development. For this reason the time of maximum thunderstorm activity is in the late afternoon hours from 4 to 6 P.M.

Jet Stream and Upper Air Divergence Higher wind velocities at the jet stream level mean that greater quantities of energy are available for transfer to midlatitude cyclones, then to smaller circulations, such as thunderstorms, and finally to the intense circulation of a tornado. The energy within a tornado must have a source and this can be traced ultimately to the heating caused by the sun. Most of the energy is transferred to the tornado from the larger circulation of the atmosphere rather than directly from solar heating in spite of the role of solar heating in thunderstorm generation. Therefore, certain regions within the larger circulation patterns are more likely to contribute to tornado development.

Tornadoes are more likely in the region from the trough to the ridge of the jet stream. The development of severe weather usually occurs just downstream from the trough of the jet stream (Fig. 9-8). One of the reasons for this is the divergence of the wind stream as it completes its cyclonic turn around the greater pressure gradient associated with most low-pressure centers in the upper atmosphere. The pressure gradient is ordinarily much less around the larger high-pressure system

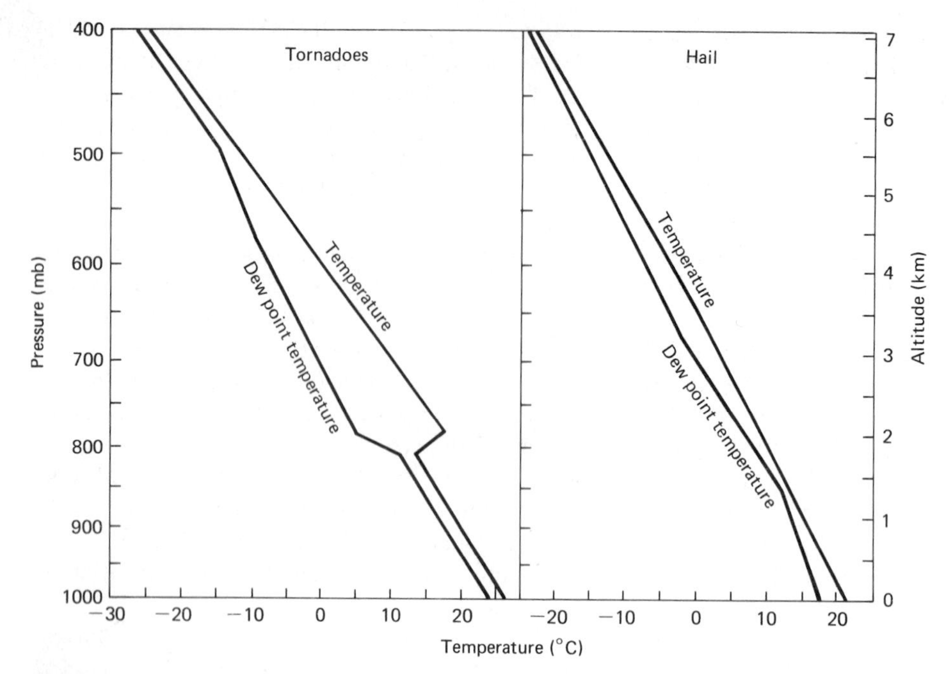

FIGURE 9-9 Vertical temperature profiles near severe thunderstorms show differences between the atmospheric environment when tornadoes occur as compared to when hail falls. The tornado-producing environment normally contains an upper air inversion beginning at about 1.5 km. Such an inversion is effective in decreasing small cumulus cloud activity and allows the atmosphere to become more unstable until a single thunderstorm breaks through the inversion and grows into a much larger storm. (From Fawbush and Miller, Meteorol. Monogr., Am. Meteorol. Soc., 1963.)

and therefore the upper air streams spread apart or diverge from the trough to the ridge.

Upper air divergence contributes to accelerations of vertical currents beneath the divergence. The divergence and outflow of air aloft is similar to a vacuum pump since it creates a deficiency of air. When this deficiency coincides with the top of a developing thunderstorm, the updrafts in it are intensified, thus improving the chances that the thunderstorm will reach the severe stage.

The circulation in the trough of the jet stream also contributes to severe thunderstorm development. As shown in Figure 7-8, if a large disk were placed horizontally in the trough of the jet stream the existing cyclonic vorticity would cause it to rotate cyclonically. Some of the cyclonic vorticity in the trough of a jet stream is transmitted to a midlatitude cyclone or thunderstorm located within it.

Upper Air Inversion An atmospheric inversion layer is very stable with little vertical motion. However, the presence of such a layer contributes to the development of a few large thunderstorms rather than many small ones. This occurs because the atmospheric layer beneath the inversion continues to heat and becomes more unstable until a cumulus cloud can penetrate the inversion. The cumulus cloud can then grow vertically while the smaller clouds are held below the inversion layer. Fawbush and Miller analyzed 75 cases of tornado activity and found great similarity in the atmospheric temperature profiles that exist when tornadoes occur (Fig. 9-9). A pronounced inversion exists at 800 mb (2 km) and the air below is almost saturated while that above is much drier. When this was compared with 68 cases of thunderstorms containing hail greater than 1.3 cm in diameter large differences were found to exist. The upper air inversion was absent for hail-producing thunderstorms. Thus the nature of the lapse rate may determine the type of severe weather associated with a thunderstorm.

A stable inversion layer has the effect of dampening initial thunderstorm development until the atmosphere is unstable enough to allow a cloud to penetrate the inversion layer. When this happens the thunderstorm is more likely to be larger and produce tornadoes since it is able to tap the humid air that has become warmer while it was located below the inversion layer. Thus, an inversion has the effect of dampening out all small thunderstorms until one is able to break through and grow into a **supercell** thunderstorm. By this limiting process the upper air inversion intensifies the development of tornado-producing thunderstorms.

High Dew Point Temperature and Moisture Tongue It is ordinarily very humid when tornadoes occur, with a high dew point temperature greater than 10°C. Sometimes a **funnel cloud** (circulation that extends downward from the cloud base like a tornado) occurs when the energy of the thunderstorm is not great enough to generate a tornado on the ground. This frequently happens when the dew point temperature (temperature where the air becomes saturated) is less than 5°C and emphasizes the importance of humidity in the air.

In the central United States air frequently comes from the Gulf of Mexico in the form of a wedge of humid air moving northward called a **moisture tongue** (Fig. 9-10). The warm maritime tropical air moves northward to the east of a midlatitude cyclone and may also be influenced by a large subtropical high-pressure area in the Atlantic. The dew point temperature within the moisture tongue is frequently high.

The moisture tongue occurs in the warm air mass ahead of a cold front. Continental polar air is behind the cold front as it advances from the northwest or west. Continental tropical air that is just as hot as the maritime tropical air but is much less humid frequently moves into the central United States from the desert Southwest and Mexico. The boundary between the cT and mT air masses forms the dry line.

The air is very hot on both sides of the dry line but the humidity of the air coming from the southwest is much less. This is an instability line since water vapor in the atmosphere decreases the weight of the air per unit volume; the molecular weight of water vapor is 18 and the molecular weight of the combined constituents of dry air is 28.9. The hot dry air coming from the southwest has the same effect as a cold front since the air is denser behind the dry line. Severe thunderstorms frequently occur along the dry line since it is an instability line.

A **low-level jet** frequently occurs along with the moisture tongue. This is a higher velocity stream of air flowing northward around and slightly toward the midlatitude cyclone in the central United States. The low-level jet occurs at about 800 mb (2 km) and has much lower wind speeds than the polar jet stream, but winds in the low-level jet are stronger than in the surrounding air. The low-level jet causes large wind shear between it and the polar jet stream above and contributes to the inflow of humid air in the leading edge of a severe thunderstorm.

Proper Wind Profile The wind profile seems to be important in developing severe thunderstorms. It is the wind relative to a moving thunderstorm that is affecting the storm. A thunderstorm that is being pushed along by 200 km/h winds typically travels at half the rate of these upper-level winds; thus, its ground speed is 100 km/h. If the thunderstorm were moving along at 100 km/h from the west and the surface winds were 20 km/h from the west, the thunderstorm would still be encountering air in front of it at the rate of 80 km/h as it was pushed through the air at low levels. The jet stream or upper-level winds ordinarily push the thunderstorm along with an average ground speed of 50

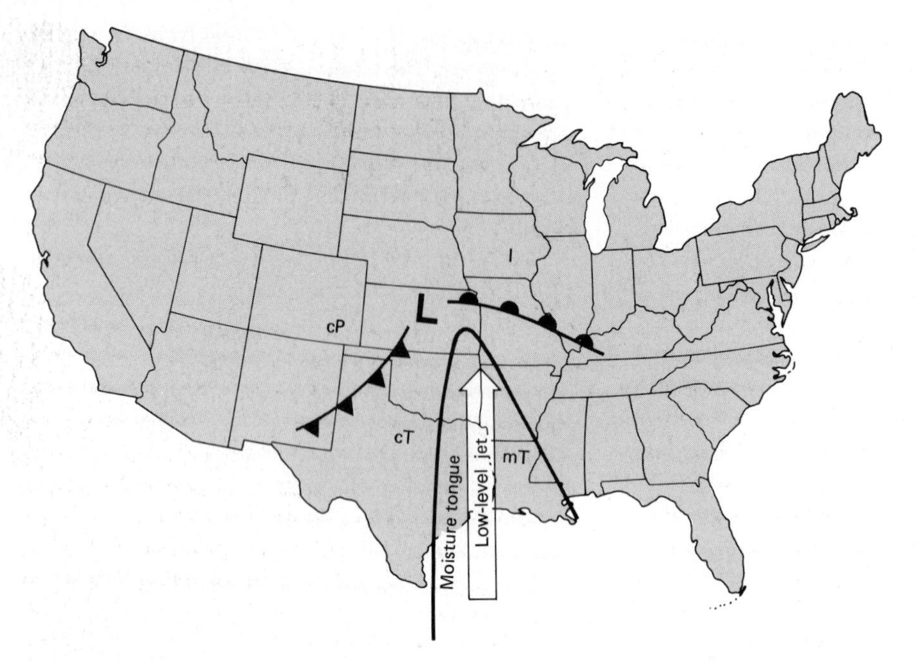

FIGURE 9-10 The moisture tongue and low-level jet develop as warm humid air is pulled northward by the low-pressure center at the surface of the earth. Water vapor decreases the density of the air making it more buoyant and easier for it to develop into a thunderstorm. The dry line, the boundary between the moisture tongue and the continental tropical air, is frequently an instability line where severe thunderstorms develop.

km/h. This forward speed generates a wind at low levels of the thunderstorm from the front side that feeds warm humid air into the lower front part of the thunderstorm, helping to generate a severe storm (Fig. 9-11).

One of the amazing things about thunderstorms is that they can develop vertically into the jet stream, which may have wind speeds of 200 km/h, and persist there in an almost vertical position rather than being blown off at the top or tilted over at an angle. The most logical explanation for this observed thunderstorm characteristic is that the storm has sufficient internal flow structure to compensate for the high-speed external winds. Severe thunderstorms are observed to thrive in strong vertical wind shear environments since they are able to tap the energy of opposing environmental winds and transfer it into an internal structure that allows the thunderstorm to become severe and generate tornadoes, lightning, and hail.

The environmental wind shear has been used as the basis for an index for forecasting severe thunderstorms. This method, developed by the author, considers the number of layers in the atmosphere that oppose the winds below the inversion layer (surface to 850 mb). If the winds striking the lower front of a moving thunderstorm are 50 km/h, for example, the number of layers (using thicknesses of 150 mb) above the inversion with winds of similar magnitude, but from the opposite direction, gives the **wind shear index.** If more layers above are opposing the low-level winds, then a better atmospheric condition exists for the development of severe thunderstorms. The geographical areas

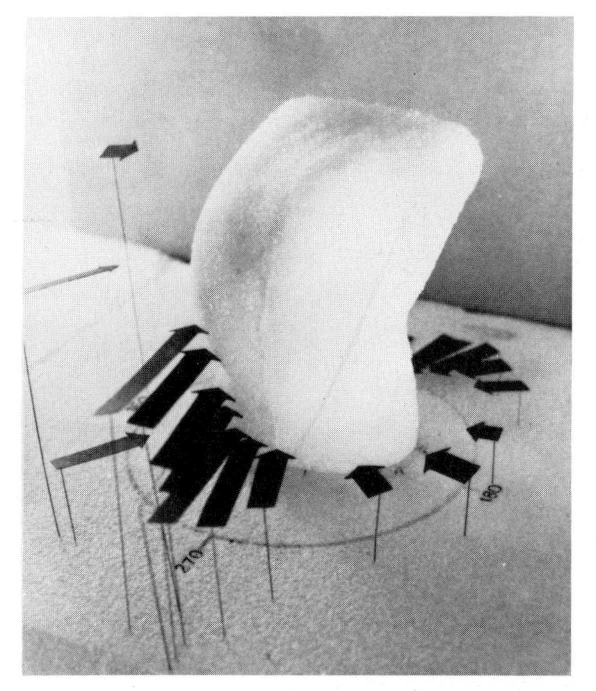

FIGURE 9-11 Wind environment relative to a thunderstorm on April 11, 1965. The length of the arrows is proportional to the relative winds at various altitudes. The illustration on the left shows the west side (270°) of the thunderstorm as the winds above 850 mb push the thunderstorm along, creating relative winds in front of the storm at low levels, as seen in the illustration on the right. (Photos by author.)

with the best wind shear environments can then be used to locate areas where severe weather is likely.

Squall Lines A seventh factor that is important in severe thunderstorm development is squall lines. Severe thunderstorms are frequently associated with a **squall line** (a line of moving thunderstorms). If a line of large thunderstorms occurs along the dry line, for example, the outflow from one storm may help feed air into the next one, allowing each to grow larger than would be possible if it were an isolated thunderstorm. There are certain preferred regions within a squall line where thunderstorms are more likely to reach the severe thunderstorm stage (Fig. 9-12). These are the southernmost cell (individual thunderstorm) of a squall line; the central cell if there is a bow in the squall line; near the center of a low pressure; or where a squall line intersects a cold front, a warm front, or a dry line. Thus severe weather is frequently associated with squall line development and preferred areas exist in an advancing squall line where thunderstorms are more likely to be severe.

SEVERE THUNDERSTORM FORECASTS

The National Severe Storms Forecast Center in Kansas City is responsible for forecasting severe weather for the entire United States. In forecasting severe weather certain characteristics of the atmosphere are used to determine where severe weather is most likely. If the location of the region

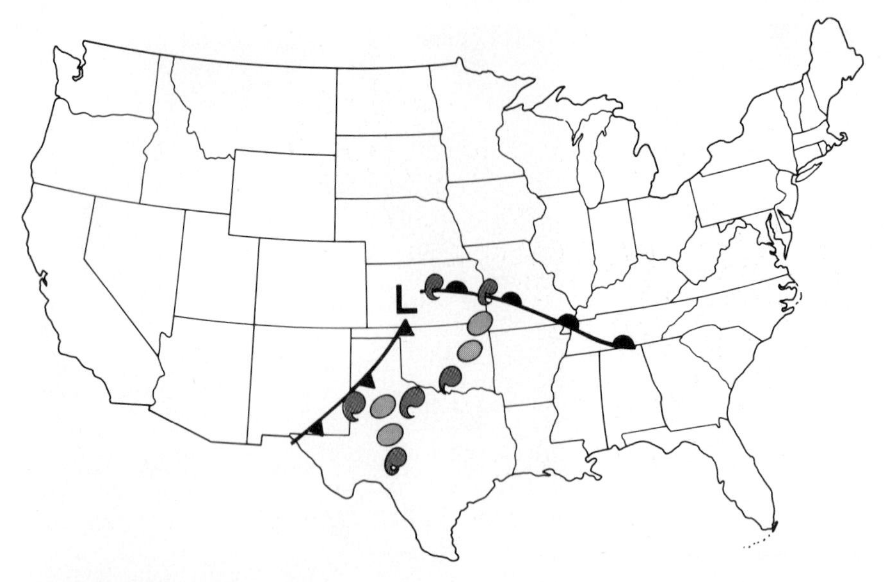

FIGURE 9-12 Thunderstorms within a squall line are more likely to be severe in particular locations, as shown by the hook-shaped radar echoes. The southernmost cell (individual thunderstorm) of a squall line is more likely to be severe, as are the cell near a bow of a curving squall line; cells within the squall line that intersect a cold front, warm front, or dry line; and cells near the center of a low-pressure area. This gives six different positions within squall lines where severe thunderstorms are likely.

of upper air divergence is known, as well as the location and movement of the other six factors that contribute to severe thunderstorm generation, then the region of probable thunderstorm activity can be isolated. The warm humid air in advance of a cold front associated with a midlatitude cyclone is an area of preferred activity. The lifted index is more likely to be more unstable farther southward in the moisture tongue. For example, the lifted index may be −12 in the southern United States and decrease northward until it becomes positive north of the midlatitude cyclone or behind the cold front. However, the location of most probable severe thunderstorm activity is determined by a combination of all the atmospheric factors rather than by a consideration of the place where one factor is the greatest. The lifted index and the moisture tongue are important if consideration is also given to the location of upper air inversions and the location where divergence within the jet stream intersects the moisture tongue in advance of a cold front. The area where all the appropriate atmospheric conditions coincide is the most probable area for severe weather activity (Fig. 9-13).

Normally the forecast area for severe weather is specified as 120 km on either side of a line between two identifiable points, ordinarily cities. The average size of a tornado watch area is 225 km by 280 km and covers an area of about 63,000 km^2. Tornado watches are issued for only those areas where proper atmospheric conditions exist for severe thunderstorm development. The specific location of squall line development cannot be forecast, but can be followed on the radar screen after development. Normally movement of an individual thunderstorm or a squall line is toward the northeast within the warm air sector of the midlatitude cyclone.

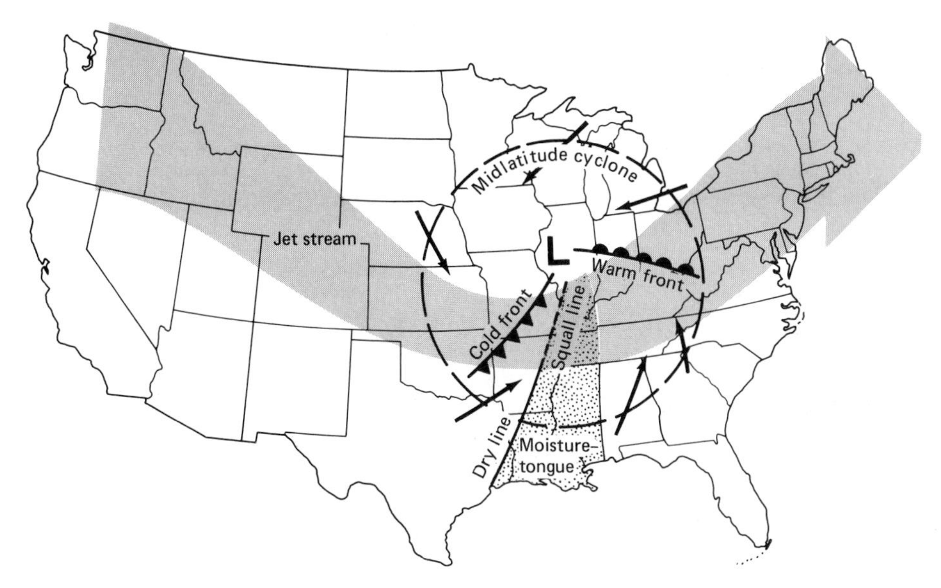

FIGURE 9-13 Various atmospheric factors combine to allow large thunderstorms to form and develop tornadoes. The most likely location of severe weather is the area ahead of the cold front where the moisture tongue is located beneath the jet stream and where a squall line is in progress.

The specific shape of the radar echo reveals information on tornado activity since the thunderstorm echo changes into a hook shape (Fig. 2-20) as a tornado is generated. After a severe thunderstorm has developed, its future movement is predictable, but forecasting its development several hours in advance is possible only by considering atmospheric conditions that are appropriate for developing severe weather. The severe weather forecast covers a large area and is not aimed at providing information on individual thunderstorms.

SEVERE THUNDERSTORM STRUCTURE

Thunderstorms occur as a single isolated thunderstorm, squall line, or **multicell thunderstorm.** Each of these three types can produce severe weather including hail and tornadoes. The actual internal structure of the severe cell may not be very different for the separate types, but their growth mechanisms are much different. New cells develop and thrive in advance (downwind) of a squall line (Fig. 9-14) while the new cells usually develop on the upwind side of a multicell thunderstorm. A strong squall line generates a well-developed **gust front** from the combined downdrafts of several thunderstorms. This is helpful in initiating the growth of new cells ahead of the squall line. The multicell thunderstorm provides a different environment for advancing the growth of new cells by blocking the strong upper atmospheric winds and forcing them around the sides of the large storm. This in turn provides a favorable growth region to the west or upwind from the mature multicell thunderstorm because the blocking action combines with the major updraft location within the western part of the

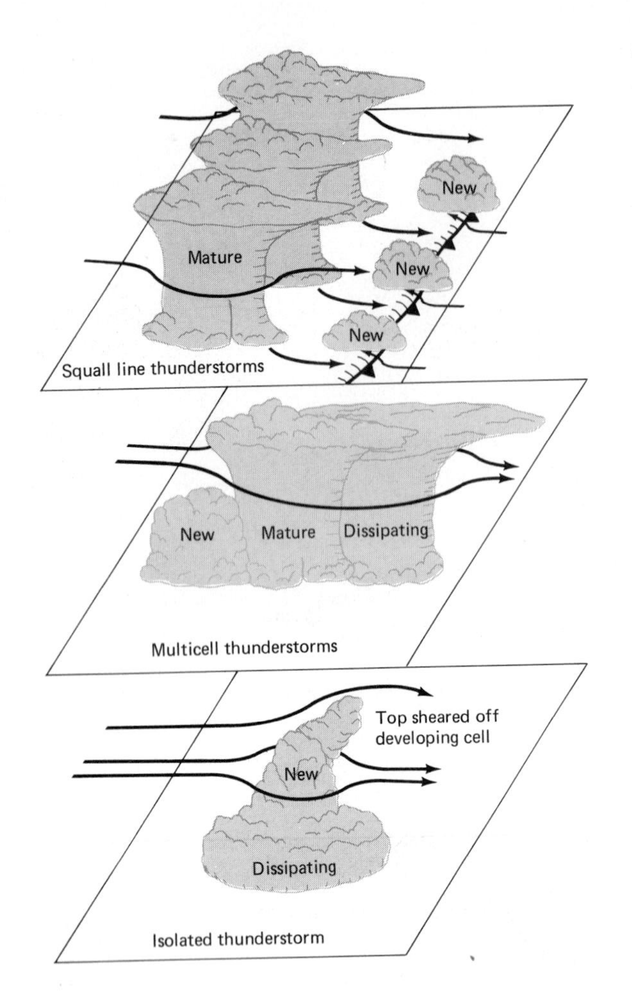

FIGURE 9-14 Severe thunderstorms can occur within squall lines, as multicell thunderstorms, or as isolated supercell thunderstorms. New cells most frequently develop downstream from the upper atmospheric winds and in advance of an older squall line. The outflow of cold air from thunderstorms within the squall line spreads outward at the ground as a miniature cold front (gust front) and helps initiate new thunderstorm activity. Thunderstorms tend to develop on the opposite side of a group of multicell thunderstorms. The blockage of upper atmospheric winds as well as the concentration of updrafts within the upwind side of the mature cell contributes to the development of thunderstorm activity. Isolated supercell thunderstorms frequently develop after several unsuccessful trials. As new cells grow into the stronger winds above, their tops are sheared off, causing dissipation and leaving the environment a little moister at this level. The next new developing cell within this location may then be able to penetrate higher in the atmosphere and develop the size and internal structure to withstand the strong upper atmospheric winds.

mature cell. The single isolated thunderstorm that reaches the supercell stage may represent the second or third bubble of unstable air that has ascended because of buoyancy, with each new bubble penetrating to a greater height than the previous one. Thunderstorm growth proceeds very irregularly, with oscillations between growth and decay until a thunderstorm reaches a stage where the internal structure approaches a steady state. A steady state thunderstorm can last for several hours, producing severe weather all the while.

The internal structure of large steady state thunderstorms has been of interest for a long time. As early as 1884 Moller suggested a model for the winds beneath a thunderstorm (Fig. 9-15). Since there were no upper air measurements at that time, the structure inside the thunderstorm was unknown, but observations at the surface gave some idea of changes in the wind direction as the storm passed; such observations allowed this 1884 model to show the area in the leading right part of the thunderstorm as the place of cyclonic rotation.

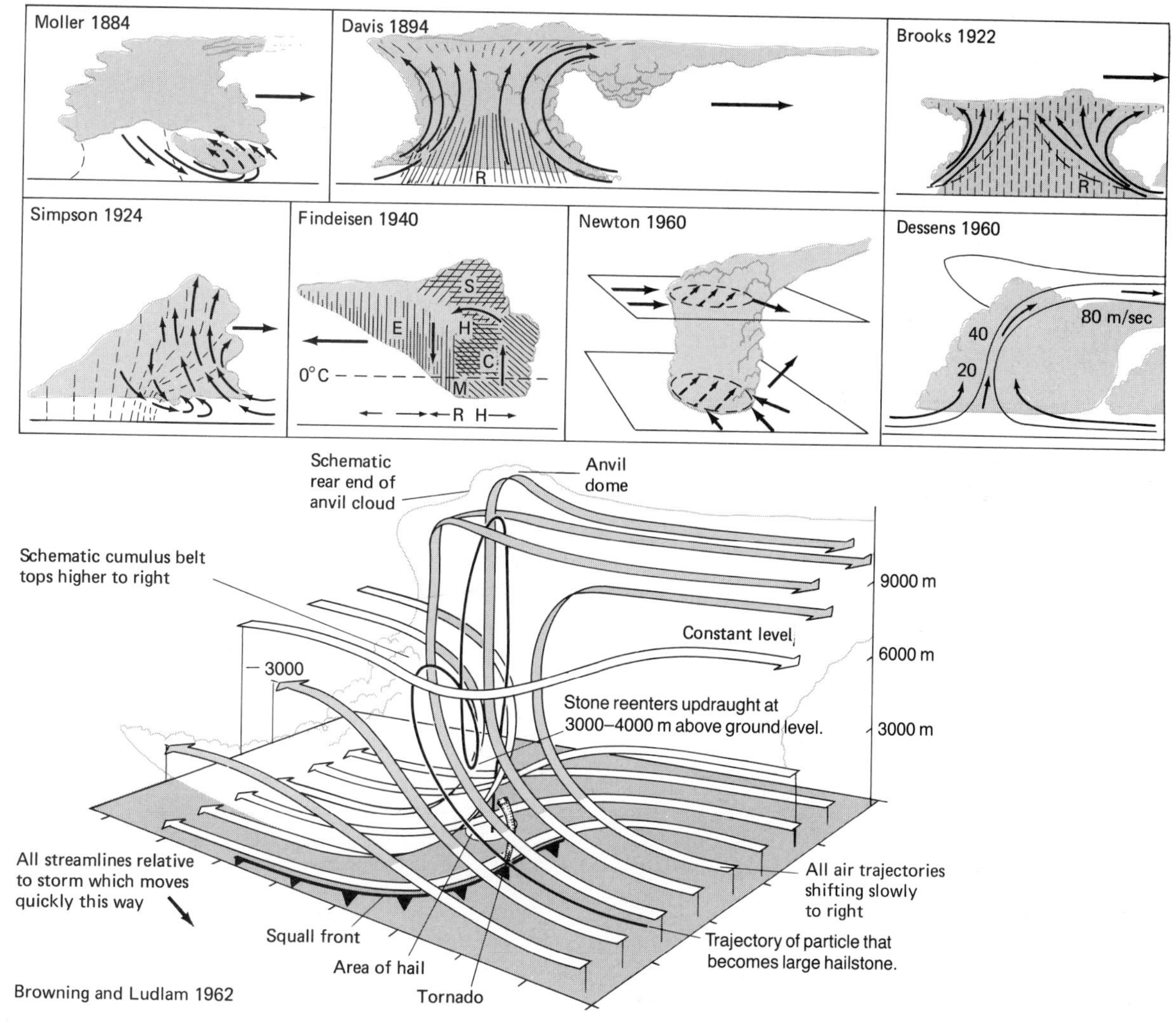

FIGURE 9-15 Various suggestions for the internal structure of thunderstorms. As early as 1884 Moller suggested that the cyclonic circulation was in the southern part of a thunderstorm moving toward the east. The model by Davis showed updrafts and downdrafts within the thunderstorm, as did the models of Brooks and Simpson. The model by Findeisen in 1940 suggests areas of condensation (C), hail (H), sublimation (S), evaporation (E), and melting (M) with suggested areas of hail and rain beneath the thunderstorm. The 1960 model by Newton shows the surrounding environmental winds with their opposing directions, but no internal structure. The Dessens model of 1960 suggests a single localized updraft of considerable velocity. The 1962 Browning and Ludlam model showed the thunderstorm to be primarily an updraft region with some downdraft in the back because of environmental air. The tornado is shown in the center with no indication of its formation mechanism. (From Meteorol. Monogr., "Severe Local Storms," Am. Meteorol. Soc., 1963.)

We now know that if a tornado develops in a thunderstorm it is more likely to be located in the southern part of the thunderstorm, with the major rain area in the northeastern part.

Various other thunderstorm models have been proposed. For example, Davis in 1894 indicated the existence of the thunderstorm anvil with mammatus formation (the development of a particular cloud formation sometimes called a tornado sky). In 1922 Brooks had the idea of a large updraft area developing rain that fell in the central part of the storm, and two years later Simpson presented a model showing an updraft and downdraft within the thunderstorm. Although the shape of the cloud was not very realistic, the updraft and downdraft indicated cyclonic circulation in the southern part of the thunderstorm, which was very realistic for that time. Others, such as Findeisen in 1940, dealt more with precipitation processes by defining areas where hail formation occurred and where condensation, sublimation, and evaporation were most common. Newton in 1960 gave a good idea of the opposing directions of the surrounding winds, with the environmental air moving the thunderstorm along and giving inflow in the front part. No indication of the internal thunderstorm structure was given, however. Dessens in 1960 postulated that an updraft exists inside the thunderstorm, similar to a chimney that does not draw well, unless it is in the jet stream where the outflow from the top is accentuated by the high wind speeds at the top of the thunderstorm. Ludlam in 1961 suggested a hail formation process with development of hail in the central part of the thunderstorm by the updrafts and downdrafts. The thunderstorm model by Browning and Ludlam in 1962 received considerable attention when it was suggested. It is essentially a two-dimensional model showing a thunderstorm with inflow at the base, with the air going up through the thunderstorm and outflow at the top through the anvil. An associated tornado is shown with no connection between it and the thunderstorm.

A steady state thunderstorm model has been developed by the author. It begins with the observation that large thunderstorms are known to maintain a vertical position in the atmosphere. If they are able to exist in a vertical position for any length of time, they must have the right internal winds to oppose the strong surrounding winds, otherwise their tops would be sheared off by the jet stream as the storm grows through it. The equations describing flow around a barrier were used to determine the internal thunderstorm structure. The resulting **double vortex thunderstorm** model was then verified by dual Doppler measurements of the winds inside a severe thunderstorm.

The thunderstorm in Figure 9-16 is moving from left to right, with the upper-level winds behind it pushing the storm along. This movement of the storm creates wind, just as a moving automobile creates wind in a direction opposite to its movement. Such a wind relative to the moving thunderstorm is generated in low levels from the front of the storm causing the air to flow into the lower part of the thunderstorm. The structure inside develops into a double vortex as the inflow below the inversion rises through the center of the thunderstorm until this **thermal updraft** (warm, moist air that rises) strikes the external winds at the back side of the storm that come from the opposite direction. As this happens the thermal updraft must either go up or be carried around the southern cyclonically rotating part of the thunderstorm or be carried around the anticyclonic rotation in the rain area in the northern part of the thunderstorm.

Additional rotation is imparted to the cyclonic rotation in the southern half of the storm by environmental airflow around the southern edge of the storm. As the cyclonic rotation develops, it simulates a tube extending from the surface up to the upper atmosphere, with the tornado connected to this tube. The anticyclonic rotation does not develop as fully because it is loaded with raindrops, and also the surrounding atmosphere has more cyclonic curvature and vorticity. Thus the inside of the thunderstorm is related to the flow of air across and around it, with the double vortex internal structure blocking the external airflow thereby causing it to go around the thunderstorm and contributing to the internal circulation. This intensifies the cyclonic rotation in the southern part of a

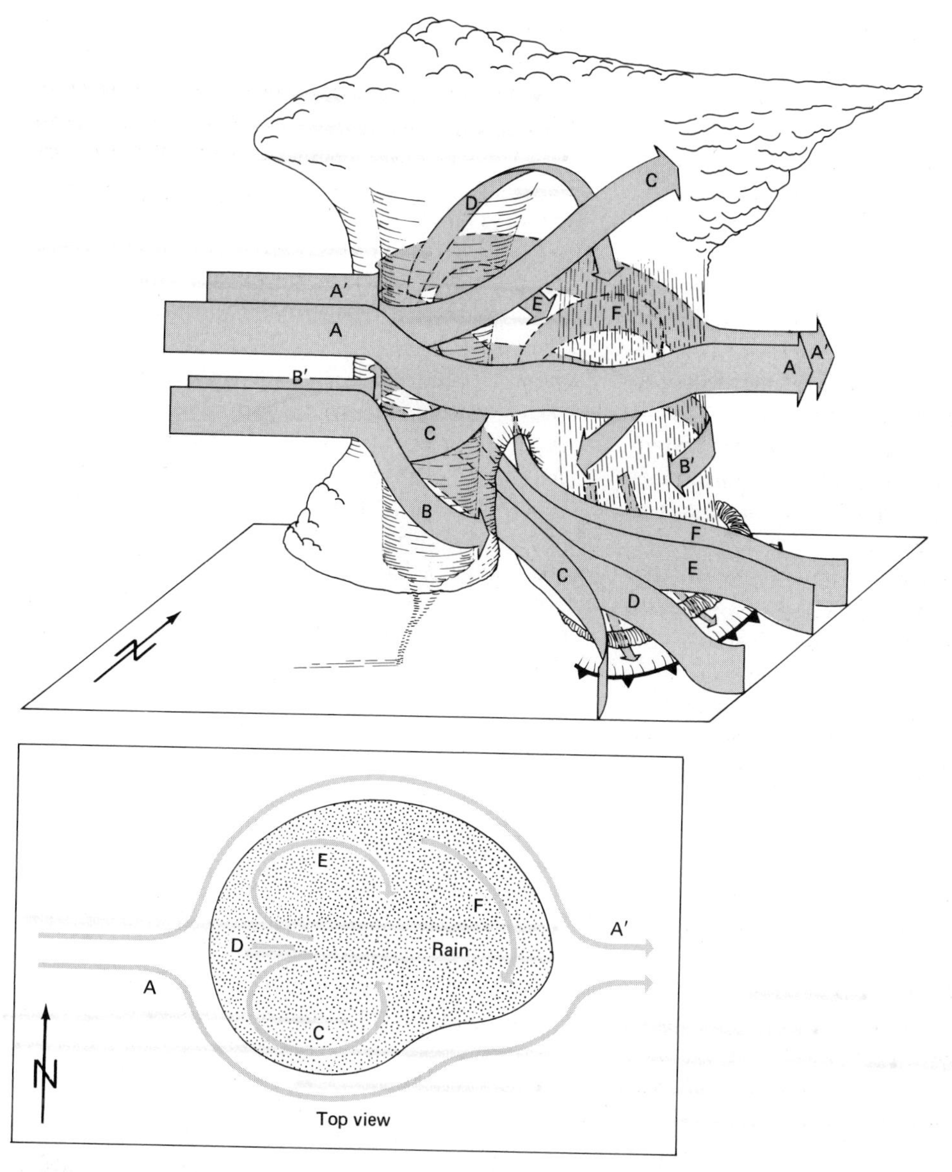

FIGURE 9-16 The double vortex thunderstorm model is based on the theory of flow around a barrier and dual Doppler radar data. The inflow of warm moist air in the front of the thunderstorm (streams C, D, E, and F) collides with the environmental airflow (A and A′), forming a cyclonic vortex to the south and an anticyclonic vortex within the northern part of the thunderstorm. The cyclonic circulation within the southern part of the thunderstorm forms a tube of rotating air that extends to a height of 8 km or more in the thunderstorm. Large tornadoes develop as this circulation reaches to the ground and smaller tornadoes may develop on the edges of this circulation. Rain occurs in the leading edge of the thunderstorm and hail takes a path shown by streamline D. This circulation is most efficient at pulling air into the base of the thunderstorm and opposing the strong winds in the upper atmosphere.

thunderstorm as well as the anticyclonic circulation in the northern part.

THUNDERSTORM GENERATION OF HAIL AND TORNADOES

We have been able to generate a large, unconfined tornadolike vortex in the laboratory by simulating a thunderstorm updraft in horizontal crosswinds. This shows that a thunderstorm that has appropriate horizontal circulation combined with an updraft will also generate a tornado by the same mechanism. The cyclonic vortex in the southern part of a thunderstorm generates tornadoes because of the horizontal winds combined with an updraft through the larger cyclonic rotation.

The hail generation region of a double vortex thunderstorm is the central part of the storm. Here the thermal updraft carries a developing hailstone upward between the vortices, perhaps even into the anvil, where its weight causes it to fall downward between the vortices and back into the thermal updraft below the inversion. Its journey then begins again as long as the updraft can support its weight.

The gust front may be the first indication of a thunderstorm at the surface. The gust front develops from the downdraft in the rain area of the storm as it reaches the ground and spreads out ahead of the thunderstorm. A **roll cloud,** which looks like a long horizontal cylinder, may develop in association with the gust front. The gust front is similar to a miniature cold front since it causes lifting of the warm humid air as it moves underneath it.

Observations of tornado damage paths can be combined with information from the laboratory tornadolike vortex to provide additional information on the nature of thunderstorms. Sometimes the damage path of a tornado is straight and sometimes it describes a series of loops on the ground (Fig. 9-17). Other thunderstorms may generate several tornadoes that stay on the ground for shorter periods of time. Several tornadoes may be on the ground at the same time or one may develop, travel a few kilometers, and then recede into the cloud as a new tornado forms in a different part of the cloud.

If a tornado is generated in the center of the cyclonic rotation of the thunderstorm (**mesocyclone**), it is likely to be on the ground for a longer period of time and describe a straight path if the thunderstorm is moving along at an average or above average speed. If the thunderstorm is moving more slowly, the tornado is more likely to create a looping damage path. The position of the vortex on the ground is unstable since the tornado is acquiring air into its central updraft core from the surface air. Many observers describe a tornado moving toward them as being like an elephant's trunk swaying back and forth. This description would correspond to a tornado producing a looping damage path.

When multiple tornadoes occur, one may be in the center of the mesocyclone with others on the edges. If a single tornado develops outside the center of the mesocyclone, it is more likely to be located on the southwestern edge of the thunderstorm. A tornado in this position is carried by the mesocyclonic circulation around to the forward part of the mesocyclonic vortex and into the rain area between the two vortices of the thunderstorm. In this position the tornado dissipates. Another tornado may develop on the southwestern edge of the mesocyclonic vortex and repeat the pattern. It is not uncommon for a single thunderstorm to produce three or more tornadoes. When multiple tornadoes occur, one vortex usually rotates around the other with the tornadoes on the edges of the mesocyclonic vortex having much shorter lifetimes than those in the center.

THUNDERSTORM RESEARCH

The generation of severe thunderstorms with an internal structure that forms lightning, hail, and tornadoes is an area of continuing research. Thunderstorm research experienced a major thrust in the 1970s as dual Doppler radar measurements became a reality. Doppler radar can be used to determine the speed of raindrops just as it can measure the speed of an automobile. When two Doppler ra-

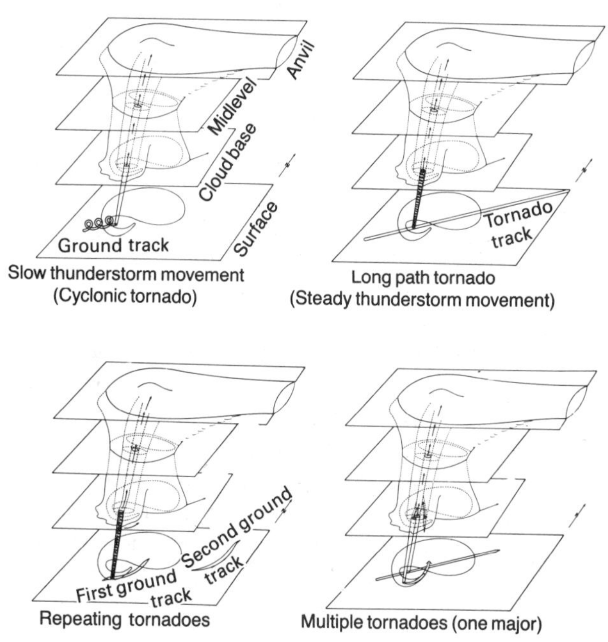

FIGURE 9-17 Tornado damage paths give information on the thunderstorm that produced them. The slow-moving thunderstorm with a tornado in the center of the southernmost cyclonic circulation (mesocyclonic vortex) produces a looping damage path while a faster moving thunderstorm with a tornado in the same position produces a long path tornado. Multiple tornadoes originate in the center of the mesocyclonic vortex or on the southern edge of the mesocyclone. Tornadoes that originate on the southern edge of the mesocyclone are carried by its circulation to the central part of the thunderstorm, where the tornado is soon dissipated. (Reprinted by permission of the publisher from J. R. Eagleman, V. U. Muirhead, and N. Willems, *Thunderstorms, Tornadoes, and Building Damage,* Lexington, Mass.: Lexington Books, D.C. Heath and Company, © 1975, D.C. Heath and Company.)

dar systems are aimed at a severe thunderstorm from different locations, the three-dimensional wind currents can be seen as they carry raindrops along. Thus the wind currents, radar reflectivity, and various other properties such as vorticity and convergence within a severe thunderstorm can be evaluated (Fig. 9-18). Dual Doppler radar measurements are made at only a few locations, including Oklahoma by the National Severe Storms Laboratory, Colorado by the National Environmental Laboratories, and Massachusetts by the Air Force Cambridge Research Laboratories. Real-time color displays of dual Doppler measurements may be of major benefit in severe thunderstorm research and operational public warning systems of the future.

The use of computers has allowed the development of complicated mathematical models for thunderstorm development and characteristics. Various computer models have been developed for specific purposes since all the various features of a severe thunderstorm, vertical winds, precipitation, lightning, hail, and tornadoes cannot be specified mathematically with the current state-of-the-art. However, **numerical cloud models** may provide an explanation of the dynamics of thunderstorm development that is difficult to obtain by any other

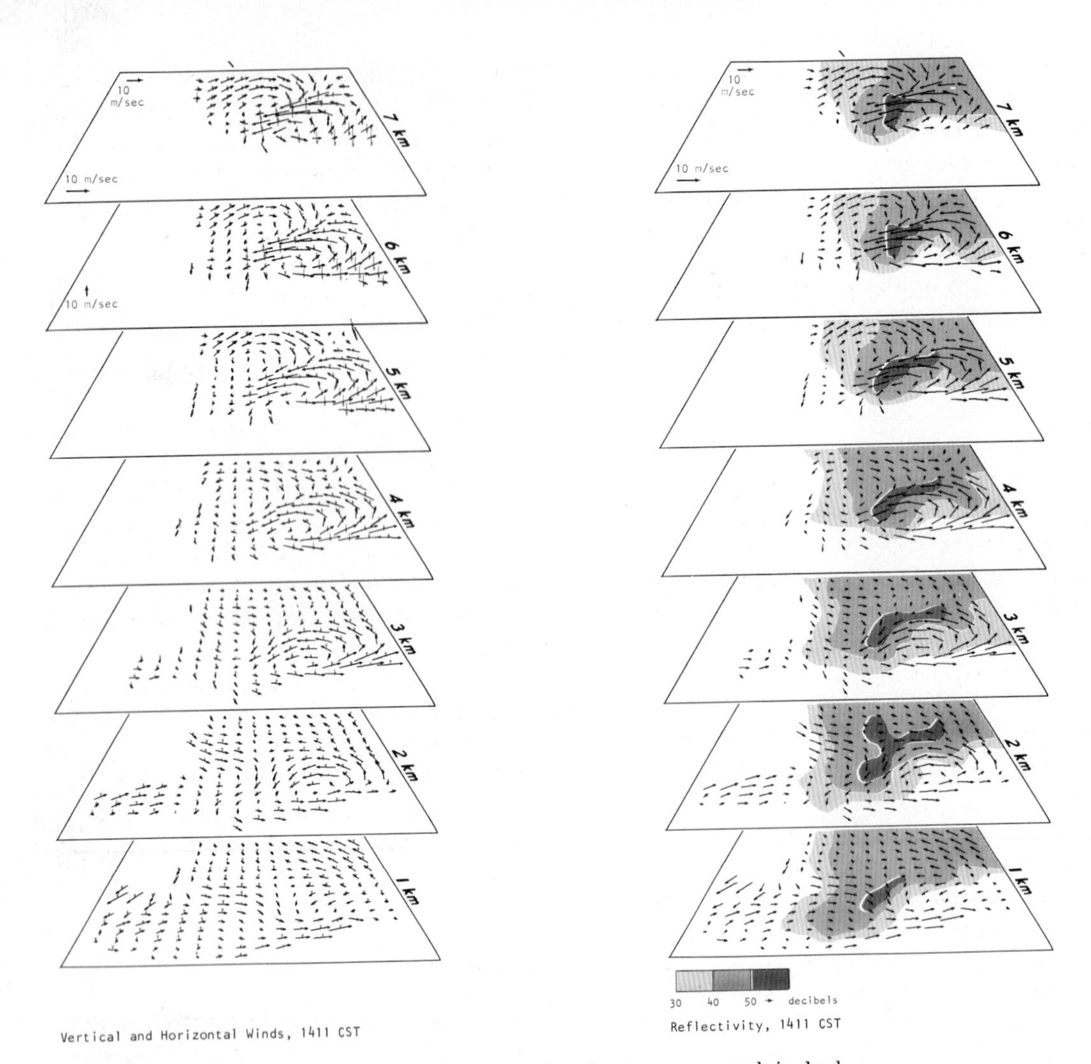

FIGURE 9-18 The latest instrument in severe thunderstorm research is dual Doppler radar. Data from two Doppler radar sets aimed at a single thunderstorm can be used to reconstruct the three-dimensional flow patterns, as shown for a storm that occurred in Oklahoma on June 8, 1974. These data revealed the development of a double vortex inside a supercell thunderstorm as a tornado extended to the ground. The radar reflectivity patterns at 2 km show the hook echo in relationship to the rotation of the mesocyclonic vortex that contributes to its shape. (From J. R. Eagleman and W. C. Lin, "Severe Thunderstorm Internal Structure from Dual Doppler Radar," *J. Appl. Meteorology*, October 1977.)

technique. A computer-generated thunderstorm model is shown in Figure 9-19 as precipitation develops above a small mountain ridge. Projections of the precipitation loading in various parts of the thunderstorm can be made every three minutes, for example. Such simulations indicate that the amount of precipitation in the lower and middle parts of a cloud has a considerable effect on the updrafts and downdrafts within a thunderstorm.

SUMMARY

Severe thunderstorms contain frequent lightning and locally damaging winds or hail 2 cm in diameter or larger. Several appropriate atmospheric conditions are required for the development of a severe thunderstorm. A midlatitude cyclone with an accompanying cold front contributes to severe thunderstorm development through surface convergence associated with the cold front or dry line and the forced lifting associated with these instability lines. An unstable atmosphere contributes to the development of severe thunderstorms by increasing the buoyancy and updraft velocities of rising bubbles of unstable air.

The jet stream contributes to thunderstorm generation by providing upper air divergence with an outflow of air that results in intensified updrafts in thunderstorms beneath it. The high speed winds of the jet stream also contribute cyclonic rotation to the thunderstorm and tornado. An upper air inversion contributes to the development of severe thunderstorms by decreasing convective activity until the air below the inversion becomes unstable enough to develop larger thunderstorms after penetration of the inversion. High dew point temperatures and the presence of a moisture tongue contribute to severe thunderstorms by providing more water vapor for condensing to release latent heat within the thunderstorm. In addition, the density of moist air is less because of the presence of water vapor. The appropriate wind profile for severe thunderstorm development consists of large wind shear between low-level winds and those above the inversion.

Severe thunderstorms frequently develop within squall lines because of the interaction in airflow between different cells within a squall line. Thunderstorms are more likely to be severe if located in specific preferred regions within a squall line. Severe thunderstorm forecasts are based on the appropriate atmospheric characteristics for the development of severe weather. Severe weather is most likely where all the various atmospheric conditions coincide in the atmosphere. This is normally a very large area.

The internal structure of severe thunderstorms frequently consists of a double vortex with cyclonic rotation in the southern half of the thunderstorm and weaker anticyclonic rotation in the northern half. Such a circulation resists the environmental wind flow and allows the thunderstorm to withstand the tremendous vertical wind shear of the environment.

Hail is generated in the central thermal updraft while tornadoes are generated in the mesocyclonic vortex because of its horizontal rotation combined with updraft. The number and nature of damage paths of tornadoes are related to the forward speed of the thunderstorm and the location of their formation in the mesocyclonic vortex.

STUDY AIDS

1. Explain the relationship between a midlatitude cyclone and severe thunderstorm development.

2. Compare the thunderstorm development effects of an unstable atmosphere containing a high dew point temperature with the occurrence of an inversion.

3. Explain the various contributions of the jet stream to severe thunderstorm development.

4. Discuss the appropriate wind profile for the development of severe weather.

5. Describe how severe thunderstorm forecasts are made.

6. Speculate on the various problems associated with the development of early models of the internal structure of a severe thunderstorm.

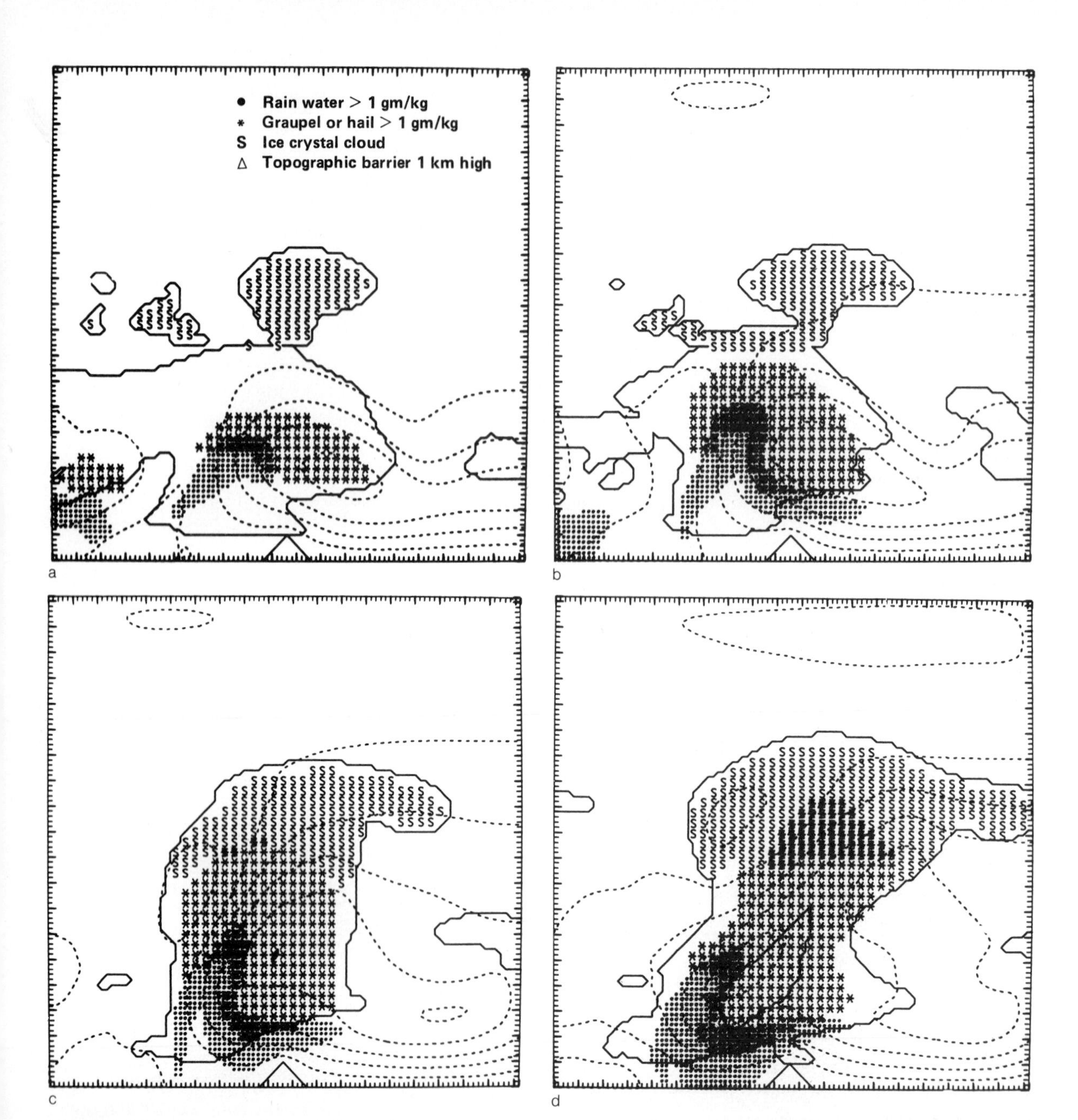

FIGURE 9-19 Numerical simulation and computer-drawn cloud and precipitation patterns within a vertical cross section of the atmosphere 20 km on a side. Cloud areas are outlined by the solid line, while the airflow is given by the dashed lines. The precipitation forms and is carried up and out from the updrafts (a), falls into the lower atmosphere (b), and is rapidly swept back into the updraft (c), causing the updraft to tilt and weaken, which allows the precipitation to fall out of the cloud (d). (From H. D. Orville et al., "The Dynamics and Thermodynamics of Precipitation Loading," preprint volume, 9th Conference on Severe Storms, Am. Meteorol. Soc., 1975.)

7. Describe the double vortex thunderstorm model.

8. What is the primary source of energy for the cyclonic vortex of a double vortex thunderstorm?

9. Discuss the generation of hail by a severe thunderstorm.

10. Describe how tornadoes are generated by severe thunderstorms.

11. Describe some of the information that tornado damage paths reveal about severe thunderstorms.

TERMINOLOGY EXERCISE

Check the glossary if you are unsure of the meaning of any of the following terms used in Chapter 9.

Air mass thunderstorms
Convective thunderstorms
Severe thunderstorm
Damaging winds
Dry line
Prefrontal wave
Supercell
Funnel cloud
Moisture tongue
Low-level jet
Wind shear index
Squall line
Multicell thunderstorms
Gust front
Double vortex thunderstorm
Thermal updraft
Roll cloud
Mesocyclone
Numerical cloud model

THOUGHT QUESTIONS

1. Describe the specific atmospheric features and their location that are required to set the stage for severe thunderstorms at your location.

2. The double vortex structure has been observed by radar. Why do you think the cyclonic rotation should be stronger than the anticyclonic rotation?

3. Describe and explain some of the difficulties in forecasting severe thunderstorms.

4. Can you speculate on methods of improving the accuracy and effectiveness of severe thunderstorm warnings?

5. Can you think of any human activities that might affect the formation of severe thunderstorms? Explain your answer.

SPECIAL TOPIC 10

FLASH FLOODS

Heavy rainfall from a thunderstorm may form puddles that flow into brooks, and these may feed a small stream, swelling it to great depths in a matter of a few hours. Such overflowing streams may inundate cropland, and cover highways and houses. Nearly 200 people are killed each year in the United States by flash floods, placing this weather event among the most threatening of all (Fig. S10-1).

Hundreds of flash floods occur each year in the United States. Damages from such floods have increased recently because of greater development of areas prone to flooding. In recognition of this problem, the National Weather Service has developed procedures for issuing flash flood watches and warnings. However, the nature of flash floods is such that warnings are frequently difficult to prepare very far in advance.

A further problem with flash floods is that many flash flood deaths are needless because some people simply ignore the warnings. A good example of this occurred in the Country Club Plaza of Kansas City on September 12, 1977. Shortly after midnight on September 12, thunderstorms over Kansas City soaked the area with about 6 in. of rain. The Weather Service issued a flash flood watch at 10:30 A.M. for that afternoon and night. The formation of another thunderstorm near Kansas City at 5:30 P.M. prompted a flash flood warning for areas surrounding Kansas City and at 7:45 P.M., the flash flood warning was extended to include Kansas City. Heavy rain began to fall in Kansas City at about 8:00 P.M. These rains swelled Brush Creek to disastrous levels. This stream can normally be stepped across as it flows along its 11-mi course, which takes it through the Country Club Plaza. This plaza was the nation's first

FIGURE S10-1 A flash flood has changed the mode of transportation from automobiles to boats. As low elevations are developed for human occupation, the flood danger increases. (Photo courtesy of U.S. Coast Guard.)

planned shopping center, started in 1922, and is now a complex of elegant boutiques, restaurants, hotels, and apartment buildings. Brush Creek's bed was lined with concrete for four miles through the plaza. It is used by bicyclists and joggers when it is dry, but on September 12, water flowed over its banks and through the Country Club Plaza at a peak velocity of 6 m/sec, or seven times the velocity needed to knock a person down and more than enough force to tumble a car.

The response of individuals in the Plaza revealed a lack of appreciation for the seriousness of the rampaging Brush Creek. It actually drew spectators who, according to police reports, interfered with rescue efforts by climbing on police cars to get a better look. Many of the 25 victims of this flood lost their lives because they failed to recognize the danger. Most of the drownings were associated with automobiles, and reliable reports indicated that motorists attempted to drive through high water when cars ahead were obviously stalled. This points to the pressing need for better education concerning flash floods as well as other severe weather reports.

SPECIAL TOPIC 11

INSIDE A THUNDERSTORM

Only one person has been known to penetrate the full depth of a severe thunderstorm. This was William H. Rankin, Lt. Col. U.S. Marine Corps. On July 27, 1959, he was forced to eject from his jet as he piloted it at 14,325 m (47,000 ft) above a well-developed thunderstorm. His survival was remarkable. No one had previously lived through the decompression effects of the extremely low atmospheric pressure at this height, only 140 mb. Yet Lt. Rankin remained conscious to experience all these effects and those of the severe thunderstorm that he was thrust into.

His first battle for survival was against the extremely cold air, −57°C (−70°F). As he fell into the large thunderstorm he soon encountered other life-threatening phenomena. The turbulence and updrafts within the thunderstorm prolonged a fall that should have lasted only 10 minutes into 40 minutes. Hailstones pelted his body. He took a terrible physical beating from the twisting, turning, and tumbling. Lightning was so close to him that he felt he could touch it. Thunder vibrated with a horror that was not meant for human ears. The drenching of rain was so torrential that he feared being drowned in midair. "How silly, I thought. They're going to find you hanging from some tree, in your parachute, limp, lifeless, your lungs filled with water, wondering how on earth did you drown!"

These thoughts and his complete description of the storm have been published by Prentice-Hall in *The Man Who Rode the Thunder*, by William H. Rankin.

CHAPTER TEN
LIGHTNING AND HAIL

THUNDERSTORMS AND THEIR UNWANTED COMPONENTS

Lightning and hail are generally unwanted components of thunderstorms. Each year lightning is responsible for starting numerous forest fires, for killing more than 100 people, and for injuring many more. Lightning strikes houses, airplanes, and rockets as they launch satellites. Such strikes frequently result in houses burning and airplanes crashing, although many survive with little damage.

Thunderstorms, by definition, all contain lightning since thunder is associated with the lightning stroke. The number of thunderstorms occurring over the world is a fantastic 44,000 each day. At any given moment about 2000 thunderstorms are present and each one produces an average of 100 lightning discharges. The distribution of thunderstorms in the United States is shown in Figure 10-1. The greatest number of thunderstorms occurs in the southeastern United States. Florida has more than 80 days with thunderstorms per year, while parts of New Mexico and Colorado experience as many as 60 per year. The drier areas of the United States are particularly plagued by the threat of fire from lightning, and many forests are burned as a result of lightning activity. This is a lesser problem in the eastern United States where there is greater humidity and moisture.

THUNDER AND LIGHTNING FORMATION

Much of the energy of a lightning stroke is dissipated through the air as shock waves, causing **thunder**. When lightning discharges, air is heated

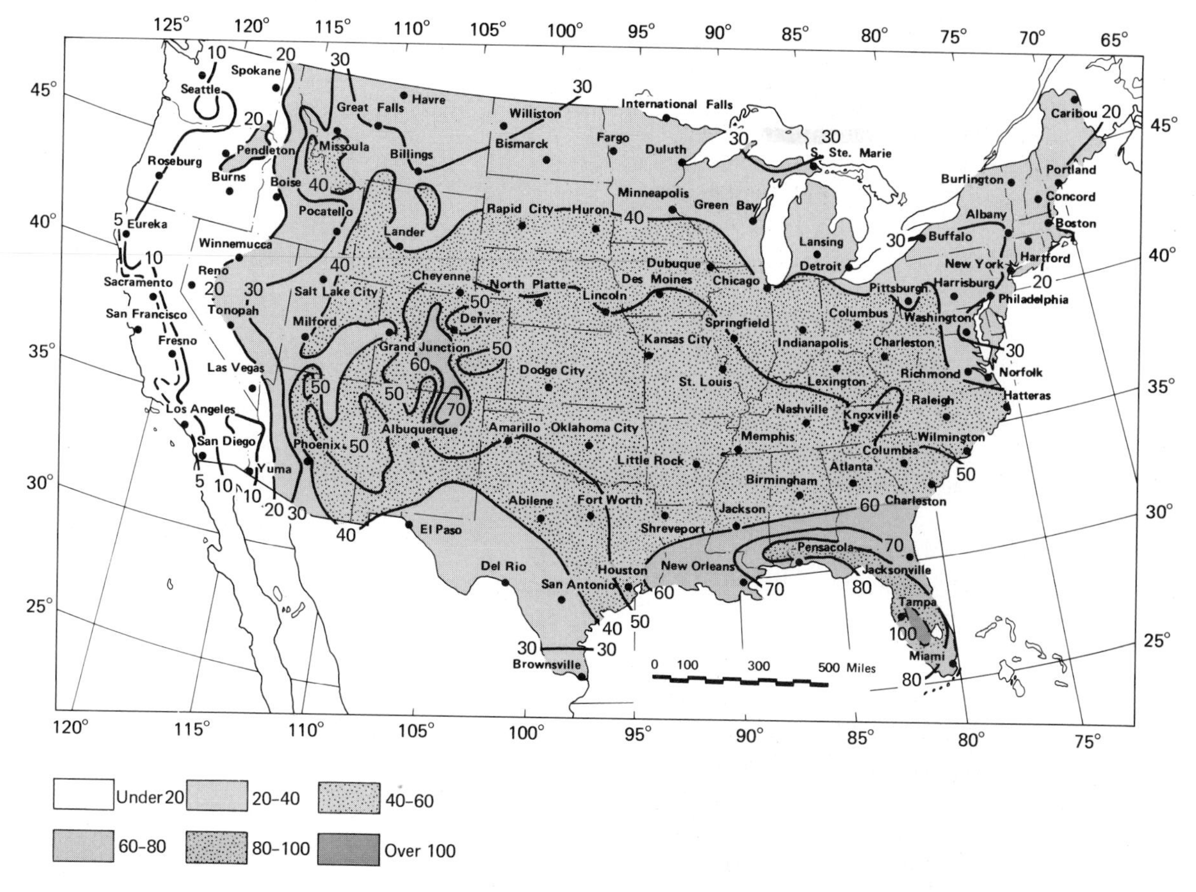

FIGURE 10-1 Distribution of thunderstorms across the United States. The numbers show the mean number of days per year with thunderstorms (Environmental Data and Information Service, NOAA.)

within a small conducting channel a couple of centimeters in diameter. During the discharge, the temperature of the air in the channel is heated to about 30,000 K. This causes a very sudden expansion of the air, producing shock waves that spread out from the discharge channel to be heard as thunder (Fig. 10-2). Sound waves travel at a much slower speed than light; light traveling at 300,000 km/sec reaches the earth from the sun in only 8 minutes. Sound waves travel at 1 km/3 sec. You can roughly determine the distance to the lightning discharge if it is within 10 km by counting the number of seconds between the lightning flash and the thunder.

The lightning process includes the formation of a conducting channel of air between the cloud and the ground. This is created when the buildup of charges within the thunderstorm increases the electrical potential between the cloud and earth's surface sufficiently that an arc of electricity bridges the gap between them.

Measurements have shown that the base of most thunderstorms is negatively charged. In fact, Benjamin Franklin determined this aspect of thunderstorms from his kite experiments. The negative charge of the lower part of the thunderstorm induces a positive charge on the earth's surface beneath it, since like charges repel each other and opposite charges attract. Thus, negative charges in the thunderstorm repel negative charges near the earth's surface, creating a positively charged earth beneath the thunderstorm (Fig. 10-3). When the thunderstorm is able to generate sufficient electrical charge separation between the earth and the thunderstorm, a discharge develops between the two that is well known as lightning.

The only resource materials for producing electricity within thunderstorms are water, ice, and wind; if we could generate electricity in this way, the materials would be cheap and our pollution problems would be solved. It is also of interest in this regard that a thunderstorm is able to generate the electricity to the point of discharge in only 20 minutes.

The process of electricity production in thunderstorms is thought to originate with the freezing of water droplets in the thunderstorm. Normally, liquid water droplets exist below the freezing level and ice crystals exist in the top part of the thunderstorm. Just above the freezing level, supercooled liquid water droplets can exist at temperatures as low as −40°C before freezing will occur spontaneously. Other ice crystals form near 0°C.

Ice crystals may contain a strong temperature gradient with perhaps 0°C near their center and −5°C on their outer surface. Thus a charge separation occurs, due to a process called the **thermoelectric effect,** as positive ions migrate within this temperature field from warmer to colder regions (Fig. 10-4). The outer edge then becomes positively charged. As ice crystals grow colder and colder, their outer surface shatters into many small pieces of ice with positive charges. The larger, central pieces of ice with negative charges fall into the lower part of the thunderstorm, while the smaller ones are carried by the updraft to the upper part of the thunderstorm to create a charge separation within the thunderstorm.

Experiments have shown that more than 20 minutes is required for a charge separation great enough to produce lightning. Therefore, another process must also be active. This is thought to be the **induced charge** on ice crystals and large drops

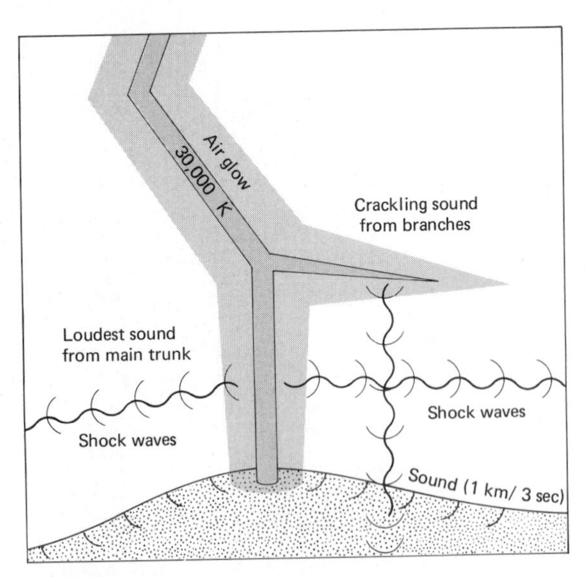

FIGURE 10-2 The lightning discharge heats an air channel to about 30,000 K, and the expanding air creates shock waves that produce sound traveling at a speed of 1 km/3 sec. Therefore, the distance in kilometers to the lightning channel can be estimated by counting the number of seconds between sighting the lightning and hearing the thunder and dividing by 3.

within the central part of the thunderstorm. As the base of the thunderstorm becomes negatively charged and the top positively charged by the thermoelectric effect, the larger drops and ice crystals within the central part of the cloud develop negative charges on top and positive charges on bottom (Fig. 10-5). This charge separation on the top and bottom surfaces of drops and ice crystals occurs because the positive charge concentration at the top of the thunderstorm repels the positive charges within an ice crystal and concentrates them on its lower surface. This charge separation is the induced charge. As smaller droplets, carried by updrafts, strike the larger ice crystals from below, they bleed off some of the positive charges and carry them along to add further to the positive

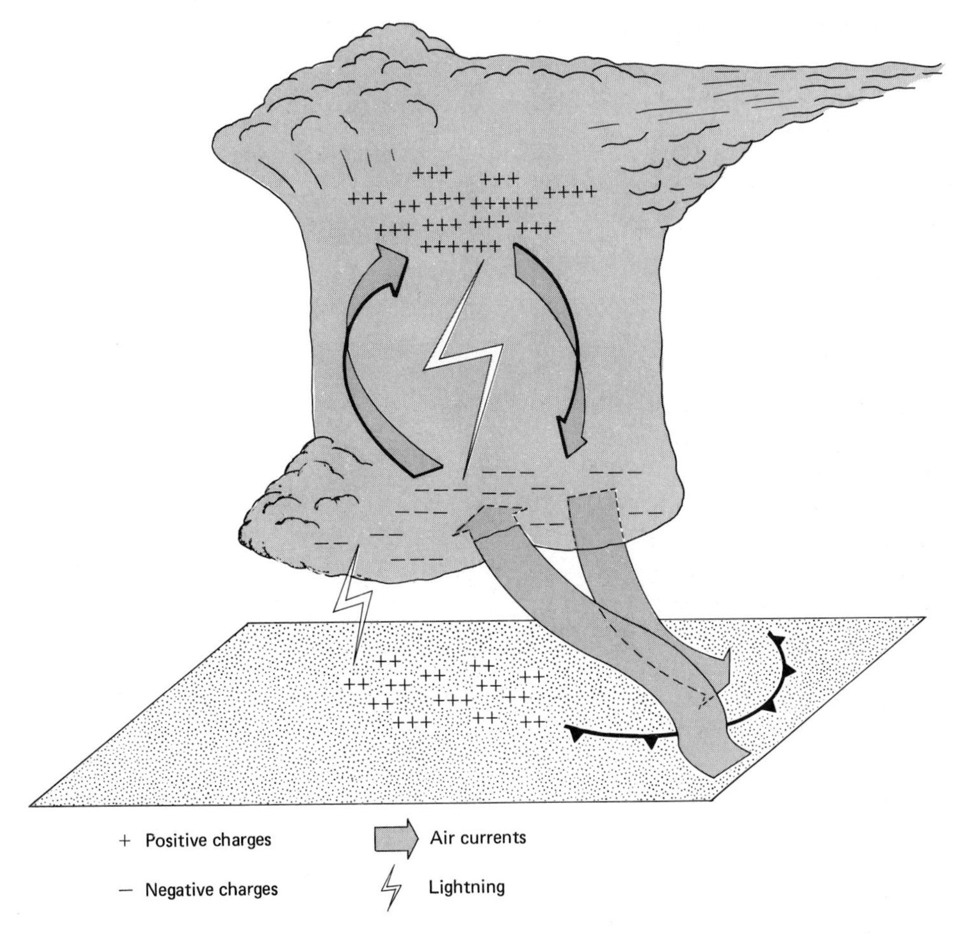

FIGURE 10-3 Thunderstorms generate large positive charges near their tops and large negatively charged regions near their bases. This induces a positive charge on the earth's surface and objects on the earth beneath the thunderstorm. When the electrical potential builds to critical values, lightning occurs either within the thunderstorm cell or between the cloud base and the ground.

charge in the upper part of the thunderstorm. The larger droplets and ice crystals with their negative charge then fall to the lower part of the thunderstorm intensifying the negative charge at the base.

Thus the production of charges because of the splitting of ice crystals starts the mechanism, with further enhancement of the charge separation due to the bleeding off of positive charges to small droplets and ice crystals in the updraft. The thunderstorm then has a charge separation built up by air currents and freezing water with a positive charge at the top and negative charge at the base, which induces a positive charge beneath it on the earth's surface.

The charge separation may be great enough to cause a visible glow of light called **St. Elmo's Fire.** It may appear as a steady glow from the masts of ships, for instance, as they become extremely

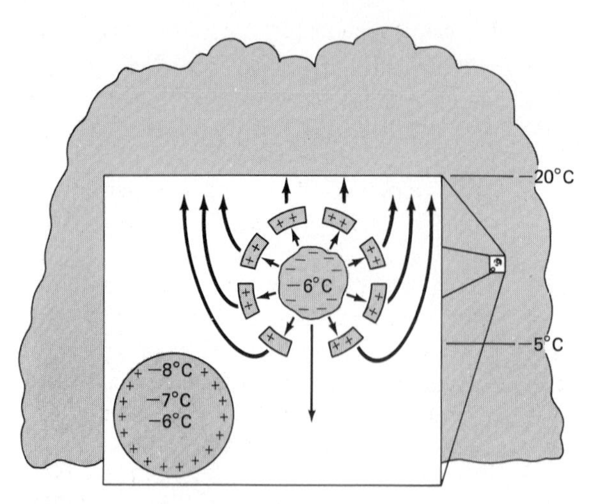

FIGURE 10-4 Charge separation in thunderstorms is initiated as supercooled water droplets and ice crystals develop a temperature gradient, as a result of which the coldest temperature, which is on the outside, creates a charge separation. The outer surfaces of ice crystals with positive charges on them may shatter into small pieces of ice as their temperature decreases. This process produces a charge separation within the thunderstorm.

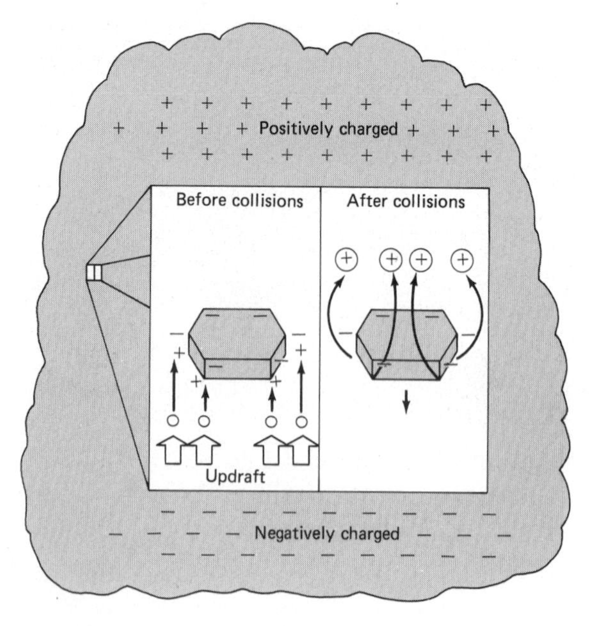

FIGURE 10-5 After a thunderstorm begins to develop a positive charge in its upper portion and a negative charge near its base, an induced charge develops on all ice crystals within the cloud in such a way that the negative charges migrate to the top of the ice crystals, leaving their bases positively charged. Within an updraft in the thunderstorm, the small cloud droplets strike the lower surfaces of the ice crystals, bleeding off the positive charges and carrying them to the top of the thunderstorm. The larger ice crystals are left with a negative charge, and as these crystals fall into the lower part of the storm, they intensify the negative charge near the thunderstorm base.

positively charged. The top of a thunderstorm may also glow, on rare occasions, if the positive charge is large enough.

FORMS OF LIGHTNING

The most frequent form of lightning occurs within the thunderstorm. The electrical discharge neutralizes the positive and negative charges at the top and base of the cloud. Lightning also occurs between clouds if clouds with opposite charges are located near each other. If the lightning simply lights up a part of a cloud, we call this **sheet lightning** since the lightning discharge itself is hidden. If the cloud is so far away that we see only some light near the horizon we call this **heat lightning.**

The most commonly observed lightning discharge is **forked lightning,** which extends from the cloud base to the ground (Fig. 10-6). A single **lightning flash** normally consists of several different **lightning strokes.** The first of these is called the stepped **leader stroke.** The leader stroke seeks the

FIGURE 10-6 Photographs of lightning can be obtained at night by opening the lens of a camera when thunderstorms are occurring. Two lightning discharges are shown in this illustration. Note that some of the forked branches appear to extend to, or very near, the ground. This lightning discharge occurred in Colombia and was photographed by Hernando Santos.

most conductive path to the ground and tries several branches before the best is located. The air becomes so hot in the conducting channel that the gaseous molecules are broken down. Oxygen and nitrogen gases are broken into ionized molecules. This charges the channel and allows the electricity to flow through much more easily.

After the stepped leader stroke extends to the ground, it is followed by a return streamer stroke to the cloud. This happens so fast that it appears as a single flash. An airglow develops around the small conducting channel, increasing its visible width considerably. An average of three strokes per lightning flash occurs, with a discharge time of about one-half second, although lightning discharges have occurred with as many as 25 strokes that last up to one second. The branches of the conducting channel also light up to give the lightning stroke its forked and branched appearance.

Air discharges sometimes occur, especially in desert regions. These consist of a lightning discharge below the cloud base, where none of the conducting channels reach the ground. Air discharges are more common in dry areas because the cloud base is normally much higher. The conducting channel may extend as far as 20 km from a thunderstorm in a horizontal direction before it finally touches the ground. This gives rise to a lightning discharge that may seem to be a **bolt from the blue.** It appears to come from blue sky, although this is not entirely true; a thunderstorm must exist nearby to propagate the flash.

Occasionally strong winds blow the lightning channel downwind fast enough to produce **ribbon lightning** (Fig. 10-7). A single conducting channel blown by the wind at the rate of 10 m/sec would move 3 m during a lightning discharge. This produces a ribbon effect as the separate strokes light up the conducting channel.

Ball lightning is an unusual form of lightning that is not well understood because it does not last very long and is usually unexpected. Ball lightning occurs near an ordinary lightning discharge, but is in the form of a ball, much like a soap bubble, and may either flow or roll downhill.

At the University of Edinburgh on June 8, 1972, a window in the Department of Meteorology was damaged by something that was thought to be ball lightning. A circular hole was found in a window to the laboratory the next morning after thunderstorms had occurred at night. The hole was 4.9×4.6 cm and the edges of the hole were smooth as if the glass had been melted. The piece of glass that fitted into the hole was also found. A metal radiator located near the window may have served as the attraction for lightning. Various discussions in the meteorological literature followed, some saying that it could not have been caused by ball lightning since only a cylinder of fire could cut such a hole.

FIGURE 10-7 Ribbon lightning occurs as strong winds blow the conducting channel along, giving a ribbon apearance to the lightning discharge. (Photo courtesy of Hernando Santos.)

FIGURE 10-8 Lightning discharges are sometimes accompanied by ball lightning, which consists of charged balls of various size that fall from the sky or roll downhill and may explode as they strike objects. (Photo by Dr. John C. Jensen, Nebraska Wesleyan University.)

However, a sphere could cut such a hole if the rate of movement were fast enough to prevent the leading edge of the ball from melting the glass while the portions of the ball at its maximum diameter, and slightly less, would be in contact with the glass for a longer time, perhaps just long enough to melt the glass and cut a circular hole.

Although there are no good theories describing the formation of ball lightning it appears to be an electrically charged mass of air (Fig. 10-8). It is usually red, yellow, or orange. Although it may vary in size from a few centimeters to a few meters, it is generally about 20 cm in diameter and surrounded by an area that is rather vague around its borders. Occasionally, it is a ball of very bright white light and has sharply defined edges. It makes a hissing or buzzing noise and foul smelling odors have been reported after its departure. Although it may last from a fraction of a second up to several minutes, it usually lasts only a few seconds. Sometimes it is observed near the base of a cloud and seems to float down to the ground until it strikes something where it usually explodes, but it may also quietly dissipate. It can move very rapidly through air or hover in one spot. Two balls of fire may be connected for a while.

Two different types of ball lightning seem to exist. The free-floating form behaves differently from the attached ball lightning although one form can change into the other. If it changes form, it is most often from the free-floating to the attached form. The floating ball is usually red and seems to avoid good conductors. This type seems to be drawn into closed spaces, such as through a window or door, and frequently down the chimney of a fireplace. The attached lightning ball is usually blindingly bright; it attaches itself to good conductors, including people.

Although many people are killed each year by lightning, it is ordinarily the common forked lightning that is responsible. One exception was Professor Richman of France, who was duplicating

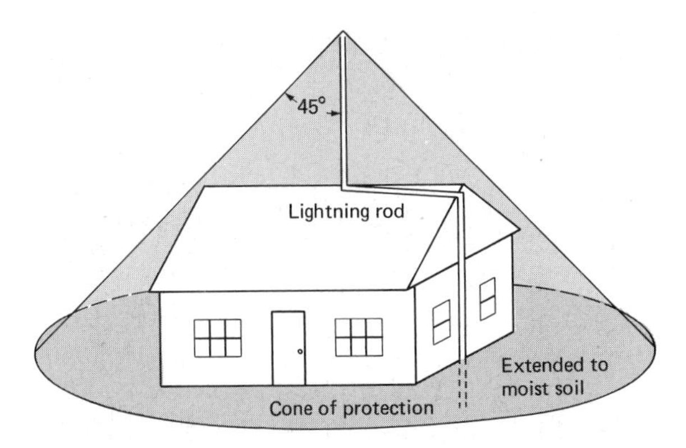

FIGURE 10-9 A lightning rod provides a cone of protection extending outward at a 45° angle from the tip of the rod. The lightning rod should be grounded into moist soil so that a lightning discharge can be conducted to the ground without damaging buildings.

Franklin's kite experiments, in the late 1700s. Ball lightning attached to his forehead killing him instantly.

LIGHTNING PROTECTION

The lightning rod is effective in protecting tall objects from lightning. Taller objects are much more likely to be struck by lightning than shorter objects. In a region with 70 thunderstorms per year, a 100-m object is likely to be struck once per year, while an object 400 m high is likely to be struck seven times. The lightning rod consists of metal rod or wire that carries the electrical discharge safely to the ground (Fig. 10-9). The zone of protection around a rod is related to the height of the rod and extends outward in a conelike shape about 45° from the tallest point. A lightning rod will protect a house only if it is tall enough so that an angle of 45° in any direction from its tip does not include any of the house. The rod itself consists of conducting metal that extends from the tip of the rod down to moist soil. if lightning strikes the rod, the charge is transmitted harmlessly to the ground. The wire does not have to be extremely large. Power line poles are grounded by wires a few millimeters in diameter.

Lightning is responsible for about 97 deaths per year in the United States. This number is similar to the number of deaths from tornadoes (118) and floods (99), while hurricanes kill fewer people (46) per year, as shown in Table 10-1. Most of the lightning deaths could be prevented. Most of these (52%) occur in the open, with 38% in houses or barns, and 10% under trees (Fig. 10-10). The most dangerous places during thunderstorms are under isolated tall trees, near open water or golf courses, along wire fences, and on hilltops and tractors. Dangerous locations inside homes are on the telephone, and near metal appliances, water pipes, sinks, and bathtubs. An automobile offers fairly good protection, as does a metal building. Inside either of these is generally a safe location.

Several people who have been struck by lightning have survived. A forest ranger holds the record for the number of lightning strikes experienced; he has lost a toenail from one strike, an eyebrow from another, and has been burned on several other occasions. Advance warning of an impending lightning strike may be provided when a person's hair stands on end. If this happens the person should immediately crouch down as low as possible and hope the discharge occurs through a taller object. It may be important for you to know that many people apparently killed by lightning can be revived if given immediate attention, including cardiopulmonary resuscitation.

HAIL DAMAGE AND DISTRIBUTION

Large hail from thunderstorms is not usually a threat to human life, although a farmer was killed by hail near Lubbock, Texas, on May 13, 1930, and a baby was killed in Fort Collins, Colorado, on July 30, 1979. However, the damage caused by **hailstorms** is generally more than $600 million annually. Much of this damage is to crops such as corn

TABLE 10-1

Deaths associated with severe weather

Year	Tornadoes Number	Tornadoes Deaths	Hurricanes Striking U.S.	Hurricanes Deaths	Flood deaths	Lightning deaths
1943	152	58	4	16	107	
1944	169	275	4	64	33	
1945	121	210	5	7	91	
1946	106	78	4	0	28	
1947	165	313	7	53	55	
1948	183	139	4	3	82	
1949	249	211	3	4	48	
1950	200	70	4	19	93	
1951	262	34	1	0	51	
1952	240	229	2	3	54	
1953	421	515	6	2	40	
1954	550	36	4	193	55	
1955	593	126	5	218	302	
1956	504	83	2	21	42	
1957	856	192	5	395	82	
1958	564	66	1	2	47	
1959	604	58	7	24	25	158
1960	616	46	5	65	32	97
1961	697	51	3	46	52	113
1962	657	28	1	4	19	120
1963	464	31	1	11	39	210
1964	704	73	6	49	100	108
1965	906	296	2	75	119	125
1966	585	98	2	54	29	76
1967	926	114	2	18	27	73
1968	660	131	3	9	31	103
1969	608	66	3	256	297	93
1970	653	72	4	11	135	111
1971	888	156	5	8	72	113
1972	741	27	3	121	554	91
1973	1102	87	1	5	148	105
1974	947	361	1	1	121	102
1975	920	60	1	21	107	91
1976	835	44	1	9	193	72
1977	852	43	1	0	210	98
1978	788	53	1	35	143	88
1979	852	84	3	22	121	63
1980	866	28	1	2	62	76
1981	783	24	0	0	84	67
1982	1033	64	0	3	24	77
Average (40 yr)	601	118	3.0	46	99	
Average (20 year)	805	96	2.1	36	131	97

Note: Data from *Climatological Data, National Summary*, and *Storm Data*.

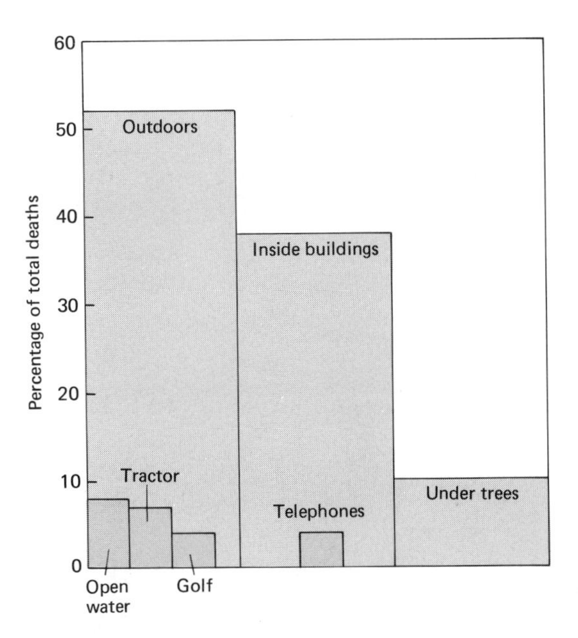

FIGURE 10-10 Locations where people are killed by lightning. More deaths occur outdoors, with 10% under trees. Other unsafe locations are near open water, and on tractors and golf courses. Telephones and metal appliances should also be avoided during thunderstorms.

FIGURE 10-11 Hailstorms are a threat to agricultural crops because they can badly damage or completely destroy a crop in a matter of minutes, as in this photograph showing winter wheat after a hailstorm. (Photo courtesy of S. A. Changnon, Illinois State Water Survey.)

and wheat (Fig. 10-11), which are not only susceptible to hail damage because of their structural characteristics, but are at vulnerable stages (young corn and mature wheat) during the time of year when hail is most likely. Wildlife also suffers from hailstorms. Animals as large as horses have been killed by hail. Automobile and airplane damage is also common as a result of severe hailstorms (Fig. 10-12).

Since hail is associated with thunderstorms, you might assume that the geographical distribution of hail is the same as for thunderstorms. This is not true, however. Although hail and tornadoes frequently occur from the same thunderstorm, the distribution of hail activity (Fig. 10-13) is different from that of tornadoes. The maximum hail frequency lies near the border of Colorado and Wyoming, where hail can be expected to occur on eight days out of the year. As you will recall, most thunderstorm activity is in the southeastern United States. This area has, however, very little hail, averaging less than one day each year.

The amount of hail that falls can sometimes be tremendous. Hail drifts more than 2 m high have been recorded. Such drifts result from hail that falls with heavy rain, where the runoff drifts the hail-

FIGURE 10-12 Hailstones are a threat to personal property and can cause considerable damage to houses, automobiles, and airplanes on the ground or flying through hail, as in this case. (Photo courtesy of NOAA.)

stones. Hail accumulations may be as much as 20 cm over an area (hail swath) of perhaps 10 by 50 km.

NATURE OF HAILSTONES

A hailstone is composed of layers of ice much like an onion is composed of layers, with some of the rings having a milky appearance and others being transparent. This difference in appearance is related to the way the ice was frozen. If it is very cold with rapid freezing, snowflakes and air bubbles are incorporated into the layer, giving the milky appearance. If the water freezes more slowly, there is more time for snow to melt and air bubbles to escape, with the result that a clear ring is formed.

Hailstones form in various shapes; small ones are commonly circular, but large ones are quite rough and knobby because of water collecting in globs on the outside as the water freezes (Fig. 10-14). Frequently, the stones are somewhat elliptical, or are **oblate spheroids.** Some are conical like an ice cream cone, particularly the intermediate sizes, while some are flattened spheres. The var-

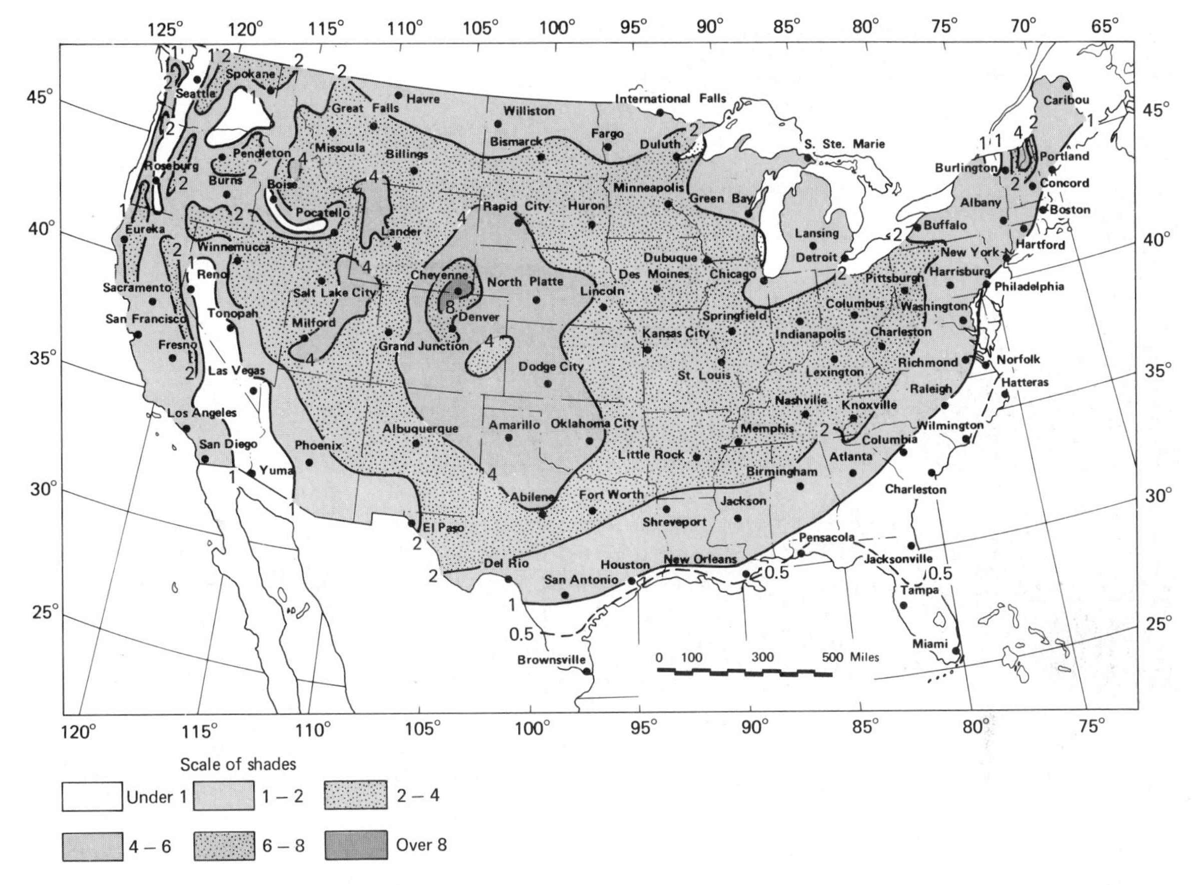

FIGURE 10-13 Distribution of hail in the United States. The numbers show the mean annual number of days with hail. (Environmental Data and Information Service, NOAA.)

ious hailstone shapes are caused by the dynamics of airflow around the ice as the stones take preferred patterns of fall through the air.

The most common hailstone size is slightly smaller than grape size. Very large hailstones of baseball or softball size are much rarer. The largest measured hailstone fell in 1970 near Coffeyville, Kansas (Fig. 10-15); it was 19 cm ($7\frac{1}{2}$ in) in diameter, 44 cm in circumference, and weighed 766 g (1.7 lbs).

The color of thunderstorms containing large hail is frequently different from an ordinary thunderstorm. The hailstorm has a distinctive greenish color because hailstones in the cloud reflect light differently from raindrops.

SYNOPTIC WEATHER FOR HAIL

The precise atmospheric conditions that lead to hail production are not well known. Many of the same factors that cause tornadoes also cause heavy hail, since thunderstorms with tornadoes are frequently hail producers. However, many thunderstorms that produce hail do not produce tornadoes. High-speed updrafts are probably more important for

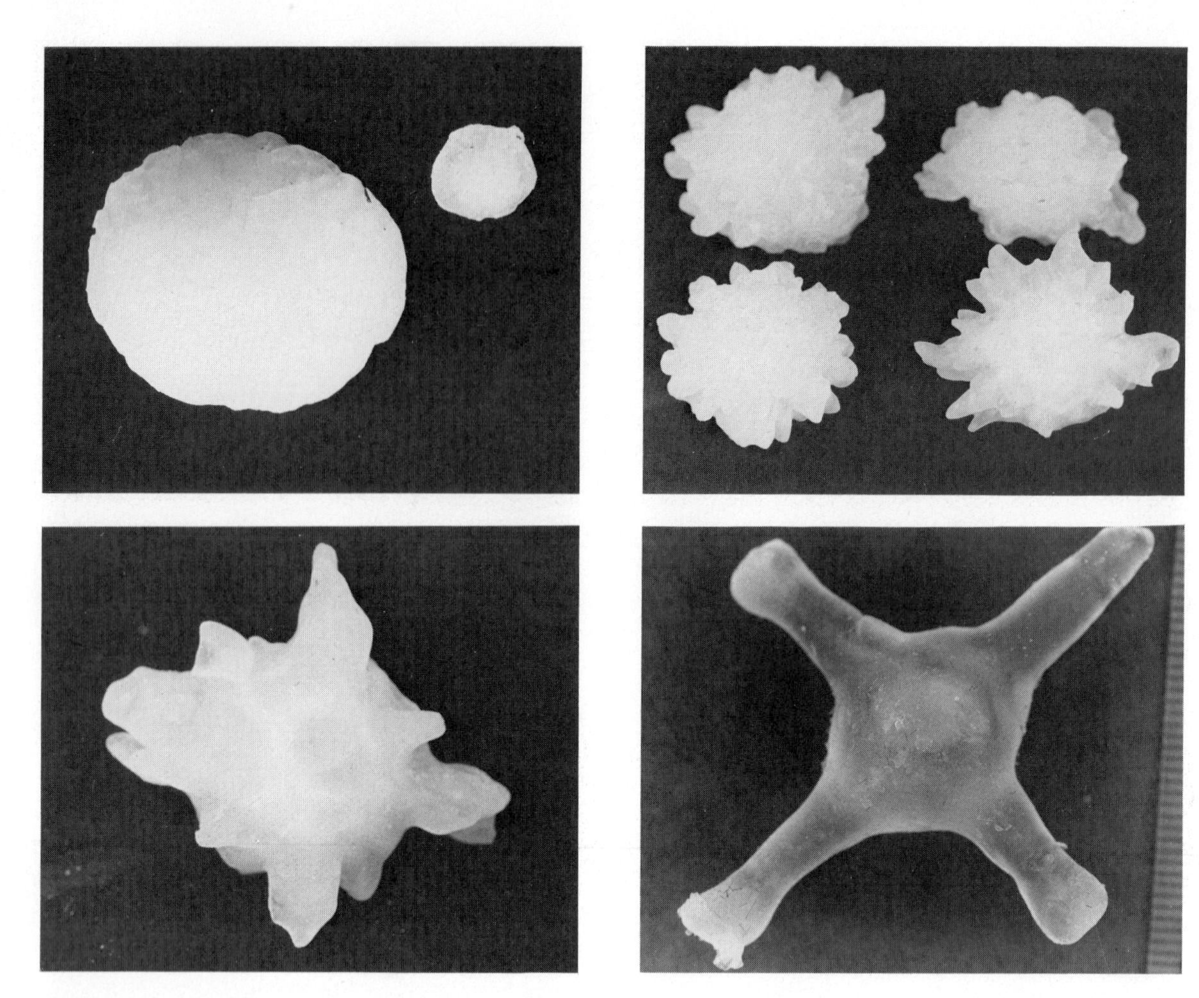

FIGURE 10-14 Hailstones range in shape from circular to very irregular. Large hailstones are seldom smooth or round. (Photo courtesy of C. A. Knight, NCAR.)

hail production, whereas strong horizontal circulation is critical in the production of tornadoes.

The speed of the thunderstorm updraft required to support large hailstones is quite fantastic. Computations show that an updraft of at least 150 km/h would support a stone of 5 cm while an updraft of more than 300 km/h is necessary to support a stone of 10 cm. Because these very strong updrafts are known to occur in thunderstorms, pilots of small aircraft do not fly through them.

The synoptic situation for hail generally includes the movement of a cold front through the region with hail formation in the warm moist air ahead of the cold front, similar to the synoptic conditions for tornadoes (Fig. 10-16). The presence of a jet stream is also a factor in hail production. Studies have shown that a greater wind shear, as determined by the difference in surface winds and winds near the 500-mb level, is directly related to heavier hailstones. We know that an upper air inversion

FIGURE 10-15 This photograph of the largest hailstone ever recorded shows the knobby outer appearance, which develops in a manner similar to the formation of icicles as liquid water collects around the falling hailstone. The illustration on the right shows the rings corresponding to different freezing mechanisms. This hailstone fell near Coffeyville, Kansas, in September 1970 and weighed 766 g. (Photo courtesy of NCAR.)

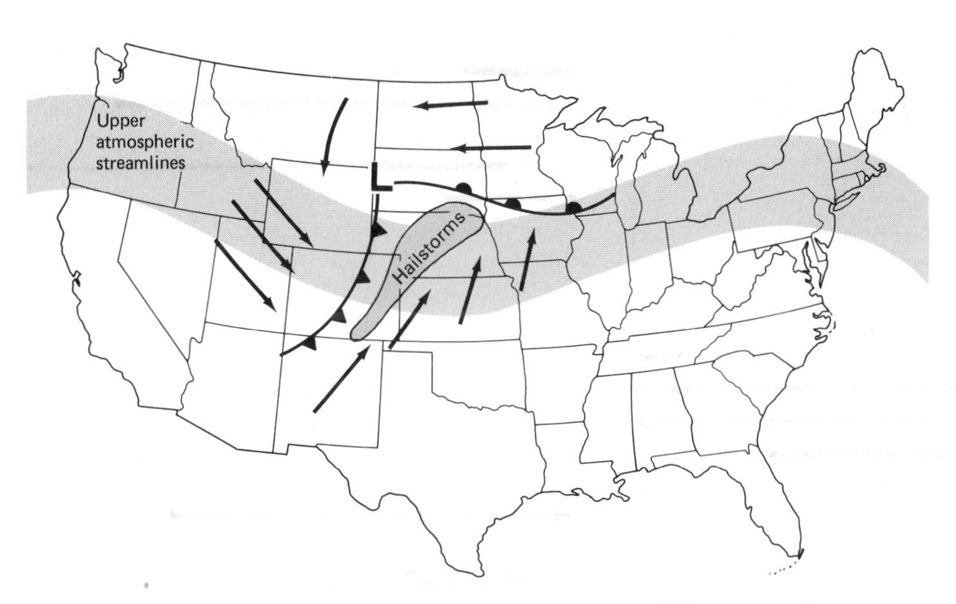

FIGURE 10-16 Hailstorms are more likely within the warm air mass preceding a cold front, especially if an area of upper atmospheric divergence occurs above the warm air mass.

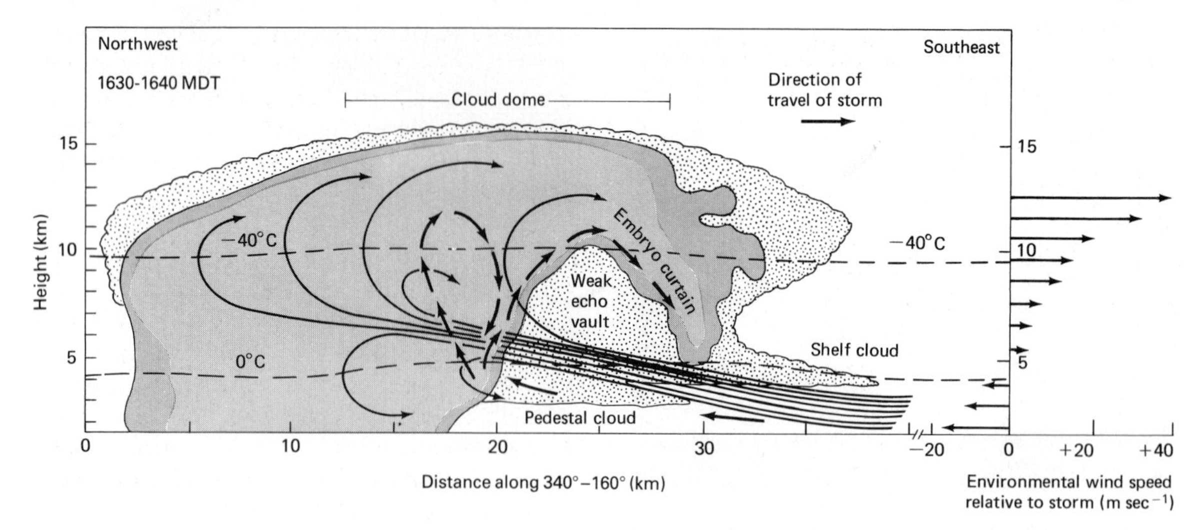

FIGURE 10-17 Vertical cross section through a supercell hailstorm oriented from northwest (340°) to southeast (160°), showing the visual cloud boundaries of the storm that occurred at 1630, June 21, 1972, near Fleming, Colorado. The radar echo is superimposed, with two levels of radar reflectivity along with arrows indicating the airflow within the storm. The short arrows show the trajectories two hailstones would follow within the updraft and near the weak echo vault. The arrows on the right of the diagram represent the environmental wind profile along the storm's direction of travel. (Adapted from Browing and Foote, 1976, *Q. J. R. Meteorol. Soc.*, London.)

is a factor in tornado development. However, studies of hailstorms have shown that no upper air inversion exists when hailstorms develop. Thus the atmospheric conditions are slightly different when hail-producing thunderstorms form.

HAIL-PRODUCING THUNDERSTORMS

Thunderstorms that produce hail may be of three different varieties. One is the isolated thunderstorm that develops into a very large storm in the warm air mass ahead of the cold front. Hailstorms also occur in squall lines in the area of propagation of new cells in advance of the squall line or in multicell thunderstorms. New growth in multicell storms generally occurs on the upwind side of the older mature cell. Many features of hailstorms are very similar to severe thunderstorms that produce tornadoes. Similarities include the presence of a shelf cloud (horizontal extension of the cloud base) at the leading edge and a pedestal cloud (lower, southern part of a thunderstorm shaped like a pedestal) associated with the mesocyclonic rotation (Fig. 10-17). A notch may be seen corresponding to the inflow of the thermal updraft. The radar echo indicates that the heaviest rain is in the leading northeast part of the thunderstorm, while the typical location of hail at the ground is underneath the central part of the thunderstorm.

Large thunderstorms are more likely to produce hail than smaller ones. An analysis of hailstorms in South Dakota showed that hail commonly occurs from thunderstorms that grow to a height of 15 km. Probabilities of the formation of hail were calculated using height of thunderstorms as the indicator (Fig. 10-18). If the height of the thunderstorm is 13 km, a 50% probability exists that the storm will produce hail. A thunderstorm growing to 15 km has a 75% probability of producing hail, while

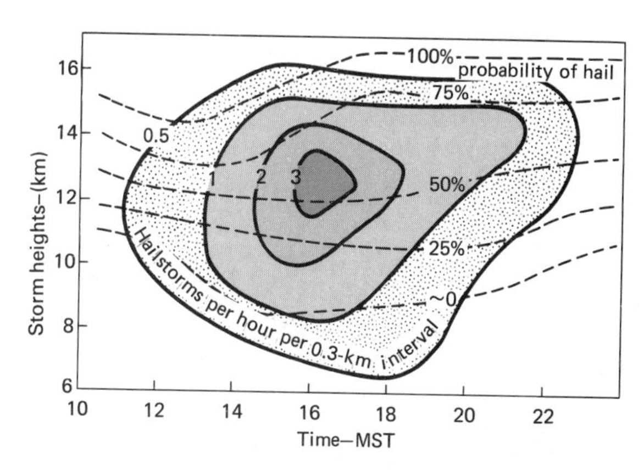

FIGURE 10-18 Distribution of hailstorms (solid lines showing number of hailstorms per hour per 0.3-km interval) with the probability (dashed lines in percent) of a storm becoming a hailer. (From A. S. Dennis, P. L. Smith, E. T. Boyd, and D. T. Musil, "Radar Observations of Hailstorms in Western Nebraska," Final Report NSF Grant, GA-1518, 1971.)

growth to 17 km results in a probability of nearly 100%. In other locations, thunderstorms of similar and greater heights (18 km in Texas) produce hail southward across the United States, while thunderstorms of much lower heights (perhaps 8 km) can produce hail in Canada.

Hail formation in thunderstorms begins as embryos within the thermal updraft form as ice around dust or other foreign material (Fig. 10-19). The hail embryos may be carried between the double vortex structure within the thunderstorm, where they rise in the updraft and are carried up toward the anvil where they drop back down until they again hit an updraft and are carried upward again until they make several trips and become too heavy to be supported (Fig. 10-20).

Pilots do not fly aircraft under the anvils of large thunderstorms, since they have learned that hail may be falling through clear air as the thunderstorm cycles its hailstones. The number of cycles

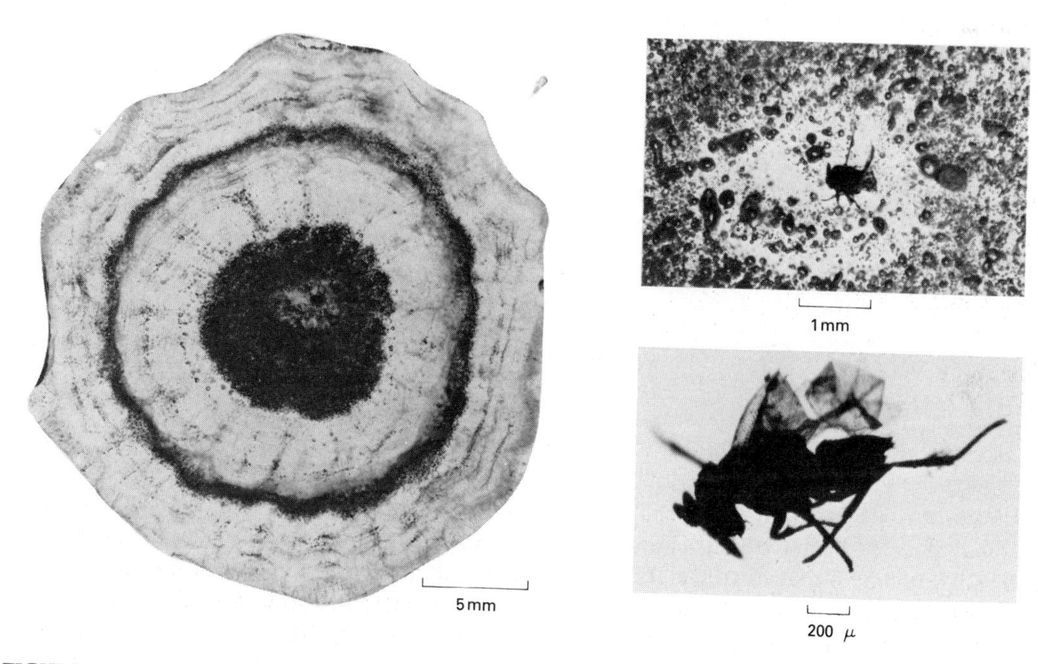

FIGURE 10-19 Hail forms around tiny embryos, in this unusual case a small insect (chalcidoidea). This hailstone fell in Norman, Oklahoma, June 14, 1975. Layers of ice form around the hail embryo enlarging it as it is carried by the air currents within a thunderstorm. (Courtesy of N. C. and C. A. Knight, NCAR.)

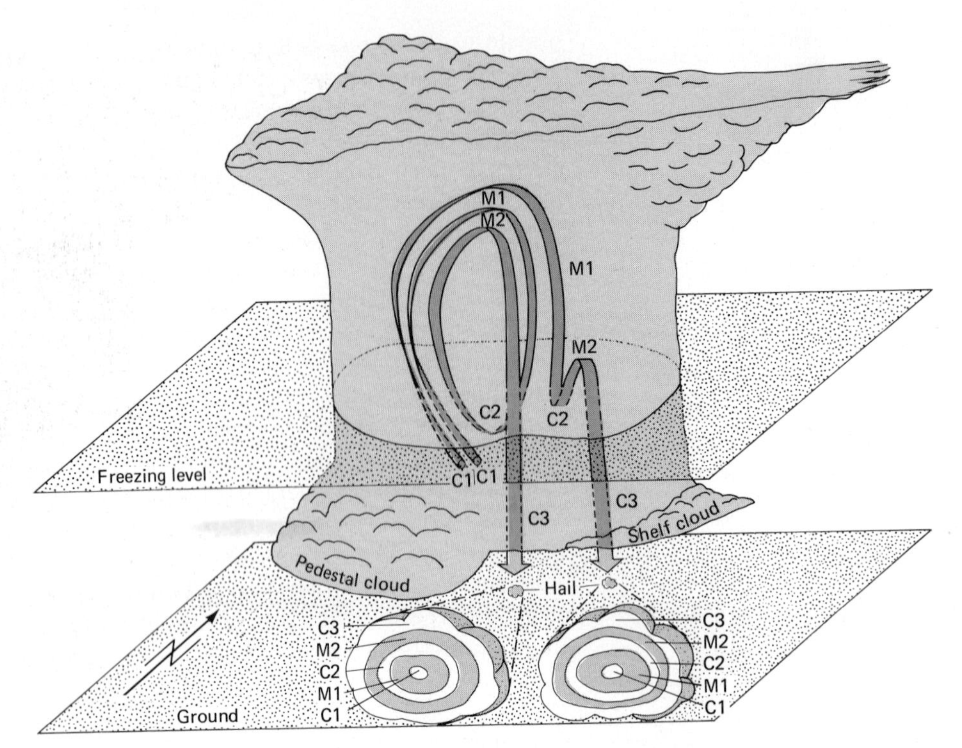

FIGURE 10-20 Hail forms in a thunderstorm as the updraft between the counterrotating vortices carries a hailstone to the upper part of the thunderstorm, where it falls out of the updraft to lower levels, where it may again be caught in an updraft or may encounter turbulence near the freezing level and fall to the ground as a smaller hailstone. Thus, the rings in hail can occur from several trips through a thunderstorm or from turbulence that carries a hailstone back and forth across the freezing level. In the diagram, *C* stands for clear and *M* stands for milky.

through the thunderstorm probably determines the number of rings of ice in a single stone. The rings are related to the number of trips across the freezing level at perhaps 1 km in the atmosphere. Some suggestions have been made that a single path through the thunderstorm is all that is required to produce large hail. In order to develop rings from a single **trajectory** through the thunderstorm, the rings would have to be produced by turbulence at the freezing level, where the hailstone would bounce across the freezing level for some time, as shown in Figure 10-20.

SUMMARY

The development of electricity in clouds is associated with raindrop formation within updrafts and downdrafts of a thunderstorm. Spontaneous freezing and shattering of supercooled water droplets helps initiate the charge separation within a thunderstorm. As the top of a thunderstorm becomes positively charged and the base negatively charged, an induced charge occurs on all ice particles and water droplets within the central part of the thunderstorm. The small cloud droplets within an intense updraft bleed off the positive charges from the lower side of larger ice pellets. The concentration of smaller cloud droplets that are positively charged in the upper part of the thunderstorm intensifies the charge separation process.

Thunder is produced by the lightning discharge as the conducting channel of air is heated to five times the temperature of the sun in a very short time. The sudden expansion of the air in the conducting channel creates shock waves that are heard as thunder.

The lightning flash consists of several strokes. The first of these, stepped leader, creates the conducting channel from the cloud base to the ground. After the conducting channel is developed, a return streamer occurs from the ground to the cloud.

Several different forms of lightning are common. The most frequent lightning discharge is within a single cloud and neutralizes the charged negative and positive regions within it. Another common type of lightning is forked lightning, which discharges between the cloud base and ground. Ribbon lightning occurs as strong winds displace the conducting channel, making the separate lightning strokes visible. Sheet lightning is seen as distant thunderstorms produce lightning that lights up part of a cloud. Heat lightning is similar, except it originates from thunderstorms that are even farther away and lights the whole sky.

Air discharges occur as the leader stroke beneath the cloud base fails to reach the ground. If the air discharge travels for several kilometers and then reaches the ground some distance from the thunderstorm, it produces a bolt from the blue. Ball lightning develops as a ball of electricity falls from the cloud base or rolls downhill.

Lightning is responsible for about 97 deaths per year in the United States. Over one-half of these deaths occur outdoors; dangerous locations include under tall trees, near fences, on a tractor with an implement in the ground, or on a golf course with a golf club over your head. Unsafe locations indoors are near telephones or metal appliances. However, properly installed lightning rods are effective in protecting buildings from lightning damage.

Hailstorms cause considerable damage to agricultural crops each year in addition to damage to personal property and animal life. Hailstorms are a great problem in the central Great Plains especially in northeastern Colorado and southeastern Wyoming.

Hailstones are composed of alternating layers of milky opaque ice and layers of clear ice. The thunderstorm environment where the hailstones originated determines the appearance of the ice layer. Opaque layers are formed as the ice freezes rapidly; clear layers form as the ice freezes more slowly. Hailstones come in various shapes. Large ones are very seldom round but are frequently knobby and elongated or conical. The largest hailstone that has been measured was 19 cm in diameter, 44 cm in circumference, and weighed more than 700 g. High-speed updrafts are required in thunderstorms to form large hailstones. An updraft of 100 km/h is required to support a hailstone of only a few centimeters diameter.

Some important atmospheric conditions that contribute to the development of hailstones are warm moist air ahead of a cold front, high wind speeds within the jet stream, large wind shear, and an unstable atmosphere. Hailstorms may occur from any of the three different types of thunderstorms: supercell, multicell, or squall line. Measurements of the height of thunderstorms show that the tallest ones are more likely to produce hail, with the average height of hail producers being 18 km in Texas and 13 km in South Dakota. Hail is most likely to fall beneath the central part of a thunderstorm, while heavy rain generally occurs beneath the leading edge of the thunderstorm, and tornadoes, if present, are more likely on the southern edge of a thunderstorm moving toward the northeast. Hailstones are produced by the updrafts and downdrafts that develop in a large thunderstorm.

STUDY AIDS

1. What is the most common type of lightning?

2. Compare some of the different types of lightning.

3. Explain the development of electricity in clouds.

4. Explain the formation of thunder.

5. Explain the nature of the lightning flash.

6. List some dangerous places during a thunderstorm together with information on how to seek better protection from lightning.

7. Discuss the layered nature of hailstones as related to formation.

8. Describe some of the atmospheric conditions associated with hailstorm development.

9. Discuss thunderstorm size in relationship to hail formation.

10. Compare the location of hail, heavy rain, and tornadoes within a thunderstorm.

TERMINOLOGY EXERCISE

Consult the glossary if you are unsure of the meaning of any of the following terms used in Chapter 10.

Thunder	Lightning stroke
Thermoelectric effect	Leader stroke
Induced charge	Air discharge
St. Elmo's Fire	Bolt from the blue
Sheet lightning	Ribbon lightning
Heat lightning	Ball lightning
Forked lightning	Hailstorm
Lightning flash	Oblate spheroid
	Trajectory

THOUGHT QUESTIONS

1. Do you think there is any relationship between the buildup of electrical charges in clouds and the precipitation formation process? Explain your answer.

2. Ball lightning is not well understood. Can you suggest ways in which it could be investigated more thoroughly?

3. Compare some of the beneficial and harmful effects of lightning.

4. Can you suggest reasons why the geographical distribution of hail is quite different from the geographical distribution of thunderstorms?

5. Can you offer suggestions for dealing with the hail problem other than crop and personal property insurance?

SPECIAL TOPIC 12

SOLAR FLARES LINKED WITH ATMOSPHERIC ELECTRICITY

Large electrical currents detected over the South Pole during a **solar flare** (disturbance on the surface of the sun) offer evidence that atmospheric electricity is directly influenced by the sun, according to a National Oceanic and Atmospheric Administration (NOAA) scientist (Fig. S12-1).

NOAA researcher William E. Cobb says that these findings sharply contradict the prevailing "global circuit theory," which describes how electricity flows from the earth to the atmosphere and back again. This theory holds that the electrical current that flows upward from thunderstorms is balanced by a weak, but widespread, air-to-earth current observed in fair weather, and that the global circuit of atmospheric electricity is controlled solely by thunderstorms.

South Pole measurements provide strong evidence to the contrary, indicating that bursts of energetic particles and radiation from solar flares strengthen this ordinarily weak return current. This would make the sun an important partner in regulating electricity in the global atmosphere.

In addition, Cobb said, it is possible that this increased electrical activity may cause the formation of more lightning strokes in thunderstorms, which in turn could affect the rain-producing efficiency of the 2000 thunderstorms estimated to be in progress over the earth's surface at any given moment.

Electrical currents were measured by balloon-borne "electrosondes" launched from Amundsen-Scott Station as part of a continuing research effort by the Commerce Department's Atmospheric Physics and Chemistry Laboratory, one of NOAA's Environmental Research Laboratories, in Boulder, Colorado, and the National Science Foundation.

Coincidentally, the first balloon was released seven hours before a solar flare occurred. While that sounding showed the atmospheric electric conditions in their normal state, subsequent ones detected a strong increase in the air-to-earth current over the next two days that finally exceeded the measuring capacity of the electrosonde at altitudes of 25 to 30 km (16 to 19 mi). From the stratosphere to the surface, Cobb reported, the measured electrical current exceeded values obtained before the flare by more than 70%.

This surge was followed by a gradual return to the weak "normal" current that flows from the atmosphere to the earth in fair weather.

These large fluctuations in electricity strongly reinforce previous findings at mountain observatories in Hawaii and Germany, and indicate that the global circuit theory should be revised to accommodate solar, as well as thunderstorm, effects.

From *U.S. Dept. of Commerce News*, by Richard E. Newell.

(Continued on next page)

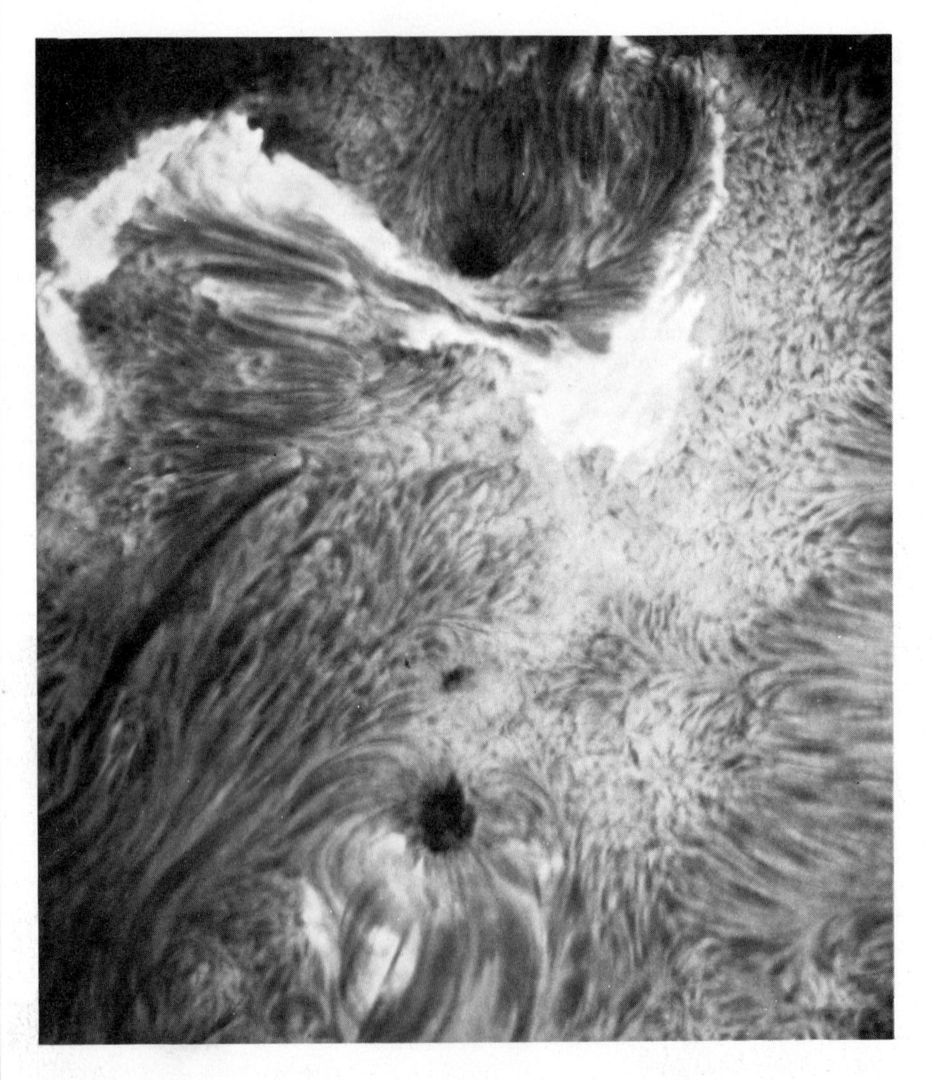

FIGURE S12-1 Photographs of the sun frequently reveal sunspot pairs (dark areas at the top and bottom) and solar flares (light area between). Evidence is accumulating that suggests that such solar activity influences our weather on earth. (Air Force Cambridge Research Laboratories.)

CHAPTER ELEVEN

TORNADOES

APPEARANCE

Tornadoes are the strongest and most powerful of all the atmospheric storms. Fortunately they are smaller than the other major atmospheric storms—hurricanes and midlatitude cyclones. The intense winds of the tornado are physically possible because they are small in size compared to the surrounding atmospheric circulation. The conservation of angular momentum can be used to gain some information on tornado wind speeds, since this is a physical principle that relates circular wind speeds to diameter of the circulation. A typical tornado is 150 m in width and stays on the ground for 10 km, while the thunderstorm that produced it may be 30 km in diameter. As the larger circulation is reduced to a smaller radius, the wind speed must increase if momentum is conserved.

A large thunderstorm that develops one tornado is likely to develop another after the dissipation of the first. Tornadoes frequently last for only a few minutes, but can be very destructive during this time. The high winds combined with the low pressure inside the tornado funnel can lift and destroy houses, cut swaths through forests, and cause strange effects, such as wheat straw penetrating wood poles, for example. Tornadoes also have considerable psychological impact on many people. It is little wonder since tales of tornado oddities are quite common. A tornado in Mountain View, Missouri, picked up a man and carried him up into the funnel. While in the air a car came by and he tried to catch it to get inside. Although he was unable to grab it, the tornado set him down in a field without harming him.

The way a tornado looks depends on the background behind it and on the debris and amount of condensed water vapor due to low pressure within

FIGURE 11-1 Tornadoes occur in a variety of shapes and sizes. They may be large in diameter with irregular edges, or they may appear as a very smooth cylinder or funnel as they extend to the ground. They frequently show a life cycle where the first stage is a vortex with a larger diameter and later stages are very thin rope-shaped funnels. Two or more funnels may be on the ground at once, as shown in the middle photograph in the second row. Occasionally the vortex is not visible between the cloud base and the earth surface, as in the second photograph from the left in the second row. Circulation of the larger cloud above is evident in three of the photographs in the bottom row. (Photo courtesy of NOAA.)

it. Tornadoes may complete a life cycle where the first stage is characterized by a large vortex with irregular edges (Fig. 11-1). Later the tornado may become very narrow until it is rope shaped. This generally corresponds to the last stage of the tornado. Sometimes in tornado and waterspout photographs distinct cylindrical layers are visible; these are caused by vertical motion through the core of the funnel with horizontally rotating air surrounding the core. The vortex occasionally has

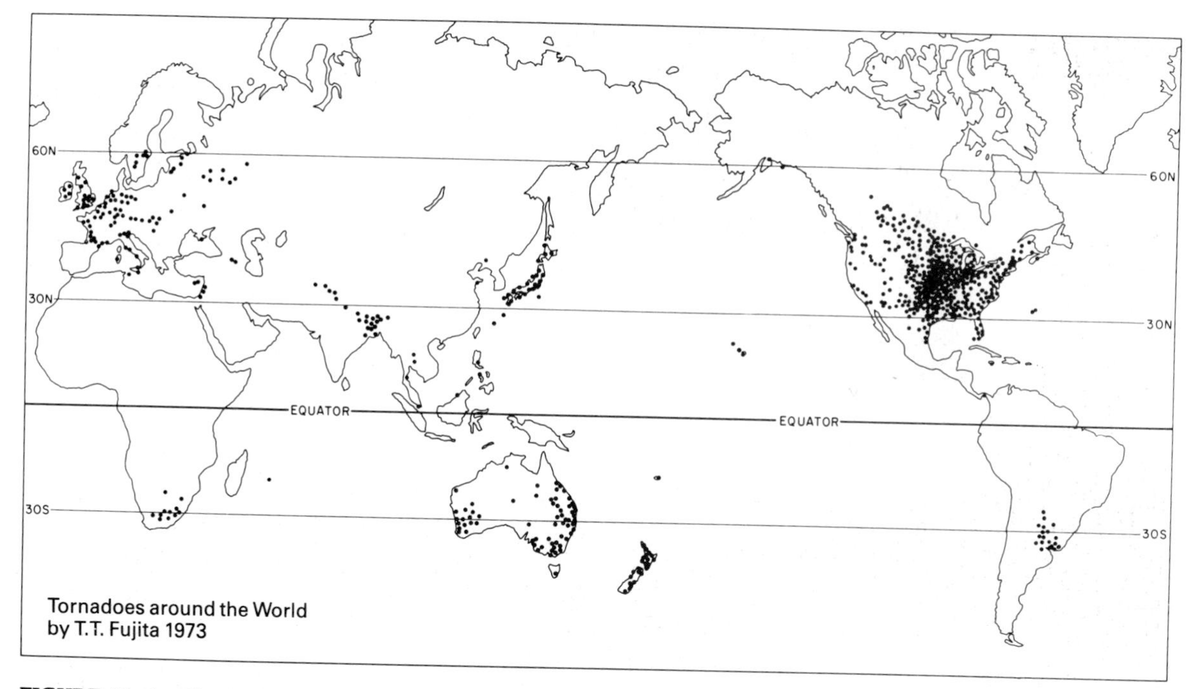

FIGURE 11-2 Distribution of tornadoes around the world expected per 4-year period. (Reprinted with permission of T.T. Fujita, University of Chicago.)

large bends near the ground as it extends downward from the cloud.

A blue sky background behind a tornado viewed from the northeast makes the funnel look dark. In contrast, if the intense rain area is behind the tornado, it may look white. This normally occurs if the observer is located south of the tornado. However, the tornado may look quite dark if it has collected debris and dust from the ground. It may also look red after passing over red soil and picking up dust.

Multiple tornadoes may develop from a single thunderstorm, as shown in Figure 11-1. The shape of a tornado is normally so distinct that it is easy to identify, although a **rain shaft** (streak of rain falling from a thunderstorm) may occasionally be confused with a tornado.

OCCURRENCE OF TORNADOES

The United States has many more tornadoes than other countries although they occur in many other parts of the world where the jet stream crosses land. Europe and Australia have tornadoes (Fig. 11-2), but the United States averages more than 700 tornadoes per year (Table 11-1), while its closest competitor is Australia with only 15 tornadoes per year, although it is almost as large as the United States. The average number of **tornado days** (any day with a tornado somewhere in the United States) is 169. Tornadoes occur in all parts of the United States (Fig. 11-3) although they are rare in the Rocky Mountain states and along the western coast as well as in Alaska and Hawaii.

TABLE 11-1

Number of tornadoes, tornado days, and deaths by states, 1953–1980 (Climatological Data National Summary)

	Tornadoes							Days	
State	Total	Average	Greatest	Year	Least	Year	Per 10,000 Sq. Mi.[d]	Total	Average
Alabama	563	20	45	1973[a]	5	1956	3.90	307	11
Alaska	1	0	1	1959	0	1980[a]	.00	1	0
Arizona	100	4	17	1972	0	1965[a]	.31	83	3
Arkansas	571	20	50	1973	2	1969	3.84	283	10
California	96	3	13	1978	0	1968[a]	.22	72	3
Colorado	460	16	42	1976	1	1959	1.58	303	11
Connecticut	41	1	8	1973	0	1980[a]	2.92	37	1
Delaware	26	1	5	1975	0	1980[a]	4.51	24	1
District of Columbia	0	0	0	—	0	1980[a]	.00	0	0
Florida	1155	41	97	1975	10	1956	7.04	737	26
Georgia	594	21	46	1971[a]	7	1960	3.60	341	12
Hawaii	17	1	4	1971	0	1978[a]	.95	14	1
Idaho	37	1	5	1967	0	1977	.16	29	1
Illinois	750	27	107	1974	4	1953	4.75	350	13
Indiana	638	23	48	1973	6	1972[a]	6.28	302	11
Iowa	766	27	54	1964	7	1956	4.86	356	13
Kansas	1212	43	97	1955	14	1976	5.26	563	20
Kentucky	226	8	34	1974	0	1953	2.00	127	5
Louisiana	546	20	55	1974	3	1955	4.02	346	12
Maine	70	3	11	1971	0	1980[a]	.75	62	2
Maryland	78	3	10	1975	0	1970[a]	2.63	62	2
Massachusetts	107	4	12	1958	0	1959	4.63	77	3
Michigan	443	16	39	1974	2	1959	2.72	258	9
Minnesota	469	17	34	1968	5	1972	1.99	282	10
Mississippi	612	22	44	1973	1	1979	4.58	323	12
Missouri	758	27	79	1973	6	1953	3.88	361	13
Montana	116	4	13	1978	0	1974[a]	.28	86	3
Nebraska	985	35	78	1975	10	1966	4.56	485	17
Nevada	18	1	4	1964	0	1980[a]	.06	17	1
New Hampshire	60	2	9	1963	0	1979[a]	2.30	54	2
New Jersey	44	2	8	1973	0	1978[a]	2.01	36	1
New Mexico	232	8	18	1972	0	1953	.68	175	6
New York	100	4	8	1978	0	1953	.72	85	3
North Carolina	328	12	38	1973	2	1970	2.22	209	7
North Dakota	473	17	52	1976	2	1961	2.39	271	10
Ohio	402	14	43	1973	3	1976	3.48	218	8
Oklahoma	1477	53	105	1957	21	1978	7.54	630	23
Oregon	24	1	3	1975[a]	0	1980[a]	.09	20	1
Pacific	1	0	1	1975	0	1980[a]	—	1	0
Pennsylvania	218	8	23	1976	0	1959	1.72	156	6
Puerto Rico	9	0	2	1979[a]	0	1980[a]	.94	8	0
Rhode Island	1	0	1	1972	0	1980[a]	.29	1	0
South Carolina	255	9	23	1973	1	1970[a]	2.93	175	6
South Dakota	683	24	64	1965	1	1958	3.17	335	12
Tennessee	316	11	44	1974	1	1962	2.67	172	6
Texas	3344	119	232	1967	32	1953	4.47	1372	49
Utah	32	1	5	1970[a]	0	1980[a]	.13	25	1
Vermont	24	1	5	1962	0	1980[a]	.89	21	1
Virginia	162	6	22	1975	1	1963	1.42	113	4
Virgin Islands	2	0	1	1979[a]	0	1980[a]	—	2	0
Washington	32	1	4	1978	0	1977[a]	.17	26	1
West Virginia	60	2	6	1980[a]	0	1960[a]	.89	46	2
Wisconsin	514	18	43	1980	0	1960[a]	3.27	288	10
Wyoming	253	9	42	1979	0	1970	.92	179	6
TOTAL: United States	20359[b]	727	1102	1973	421	1953	2.01	4743[c]	169

Deaths		
Total	Average	Per 10,000 Sq. Mi.[e]
202	7	39
0	0	0
3	0	0
122	4	23
0	0	0
2	0	0
4	0	8
0	0	0
0	0	0
52	2	9
72	3	12
0	0	0
0	0	0
129	5	23
205	7	56
54	2	10
162	6	20
101	4	25
88	3	18
1	0	0
1	0	1
99	4	120
231	8	40
74	3	9
316	11	66
120	4	17
0	0	0
49	2	6
0	0	0
0	0	0
0	0	0
3	0	0
2	0	0
22	1	4
21	1	3
147	5	36
177	6	25
0	0	0
0	0	0
8	0	2
0	0	0
0	0	0
24	1	8
8	0	1
74	3	18
371	13	14
0	0	0
0	0	0
16	1	4
0	0	0
6	0	1
1	0	0
56	2	10
2	0	0
2325	98	6

[a]Also in earlier year(s)

[b]Corrected for boundary-crossing tornadoes

[c]Tornado days for country as a whole

[d]Mean annual tornadoes per 10,000 square miles

[e]Number of deaths per 10,000 square miles

Tornadoes that travel across shorelines are called **waterspouts** over water. However, the typical waterspout in the Gulf of Mexico or along the eastern coast of the United States occurs from small thunderstorms, with tops only 8 km high or less (Fig. 11-4). Waterspouts actually have three different formation mechanisms. They occur (1) as land-based thunderstorms move over water, (2) from smaller cumulonimbus clouds that develop over oceans or large lakes, and (3) in thunderstorms within the rainbands of hurricanes.

The most frequent type of waterspout develops from small thunderstorms that may be only half the size of a Great Plains thunderstorm that produces a tornado. The generation mechanism of waterspouts from such small thunderstorms has not been entirely explained, but is made possible by the warm humid air over the water combined with large updrafts in small cumulus clouds above. The typical waterspout is not as destructive as a typical tornado, but nevertheless may do severe damage. For example, on February 7, 1971, a single waterspout moved across the shoreline into Pensacola, Florida, and caused $3 million worth of damage. Waterspouts are probably more frequent in the Florida Keys than anywhere else in the world, with smaller numbers occurring in other locations in the Gulf of Mexico and along the eastern coast of the United States (Table 11-2).

The topography and pressure patterns combine to make the central and eastern United States the playground of tornadoes (Fig. 11-5). The Rocky Mountains form a barrier for western surface winds. The frequent existence of a large high-pressure area over the Atlantic at about 30° N latitude causes south winds from the Gulf of Mexico to carry warm, humid air into the Great Plains. The jet stream flowing over the Rocky Mountains from the west is much less humid, providing an extreme contrast in density between these two distinct air masses. Troughs in the jet stream are more likely to occur on the leeward side of the mountains because of the dynamics of flow over them, thus favoring midlatitude cyclone formation and resulting tornado activity there.

The location of tornadoes in the United States follows the seasonal migration of the jet stream

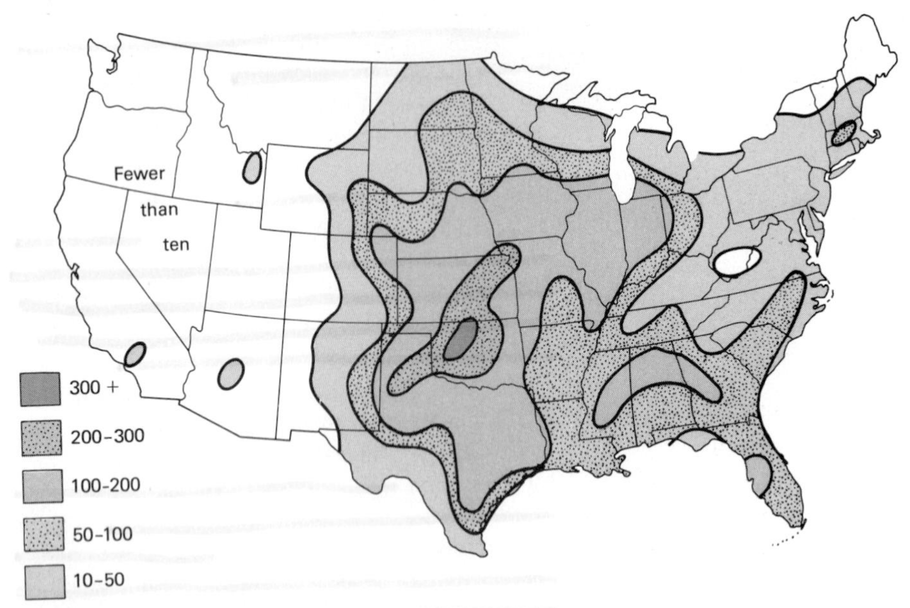

FIGURE 11-3 Number of tornadoes in the United States 1955–1967. (From M. E. Pautz, ESSA Technical Memo WBTMRCSE 13, Dept. of Commerce, 1969.)

FIGURE 11-4 A waterspout is a vortex similar to a tornado. More than one can be formed by a single thunderstorm. (Photo courtesy of NOAA.)

(Fig. 11-6). For this reason, during the winter tornadoes are more likely to occur in the southern United States along the Gulf. The jet stream provides the energy and vorticity that is helpful in developing severe thunderstorms and tornadoes. It is also located at the boundary between warm air masses from the south and cold air masses from the north. The central United States is the area of maximum tornado activity during the spring, while tornadoes are more likely in the north central United States during the summer, as the jet stream moves toward the Canadian border.

Statistics on the seasonal distribution of tornadoes for the whole United States show that more cornadoes occur in the month of May than in any other month (Fig. 11-7). Tornadoes have occurred in each of the other months, but these are generally displaced geographically, so most of them are located near the jet stream as it moves with the seasons and short-term fluctuations.

TABLE 11-2

Locations where waterspouts are frequent along U.S. coastlines, based on reports from 1959–1973

Location	Waterspouts per year per 1000 km²
Florida Keys	43.8
Palm Beach, Fla.	3.1
Tampa Bay, Fla.	2.4
Ft. Lauderdale, Fla.	2.4
Corpus Christi, Tex.	2.3
Greater Miami, Fla.	2.2
Pensacola, Fla.	1.8
Mississippi Sound, Mass.	1.2
Port Arthur, Tex.	1.2
Galveston Bay, Tex.	0.9
Mississippi River Delta	0.6
Ft. Myers, Fla.	0.5

Source: From J. H. Golden, "An Assessment of Waterspout Frequencies along the U.S. East and Gulf Coasts," *J. Appl. Meteorol.*, vol. 16, no. 3, 1977.

The number of days with tornadoes, tornado days, is greatest in June although more individual tornadoes occur in May. A tornado occurs somewhere in the United States on an average of 26 of the 30 days in June or on 87% of the days. Tornadoes occur on 23 days in May on the average. Most tornadoes occur between 2 and 9 P.M., with most occurring at 5 P.M. Each year an average of 98 people are killed by tornadoes, several thousand are injured, and several millions of dollars worth of damage occurs (Fig. 11-8).

Outstanding tornadoes include the tri-state tornado that developed in Missouri on March 18, 1925, and moved through Illinois and Indiana. This tornado was on the ground for more than 325 km, causing 689 deaths and injuring almost 2000 people. More fatalities resulted from the fact that the tornado occurred before the first tornado forecasts (made in 1948) as well as before the current modern communication networks. Another tornado completely destroyed the city of Udall, Kansas, on May 25, 1955, at 10:30 P.M., killing 80 people and injuring 270 others.

A few single tornadoes have caused over $100 million worth of damage. The first of these was in Topeka, Kansas, on June 8, 1966, when 16 people were killed. On May 11, 1970, a tornado struck Lubbock, Texas, causing damages of more than $100 million and killing 26 people. A third tornado causing such extensive damage occurred in Omaha, Nebraska, on May 6, 1975, although only 3 people were killed. One of the most destructive was the tornado in Wichita Falls, Texas, on April 10, 1979, which killed more than 40 people and caused great monetary loss.

The two worst tornado days were April 11, 1965, and April 3, 1974. On April 11, 1965, there were 256 deaths from 47 tornadoes over five states, with over $200 million worth of damage resulting. However, the worst outbreak of tornadoes

FIGURE 11-5 Because of the pressure patterns and topographic features, the region of most frequent tornado activity is the central and eastern United States. Upper atmospheric winds are frequently from the west, bringing cooler air over the mountains, as indicated by the direction of the thunderstorm anvils shown in this satellite photograph. The low-level winds in the eastern two-thirds of the United States are frequently from the south as the air circulates anticyclonically around a large high-pressure system centered eastward over the Atlantic Ocean. Since troughs in the jet stream tend to occur on the leeward side of mountains, the central and eastern part of the United States is an ideal location for the occurrence of all the atmospheric factors required for severe thunderstorm formation. (Photo courtesy of NASA.)

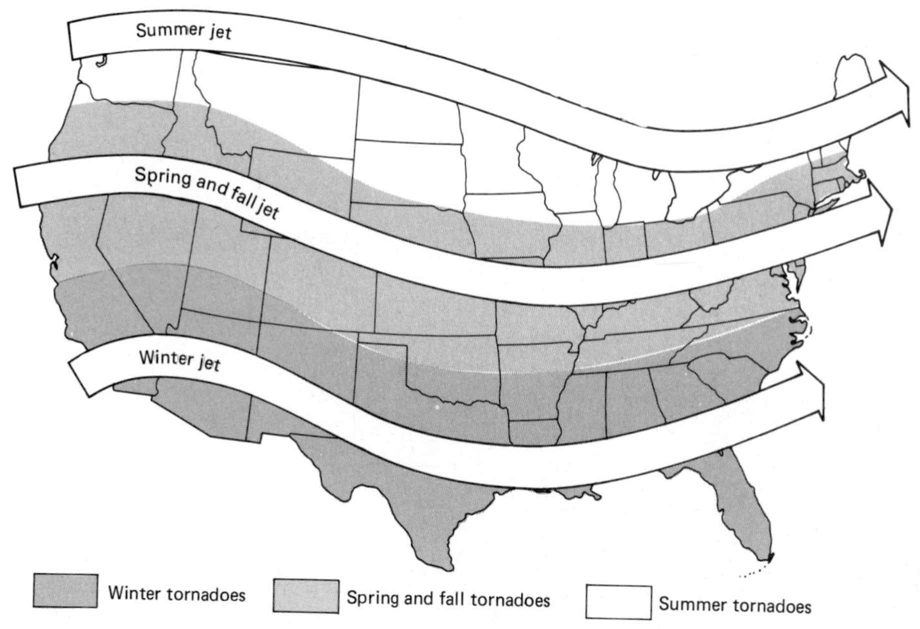

FIGURE 11-6 The jet stream migrates with the seasons, causing the areas of greatest tornado activity to shift to the south as temperatures decrease during the fall and winter months. During the spring and summer, as the jet stream and the boundary between the cold and warm air masses shift northward, the location of most likely tornado activity also shifts northward.

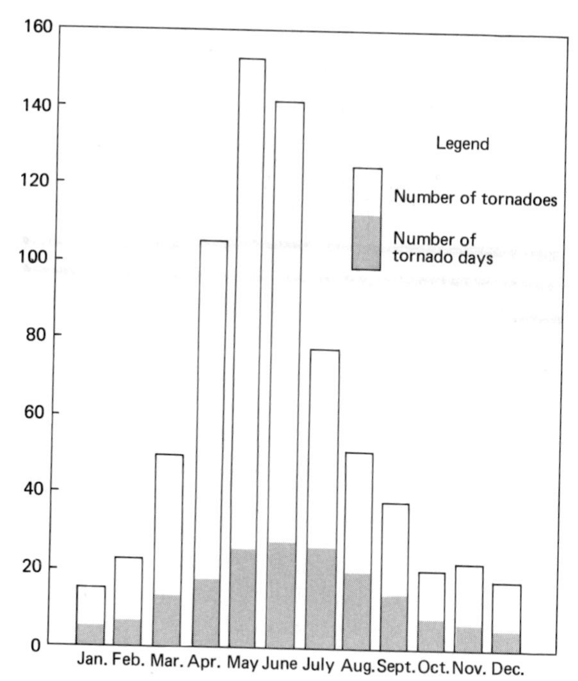

FIGURE 11-7 Monthly distribution of tornadoes in the United States based on more than 20,000 tornadoes that occurred from 1953 to 1980. More tornadoes form in May, while June has more tornado days than any other month. (National Oceanic and Atmospheric Administration, *Climatological Data National Summary,* Annual 1980.)

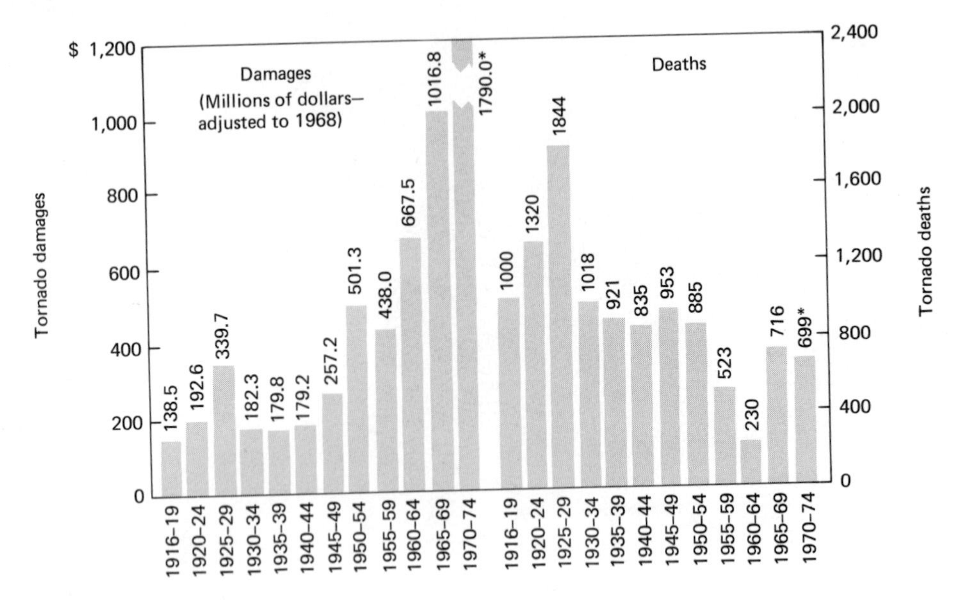

FIGURE 11-8 The trend of damage from tornadoes shows a continual increase as more of the surface of the United States becomes occupied by buildings or other structures of value, while the trend in number of deaths from tornadoes has shown a gradual decrease as better warning systems are put into operation and more accurate severe weather forecasts become available. (*Bull. Am. Meteorol. Soc.*, vol. 56, 1975.)

in history occurred on April 3, 1974, when 148 tornadoes developed over 11 states killing 329 people and leaving damages amounting to about three-quarters of a billion dollars.

TORNADO FORMATION AND MOVEMENT

The formation of tornadoes depends to a very great extent on the existence of the appropriate atmospheric conditions. These conditions were discussed in Chapter 9 on severe thunderstorms. In the central and eastern United States, tornadoes are normally associated with a cold front that is moving into the region, and the most likely location of tornado formation is ahead of the advancing front in the moist unstable air coming from the south or southwest. An inversion normally exists beneath an area of upper air divergence.

The synoptic weather patterns for some of the hundred-million-dollar tornado outbreaks were used in previous illustrations. It may be informative now to compare these figures. Figure 2-5 shows the synoptic map for April 3, 1974, and Figure 4-14, the one for May 6, 1975, while Figures 1-23, 7-14, and 7-19 show satellite photographs and the synoptic map for a large tornado in Emporia, Kansas, that occurred on June 8, 1974. Comparison of

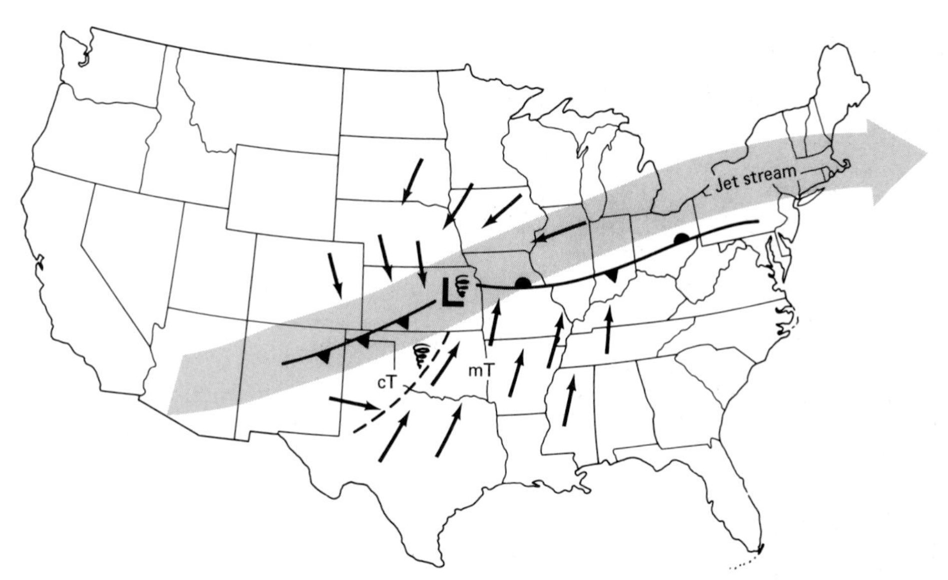

FIGURE 11-9 The synoptic weather patterns corresponding to the Topeka tornado on June 8, 1966, at 7:00 P.M. This tornado occurred near the center of the frontal cyclone, while another tornado occurred in south-central Oklahoma along the dry line separating the maritime tropical air mass from the continental tropical air mass. Surface winds converged along the dry line and near the center of the midlatitude cyclone. Upper atmospheric patterns showed diffluence at 500 mb over eastern Kansas, with the strongest jet stream flowing over Kansas. More tornadoes would have been likely if the frontal cyclone center had been a little farther northward, allowing it to be closer to the upper atmospheric low-pressure center and allowing the jet stream to carry the cold front and dry line along at a greater rate of speed. The Topeka tornado, however, was the first single tornado to cause more than $100 million worth of damage.

these figures shows where the cold front, low pressure center, and the like, were located in comparison with the large cloud patterns and tornadoes, shown in the satellite photographs.

The June 8, 1966, Topeka tornado mentioned in the last section was an example of a tornado that occurred near the center of a large midlatitude cyclone. Two tornadoes occurred on June 8, 1966, one in Topeka and the other in central Oklahoma near a cold front that extended southwestward from the center of the midlatitude cyclone (Fig. 11-9). On June 8, 1966, the upper air patterns showed considerable streamline diffluence at the 500-mb level. This is one of the factors that is helpful in creating extra large thunderstorms, since upward vertical motion in the atmosphere beneath the diffluence is intensified by the lack of air above. Photographs of the Topeka tornado (Fig. 11-10) showed that it was quite large, covering about 1 km on the ground. Houses in its path were almost completely destroyed.

Because of the importance of the jet stream with the winds aloft pushing thunderstorms along, the paths of most tornadoes are related to the upper air currents. The **damage paths** of tornadoes show that more come from the southwest than from any other direction. Damage path lengths greater than 300 km are rare; those greater than 200 km are

FIGURE 11-10 The 1966 Topeka tornado had a diameter of about 1 km and, as shown in this photograph, its strong winds carried large amounts of debris along with them. (Photo courtesy of Topeka *Capital Journal*.)

slightly more common. Tornadoes occasionally develop behind a cold front where the surface and upper atmospheric winds are from the northwest, resulting in tornado damage paths from the northwest. On one occasion a short damage path was produced from east to west as a stationary thunderstorm enlarged in the absence of a jet stream and carried an associated tornado toward the west.

The path of a tornado on the ground is influenced not only by the direction of the winds in the upper atmosphere but also by the circulation within the individual thunderstorm. The mesocyclone, or the cyclonically rotating half of the thunderstorm, may carry the tornado funnel along with it as it rotates. Such a tornado occurred near Lawrence, Kansas, in 1964. The tornado was about 3 km from the city when it curved to the north; this indicated that the tornado was not in the center of the larger mesocyclone in the thunderstorm. This is not an unusual path for tornadoes; numerous others have been observed to have similar paths, as indicated in Figure 9-17.

A cloud larger than the funnel, called the **collar cloud,** which is located at the base of the thunderstorm and corresponds to the mesocyclonic rotation, may be visible. Occasionally, more than one funnel extends from the collar cloud of a single thunderstorm. A funnel that forms on the edge of the mesocyclone will dissipate sooner than one formed in the center. Figure 11-11 shows two large tornadoes that occurred during the April 11, 1965, tornado outbreak. When two tornadoes occur as in this photograph, they do not travel side by side for any great distance, but one of the tornadoes rotates around the other one. Three funnels were on the ground at the same time in Kansas in 1973, as shown in one of the photographs in Figure 11-1. The photographer ran out of film after taking this photograph but reported that five funnels occurred shortly afterward.

In addition to being formed by large thunderstorms, tornadoes can be generated by small thunderstorms that produce waterspouts that strike land, by large thunderstorms within hurricanes, and there is some indication that dust whirls in deserts can become tornadoes if they develop on the ground at just the right time, that is, as an updraft within a developing cumulus cloud moves over the dust whirl. The most usual type of formation mechanism is, however, the large thunderstorm.

TORNADO CHARACTERISTICS

One of the damaging features of a tornado is the large quantity of debris that it picks up. Photographs of the Dallas tornado in 1957 and the 1966 Topeka tornado show flying pieces of wood and other debris carried by the strong winds (Fig. 11-10). Such debris can cause tremendous damage if it is traveling a few hundred kilometers per hour.

The maximum wind speeds in tornadoes are uncertain but may occasionally reach more than 400 km/h, or even the speed of sound. It is thought that about half of all tornadoes, however, have wind speeds of less than 300 km/h. The noise produced by tornadoes is one of their warning characteristics. This extremely loud noise has been described as the noise of many freight trains or jet airplanes and may mean that supersonic winds occur somewhere in the thunderstorm.

FIGURE 11-11 These two large tornadoes were on the ground at the same time during the tornado outbreak on April 11, 1965. These twin tornadoes occurred east of Elkhart, Indiana, about 6:30 P.M. and were photographed by Paul Huffman of the Elkhart Truth newspaper just before they struck Dunlap, Indiana, causing deaths and heavy damage.

The high wind speeds within a tornado make it possible for a piece of wood to penetrate metal, wood, or glass. In Kansas, wheat straw frequently penetrates telephone poles during tornadoes. Such penetration is more likely on the side of the telephone pole facing the approaching tornado. Tornado winds are also strong enough to pick up large boards and tree limbs and cause them to pierce the walls of houses (Fig. 11-12). The sides of houses most likely to be hit by flying debris are those on the south and west.

Observations of hundreds of houses damaged by tornadoes coming from the southwest consistently show indications of high winds with debris penetrating on the south and west sides, but not on the east and north sides. Aerial photographs taken

FIGURE 11-12 One of the hazards associated with tornadoes arises from the debris and other objects that are carried by the high winds of the tornado. The winds are of sufficient strength to cause wood to penetrate glass, metal, and other wood objects. In this case boards penetrated the south wall of a house. When the tornado is coming from the southwest, damage from flying objects is most common on the south and west walls. (Photo courtesy of author.)

after the April 3, 1974, tornadoes showed scattered debris and tree fall directions in the same direction the tornado was moving, thus indicating that the most damaging winds were in one preferred direction. Circular suction marks from weaker winds may be left on bare ground on the north side of the damage path, but the most damaging winds are concentrated on the southern edge of the tornado.

The higher wind speeds in the direction of tornado travel are caused by the vortex dynamics and by the translational (forward) speed of the tornado. The forward speed of the tornado (typically 50 km/h) adds speed to one side of the tornado while subtracting it from the other, in the manner previously described for a hurricane. However, this is less important for a tornado since the wind speeds are so much greater. The nature of a vortex requires that it dissipate its energy in a specific way at the ground. If the vortex were perfectly symmetrical, the surface winds would flow into it equally from all sides. Since this is not possible in nature because of surface roughness and the movement of the tornado, the surface winds instead converge toward the vortex from certain preferred directions. This causes **inflow streaks** and greater winds from behind the tornado than from the front (Fig. 11-13), and results in the greatest damaging winds occurring in the direction of tornado movement.

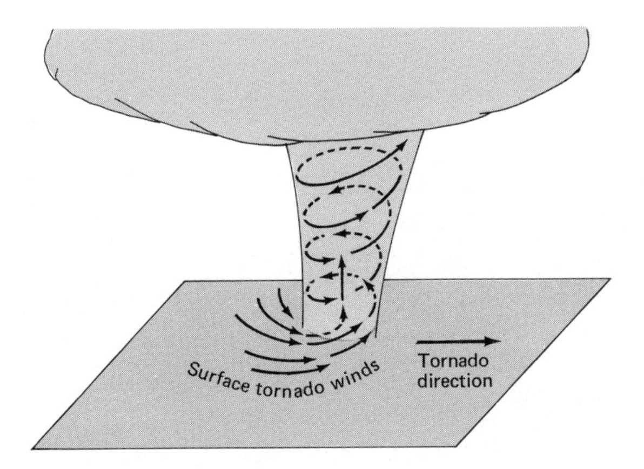

FIGURE 11-13 A perfectly formed stationary tornado funnel would have equal inflow from all directions near the ground. As a tornado moves over the earth's surface, however, the preferred direction of inflow is from the trailing edge of the vortex, where inflow streaks of high-velocity winds are created near the ground in the direction that the tornado is moving. Observations of tornado damage verify both the existence of such inflow streaks and the fact that the most damaging winds are in the direction that the tornado is moving.

TORNADO-PRODUCING THUNDERSTORMS

Winds inside the Thunderstorm Tornadoes, like hail, may occur from isolated supercell thunderstorms, from multicell thunderstorms, or from within a squall line. The general wind environment associated with the development of a severe thunderstorm is normally characterized by opposing winds above and below the 800-mb level. Figure 9-14 shows a thunderstorm moving in the direction of the winds in the upper layers of the troposphere. The winds in the lower part of the thunderstorm, relative to its movement, are from the opposite direction. As a result of these opposing winds in the atmosphere, a large wind shear exists. These opposing winds, together with the airflow around a large thunderstorm, then initiate a double vortex structure inside the thunderstorm. The southern half of the typical severe thunderstorm rotates cyclonically while the northern half may rotate anticyclonically, depending on the rain interference.

Interaction with the Jet Stream The jet stream is important in supplying a source of energy for the tornado as well as allowing outflow at the top of the thunderstorm, although tornadoes occasionally occur in the southern United States without a strong jet stream. It has been demonstrated that a tornadolike vortex can be generated in the laboratory by initiating horizontal circulation of air and providing outflow through the upper center of the rotation. The jet stream, with its high-velocity airflow, and the lower atmospheric pressure at the top of the thunderstorm create an upward pull through the vertical cyclonic "tube" generated by rotation, thus providing the thunderstorm with the required horizontal rotation combined with a central updraft. The tornado, then, has a structure and generation mechanism similar to that of the midlatitude cyclone, while the structure of the hurricane and dust whirl are comparable.

General Observations Tornadoes typically occur in the southern part of a thunderstorm that is moving toward the northeast. The major rain area is located beneath the northeastern leading part of the thunderstorm, which may also contain anticyclonic rotation. Several observers of severe thunderstorms with approaching tornadoes have described two clouds colliding at the leading edge of the thunderstorm (Fig. 11-14). If the observer were located in front of the thunderstorm, where both the anticyclonic and cyclonic rotation were visible, it would appear that the two clouds were fusing together. Other observers of approaching tornadoes have described them as swinging back and forth like an elephant's trunk. This observation corresponds to slow-moving tornadoes that are creating looping damage paths on the ground.

Development of Low Pressure The surface atmospheric pressure is higher in the northern rain area of the thunderstorm than in surrounding locations. The tornado in the southern part of the thunderstorm is surrounded by a region of low pressure. In addition, the center of a tornado is characterized by extremely low pressure, which develops as the spinning action of the mesocyclonic vortex throws the air outward, away from the center of rotation, and effectively creates a tube that may extend up to 8 km, where the atmospheric pressure is only 350 mb. If the tube extending to this level were a perfect tube, the lowest pressure inside would be 350 mb. The actual surface pressure is probably higher than this, however, since it would be impossible to create a perfect tube from the cyclonic rotation.

FIGURE 11-14 This photograph of the southeastern portion of a severe thunderstorm shows the forward edge of the storm. A ground observer located beneath the storm would see the portion of the cloud in the cyclonic circulation, on its collision course with the rest of the thunderstorm. (Photo courtesy of David Jones.)

Thunderstorm Movement and Tornado Development Measurements of the actual winds in the atmosphere may be used to calculate the direction of movement of thunderstorms. Then, based on this movement and on observations of thunderstorm echoes on the radar screen, the most severe thunderstorms can be identified even before they reach the severe stage. The most severe thunderstorm moves in a direction that gives it the greatest opposing relative winds above and below 800 mb.

Investigations of the speed of past severe thunderstorms in relation to the mean winds (average winds) from the surface to the top of the thunderstorm have shown that a certain preferred speed and direction exist for the most severe thunderstorms. Thunderstorms are more likely to be severe if their forward speed is slower than the mean winds (Fig. 11-15). The relationship between speed and direction of movement in a group of past severe thunderstorms showed frequent movement of thunderstorms to the right of the mean wind with a forward speed of only 50% of the mean wind. Another preferred movement was slightly to the left of the mean wind at about 75% of the mean wind speed.

Splitting Thunderstorms Radar observations of thunderstorm echoes show that single thunderstorms occasionally split into two individual thunderstorm cells that become severe. When this happens, the southern cell moves to the right and the other one to the left of the mean wind. This movement indicates rotation within the two cells, giving further evidence of a double vortex structure within thunderstorms, since a cyclonically rotating cell should move to the right of the mean wind and an anticyclonically rotating cell should move to the left because of forces resulting from the rotation.

Satellite Information on Thunderstorms The Skylab satellite obtained photographs of cloud patterns that are interesting in relation to severe thunderstorm formation. The Missouri River had no clouds over it on September 18, 1973, while surrounding areas were covered with developing cumulus clouds (Fig. 1-22). A temperature difference between the Missouri River valley and the sur-

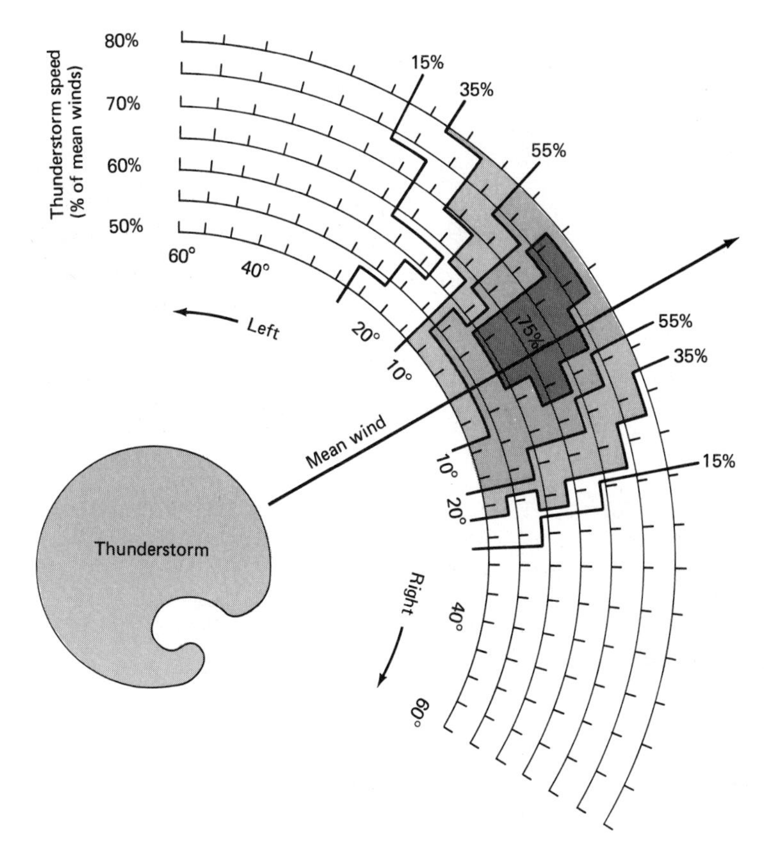

FIGURE 11-15 An analysis of the atmospheric environment corresponding to the time of tornado formation has shown that the ideal wind environment for developing severe thunderstorms exists when the thunderstorms move more slowly than the mean winds. Severe thunderstorms also frequently move to the left or right of the mean wind direction. The arcs in this illustration from 50% to 80% give the storm speed in percentage of the mean wind speed. The number of degrees to the left or right are shown along the inner arc. The percentages along the outer arc give the occurrence of thunderstorms with these particular speeds and directions.

rounding country affected cumulus cloud formation. This example demonstrates the relationship between surface characteristics and cloud formation, although the relationship is not always simple—in this case, for instance, the water temperature was warmer than the land. The warmer earth temperature in the late afternoon corresponds to the occurrence of more tornadoes from 4 to 6 P.M. since greater solar heating contributes to warmer air temperature with the greater likelihood of more buoyant clouds.

The Florida peninsula illustrates very well the difference in heating rates between land surfaces and water. Evidence of the resulting difference in temperature can be seen on satellite photographs, which show clouds forming over the land more frequently than over the water during the daytime. Figure 1-1 shows numerous small cloud formations

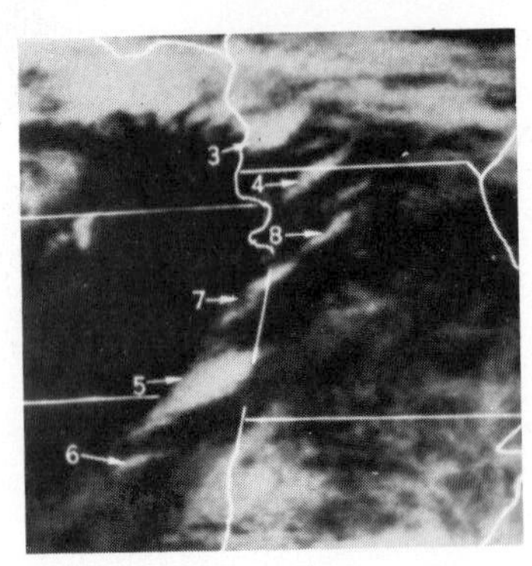

ATS-3 Frame 33
2109 GMT May 5, 1971

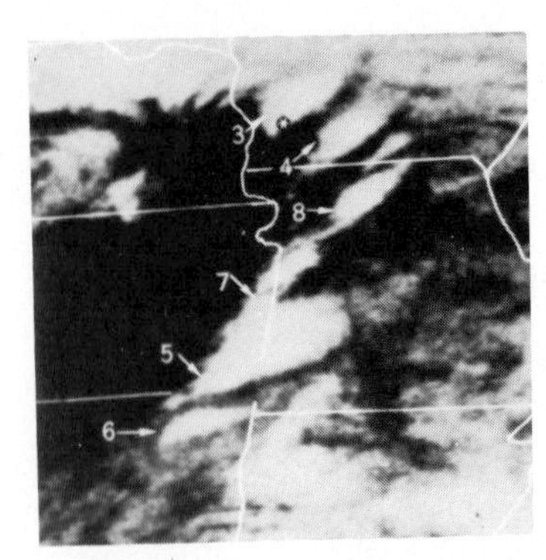

ATS-3 Frame 36
2147 GMT May 5, 1971

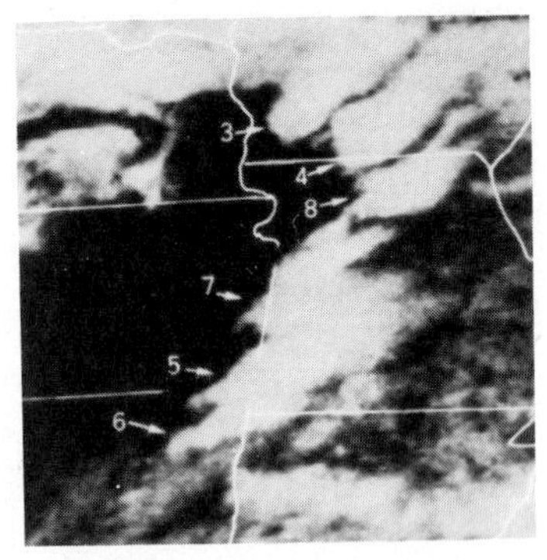

ATS-3 Frame 39
2229 GMT May 5, 1971

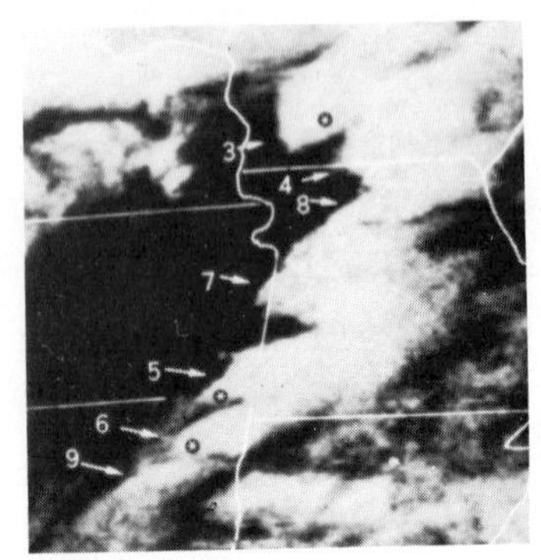

ATS-3 Frame 42
2307 GMT May 5, 1971

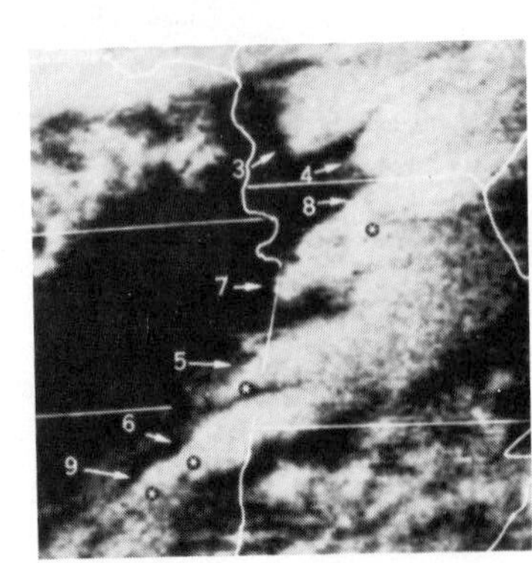

ATS-3 Frame 45
2347 GMT May 5, 1971

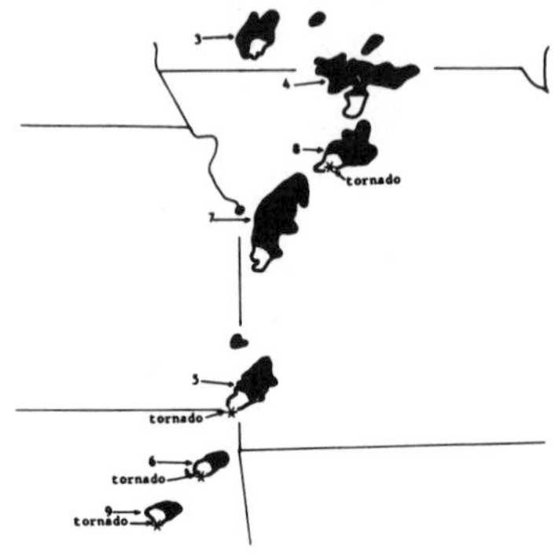

Radar Echoes
2347 GMT May 5, 1971

FIGURE 11-16 This series of ATS photographs taken on May 5, 1971, shows a developing squall line across Oklahoma, Kansas, and Missouri. These thunderstorms developed several tornadoes, including a devastating tornado in Joplin, Missouri. A series of ATS photographs like this, showing rapid thunderstorm growth, is useful as one indicator of tornado activity. (From *Applications of Meteorological Satellite Data in Analysis and Forecasting*, Suppl. No. 2, National Environmental Satellite Service, 1973.)

that have developed over the land while fewer clouds exist over the ocean.

Satellites are also useful in monitoring the rate of thunderstorm growth, as shown in Figure 11-16, as a means of identifying prospective severe thunderstorms. Evidence indicates that a thunderstorm that grows more rapidly than others is more likely to develop a tornado. Successive satellite photographs of thunderstorms can be used to monitor this growth.

TORNADO DETECTION AND CONTROL

Many people are surprised to learn that in this age of space exploration the most reliable indication of tornadoes is still visual sighting of funnels, such as those in Figure 11-1. Because of this, storm spotter organizations remain very important. Many communities have a large storm spotter group that is on duty during severe thunderstorm watches. Tor-

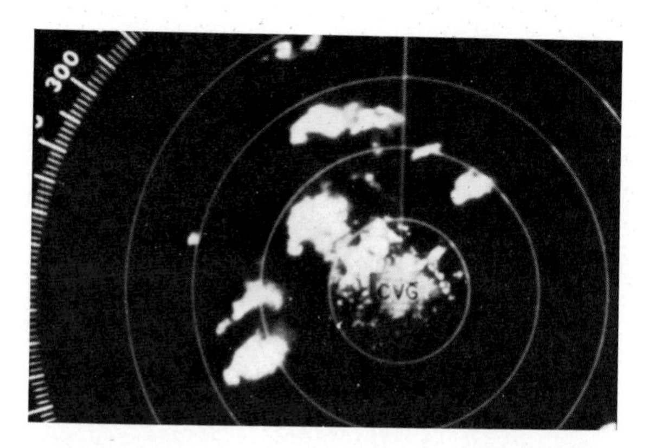

FIGURE 11-17 Many of the thunderstorms that occurred across the eastern United States on April 3, 1974, became severe. This radar echo obtained at 3:20 P.M. CDT shows three echoes near Cincinnati, Ohio, that have developed hook echoes. Several tornadoes occurred from each of these thunderstorms. The echo to the northeast of Cincinnati does not show the characteristic hook shape at this time, but soon after this picture was obtained, a hook echo and large tornado developed that destroyed much of Xenia, Ohio. (Photo courtesy of NOAA.)

nado warnings are issued after a tornado is spotted by observers or a hook echo develops on the radar screen. Some special cloud formations indicate severe turbulence and are associated with tornadoes. A **mammatus cloud** formation indicates turbulence and is frequently observed before and after tornadoes.

Operational tornado detection devices are all somewhat unreliable. The best indication of severe weather by remote sensors is a hook-shaped thunderstorm echo on the radar screen. Radar operates by sending out a radiation signal that is reflected off of and returned to the radar system by raindrops in clouds. When the returned signal is displayed as a horizontal slice through the lower part of the thunderstorm, a hook-shaped echo indicates that the proper internal thunderstorm structure now exists for the development of a tornado (Fig. 11-17). Several **hook echoes** were observed on the radar screen at Cincinnati on April 3, 1974. The hook shape is caused by mesocyclonic circulation, which carries raindrops along with it. The primary rain area in the leading northeastern part of the thunderstorm, as well as the lesser amounts of rain spiraling around the mesocyclone, both return the radar signal.

The hook echo is visible only if the radar is pointed toward the lower third of a thunderstorm. Different shapes of hook echoes occur from different thunderstorms. The pointing angle of the radar also affects the return since radar aimed above the lower third of the thunderstorm may give an echo that is circularly shaped, with a hole in it corresponding to the intense cyclonic circulation in the southern part of the thunderstorm. The dependability of radar for tornado detection is also diminished by the fact that smaller tornadoes occur from thunderstorms that do not have a hook-shaped echo. Fewer than one-third of severe thunderstorms may form a hook echo on the radar screen. However, radar is always useful for locating the largest thunderstorms and documenting their movement, and it is the largest thunderstorms that are most likely to produce tornadoes.

Experimental **tornado detectors** are being developed that are similar to the television detection method. The tornado detection method by television consists of turning to Channel 13, darkening the screen, then switching to Channel 2. If the screen glows, then a tornado may be within 30 km. This detection method is thought to work because of the electrical activity of a severe thunderstorm at frequencies close to that of Channel 2. Experimental tornado detectors operate on a similar basis and are sensitive to the 3 MHz of energy emitted by severe thunderstorms. Approximately 75% of the

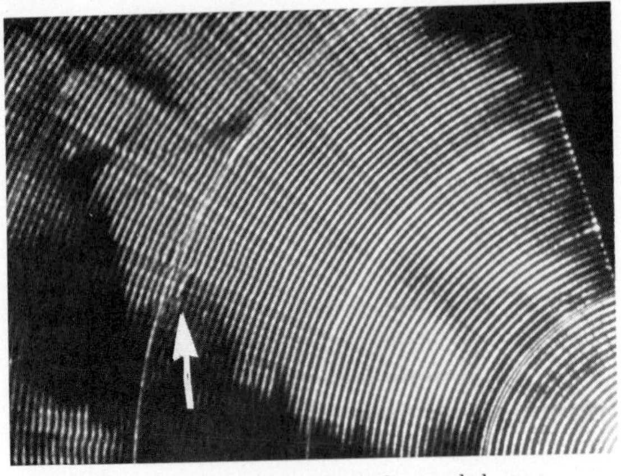

1535 CST—3 min. prior to tornado touchdown

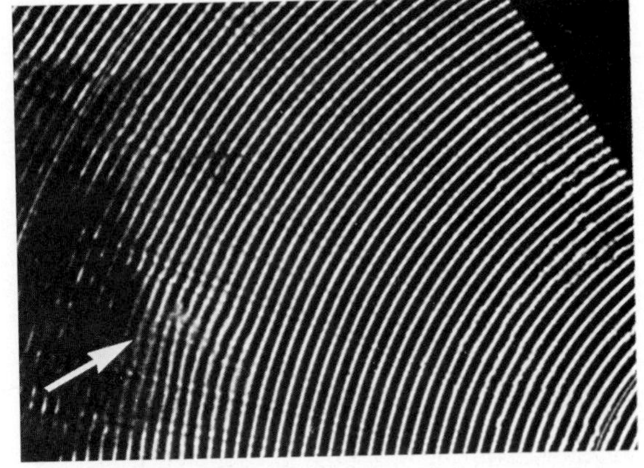

1554 CST—Tornado on ground

FIGURE 11-18 Tornado detection by Doppler radar may be possible by means of the Plan-Shear Indicator, as shown in these photographs taken at 1535 CST at a height of 3.3 km and at 1554 CST near the cloud base. The Union City, Oklahoma, tornado touched down at 1538 CST and is indicated by the deviations from the normally circular arcs. This is one of the new techniques that is being developed for tornado detection. (Courtesy of R. J. Donaldson, Jr., Air Force Geophysics Laboratory.)

thunderstorms with tornadoes emit energy at this frequency. However, the false alarm rate of this experimental tornado detector is also high since many thunderstorms without tornadoes also emit energy at this frequency. These methods are based on **sferics** (a contracted form of their previous designation of *atmospherics*), which are electrical discharges within thunderstorms. Electrical discharges in thunderstorms that produce radio waves received as static on radio receivers are similar except for a slightly longer wavelength.

Another experimental tornado detection instrument is Doppler radar. A Doppler radar measures the velocity of raindrops carried by the wind currents in a thunderstorm. Research with Doppler radar at the National Severe Storms Laboratory indicates that the mesocyclonic circulation in the southern part of a thunderstorm can be distinguished about 30 minutes prior to tornado touchdown, a fact previously suggested by the double vortex thunderstorm model. Several Doppler radar observations indicate that the mesocyclonic circulation develops in middle levels of the thunderstorm and gradually extends to the ground as a tornado.

Operational use of Doppler radar will require solving such problems as processing the detailed radar data and displaying the relevant information in a manner that allows rapid interpretation. One method of displaying the data is shown in Figure 11-18. A vortex is indicated by deviations from normal arcs. Other problems include developing operational Doppler radar with appropriate resolution as well as developing the capability to sort out ordinary wind shear from the mesocyclone or tornado vortex within a thunderstorm.

There are no tornado control experiments in practical operation. Some suggestions have been made, however. One of these, which requires more theoretical as well as experimental work, suggests using airplanes to generate contrails over areas where tornado formation is likely. The objective would be to create cirrus clouds over a tornado watch area so that the ground and lower atmosphere heat less from sunlight. Lower atmospheric temperatures reduce the possibility of tornadoes since the air is less buoyant and less likely to develop into a large thunderstorm.

An interesting observation of a waterspout was made that may offer clues to tornado control. As a lightning discharge passed through the vortex, the vortex quickly broke into pieces and then disintegrated. If this occurrence were better understood, perhaps specific thunderstorm modifications could be made that would decrease the impact of tornadoes.

FIGURE 11-19 Laboratory tornadolike vortex generated by simulating the wind currents in a severe thunderstorm containing a double vortex internal structure. The vortex was created by a fan at the top to simulate the updraft within the mesocyclone and two other fans mounted perpendicularly to the first fan for simulating the opposing winds from the back and front of a thunderstorm near its base. The vortex is made visible by the layer of dry ice on the floor. The author is making temperature measurements near the vortex, although more sophisticated equipment is necessary for detailed measurements. (Photo courtesy of University of Kansas Public Relations.)

LABORATORY TORNADOES

Since it has been impossible to obtain detailed information on tornadoes in the atmosphere, because of their destructive power, their duplication in the laboratory is of considerable interest. Laboratory vortices are not difficult to generate. The difficulty is in creating a vortex that is similar to a tornado without restricting boundary conditions such as occur inside a cage or box. A realistic appearing **tornadolike vortex** was generated in the author's laboratory, as shown in Figure 11-19, by using a large fan to supply an updraft between horizontal air currents generated by two other fans mounted near the ceiling of the laboratory. The combination of horizontal air currents and updraft through the center generates a tornadolike vortex that extends from the ceiling to the floor. Certain tornado characteristics can be investigated by such a vortex. Since the vortex is not confined within a cage or box, movement across the ground can be simulated in a realistic manner.

Several different photographs of tornadoes and waterspouts have shown a visible central core. Circulation shells, corresponding to the horizontal rotation around the core updraft, appear to surround a central core. The laboratory vortex has a similar appearance, with a central updraft core surrounded by horizontal circulation. Observations of the vortex show that the horizontal rotation increases and the diameter of the vortex decreases as the speed of the updraft is increased. Air is drawn into the base of the vortex from great distances, with inflow streaks of air entering the vortex at the surface. Observations of tornado-damaged areas show similar wind streaks caused by very strong winds entering the base of the vortex. This causes houses in particular locations to be damaged more severely than others nearby.

The center of the tornado is characterized by very low pressure. Accurate measurements of the pressure inside a tornado are not available because of the response time of barometers and the fact that they are located inside buildings. Some measurements have been made, however—for example, a pressure of 996 mb for the Lubbock tornado and of that shown in Figure 11-20 for the Topeka tornado. These are almost certainly not the lowest pressures within these respective tornadoes. Building damage indicated much lower pressure. Measurements from the vortex in the laboratory have shown a core pressure as low as 500 mb.

The laboratory vortex provides a very good simulation of the surface interaction of a tornado. Figure 11-21 shows the path of the simulated tornado after it has traveled from left to right, with an obviously large component of airflow in the direction the vortex was moving and a lesser indication of circular airflow patterns. This pattern is very similar to the treefall patterns observed after tornadoes.

A slow-moving tornado simulated in the laboratory produced a looping damage path similar to that of the Lubbock tornado. If a longer tornadolike vortex is generated, the vortex is more likely to skip along the surface. This indicates that **skipping tornadoes** are probably related to the strength of the thunderstorm. If the thunderstorm has enough energy to generate sufficient circulation to extend the tornado to the ground, then it can remain on the ground as it goes over hills, valleys, or bluffs.

TORNADO SAFETY

If tornadoes can't be controlled, their effects should be decreased if possible. One way to accomplish this is to use all the information available on tornadoes when designing houses and when seeking shelter from tornadoes. Tornadoes, as shown in Figure 11-3, occur more often in the central and eastern United States, and the combination of large numbers of tornadoes and denser population should indicate the distribution of potential casualties. However, statistics show that these do not correspond; rather, more deaths occur in the southern United States than in any other location (Fig. 11-22). One reason for this may be the general misconception that the safest location is in the southwestern part of a house. This wrong information combined with the almost total absence of basements in the South undoubtedly causes more casualties there.

Houses are damaged by tornadoes in very systematic ways. Observations of hundreds of tornado-damaged houses have been made in various states by the author to determine which areas are safest during a tornado. In houses with poured concrete basements, the floor frequently remains over the foundation if it is properly bolted to it. Any location away from windows is usually safe in such houses. Houses are not always bolted down, however, because many builders assume that the weight of the house is sufficient to secure it to the foundation. If it is not bolted properly, the northeast part of a house is preferable to the southwest because the whole house may be shifted to the northeast, causing the southwest part to fall into the basement.

One of the factors evident from investigating tornado-damaged houses was that the walls tended to fall outward because of the overpressure developed inside during the tornado. The major winds in a tornado are most frequently from the

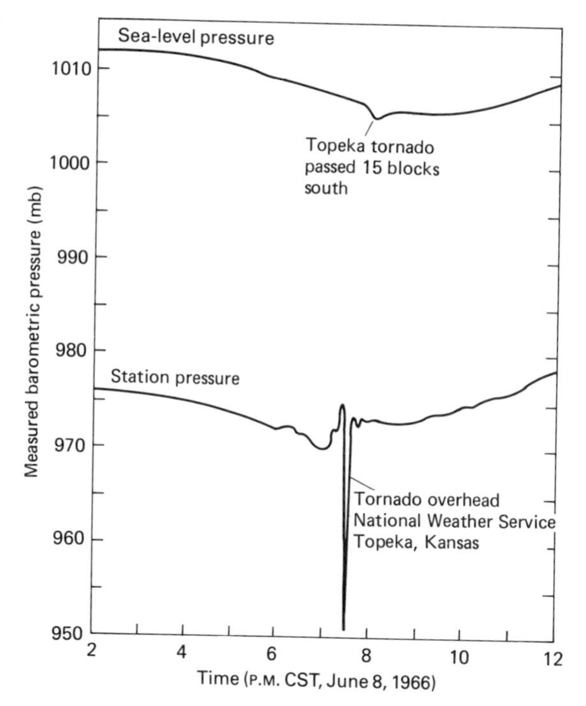

FIGURE 11-20 Barometric pressure variations measured on June 8, 1966, in Topeka, Kansas, as a tornado passed through this city. The barometer recording the top trace was set to measure sea-level pressure. Although the tornado passed over the instrument at the National Weather Service, the actual lowest pressure in the tornado was probably not measured because of the response time of the instrument inside a building. (Data from Topeka National Weather Service.)

FIGURE 11-21 The laboratory tornado provides information on the surface inflow of air at the base of the vortex, as indicated here by its movement over a viscous liquid applied with up and down brush strokes before the laboratory tornado was moved over the surface from left to right. Indications of strong airflow in the direction that the vortex was moving are shown, but there is only a slight indication of circular rotation. The major surface winds are concentrated in the right half of the vortex trail, with more indication of the circular rotation on the left half. This same damage pattern is revealed by field investigations of tornadoes, which show the most damaging winds to be in the direction a tornado is moving. (Photos courtesy of author and NASA.)

same direction the tornado is moving. Therefore, if any walls fall inward, they are usually located on the southern or western sides. Figure 11-23 shows tornado damage that occurred on June 8, 1974, in Emporia, Kansas. The tornado struck a shopping center where concrete blocks were used as the building material. The photograph shows the weak nature of this building material as well as the remains of an acoustical tile ceiling where metal strips were used to hold the tile.

The roof-wall connection is frequently a weak part of a building, allowing the whole roof to be

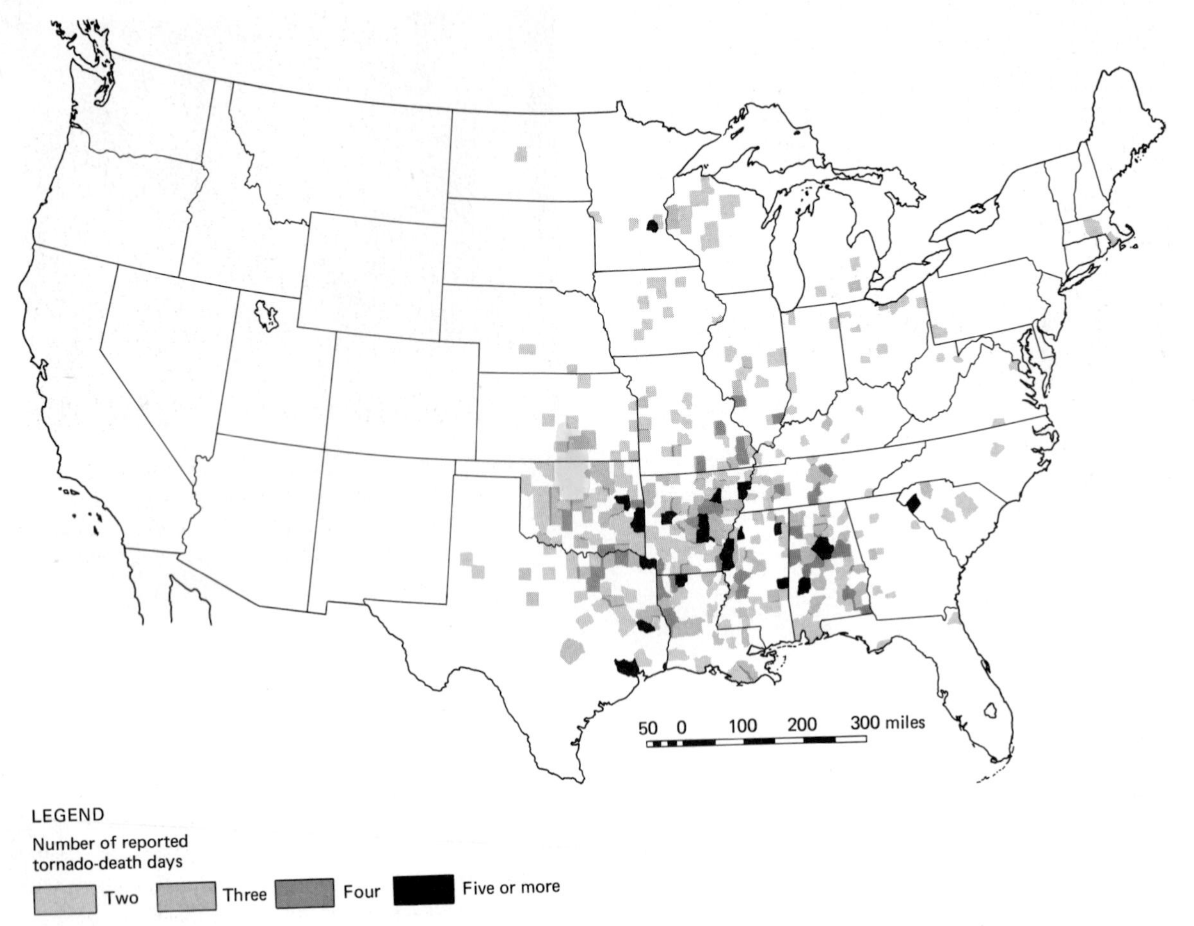

FIGURE 11-22 Number of days with recorded tornado deaths by county for the time period 1916–1953. The distribution shows that tornado deaths are more frequent in the southern United States than in other locations. The combination of the distribution of actual tornado occurrence and numbers of people located in various states would seem to predict, however, that more tornado deaths should occur in the central and eastern United States. (From U. S. Linehan, "Tornado Deaths in the United States," U.S. Government Printing Office, 1957.)

blown away. The wall-foundation joint is also a potentially weak area. Figure 11-24 shows a house that had a poured concrete foundation with no bolts anchored in the concrete to hold the walls in place. Such houses are frequently pushed in the direction the tornado is traveling.

Observation also revealed a considerable accumulation of debris on the southern and western sides of houses in the tornado damage path, with some of it penetrating walls and windows.

Observations from many different tornadoes in several different states have consistently shown that the southwest part of a house (including that part of the basement and first floor) is the worst place to seek shelter from a tornado. The percentage of various locations that are unsafe in houses is

FIGURE 11-23 The tornado that struck Emporia, Kansas, on June 8, 1974, passed through a shopping center, and some of that damage is shown here. Concrete blocks and other mortar structures are frequently weak, as are acoustical tiles held together by metal strips. (Photo courtesy of author.)

FIGURE 11-24 This house was moved several meters in the direction the tornado traveled. This type of damage occurs to houses that are not bolted securely to the poured concrete foundation. (Photo courtesy of author.)

shown in Figure 11-25. The safest location in basements is in the northeast. This is twice as safe as southern locations. The most submerged part of walkout basements is safest. Investigations of the first floor of damaged houses also showed the northeast and central parts to be much safer than other parts. Based on damage statistics gathered from several different states after tornadoes traveling from the southwest hit the area, the southwest part of 72% of the houses was found to be unsafe, while the northeast and central parts of 42% of the houses were unsafe.

Since these studies were completed by the author, the National Weather Service no longer promotes the southwest part of the house as the safest location, but now advises that the central part of a house is the safest location. A problem still exists, however, since most people still believe the southwest part of a house is safest. Surveys conducted by the author have shown that about 60% of the people in tornado areas still think that the southwest corner of the basement and of the first floor of houses without basements is safest from tornadoes. If more than half the people in these areas seek shelter from tornadoes in the southwest part of houses, where debris is likely to come through the walls and the walls are more likely to fall inward from the high wind speeds, injuries are likely to be widespread.

Other unsafe locations include mobile homes, since their construction provides very little protection from the winds in a tornado. Mobile home parks should provide underground storm shelters for residents.

If caught in open country with an approaching storm the best thing to do is to move at a right angle from the direction of the tornado. If it is not possible to get out of its path, a depression or ditch is better than level ground since winds traveling at speeds of 300 km/h in a tornado cannot make sharp turns. Therefore debris will go over ditches or depressions. A culvert or tunnel under a bridge also offers some protection from the tornadic winds.

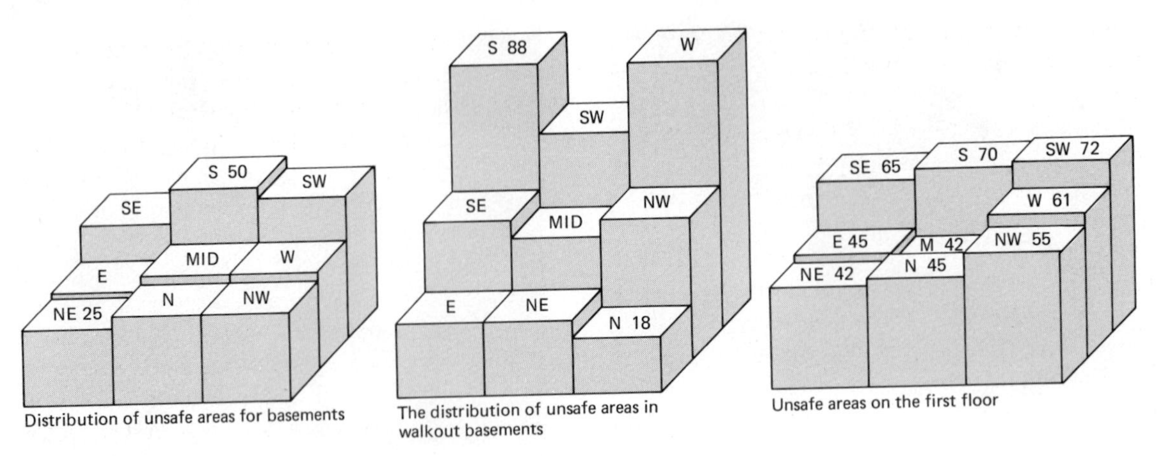

FIGURE 11-25 Investigations of hundreds of houses damaged by tornadoes in many different states have revealed systematic patterns of safe and unsafe locations in various parts of houses. Shown here is the distribution of unsafe locations in various sections of houses on the first floor, in walkout basements, and in submerged basements. In submerged basements the safest locations were downwind from the approach direction of the tornado, in the northeast part of the basement. Most unsafe locations were along the south and southwest, with 50% of the basements unsafe in the south and only 25% of them unsafe in the northeast. For walkout basements the safest locations were those that were the most submerged below ground, with 88% of the houses being unsafe in the south and only 18% of them unsafe in the northern part of the house. On the first floor of houses without basements 72% of the houses were unsafe in the southwest part while only 42% of the houses were unsafe in the northeast and central parts of the houses. Thus the southwest part of a house is the worst place to be located when a tornado is approaching from the southwest—the typical approach direction of tornadoes. (From J. R. Eagleman, V. U. Muirhead, and N. Willems. *Thunderstorms, Tornadoes and Building Damage,* Lexington, Mass.: Lexington Books, © D.C. Heath & Co., 1975.)

SUMMARY

Tornadoes are the most powerful of all atmospheric storms. They may complete a life cycle that includes the formation of a large vortex followed by a smaller rope-shaped vortex and retraction into the cloud. What a tornado looks like is determined by its background and by the type and amount of debris that is picked up from the surface.

The United States has many more tornadoes per year than any other country. Tornadoes are responsible for an average of 98 deaths per year and several million dollars worth of damage. The area of greatest tornado activity migrates with the jet stream, with more tornadoes occurring in May than in any other month. An average of 87% of the days in June have tornadoes somewhere in the United States.

A single tornado killed 689 people in 1925 and four other recent single tornadoes have each resulted in more than $100 million worth of damage. The path of a tornado on the ground is determined by the upper atmospheric winds, which push the thunderstorm along, as well as by the circulation within the mesocyclone of the thunderstorm. Tornadoes are characterized by high wind speeds and low pressure inside the funnel. One of their damaging features is the large amount of debris that is carried along by the high winds. Major damaging winds occur in the direction that the tornado is moving.

Tornadoes develop from the largest thunderstorms that are created in an unstable atmosphere with a strong wind shear environment. Recent thunderstorm models and dual Doppler radar measurements show that the tornado vortex is generated in midlevels of the thunderstorm and extends down to the surface. Tornadoes occur within the mesocyclone in the southern part of a thunderstorm that is moving toward the northeast. Severe thunderstorms normally move more slowly than the mean wind in the atmosphere.

Tornado warnings are ordinarily issued after the development of tornadoes; visual sightings are still the most reliable method of detecting tornadoes. Weather stations have conventional radar sets that are useful in tracking individual thunderstorms and in watching for the development of very large thunderstorms. About one-third of the tornadoes are associated with hook-shaped echoes on the radar screen. These occasionally develop soon enough that tornado warnings can be issued on this basis. Other experimental tornado detectors are being investigated including sferics detectors and Doppler radar. Doppler radar can be used to locate the mesocyclonic circulation in midlevels of thunderstorms prior to tornado development. There are indications that this circulation may provide about a 30-minute warning before a tornado develops. The Doppler radar design and data display must be perfected further before this system can be made operational.

No tornado control experiments are in operation although it has been suggested that aircraft could be used to generate contrails over tornado watch areas to reduce the ground temperature and decrease the likelihood of tornado development.

Laboratory tornadoes have been generated to gain more information on various aspects of tornadoes. Damage paths are either looping or straight depending on the speed of movement of the thunderstorm and the location of the tornado within the mesocyclone. Severe damage occurs in the form of wind streaks as surface winds enter the base of the vortex. The laboratory model also indicates that skipping tornadoes are the result of weak support of the tornado within the thunderstorm.

Greater tornado safety can be achieved by proper information on seeking shelter from tornadoes and on better designed houses. Houses are frequently weak at the roof-wall connections and wall-foundation connections. The walls tend to fall outward due to the low pressure within the tornado vortex except on the south and west sides of houses, where strong winds frequently blow the walls inward. The safest locations in houses are small rooms in the northeast and central parts. The most unsafe location is the southwest part of houses, contrary to the belief of many people.

STUDY AIDS

1. Explain why tornadoes have various shapes and colors.

2. Compare tornado occurrence in the United States with other countries and comment on the differences and reasons for them.

3. Discuss the seasonal distribution of tornadoes in the United States.

4. Compare the synoptic weather patterns for some of the major tornado outbreaks in the past.

5. Explain some of the characteristics of tornado damage paths.

6. Use the law of conservation of angular momentum (vr = constant) to calculate the wind speed (v) in a tornado of 300-m radius (r) when the mesocyclonic vortex contains a wind speed of only 50 km/h with a diameter of 8 km.

7. Discuss the direction of the most damaging winds within a tornado.

8. Discuss the generation of tornadoes by thunderstorms.

9. Discuss tornado detection and control.

10. What information has been obtained from laboratory tornadoes?

11. How might Dorothy or her family have prevented her journey to see the Wizard of Oz?

TERMINOLOGY EXERCISE

Check the glossary if you are unsure of the meaning of any of the following terms used in Chapter 11.

Rain shaft
Tornado day
Waterspout
Damage path
Collar cloud
Inflow streaks
Mammatus cloud
Hook echo
Tornado detector
Sferics
Tornadolike vortex
Skipping tornado

THOUGHT QUESTIONS

1. Why do most tornadoes rotate cyclonically while dust devils rotate in either direction?

2. Tornadoes have sometimes been reported to glow in the dark. Can you suggest reasons why this might happen?

3. Describe a typical sequence of weather events as a severe thunderstorm containing a tornado moves over your location.

4. Is it possible for a single frontal cyclone to produce blizzard conditions and tornadoes at the same time? Explain.

5. Conduct a survey and determine which location in houses most people consider to be the safest. Do statistics on safest location in houses agree with the results of your survey?

TORNADO MEASUREMENTS: TOTO

The progress of many scientific problems is related to the availability and accuracy of measurements. No accurate measurements have been made inside a tornado. Evidence indicates that the pressure is extremely low, but this has not been verified. Now an instrument is being tested that may obtain some measurements inside a tornado (Fig. S13-1).

The Wave Propagation Laboratory of NOAA has developed a prototype instrument package that has been named TOTO after Dorothy's Toto from the *Wizard of Oz* and also to indicate Total Tornado Observatory. The device is designed for rapid deployment in the path of a tornado by tornado chase teams, and it is designed so that the sensors turn on automatically when it is lifted into a vertical position. The device is equipped to measure the temperature, wind speed and direction, pressure, and corona discharge inside a tornado.

It has been successfully placed under the rotating mesocyclone of a tornado-producing thunderstorm, where measurements have been made. Future attempts will be made to place it inside a tornado.

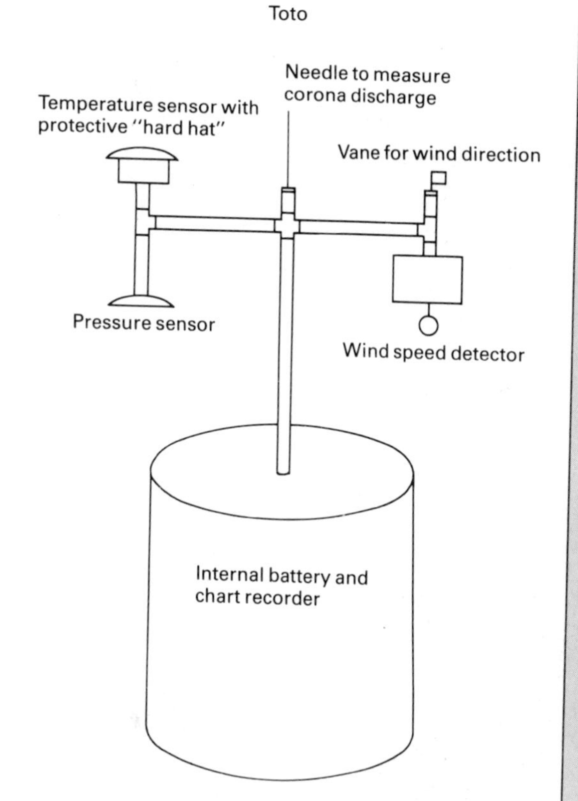

FIGURE S13-1 *Toto* is the instrument package designed to obtain measurements of atmospheric conditions inside a tornado.

SPECIAL TOPIC 14

TO AVOID THE TORNADO TURN LEFT AT THE RAINBOW

Few subjects are surrounded by as much mystery and misunderstanding as tornadoes. It is not uncommon for people from other states to believe that tornadoes occur predominantly in Kansas, perhaps because of the general association of Dorothy and Toto with Kansas. The fact is, however, that many people have lived in Kansas all their lives and never seen a tornado.

Another misconception is that tornadoes are random events rather than associated with some fairly specific atmospheric conditions. Further, when tornadoes are developed by an individual thunderstorm that has been favored with all the right atmospheric surroundings, it is attached to its parent thunderstorm in a specific way.

The upper-level winds of the parent thunderstorm are typically from the southwest and move the thunderstorm along with them. The heavy rain area is concentrated in the leading northeastern part of the thunderstorm, while hail falls mostly beneath the central part and tornadoes develop on the southern edge.

In order for a rainbow to be seen, the sun must be behind the viewer. Thunderstorms that form tornadoes typically grow in the afternoon hours. Combining these two facts means that if you see a rainbow within a severe thunderstorm, it is most likely within the rain area as you are looking in an easterly direction. Therefore, the tornado-prone region of the thunderstorm would be on your right, on the south side of the thunderstorm. If you were driving a car toward the thunderstorm you would want to turn to the left and not the right in order to avoid a possible encounter with the tornado.

PART FOUR

APPLICATIONS AND SELECTED TOPICS

CHAPTER TWELVE
WEATHER MODIFICATION

EARLY IDEAS AND EFFORTS

Weather modification is a subject that has been of interest for a long time. Since the general public does not thoroughly understand weather processes, various wrong ideas have been common. The observation was made during wars that rain always followed battles. Some thought the noise from cannons and guns was affecting the weather. Others suggested that preparations for battle were always made during fair weather, resulting in increased chances for fog or rain during the battle.

Observations of increased rainfall on the windward side of mountains and drier conditions on the leeward side indicated to some that it might be possible to build structures tall enough to interfere with the flow of air sufficiently to cause precipitation. If this were possible, however, more precipitation would fall near New York City than in many other places, but an analysis of surrounding precipitation patterns does not show any clear effects from buildings.

James Espy, a Pennsylvania meteorologist who wrote extensively between 1838 and 1857, suggested, "If masses of timber to the amount of 40 acres for every 20 miles should be prepared and fired every seven days in the summer on the west of the United States in a line of 600 to 700 miles north to south, then it appears highly probable from the theory that a rain of great length from north to south will commence on or near the line of fires. That the rain will travel towards the east and will not break up until it reaches far out into the Atlantic Ocean. And it will rain over the whole country east of the place of beginning, that it will rain only a few hours in any one place, that it will rain enough and not too much in one place." This suggestion remained only that—it was never actually tried.

APPLICATIONS OF WEATHER MODIFICATION

A lot of thought has been given to weather modification by various segments of society because the idea is very intriguing. If we could actually control the weather, however, we would face enormous problems. For example, the farmer might want rain at the same time someone else, planning outdoor activities, wanted sunshine. Weather modification has not yet advanced to this stage, but there is evidence that it is possible to modify the weather in certain specific ways. Many weather modification efforts are concerned with increasing the amount of precipitation. Others endeavor to dissipate clouds over airports. Fog also hampers aviation activities and may be dissipated under the right conditions. Other weather modification activities concern decreasing the impact of hurricanes and decreasing or eliminating lightning and hail.

The two weather modification activities most likely to be applied over widespread areas are precipitation augmentation from summer convective clouds and orographic cold cloud seeding for **precipitation management** during the winter. The regions of most likely application are those that need additional water for agriculture, industry, municipalities, and hydroelectric power generators (Fig. 12-1). At least occasional application of weather modification technology can be expected over a large part of the United States in the future.

CLOUD SEEDING AGENTS

The usual cloud seeding materials used for the augmentation of precipitation, as well as for other purposes, are dry ice or silver iodide for cold clouds and salt for warm clouds. Other compounds are being developed that have a similar structure to ice crystals and are also environmentally more appealing than silver iodide. Silver iodide is composed of silver and iodine, and there is some concern about the environmental impact of large quantities of these materials accumulating on the ground when clouds are seeded. Measurements have been made to determine the change in concentrations of silver in farm ponds, rivers, and so on in areas where cloud seeding with silver iodide is practiced. These have generally shown very minor concentrations. If cloud seeding with silver iodide became widespread, it would result in temporary local concentrations of silver in precipitation of the same order of magnitude as the natural concentration in surface waters.

Dry ice is composed of solidified carbon dioxide gas, and there is little objection to its use in the quantities required for cloud seeding. The temperature of dry ice is −78°C, which is so cold that a very small pellet of dry ice will cool the surrounding air sufficiently to cause the formation of many millions of ordinary ice crystals. This observation by Vincent Schaeffer in 1946, along with the discovery by Bernard Vonnegut that silver iodide crystals would induce supercooled droplets to freeze, provided the basis for most modern weather modification activities.

CLOUD SEEDING METHODS

Cloud seeding with silver iodide is usually performed by burning silver iodide flares. The resulting smoke particles furnish the nuclei for ice crystal formation. The crystalline structure of silver iodide is similar to an ice crystal, which is a six-sided hexagonal-shaped crystal. The silver iodide is similar enough in shape so that ice nucleation (initial transfer of vapor to solid) occurs on the smoke particles of silver iodide (Fig. 12-2). Since no actual cooling is associated with the introduction of silver iodide particles, this material is effective only if the cloud temperature is appropriate. The effective temperature range for seeding with silver iodide is from −4°C to −30°C, with the optimum temperature being −12°C. Cloud regions with a temperature warmer than −4°C are not sufficiently supercooled, while cloud regions where the temperature is lower than −30°C usually already contain only ice crystals since the temperature at which spontaneous nucleation occurs is −40°C.

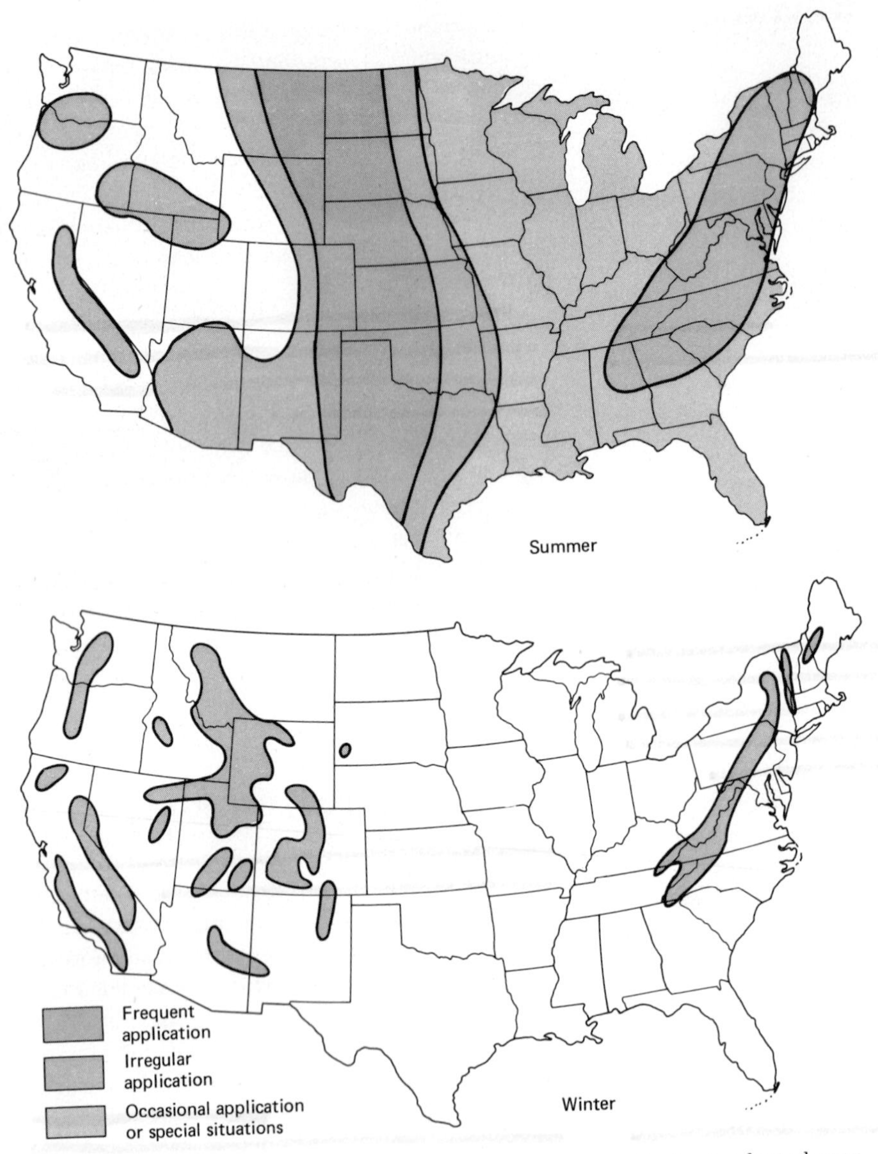

FIGURE 12-1 Future weather modification can be expected to be conducted over large areas of the United States. Areas of most likely application of summer convective precipitation management (top) are those that need additional water due to intensive agriculture, greater industrialization, and population concentration. Summer convective precipitation management can be expected to be frequently applied through a band extending from Texas to North Dakota and Montana. Many of the states within this area already have operational weather modification programs. The geographical areas containing mountain crests suitable for winter or orographic precipitation management applications are not as widespread, but the results of this weather modification activity have been so encouraging that winter precipitation management programs can be expected to become more important in the future. (From *Bull. Am. Meteorol. Soc.*, vol. 58, Nov. 6, 1977.)

FIGURE 12-2 When viewed through a microscope, silver iodide crystals show that they are similar enough in shape to six-sided ice crystals to serve as deposition nuclei for the formation of ice crystals in a supercooled environment. Silver iodide crystals are most efficient in causing deposition when they are coated with impurities, such as occurs during combustion processes. (Courtesy of Roger Cheng, Atmospheric Sciences Research Center, State University of New York at Albany.)

FIGURE 12-3 The most effective way of applying cloud seeding agents is by direct delivery into the cloud from aircraft. The updraft area near the base of a cloud may be seeded by flying through it, or the top of a different cloud formation may be seeded by flying near the top of the cloud deck. (Photo courtesy of NOAA.)

Cloud seeding with dry ice is performed by grinding the dry ice into small particles and dispersing them through the cloud from an airplane. Silver iodide is also most effective if it is applied from an airplane, normally from a flare attached to the wing of the airplane, which disperses it into the cloud (Fig. 12-3). Surface generators are sometimes used to seed clouds with silver iodide smoke, but these, in general, are less effective than delivery of the seeding agent directly into the cloud.

A typical seeding rate for dry ice is 5 g/km of travel, while the seeding rate of silver iodide may be 25 g/km. Data from cloud modification experiments indicates that the seeding rate is important in the cloud seeding process. The first thing that happens when a flat cloud top, or deck, is seeded from an airplane is that a bulge develops in the cloud deck. The cloud visibly expands as the latent heat of fusion (335 J for every gram of ice crystals) is released. Next a depression develops in the cloud where the ridge occurred initially (Fig. 12-4). The actual appearance is related to atmospheric air currents. If there is convective activity, holes may develop in the cloud; otherwise a depression in the cloud develops along the aircraft flight line. The cloud deck depression results from conversion of supercooled water to ice crystals that fall to a lower part of the atmosphere.

CLOUD DISSIPATION

It may seem paradoxical that cloud seeding can be used to both increase precipitation and dissipate clouds. Important factors contributing to the results of cloud seeding, in addition to cloud temperature, type of seeding agent, and method of delivery, are the cloud thickness and water content, and the amount of seeding agent applied. The object of cloud or fog dissipation experiments over airports is to initiate the growth of additional ice crystals at the expense of liquid water droplets by

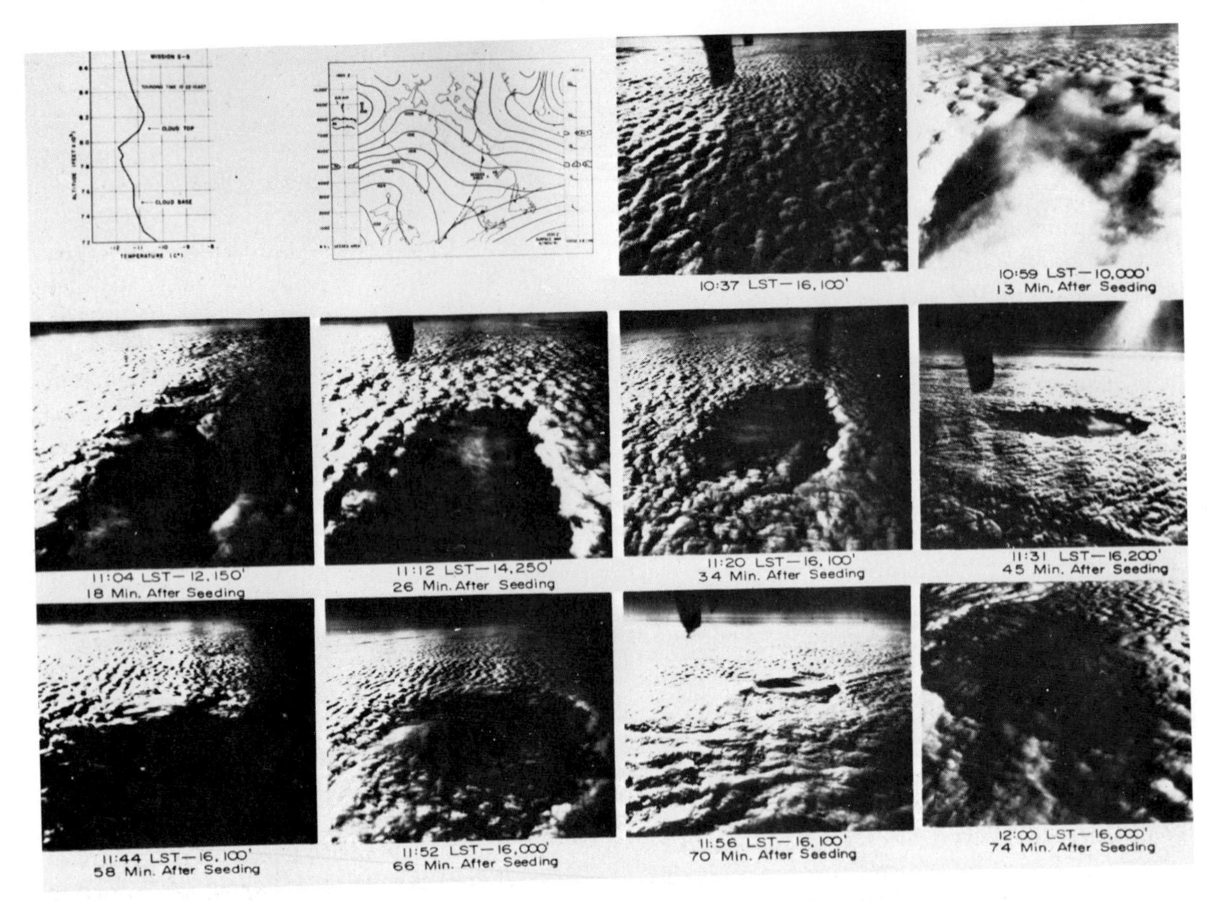

FIGURE 12-4 Sequence of photographs showing the visual modification of an altostratus cloud deck 200 m thick that was seeded with dry ice pellets. The location of the seeding experiment and the temperature through the cloud layer are shown in the two diagrams above. The first photograph shows the altostratus cloud deck before seeding, at 10:37 local standard time. The second photograph, at 10:59 local standard time, shows the cloud deck after seeding as the ice pellets cause a fuzzy appearance because of ice crystal formation. The next visual stage occurs as the ice crystals fall to a lower part of the cloud, leaving a hole or cleared strip through the cloud deck. (Photos by Air Force Cambridge Research Laboratories, Weatherwise, June 1968.)

adding dry ice pellets. Ice crystals, if present at the same temperature as liquid droplets, will have a lower vapor pressure and slower evaporation rate. Therefore the ice crystals are able to grow since they can gain mass, while liquid water droplets are not since they lose mass at the same temperature and relative humidity. It has been observed that both ice crystals and supercooled liquid water droplets may coexist in clouds with temperatures between 0°C and −40°C. If more of the supercooled water droplets are converted to ice crystals through cloud seeding, they may be able to grow fast enough to fall to a lower elevation, where they will melt and evaporate, thus dissipating the cloud.

Some dramatic cloud dissipating experiments have been performed. Cloud covered areas of 200 km² or more have been cleared in an hour. After seeding with dry ice, the cloud layer may simply vanish, resulting in a very impressive operation.

Whether clouds can be cleared or not depends on their temperature and thickness. The temperature must be between −10°C and −25°C for successful results. If the thickness of the cloud deck is less than 500 m, then it may be possible to clear it. In general, though, it is not possible to clear a cloud deck thicker than this.

PRECIPITATION ENHANCEMENT

Warm Clouds The usual purpose of cloud seeding is to increase the amount of precipitation. Most cloud seeding is directed toward supercooled clouds, although there is some emphasis on seeding warm clouds. Warm clouds may be seeded with small quantities of finely ground salt or with a water spray. This type of seeding is done to increase the number of hygroscopic (water absorbing) nuclei and to provide a range of droplet sizes. An ordinary cloud may not be very efficient in precipitating out all of the water that is present. Therefore, even though there is not a large deficiency of liquid condensation nuclei in the atmosphere, the addition of more condensation nuclei may increase the range of drop sizes and make the cloud more efficient. In general, warm cloud seeding has had less success than supercooled cloud seeding. This is probably because there are already a large number of condensation nuclei for liquid water in the atmosphere from natural and artificial sources.

One precipitation enhancement experiment consisted of dumping a few thousand liters of water on top of a developing cumulus cloud. Radar observations indicated increased return from the cloud, although it was uncertain whether the water showing up on the radar scope was the water that was poured over the cloud or was actually from increased raindrop formation. The idea was to supply larger droplets within the cloud to serve as agents in the coalescence process (explained in Chapter 6).

Cold Clouds The most productive cloud seeding efforts are probably those involving supercooled clouds. By using silver iodide or solidified carbon dioxide, it is possible to take advantage of the lower vapor pressure of ice crystals, in comparison to supercooled water droplets of the same temperature, so that ice crystals grow by natural diffusion.

Numerous natural sources of condensation nuclei keep the atmosphere well supplied. Many particles of various sizes are present in the atmosphere. These come from the ground as winds pick up clay particles, and from the oceans as the winds pick up salt particles evaporated out of tiny droplets left in the air by wind and wave action. Industrial activity and automobiles in cities also add particles to the atmosphere. Therefore, a tremendous number of nuclei are ordinarily present in the atmosphere, although the number and type vary over cities, land, and oceans (Fig. 12-5). More condensation nuclei are present over some cities than over others. In general, more condensation nuclei are in the atmosphere over cities than over open land, and the fewest number are over oceans. Many of the particles are hygroscopic in nature. This means they attract and absorb water vapor just as ordinary salt does.

Two separate raindrop formation processes are involved in precipitation modification. One of these results from a difference in the vapor pressure and expresses the rate that a droplet will grow because of the diffusion of water vapor to that droplet. If liquid water and ice crystals are present together, the ice crystal will have a lower vapor pressure and will grow in an environment where a liquid water droplet cannot grow (Fig. 12-6). The rate that it grows is proportional to the difference in vapor pressure between that of the ice crystal and the vapor pressure of the surrounding atmosphere. If a large difference exists, the growth rate is rapid; if the difference is small, the growth is slow.

The other raindrop growth process is the coalescence process. Many small droplets exist in a cloud if the water content is very great. The fall velocity

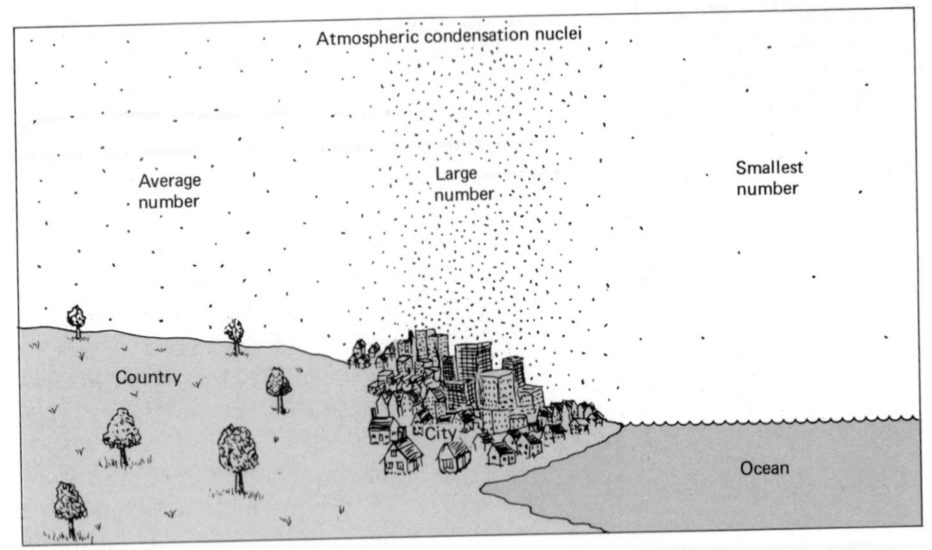

FIGURE 12-5 There is generally an abundance of atmospheric condensation nuclei for the formation of liquid water droplets. More condensation nuclei occur in some locations than others, however. They are most abundant over cities and least abundant over oceans; however, many of the condensation nuclei over oceans are salt particles, which are very effective nuclei for the formation of liquid droplets. The optimum concentration of condensation nuclei for the formation of precipitation is not known. However, it is likely that too many condensation nuclei will reduce the amount of precipitation just as a deficiency of condensation nuclei will do.

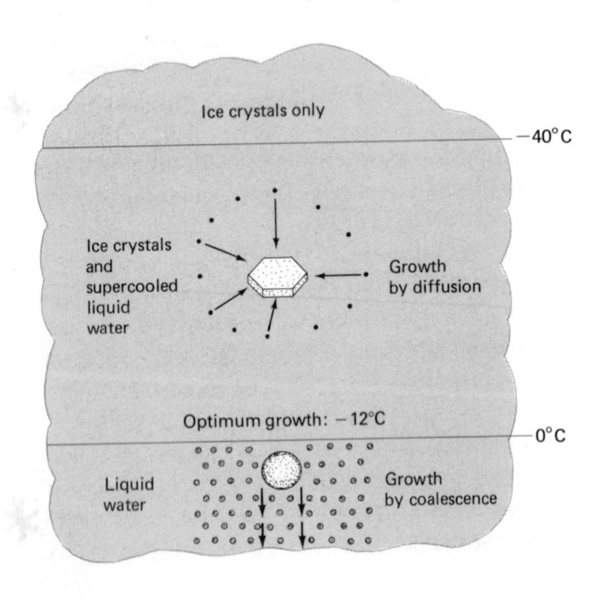

FIGURE 12-6 Precipitation grows in clouds by diffusion in the region above the freezing level. Diffusion of water vapor causes it to collect on ice crystals more rapidly than on liquid water droplets in the same region because of the difference in vapor pressure. The optimum temperature for such growth is −12°C. Below the freezing level raindrop growth may continue by the coalescence process, whereby large cloud droplets gather up smaller droplets that coalesce with them.

FIGURE 12-7 This photograph shows the conversion of an altocumulus cloud layer to cirrus clouds. The unique hole in the altocumulus cloud layer may have been produced naturally as ice crystals from a thin cirrus layer above mixed with the altocumulus layer because of atmospheric turbulence, producing natural cloud seeding. The hole in the altocumulus cloud deck was produced as the supercooled liquid water droplets were converted to ice crystals with the formation of cirrus type clouds. Such holes in cloud decks can be produced by spot cloud seeding and even by the penetration of an aircraft with enough associated turbulence to cause a mixing of ice crystals from above with the supercooled liquid water in a lower cloud deck. (U.S. Air Force photo.)

of large drops is much greater than that of small drops. Therefore, large drops will overtake many small drops within a cloud with a high water content. The larger drops will coalesce with enough smaller ones to grow rapidly in a short period of time. The total growth of a raindrop is determined by the combination of both the diffusion and the coalescence processes.

A natural process occurs in the atmosphere that is similar to cloud seeding. This is set up by a **releaser cloud** (cirrus) composed of ice crystals, with a **spender cloud** (altocumulus) beneath it (Fig. 12-7). With a high water content cloud above the freezing level and a cirrus cloud composed of ice crystals higher in the atmosphere, vertical mixing causes natural cloud seeding. The ice crystals that are mixed into the altocumulus may grow rapidly if the amount of supercooling of the liquid water droplets is large. The greatest growth region is where the supercooling of the liquid water is greatest, and the region may spread rapidly after rapid growth of ice crystals is initiated by atmospheric turbulence or, in some cases, the turbulence accompanying the passage of a jet aircraft.

Seeding Rates Some experiments have indicated that the rate of cloud seeding is important in determining the outcome. If the atmosphere is seeded with an overabundance of ice crystals or silver iodide, too many droplets may result or the cloud may become glaciated as in hail suppression experiments. Above the freezing level ice crystals grow primarily by diffusion; below the freezing level droplets grow primarily by the coalescence process. There are indications that a cloud seeded with 1000 ice crystals per cubic meter will see a very slow growth rate in the diffusion region, while a much faster growth rate will occur in the lower atmosphere, where coalescence occurs. If 10,000 ice crystals per cubic meter are used, the growth rate and the diameter of the drops will be much smaller. In fact, so many tiny droplets can develop that they may all evaporate before reaching the surface. So the seeding rate is important, although not enough experiments have been conducted to give complete information on this critical rate.

Weather Modification Evaluation It is very difficult to evaluate precipitation enhancement programs. It would be simple if two clouds developed in the atmosphere that were exactly alike throughout their life cycle; then one could be seeded while the other would serve as a control for comparison. Since the atmosphere does not generate identical clouds, other methods of evaluating the effects of artificial cloud modification must be used (Fig. 12-8). A statistical approach, for example, may be

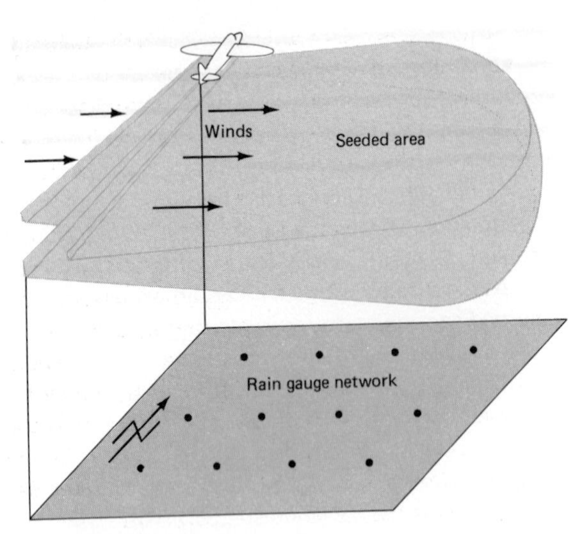

FIGURE 12-8 Evaluation of cloud seeding experiments is difficult because of the natural variability of clouds and precipitation at the ground. A statistical approach may be used where the area that is seeded is carefully determined by the aircraft flight path combined with the environmental winds through this region of the atmosphere. The amount of rainfall beneath the seeded area is then compared with the amount of rainfall outside the seeded area based on measurements obtained from a rain gauge network. Several such experiments provide a statistical answer to the effect of cloud seeding.

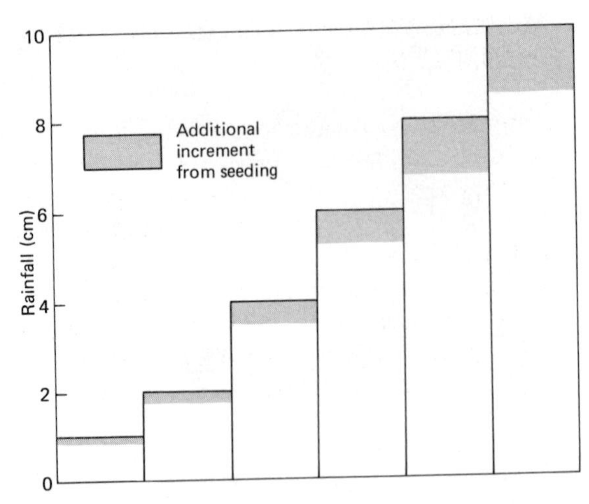

FIGURE 12-9 Generalized diagram indicating the amount of additional precipitation that can be expected from cloud seeding operations if the operations are performed under appropriate conditions. It is generally thought that a 10% to 20% increase in rainfall is likely. This means that the additional increment from seeding is proportional to the amount of rainfall that could be expected naturally, making it difficult to obtain large amounts of additional rainfall from seeding during extremely dry weather.

used by seeding a certain area and noting the winds at the height of the aircraft so the dispersion of the seeding agents after a certain period of time will be known. The effect of seeding is then determined by comparing the rainfall within the seeded area to the rainfall outside the seeded area for a number of different cases. Some of these experiments indicate that the amount of precipitation can be increased 10% to 20% by seeding under the right conditions (Fig. 12-9).

Considerable concern has been raised over the **sponge effect.** A sponge can hold only a finite amount of water. When that is squeezed out, the sponge is dry. A common assumption is that augmentation of precipitation in one area must result in its diminution in another area because the atmosphere is drier. If cloud seeding, when the natural rate of precipitation was constant through several hours, resulted in an immediate increase for the first hour followed by a decrease to a rate much smaller than the natural rate for the next hour, it would be possible to change the distribution of rainfall. One county would be able to seed the clouds and get more rainfall at the expense of the counties downwind. This would be possible if all the water were extracted from the atmosphere. Current cloud seeding operations, however, do not indicate that this is happening. In fact, increases in precipitation to distances of 400 km downwind have been seen.

MAJOR WEATHER MODIFICATION PROJECTS

A cumulus cloud modification project was conducted in Florida for several years by characterizing days as rainy or fair. A rainy day was one with over 10,000 km² of showering cumulus clouds within 160 km of Miami; a fair day was one with less. Cloud seeding on fair days resulted in a sevenfold enhancement of rainfall while the rainy day seedings diminished precipitation by 20%. If this experiment is typical, then on days when it is raining naturally, seeding reduces the amount of precipitation, while cloud seeding on fair days may increase the precipitation by as much as seven times. One explanation for this is that the overabundance of ice crystals on rainy days, when the natural processes are operating effectively, may cause the water in clouds to be converted to many small droplets that evaporate before reaching the ground.

A significant amount of funds has been expended for cloud seeding and other weather modification efforts. In the mid-seventies, the amount spent for precipitation augmentation and redistribution was $5 million per year; hail suppression, $4 million; reduction of winds in hurricanes, $2 million; fog dispersal, $2 million; and lightning modification, $1 million. Three of the major agencies that funded research and operational weather modification programs were the National Science Foundation, the National Oceanic and Atmospheric Administration, and the Bureau of Reclamation. In addition, the Navy and the National Aeronautics and Space Administration also expended funds for research on weather modification.

Some states, such as South Dakota, have tried operational cloud seeding programs where the clouds are seeded during dry weather in an effort to increase the amount of rainfall. Groups of counties have also combined funds to pay for operational cloud seeding. Five counties in Kansas each contribute several thousand dollars each year to pay for **operational cloud seeding** during the summer when droughts occur. Operational cloud seeding is aimed at results rather than experimentation.

A major federally funded project is the **Colorado River Basin Project,** which is supported by the Bureau of Reclamation. An area in the San Juan Mountains in Colorado of about 5000 km² is the site of cloud seeding to increase the snow pack in the mountains during the wintertime. Only days when the temperature is between −12°C and −23°C are chosen. The results have been favorable and indicate that about a 30% increase in the snow pack results from the seeding operations. Calculations indicate that the value of the stored water is about $100 million, since much of the water supply for that area comes from the snow that melts after the winter season. This project is one of about fifteen research programs of the Atmospheric Water Resources Program called Project Skywater.

Another weather modification project is the **High Plains Project.** This program is being developed by the Environmental Research Laboratories of NOAA, in cooperation with the Bureau of Reclamation. This project is concerned with trying to enhance the precipitation in the Great Plains area. The sites that have been chosen are Miles City, Montana, and Colby, Kansas. Some theoretical studies of cloud models will be combined with actual field experiments with radar and other instruments to determine the effects of cloud seeding.

Another cumulus cloud modification project is being conducted in Florida by the Environmental Research Laboratories of NOAA. This project is also concerned with mathematical models of the buoyancy of cumulus clouds, with calculations related to obtaining more precipitation from them. The data have not been very conclusive, but in 1971, when southern Florida had a severe drought, sufficient political pressure was generated that NOAA was asked to send airplanes to seed the clouds and break the drought. There was some indications that their efforts were successful.

The hurricane modification project is **Project Stormfury.** In this project, cloud seeding is used to try to decrease the winds in certain hurricanes. The first hurricane to be seeded was Hurricane Esther in 1961. The most noticeable result when Hurricane Esther was seeded was a change in the direction of the storm. The hurricane had been traveling in a typical northern direction prior to seeding, then it made a loop and struck the east coast of the United States (Fig. 8-12). No evidence has been acquired from other storms that would indicate that a change in direction is likely because of seedings. Other hurricanes have traveled in the same kind of looping path naturally. Even though there was no real indication that the seeding caused the change in direction, hurricane seeding was off to a slow start because of possible political ramifications. Specific areas are now defined where hurricane seeding can be conducted.

The next hurricane to be seeded was Hurricane Beulah in 1963. Measurements were made before and after seeding. There was no dramatic change in direction but a large reduction in wind speeds occurred after the second seeding. The pressure in the eye also rose after the second seeding, indicating a weaker storm.

The seeding of two different regions in hurricanes has also been tried. Either the eyewall is seeded in an effort to increase its diameter and weaken the intensity of the storm, or seeding is performed some distance out from the major, vertically ascending air surrounding the eye (Fig. 12-10). The idea in this case is to cause more vertical motion and cloud formation from the surface inflowing air before it gets to the tight band around the eye. This would reduce the band of highest velocity winds just outside the eye of the hurricane.

The greatest success in seeding hurricanes came from Hurricane Debbie in 1969. Seeding of this hurricane on August 18 was followed by a 31% reduction in the winds. The hurricane was not seeded on August 19 and the winds in the hurricane regained most of their original intensity (Fig. 12-11). It was seeded again on August 20 and the maximum winds were again reduced by 15% of their preseeding value. These results were quite favorable.

Hurricane Ginger was seeded in 1971. This hurricane was not a very good candidate for seeding. Hurricanes have some natural oscillations of wind speed and pressure in the eye. The results of seeding Hurricane Ginger were very inconclusive because no changes occurred beyond the natural changes of the storm. Project Stormfury was suspended in 1973, but was reactivated in 1976 with the purchase of additional airplanes having instrumentation capable of continuous measurements for more extended periods of time within hurricanes. Hurricane seeding is anticipated again during the 1980s.

There are no national projects to reduce the winds associated with tornadoes, thunderstorms,

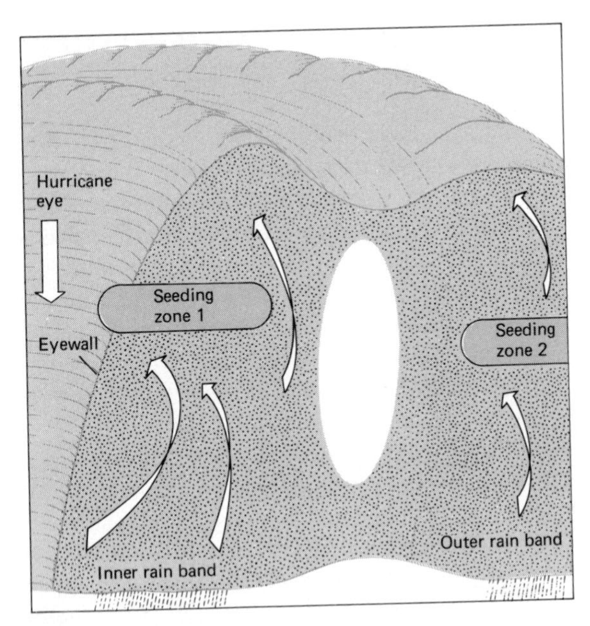

FIGURE 12-10 Two different approaches to hurricane modification have been attempted. The winds spiraling around the eye of the hurricane are in delicate balance with the decreasing pressure toward the center of the eye. The first seeding zone involves aircraft penetration and seeding near the eyewall to interfere with the pressure–wind balance. If the eyewall can be made larger, then the winds surrounding the eye will decrease in intensity, resulting in beneficial effects. Another approach to hurricane modification involves seeding the updrafts located further away from the eye in an attempt to intensify them and decrease the moisture content of the air before it reaches the inner rainbands. This should contribute to weaker winds in the inner rainband, which contains the strongest winds in the hurricane.

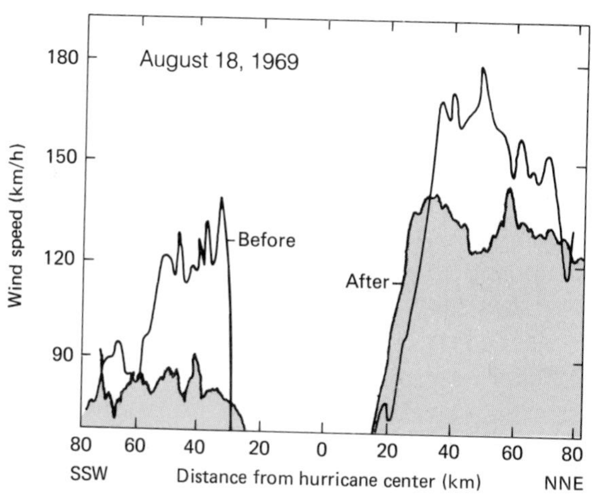

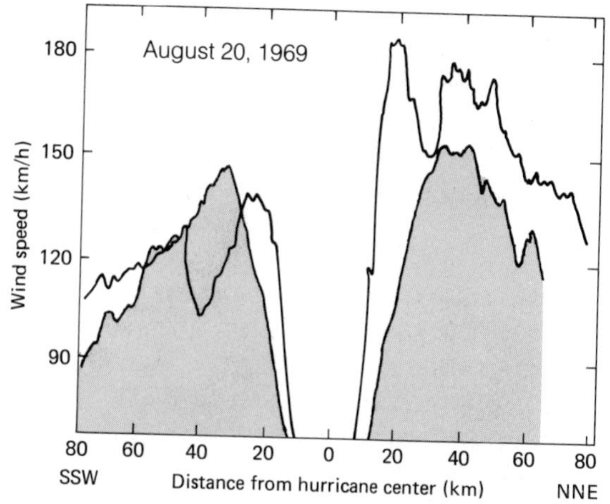

FIGURE 12-11 The wind speeds through Hurricane Debbie were measured on August 18, 1969, by aircraft penetration across the storm from southsouthwest to northnortheast at a height of 3.6 km. Wind speeds measured four hours after seeding showed a significant reduction. Two days later the winds had regained their strength, but again decreased six hours after seeding, as indicated by the aircraft measurements. (From R. C. Gentry, *Science,* 1970, 168: 473–475, copyright 1970 by the American Association for the Advancement of Science.)

or squall lines. The major effort is to try to better understand the causes of these storms so that experiments can be planned to modify them.

The National Hail Research Experiment (NHRE) was conducted by the National Center for Atmospheric Research (NCAR) in northeast Colorado in cooperation with several universities. It started in 1972 and was scheduled to proceed for five years with the object of developing techniques to reduce damage from hail. The beneficial competition concept was used in the hail modification experiments, (by the introduction of sufficient nuclei to cause great competition for growth of large hail.) The experiment used many instruments to make measurements in hailstorms created with and without cloud seeding in order to get additional information on these very important storms. Part of the stimulus for this program came from the very encouraging reports from field experiments and operational programs in Russia. However, the NHRE program was suspended in 1974 with inconclusive results. Such a program may be revived again after careful replanning.

Another weather modification effort concerns **lightning suppression.** The stimulus for this project has come from the many forests that are destroyed each year by lightning (Fig. 12-12) and from the hazard lightning represents during satellite launches. The Forest Service and Department of Agriculture conducted experiments in the 1960s that indicated that some of the electricity in clouds could be reduced. Probably the most successful experiment was performed in 1972 by the Environmental Research Laboratory on thunderstorms seeded with fine aluminum fibers. Measurements of the electrical charges within these thunderstorms showed that the charges were almost eliminated in about 10 minutes.

Fog modification efforts are also in progress. The Federal Aviation Administration is the leading sponsor of this effort since fog is a major problem around airports. The cost of fog to the airlines is

FIGURE 12-12 About 7000 forest fires are started each year by lightning discharges. Many of these are small, but others become quite large and cover large sections of forests. This forest fire in Boise National Forest, Idaho, burned for five days over 7000 acres. (Photo courtesy of U.S. Forest Service.)

FIGURE 12-13 Helicopters were used to remove a warm fog engulfing a bridge across Smith Mountain Lake, Virginia. Two helicopters flew slowly across the top of a 100-m deep fog layer, where the surface visibility was only about 30 m. The downwash action from the rotor of the helicopter forced dryer air from above to descend and mix with the fog layer, causing the fog to evaporate. The first photograph, taken at 8:13 A.M., shows the amount of clearing after one helicopter was halfway across the fog layer. The second photograph, taken at 8:21 A.M., shows a cleared strip about 600 m wide. This clearing remained over the bridge until dissipation of the surrounding fog about 45 minutes later. (Air Force Cambridge Research Laboratory, Bedford, Massachusetts, courtesy of V. G. Plank, Weatherwise, June 1969.)

over $15 million per year, with an additional cost of more than $25 million to passengers involved in fog-related problems.

About 95% of fogs in the United States are warm fogs. Efforts to eliminate warm fogs have not been as successful as for cold fogs. Heat has been supplied to evaporate fog in some locations. Helicopters have also been used to mix the air above airports to eliminate warm fogs. Some of these methods are able to evaporate or disperse fog (Fig. 12-13), but most are not effective for very long. Warm fogs are largely in the uncontrolled category.

Cold fog is easier to eliminate. If the temperature is colder than −5°C, the supercooled fog droplets can be seeded with dry ice or silver iodide to convert them to ice crystals that fall out of the lower atmosphere. Several dozen airports, such as Salt Lake City, Medford, Spokane, Boise, and Portland, have continuing fog seeding programs. The fog modification programs are similar in concept to cloud dissipation operations since ice crystals grow at the expense of liquid water droplets to become large enough to fall to the ground or into a warmer atmospheric layer, where they melt.

LEGAL ASPECTS OF WEATHER MODIFICATION

As certain weather modification activities became a reality in the early 1950s, they brought many complex legal questions. Most of these have remained controversial and may continue to do so until more experiments have been conducted and a more complete understanding of weather modification results is possible. Legal aspects of weather modification may be local, national, or international. One issue is the ownership of the atmosphere. Questions have been raised as to whether it belongs to the person who owns the land beneath it, to the state that is located beneath the atmosphere, or to the nation over which the atmospheric winds travel.

Weather is international in scope. Therefore, weather modification may have important international ramifications. Specific international legal codes are very developmental in nature. The United Nations Conference on the Human Environment that met in Stockholm in 1972 urged the establishment of an advisory committee to consider the weather modification problems that have potential international concern. It is conceivable that serious problems may arise in the absence of an international legal policy concerning weather modification. Internationally, the most active nations in weather modification programs are Russia followed by the United States. More than a dozen other nations are involved in weather modification activities to a lesser degree.

In the United States, there are no federal regulations on weather modification activities. In 1970, the Environmental Science Services Administration of the National Oceanic and Atmospheric Administration (NOAA) was authorized to establish a mandatory reporting system for all weather modification activities in the country. This reporting system, however, was not designed to represent comprehensive weather modification action at the federal level.

Much of the legal responsibility for weather modification rests with the particular state involved. Slightly more than half of all states have passed **weather modification regulations.** These state regulations are not uniform, and in some cases they are even contradictory. The state of Maryland prohibited any form of weather modification for the 2-year period starting in 1965. Pennsylvania has a law granting to the individual county the option to prohibit weather modification if it is considered detrimental to that county's welfare. Some states, such as Kansas, North Dakota, Wisconsin, and Oregon, have laws that attempt to regulate weather modification activities through licensing. Some of these also provide for collecting and evaluating data that result from the modification activities. States such as Colorado, Louisiana, and New Mexico have regulations that prohibit certain weather modification activities within their boundaries that might conceivably affect another state. On the other hand, some states claim sovereignty over the moisture that is in the atmosphere above.

Legal cases of individuals who have allegedly suffered damages because of cloud seeding operations represent a problem for the legal system. In general, plaintiffs are unable to establish cause and effect relationships that would allow a positive decision. Because of the unpredictable nature of the atmosphere and the difficulty of obtaining complete and accurate measurements during weather modification activities, it is often difficult to specify the exact cause of damages. Most of the few lawsuits involving weather modification have been decided in favor of the weather modifiers.

The complex interactions that arise because of weather modification efforts emphasize the necessity for cooperation among legislators, scientists, politicians, and the general public. The potential benefits from properly applied weather modification activities appear to outweigh the possible disadvantages to such an extent that legislation should not be enacted that would prohibit well-planned programs.

Weather modification has apparently been used for military and political purposes. Certainly, the Department of Defense is one of the larger funders of weather modification operations and research. There are indications that artificially induced rainfall was produced by United States military operations to break up unwanted demonstrations and muddy the Ho Chi Minh trail, an important supply route, during the war in Vietnam. There is also some indication that cloud seeding operations were conducted within the wind currents that carry moisture to Cuba and that the operation brought erratic weather to the island in 1969 and 1970, resulting in the sugar crop falling far short of goals that had been set.

Hopefully, future weather modification activities will be conducted only in the best interests of the general public. It is clear that the responsibility for weather modification activities should not be left entirely to one group such as scientists, legislators, or political leaders. It should be a common expectation that weather modification activities that could have such tremendous benefits not be allowed to deteriorate to the realm of warfare, where the battle could be won by those countries with the best scientist or the money to buy operational weather modification capabilities. In such a situation it is possible—because of adverse climatic effects—that no one would win.

SUMMARY

Weather modification has been of interest for centuries, although it is only in recent times that the capability of actually modifying the weather in certain specific ways has become a reality. Weather modification programs are in operation to increase the amount of precipitation, to dissipate clouds and fog over airports, to decrease the winds in hurricanes, and to suppress lightning and hail. Most of these weather modification activities are based on the discovery that silver iodide or dry ice crystals cause the conversion of supercooled water droplets to ice crystals. Silver iodide has a crystalline structure very similar to the ice crystal, while dry ice forms additional ice crystals because of its very cold temperature.

Cloud seeding is most effective if the seeding agent is delivered directly into the cloud rather than injected into the atmosphere at the surface. Thin cloud layers less than half a kilometer thick can be dissipated by heavy cloud seeding if their temperature is about $-20°C$. The major weather modification effort is to increase the amount of precipitation. Although warm clouds may be seeded with finely ground salt particles, most cloud seeding efforts involve the seeding of supercooled clouds with silver iodide or carbon dioxide crystals. Measurements indicate that the atmosphere normally contains an abundance of condensation nuclei for liquid water although there is frequently a deficiency in deposition nuclei for the formation of ice crystals.

The raindrop formation mechanism in cloud modification activities is the growth of ice crystals at the expense of liquid water droplets. Further growth then occurs by the coalescence process. Natural cloud seeding may occur between a cirrus and lower cloud deck, if mixing of ice crystals within the cirrus cloud causes supercooled droplets within the water cloud to form ice crystals.

The number of ice crystals or amount of silver iodide introduced into a cloud probably affects the outcome, although it is difficult to evaluate the results of cloud seeding experiments because of the natural variability of precipitation. The seeding of supercooled clouds with silver iodide seems to be more effective than seeding warm clouds with salt particles, as would be expected from measurements of condensation and deposition nuclei. Many precipitation modification experiments are based on a statistical approach, with comparisons made of the amount of rainfall within the seeded area and that outside the seeded area. Reliable results require a large number of cases because of the variability of rainfall. Generally the results indicate that increases of 10% to 20% in the amount of rainfall is possible by seeding under the right conditions.

Some of the major weather modification projects include the Colorado River Basin Project, where cloud seeding is performed to increase the winter snow pack in the mountains; the results have been quite favorable. Project Stormfury is the hurricane modification program, where cloud seeding has been conducted to decrease the winds in certain hurricanes. The greatest success in seeding hurricanes came from Hurricane Debbie in 1969, when winds were reduced after seeding on two separate days.

Hail suppression research has been conducted by various universities at the National Center for Atmospheric Research. The results of hail suppression activities have been much less impressive than was anticipated from hail suppression reports from other countries. A greater understanding of the hailstorm structure is required before storms can be seeded with the right amount of seeding agent. Further studies are required before hail suppression activities can become operational on a large scale.

Fog modification efforts are generally more localized because of the nature of fog. Cold fogs are commonly seeded with silver iodide at many airports. Warm fogs are less responsive to modification efforts, but techniques are being developed that may help eliminate them in the future.

Lightning suppression experiments indicate that electrical charges can be reduced in thunderstorms by seeding them with fine aluminum fibers, thereby decreasing the amount of lightning.

Weather modification activities have been accompanied by complex legal questions. The legal aspects of weather modification may be local, national, or international. No federal regulations exist in weather modification activities although a mandatory weather modification reporting system has been established. More than half of the states have weather modification regulations, although they vary from state to state. Legal cases of individuals who have allegedly suffered damages as a result of weather modification have, in general, been decided in favor of the weather modifiers because of the difficulty in establishing cause and effect relationships. Weather modification efforts require the cooperation of legislators, scientists, politicians, and the general public if such activities are to result in maximum benefits for society.

STUDY AIDS

1. Describe some of the problems that could develop if weather modification on a large scale were possible.

2. Explain the theories behind the seeding of cold and warm clouds.

3. Explain how cloud seeding may be used in one case to dissipate clouds and in another case to enhance precipitation amounts.

4. Discuss the natural occurrence of condensation and deposition nuclei.

5. Discuss the raindrop growth mechanisms in a cold seeded cloud and compare these with natural raindrop growth.

6. Discuss the evaluation of the results of cloud seeding experiments.

7. Explain the sponge effect.

8. Compare the objectives and activities of several of the major weather modification programs.

9. What are some of the problems in dispersing fogs?

10. Discuss the legal aspects of weather modification.

11. Discuss the use of weather modification for the benefit of society, including your opinions as to who should make the decisions and the activities that should be encouraged.

TERMINOLOGY EXERCISE

Check the glossary if you are unsure of the meaning of any of the following terms used in Chapter 12.

Precipitation management
Releaser cloud
Spender cloud
Sponge effect
Operational cloud seeding

Colorado River Basin Project
High Plains Project
Project Stormfury
Lightning suppression
Fog modification
Weather modification regulations

THOUGHT QUESTIONS

1. Do you think weather modification activities are being practiced as widely as they should be? Explain your answer.

2. Discuss the role of legislators, scientists, and others such as politicians in determining weather modification activities.

3. Which particular aspects of weather modification would you prefer to receive the greatest amount of attention? Give your reasons for this.

4. Explain how it may be possible to use similar cloud seeding techniques to both dissipate clouds, and in other cases, increase the amount of precipitation from clouds.

5. Discuss some of the many ramifications of seeding a hurricane just prior to its striking the United States.

SPECIAL TOPIC 15

NOAA'S DROUGHT WARNING PROGRAM

An early warning system has been developed to alert U.S. foreign aid officials to potentially severe drought and food shortages in developing nations, the National Oceanic and Atmospheric Administration (NOAA) has announced.

The Agency for International Development (AID), which requested the system, is using it to monitor and evaluate potential and actual drought-caused conditions in the Caribbean and sub-Saharan Africa. Systems also are being developed for southern Africa and portions of Central and South America and southern Asia.

Paul F. Krumpe, science advisor for AID's Office of Foreign Disaster Assistance, said the system has the potential for providing information 30 to 60 days prior to harvest of a crop, as well as indicating potential shortages that may not occur until three to six months after harvest.

NOAA's Environmental Data and Information Service began providing weekly weather assessments late in 1977, according to Louis T. Steyaert of the Center of Environment Assessment Services. Steyaert said qualitative crop conditions are integrated with weekly weather assessments.

The operational early-warning system depends heavily on computer assessment models, Steyaert said, but data limitations often make it impossible to develop traditional climate/crop yield models. Therefore, analysis is conducted on the interrelationship of climate, subsistence agriculture, abnormal food shortages, and soil erosion.

The risk of food shortages due to drought is compounded sometimes by land-management practices that decrease agricultural productivity, according to Steyaert. Some countries have experienced food productivity shortages during a drought while neighboring nations have not been affected. Deforestation and poor land-management practices have led to severe soil erosion, and this has increased the risk of potential food shortages.

Erosion reduces the water-holding capacity of the soil and lowers potential soil moisture reserves, the NOAA scientist explained. This increases the vulnerability of crops to short-term dry spells. A corn yield model used to investigate these "pseudo-drought" conditions in one country showed the simulated yield for eroded soil was 30% less, and four times more variable, than that for uneroded soil.

This analysis suggests "pseudo-drought" is probably the key factor to drought-related food shortages in some countries, Steyaert concluded. Thus, these nations can expect to be more severely affected by short-term dry spells than their neighbors.

From William J. Brennan, *U.S. Dept. of Commerce News.*

CHAPTER THIRTEEN

ATMOSPHERIC OPTICS AND ACOUSTICS

ATMOSPHERIC DISPLAYS

Because white sunlight can be separated into a number of different colors as it passes through an atmosphere containing a variety of tiny prisms and spheres, it is not surprising that a variety of intriguing atmospheric displays occur. Cirrus clouds contain millions of ice crystals that are not all randomly oriented but assume particular positions determined by their shape and their interactions with air currents as they fall through the atmosphere. The preferred orientation of hexagonal plates is to have their flat sides horizontal; the preferred orientation of the longer prisms is to have their long axis horizontal. Atmospheric optical effects produced as sunlight passes through ice crystals include haloes, sundogs, and sun pillars.

Spherical raindrops in clouds produce such optical phenomena as rainbows and coronas. Rainbows provide a link between artists, poets, and scientists because of their natural beauty and scientific interest in the processes involved in their formation. Their main features can be explained by simple geometrical optics, but some of their detail requires deeper scientific explanations. The brightest arc after a rain shower in the afternoon, or before a shower in the morning, is the primary rainbow. Although its colors vary in brightness, they always follow the sequence of violet (innermost) blending into shades of blue, green, yellow, orange, and red. Higher in the sky a fainter, secondary rainbow may appear with the colors in reverse order. The area between the two is darker than the area enclosed by the primary rainbow, which may be distinctly lighter. Occasionally faint bands of pink and green alternate on the inside of the primary rainbow. These are called **supernumerary arcs** and are usually clearest near the top of the bow.

REFLECTION, REFRACTION, AND DIFFRACTION OF LIGHT

The various atmospheric optical effects can be explained by using the basic properties of light combined with the physical characteristics of the atmosphere, ice crystals, and raindrops. As a background for understanding these phenomena the **reflection, refraction,** and **diffraction** of light become important (Fig. 13-1). As light traveling through the atmosphere strikes water, for example, some of it is reflected at the same angle as the incident ray while the transmitted light is refracted at a different angle. The path of the reflected light is determined entirely by geometry, but the path of the refracted light involves the properties of the medium as well. Since light travels more slowly in denser mediums, it is bent as it enters a denser material. The ratio of the speed of light in a vacuum to its speed in the transparent medium is called the **refractive index.** The refractive index of water is 1.33 while that for air of average density is approximately 1.0. The amount of refraction in a substance can be determined by a law developed by Willebrord Snell in 1621 and called Snell's law (see Appendix A).

Light is diffracted by particles with a diameter similar to the wavelength of the light. Light waves can be compared with waves on water. For example, as with water waves, constructive interference occurs when two crests combine and destructive interference occurs when a crest and a trough arrive at a point at the same time. A light beam has tiny wavelets at its edge that are not very noticeable until it passes through a small hole or around a cloud droplet, as in Figure 13-1. The wavelets on the edge of the beam (cloud droplet shadow in this case) then spread outward, with constructive interference producing streams of brighter light called diffracted light. Diffracted light produces some of the atmospheric optical effects.

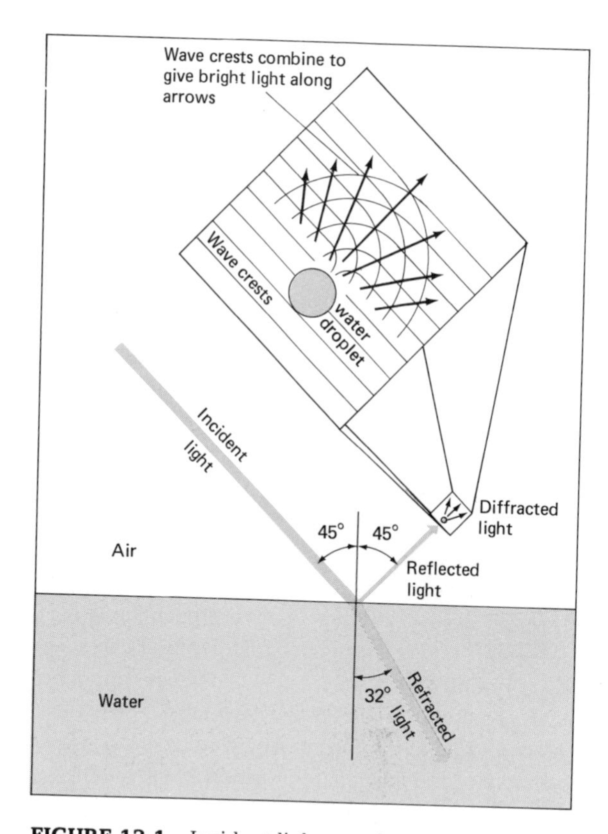

FIGURE 13-1 Incident light may be reflected, refracted, or diffracted as it passes from one material to another. Light is reflected at the same angle as the incident light. However, the angle of refracted light depends on the density of the material. Water is much denser than air, so light that is refracted as it enters water is sharply bent toward a line normal to the surface of the water. Diffraction of light can occur from reflected light or incident light. As light passes around any small object, wavelets spread from the shadow of the object, where the crests of these small waves combine, giving diffraction patterns of light spreading outward from the object. The intensified light is indicated by the arrows extending outward from the water droplet in the figure.

ATMOSPHERIC BENDING OF LIGHT RAYS

Some interesting optical effects are produced in the atmosphere by changing lapse rates. The lapse rate is determined by the temperature structure of the atmosphere and this is related to the density of the air. As previously discussed, light rays are bent if they pass from one substance into another substance of differing density. Light rays entering water, for example, are bent toward a line perpendicular to the water surface, with the specific amount of bending determined by the density of water. Therefore objects beneath the water surface are not located where they appear to be. A similar effect occurs in the atmosphere. Light rays are gradually bent in the atmosphere by air of varying densities, causing some interesting optical effects. Density differences are always present in the atmosphere unless one specific lapse rate exists throughout the atmosphere. This specific lapse rate is called the **autoconvective lapse rate,** and it corresponds to the exact temperature change with height needed for constant atmospheric density. If the temperature decreases exactly 34.2°C/km (0.3°C/10 m), no density differences exist in the atmosphere between various layers of air. For all other lapse rates, the density either increases or decreases with height. Since the measured lapse rate is never exactly equal to the autoconvective lapse rate for a very thick atmospheric layer, light rays are always bent as they travel through the atmosphere.

The average temperature profile in the troposphere is 6.5°C/km, which results in a slight decrease in density with height. This is a stable atmospheric density structure with the denser air below the less dense air. An atmosphere with decreasing density with height produces bending of the light rays with a curvature in the same direction as the earth's curvature (Fig. 13-2). The amount of bending depends on the specific lapse rate. A lapse rate of 6.5°C/km produces one amount of bending while lesser lapse rates, including inversions, produce even greater bending in a similar direction. Stronger inversions represent more rapid increases in temperature with height and result in denser air below air that is less dense.

Lapse rates greater than the autoconvective lapse rate of 0.3°C/10 m (34.2°C/km) develop in a thin atmospheric layer due to intense solar heating of the earth's surface. This commonly occurs in desert regions. The heat input must be sufficient to surpass the heat lost by convection and terrestrial radiation in order to develop a lapse rate that represents an increase in density with height. Atmospheric density increases with height cause bending of light rays with a curvature opposite to the direction of the earth's curvature (Fig. 13-3). Various optical effects are possible in the atmosphere as light rays are bent in varying degrees in either direction. The atmosphere produces an effect similar to that of light entering water, which is bent sharply, except that the bending is gradual. The amount and type of bending depend on the density change with height and this is determined by the temperature of the air.

In summary, the atmospheric density *decreases* with height for all lapse rates less than the autoconvective lapse rate. Inversions produce very rapid decreases in density with height while the average lapse rate produces a smaller decrease. The atmospheric density *increases* with height for all lapse rates greater than the autoconvective lapse rate. Such lapse rates require strong surface heating and are possible only in shallow layers of the atmosphere near the earth's surface. A shallow layer up to a person's eye level or less is all that is required to produce atmospheric optical effects.

POSITION AND SIZE DISTORTIONS

Objects may appear to be in a different position or have a different size because of atmospheric density changes. Atmospheric optical effects related to a change in position are called looming and sinking. These occur as objects appear to be higher (looming) or lower (sinking) than their actual location because of the specific atmospheric density profile. It is also possible for objects to appear larger or

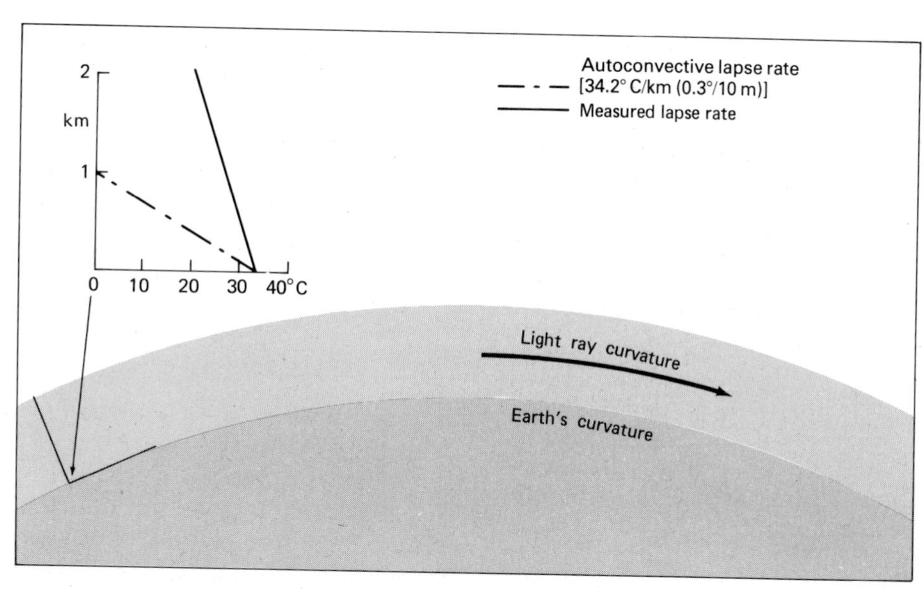

FIGURE 13-2 When the atmospheric lapse rate is less than the autoconvective lapse rate, the density of the air decreases with the height above the surface. This causes light rays to bend with a curvature in the same direction as the curvature of the earth.

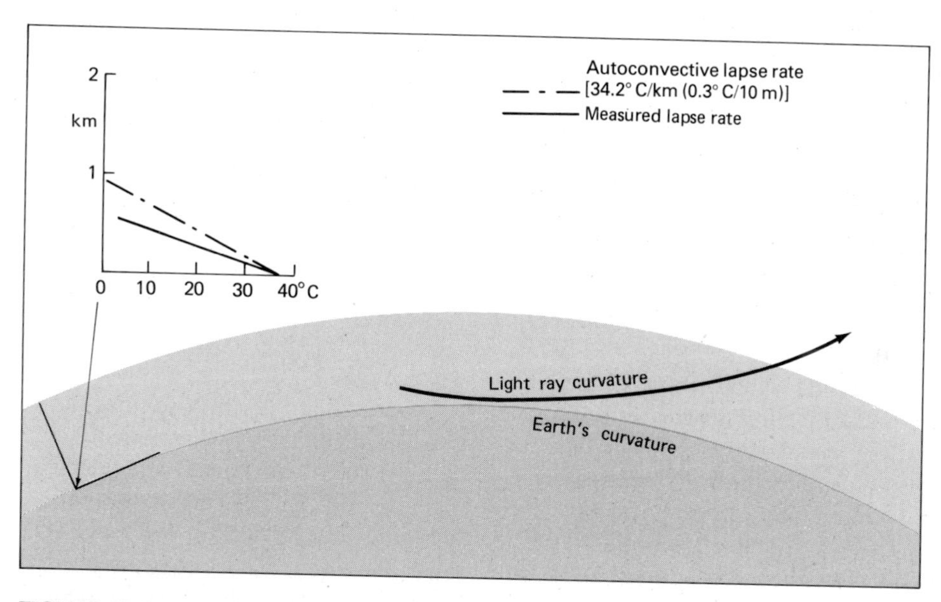

FIGURE 13-3 If the measured atmospheric lapse rate is greater than the autoconvective lapse rate, the density of the air increases with height through a thin layer near the surface and light rays are bent with a curvature in the opposite direction to the curvature of the earth.

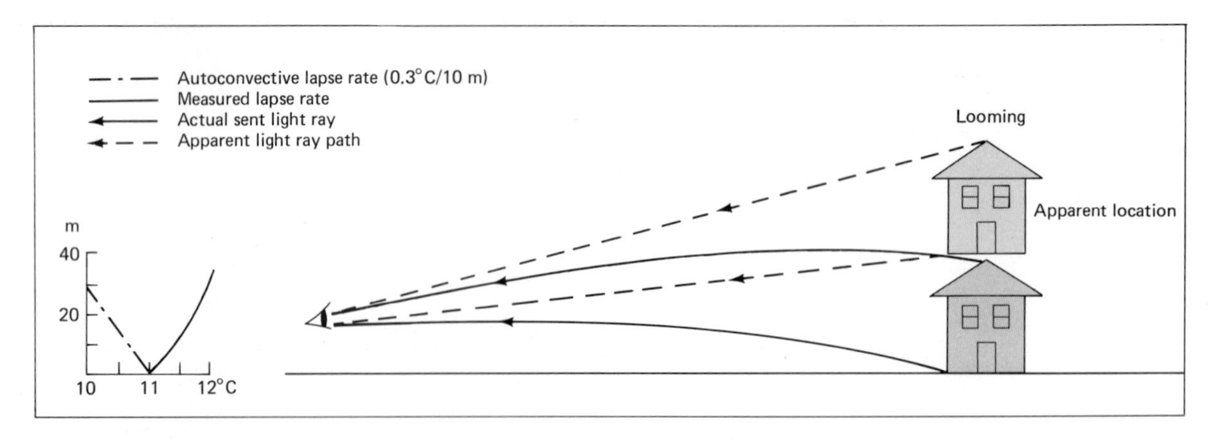

FIGURE 13-4 Looming occurs when the lapse rate is less than the autoconvective lapse rate, such as when an inversion exists. Bending of light rays with a curvature greater than the earth's curvature causes objects to appear at a higher elevation.

smaller than their actual size. This is called towering and foreshortening. Towering objects appear to be taller while foreshortened objects appear to be smaller than they really are.

Looming Position changes occur in the atmosphere when the atmosphere contains either strong inversions or large lapse rates. It is common for the lapse rate to change between day and night, with inversions frequently occurring near the surface during the nighttime and large lapse rates usually occurring during the daytime. These are related to the radiation balance, with loss of surface heat at night and gain in energy at the surface during the daylight hours producing very large hourly changes in the lapse rate. Other atmospheric conditions, such as the advection (horizontal transfer) of warm air over a colder surface, may also generate strong inversions. Inversions are common during the daytime over water surfaces for this reason.

Inversions and all other lapse rates that are less than the autoconvective lapse rate cause light rays to bend with a curvature in the same direction as that of the earth, causing displacement of objects (Fig. 13-4). Objects appear to be higher than they really are—that is, a superior image is produced. A person looking at an object some distance away has no way of discerning whether light rays are bent. This optical effect, called **looming,** is most pronounced with strong surface inversions.

Sinking The appearance of objects at a lower elevation, **sinking,** occurs when the atmospheric lapse rate becomes greater than the autoconvective lapse rate. Such lapse rates develop from surface heating in dry hot regions under sunny skies. Large lapse rates cause bending of the light rays in a direction opposite the earth's curvature. Objects appear to be at a lower elevation—that is, an inferior (lower) image is produced due to this bending of the light rays (Fig. 13-5). Topographic features of the earth's surface are also perceived to be lower than they actually are with this type of lapse rate, thus changing the whole view.

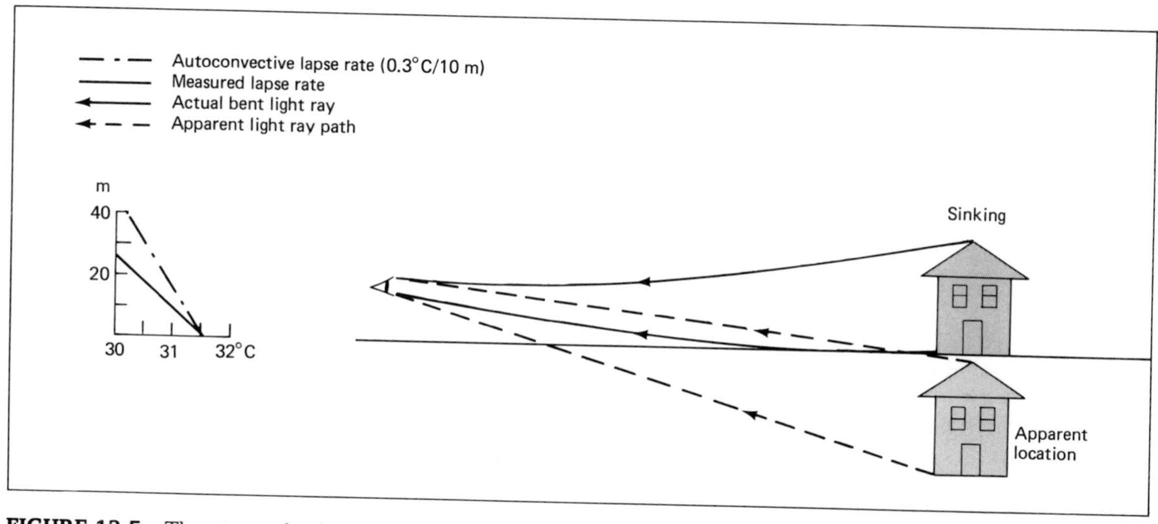

FIGURE 13-5 The atmospheric optical effect of sinking occurs when the measured atmospheric lapse rate is greater than the autoconvective lapse rate. This causes bending of light rays in a direction opposite the curvature of the earth, causing objects to appear at a lower elevation.

Towering Other optical effects include changes in the size of objects. This occurs in general where changes in atmospheric density occur in a short distance between the thin surface atmospheric layer below eye level and the atmosphere above. A decrease of lapse rate with height through both layers is required. Objects may appear to be much larger—that is, a **towering** optical effect may be produced. Towering may occur as surface heating warms a shallow layer close to the earth, producing a contrast with the atmospheric layer above that contains a much smaller temperature decrease with height. This temperature profile (Fig. 13-6) causes the light rays to bend with a curvature opposite that of the earth's in the shallow surface layer, with opposite bending of the light rays in the atmospheric layer above. This results in an apparent stretching of objects and produces the towering optical effect. Such optical illusions can be spectacular, as small objects become castles rising in the air and gentle slopes become cliffs rising dramatically along shorelines.

Foreshortening Objects may also appear to be shorter, producing the optical effect called **foreshortening.** The temperature profile required for this effect consists of a shallow surface inversion with a more normal temperature decrease with height above eye level (Fig. 13-7). The density of the air in the shallow inversion layer decreases very rapidly up to eye level. The layer above contains air that decreases less rapidly in density with height. Light rays both above and below eye level are bent in the same direction as the earth's curvature, but in the shallow surface layer they are bent much more than they are above eye level. Objects appear to be squeezed together and smaller in size. In general, an increase of lapse rate with height is required for foreshortening.

These various atmospheric optical effects can be observed by using a stationary time lapse movie camera to obtain pictures of the same object throughout the day as the lapse rate between the object and camera changes. Objects appear to go up and down because of atmospheric changes.

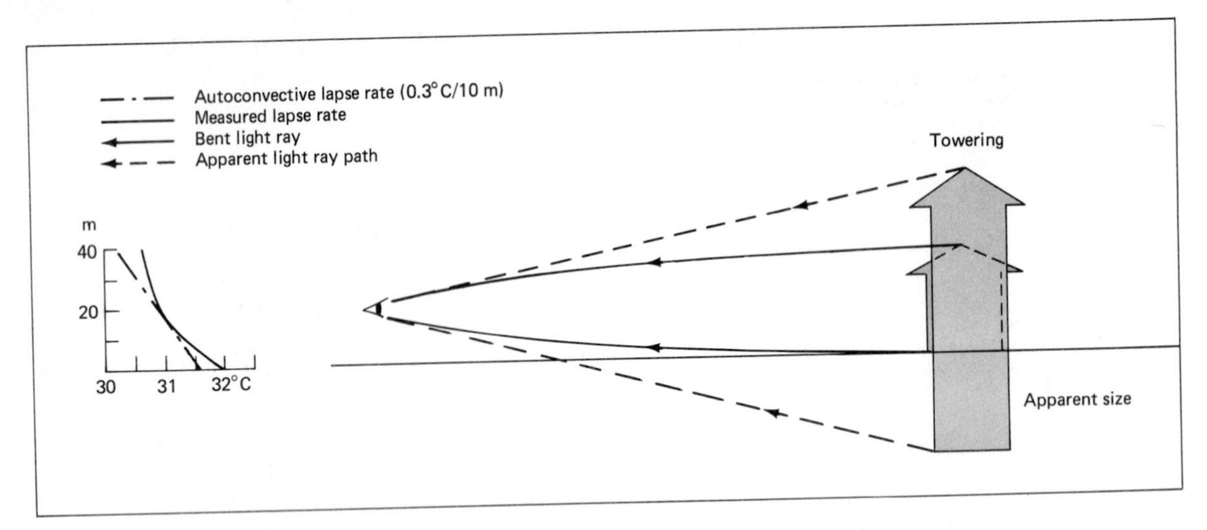

FIGURE 13-6 The atmospheric optical effect of towering occurs with atmospheric lapse rates greater than the autoconvective lapse rate through a thin layer near the earth's surface, with lesser lapse rates above eye level. This may cause extreme vertical stretching of objects and make them appear much larger than they are.

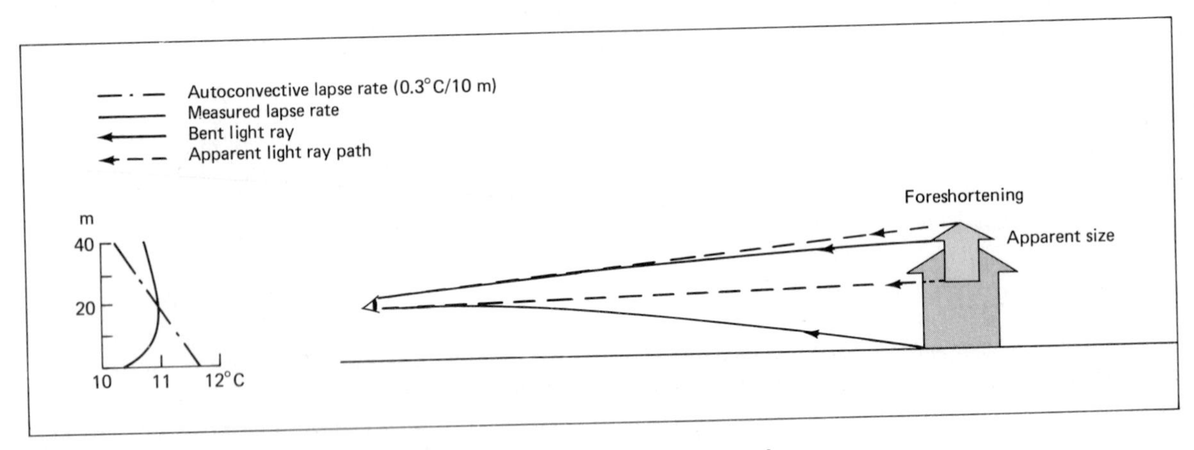

FIGURE 13-7 The atmospheric optical effect of foreshortening occurs when a shallow surface inversion exists, with a normal decrease in temperature with height above eye level. This causes objects to appear much smaller than they really are.

MIRAGES

The previously discussed optical effects may all be considered forms of mirages. However, several optical effects are more commonly known as mirages. The **desert mirage** commonly occurs as a result of very strong bending of light rays in a direction opposite to the earth's curvature. These atmospheric conditions are similar to those required for the towering and sinking optical effects, but lapse rates for desert mirages are greater. The desert mirage causes the blue sky to appear to be on the desert

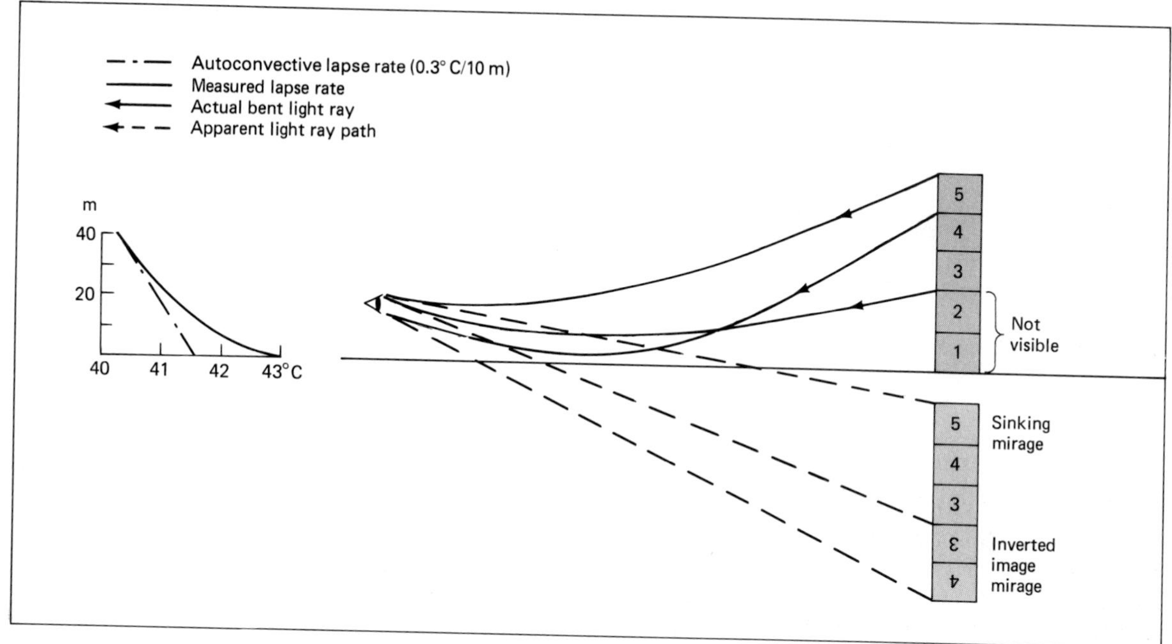

FIGURE 13-8 Mirages and inverted images occur as the lapse rate near the earth's surface becomes much greater than the autoconvective lapse rate, with changes to a smaller lapse rate with height above the surface. This situation causes extreme bending of light rays, with a curvature opposite that of the earth's, through the very thin surface layer. Each higher atmospheric layer has a slightly different effect on the light rays. This may make it impossible for light from parts of an object, such as the blocks labeled 1 and 2, to reach an observer's eye. The variation in bending of the light rays may also cause objects to be inverted, which occurs over water or on hot days in a desert, as shown by blocks 3 and 4.

floor. The image of the sky shimmers because of heat waves and gives the appearance of a water surface.

The intense heating in deserts may cause inverted images of other objects besides the sky, due to very large lapse rates near the ground, as in Figure 13-8. The larger lapse rates cause greater bending of the light rays and result in several possible effects such as inverted images and disappearing portions of an object. In fact, a person walking into a hot desert frequently appears to be walking into a pool of water, with his legs disappearing first, followed by the rest of his body (Fig. 13-9). This effect, as well as inverted images over water, are possible because light may have more than one possible path from an object to the viewers' eyes.

A similar **highway mirage** commonly occurs on sunny days. The heated surface of the highway produces a very thin atmospheric layer with a lapse rate greater than the autoconvective lapse rate and a mirage is generated on the highway as light rays are bent. The highway appears to be covered with water a short distance ahead. Above and within the water mirage an inverted image of the highway also commonly occurs (Fig. 13-10). If the highway is straight beyond the mirage, the driver is less affected than when a curve in the highway occurs just above the mirage. In this latter case the inverted image of the curve gives the appearance

FIGURE 13-9 Mirages can make travel through deserts treacherous. This mirage is apropriately named Silver Dry Lake and is in the Mojave Desert in California. (Richard W. Brooks/ Photo Researchers.)

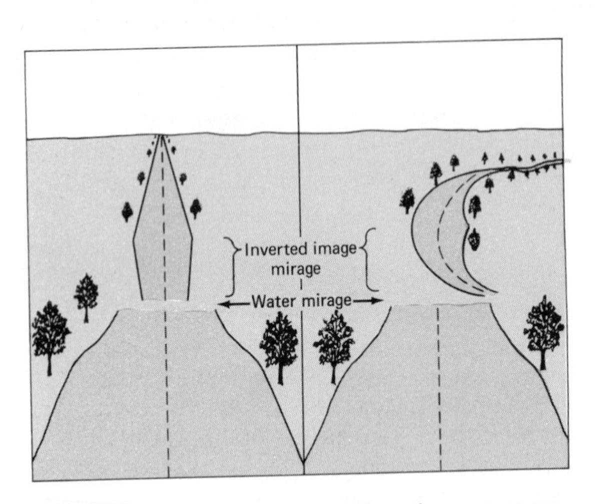

FIGURE 13-10 The highway mirage is common on hot summer afternoons. The highway ahead may contain a water mirage as light is bent from the blue sky overhead and may also have an inverted image mirage because of the large lapse rate near the surface of the highway. If the highway ahead is curved, the mirage may give the image of a side road entering the highway. Distortions of oncoming cars are a greater hazard than is commonly recognized.

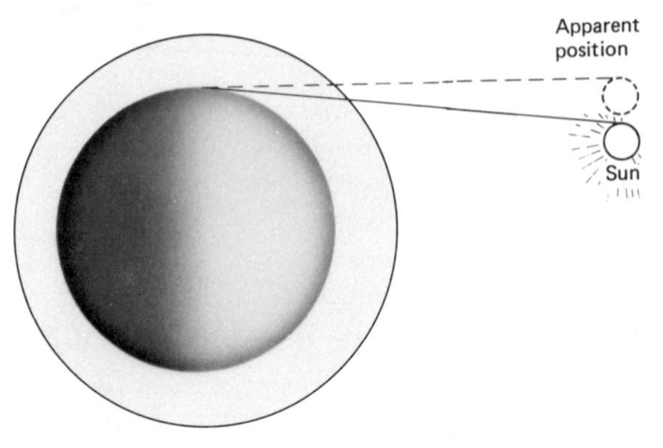

FIGURE 13-11 The position of the sun or moon appears to be slightly higher above the horizon than it actually is because of the bending of light rays through the earth's atmosphere. The average atmospheric lapse rate causes bending of the light rays with a curvature in the same direction as the earth's curvature. This may cause the sun to set a week or more later than anticipated near the Arctic Circle.

of an adjoining side road. Such highway mirages are a greater hazard in automobile driving than is commonly recognized as they make details of the highway and traffic ahead more uncertain.

DISPLACEMENT OF THE SUN, MOON, AND STARS

Atmospheric bending of light causes the stars, sun, and moon to be displaced. The whole atmospheric layer bends light rays slightly, with a curvature in the same direction as the earth's curvature. This causes the sun to appear to be higher than it really is at sunrise and sunset (Fig. 13-11). The sun ordinarily sets two minutes later and rises two minutes earlier than it would without bending of the sunlight. The specific amount of bending varies with the actual atmospheric lapse rate. Near the Arctic Circle, where the sun stays near the horizon for long periods of time, it shines for as long as two weeks more than would be possible without atmospheric bending of the light rays. Strong surface inversions, by causing much greater than normal bending of light in the same direction as the earth's curvature, are most influential in producing this optical effect.

GREEN FLASH

Sunlight is composed of the various visible colors, each of which has a different wavelength that is bent a different amount from the others. Shorter wavelengths are bent and scattered much more than the longer wavelengths in sunlight. The sky is blue because the shorter wavelengths of blue light are scattered over the sky, and the sun is red at sunset because the longer wavelengths of red light are not bent or scattered as much by the same atmospheric conditions. Selective bending of the different colors also causes other optical effects. As the sun sets or rises, the shorter wavelengths of blue and green light may be visible longer since their bending by the atmosphere is greater.

Occasionally a distinct green color is visible at the top edge of the sun just as the sun sets or rises. This is the optical effect called the **green flash.** It was once thought that this was an optical illusion created by some sort of false color, but it has been shown to be a real atmospheric effect since it has been photographed and can be explained by selective bending of shorter wavelengths. The green flash lasts for less than a second and is usually observed over oceans or other flat surfaces.

RAINBOWS

Rainbows are caused by light rays bending in a different way. Whereas the previously described effects are caused by atmospheric bending of light, rainbows are the result of the refraction and reflection of light rays by liquid water droplets. Light must enter a liquid water droplet from the air to produce a rainbow. The entering light is sharply bent since the density of air and water are quite different. The nature of the refraction of light in water is such that only the sunlight that enters a raindrop at certain specific angles reaches the observer's eye in bright bands of light separated into colors.

One internal reflection occurs inside a spherical raindrop to produce the primary rainbow, while refraction bends the light as it enters and as it leaves if the proper angle exists between the sunlight reaching the raindrop and the observer (Fig. 13-12). The resulting angle after the two refractions and one reflection is only slightly different for the various visible colors. The longest wavelengths (red light) are refracted with the largest angle, 42° 18′, to produce the primary rainbow while the shortest wavelengths (violet) are refracted at 40° 36′. The primary rainbow is separated into the various colors between these two angles, with each color forming a complete arc at a constant angle from the observer.

The colors in a rainbow are arranged in order according to their wavelengths (Fig. 3-3). The violet light in a rainbow, for example, is visible at an exact angle of 40° 36′ in all directions from the eyes of the observer if the cloud is large enough to provide the water droplets to refract the sunlight back to the observer. Portions of the primary rainbow less than the whole arc are visible if liquid raindrops cover only part of the arc.

Two internal reflections of light instead of one may occur in spherical drops. This produces a **secondary rainbow** at an angle of about 50° as compared to 42° for the primary rainbow. The path of the sunlight must cause it to strike the raindrops from a specific angle (Fig. 13-13) that causes two internal reflections before it emerges to reach the observer. This situation causes the secondary rainbow to have a larger radius of curvature than the primary rainbow. The different colors are refracted at different angles, as in the primary rainbow. The secondary rainbow has fainter colors because some of the light is lost with each internal reflection. The secondary rainbow is not always visible when a primary rainbow occurs.

The primary and secondary rainbows occur as a cloud population of spherical drops concentrate the light in bands at 42° and 50° from the observer. The geometry of the drops and the refraction angles are such that if all the incident angles are considered,

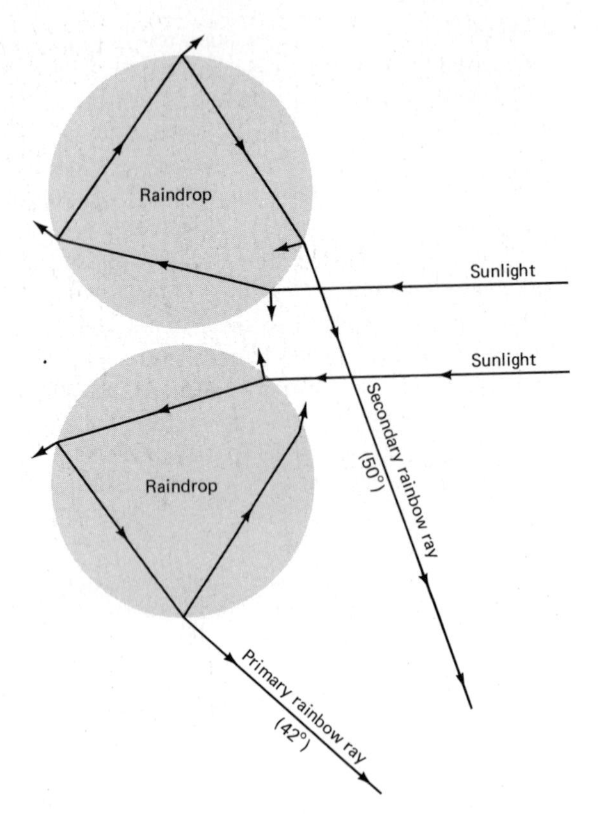

FIGURE 13-12 Primary and secondary rainbows are produced as sunlight enters raindrops and is reflected either one or two times inside the liquid water droplet. Each color of the primary and secondary rainbows is refracted and reflected at a slightly different angle, causing the colors to be separated and making them visible. The primary rainbow ray is created as sunlight enters a raindrop, is refracted on entering, is reflected one time from the back of the raindrop, and is refracted again upon leaving the raindrop at an angle of about 42°. Thus the sun has to be in back of the observer and a rain cloud must be so located as to give a 42° angle for the primary rainbow and a 50° angle for the secondary rainbow. A small amount of the sunlight is reflected each time the light rays are bent. This loss of light due to the additional internal reflection of the light causes the secondary rainbow to be fainter than the primary rainbow.

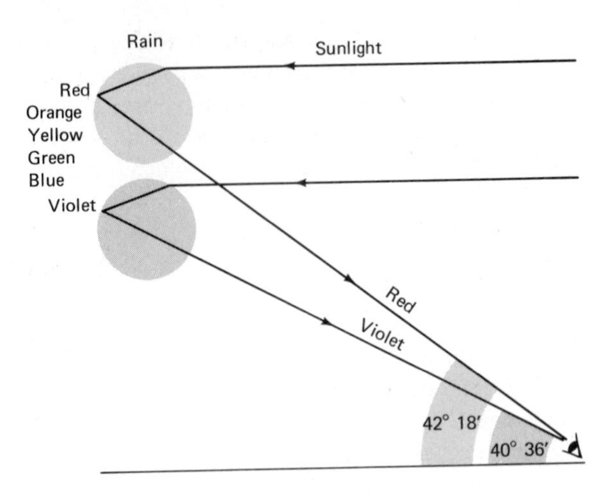

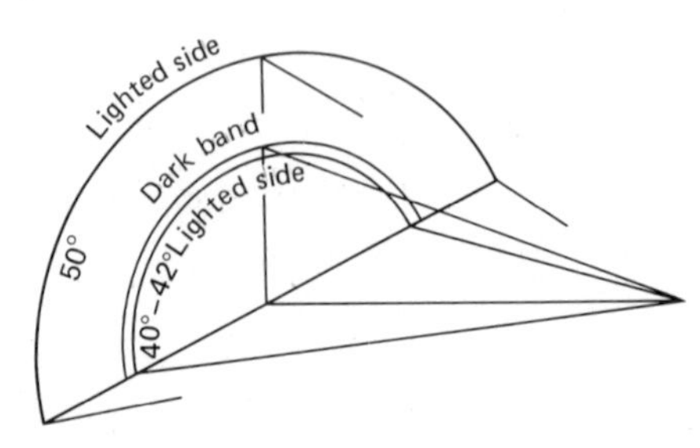

FIGURE 13-13 The primary rainbow is formed as red light is returned to the observer at an angle of 42° 18′ while violet light is returned at an angle of 40° 36′. The other colors are arranged in order from the longest wavelengths to the shortest so that red is followed by orange, yellow, green, blue, and violet. The areas inside the primary rainbow and outside the secondary rainbow are lighter than the area between the two since some of the light rays are directed there.

no light is refracted between 42° and 50° and thus the dark band is created between the primary and secondary rainbows. It is possible for light to strike raindrops at an angle that scatters at least a small amount of light at all angles except from 42° to 50°. This explains the lighter sky below the primary rainbow and above the secondary rainbow.

The supernumerary arcs on the lighted side of the primary rainbow are the result of diffraction. Two light rays refracted by a raindrop may have paths that differ by a very small amount. Constructive and destructive interference produces a series of alternating bright and dark bands near the primary rainbow on the lighted side. The arcs form a series of bands that may alternate between pink and green. On rare occasions supernumerary arcs can also be observed on the lighted side of the secondary rainbow; these are formed in the same way: from light that was reflected twice within raindrops.

The diffraction of light to produce supernumerary arcs is more likely to occur from small droplets. Therefore such arcs commonly appear near the top of the rainbow arc as diffraction occurs from small droplets higher in the cloud. Larger raindrops near the ground produce a better separation of colors, making the rainbow most vivid as it reaches the ground.

FIGURE 13-14 Coronas occur as light is diffracted and dispersed by small liquid water droplets. The result is rings of varying size around the moon or sun when these are covered by a thin layer of clouds.

CORONAS AND GLORIES

Coronas are atmospheric optical effects produced as light passes between and around small liquid water droplets. As light shines between liquid water droplets, diffraction of the sunlight or moonlight separates the colors into alternating circles of bluish and light reddish colored bands. This separation of light into colored circles around the sun or moon by diffraction is called a **corona** (Fig. 13-14). Coronas have in the past been a source of folklore weather forecasts, with the size of the corona rings being related to the amount of expected precipitation. The size of the droplets producing diffraction determine the diameter of the corona ring, with large drops producing small rings. Coronas can be seen more often around the moon when a thin layer of clouds is present, but they can sometimes be seen around the sun in a reflection from still water or when a thicker cloud blocks the sun from view.

A similar optical effect may be produced by the diffraction and refraction of light called a **glory.** Glories are rings of light that can be seen when light is bent 180° due to refraction and reflection by liquid water drops, and is then diffracted. Thus, glories have similarities to both rainbows and coronas. They are seen when you turn your back to

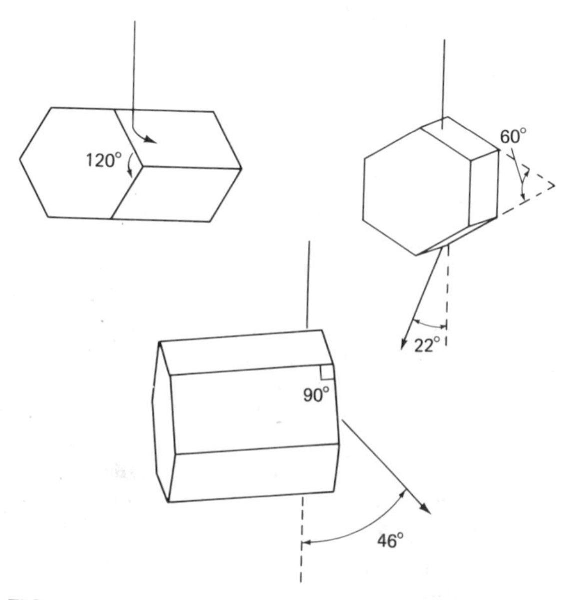

FIGURE 13-15 Light is refracted by ice crystals at an angle of 22° if it passes through 60° prisms; it is refracted at an angle of 46° if it passes through prisms with a 90° angle. Thus the 60° and 90° ice crystal prisms cause various atmospheric optical effects.

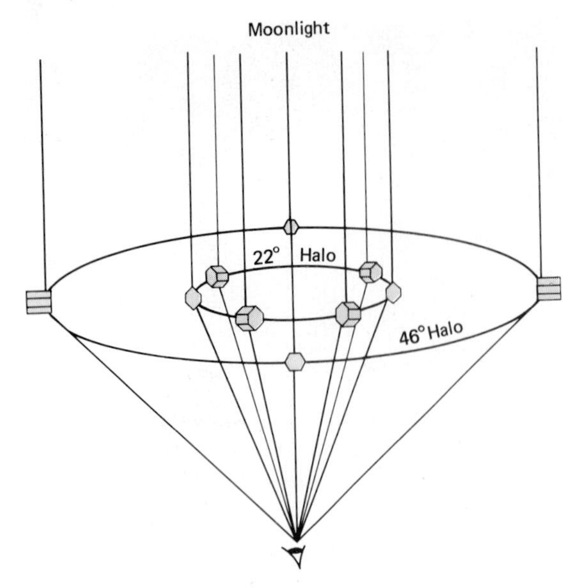

FIGURE 13-16 A circular 22° halo may form around the moon as horizontally oriented ice prisms refract the moonlight as it passes through 60° angles within the prisms. Circular 46° halos may form as the moonlight passes through 90° angles within the ice crystals.

the sun and look at the shadow of your head when heavy dew is on the ground. They are also commonly observed from airplanes around the airplane shadow as it falls on clouds with liquid droplets.

Ordinary refraction of light as it enters a drop, together with one internal reflection, cannot bend light the required 180° to produce a glory. The light entering water droplets is refracted upon entering and is reflected from the back side of the droplet as in the production of a primary rainbow, but instead of a simple refraction as the light leaves the drop, it travels along the surface of the drop for a short distance as a surface wave and is then refracted from the droplet in a path directly back toward the sun. Diffraction of sunlight then produces the rings of light seen as the glory.

HALOES

High cirrus clouds are composed of ice crystals that may act as prisms for light. Ice crystals are six-sided and thus similar to ordinary prisms used in the laboratory to separate sunlight into various colors. The nature of prisms causes them to refract light at specific angles depending on the angle of the prism. Sunlight may pass through different parts of ice crystals, resulting in two possibilities for refracting light (Fig. 13-15). A third angle of 120° exists at the edges of the six-sided crystal, but the refraction characteristics of sunlight do not allow it to pass through ice prisms with this angle.

One path takes sunlight through the sides of an ice crystal where an angle of 60° is encountered. A 60° prism refracts light at an angle of 22°. Therefore cirrus clouds with their ice crystals may bend light 22° as it passes through 60° angles. Another possible path of light through ice crystals takes it through one end of a prism or plate where a 90° angle is encountered. This results in a bending of the light rays at an angle of 46°.

Thus cirrus clouds above an observer may produce a 22° or 46° **halo** (circle of light) around the sun or moon as the light shines through randomly oriented ice crystals (Fig. 13-16). The 22° halo arises as refraction occurs through the 60° angle formed by the sides of the ice crystal, while the 46° halo is generated by light passing through the 90° angle at the ends of the ice crystals. These latter angles are formed less perfectly and thus the 46° halo occurs less frequently than the 22° halo. The area inside the inner circle of the halo is dark because no light is bent at angles less than 22°; it is lighter outside since bending at greater angles is possible. The longer wavelength red light is bent less, so it forms the inner edge of the halo.

SUNDOGS

Another optical effect called **sundogs** is produced by ice crystals in cirrus clouds particularly near the time of sunset or sunrise. As the sunlight comes through horizontally oriented hexagonal plates, and through the 60° angle of the ice crystals, bending occurs at 22° angles (Fig. 13-17). This creates an image of the sun 22° to the left and 22° to the right of the sun known as a sundog or mock sun. Sometimes sundogs are simply bright spots in thin cirrus clouds although separation of colors does occur, showing that sundogs are really intense portions of haloes produced by horizontal ice crystals.

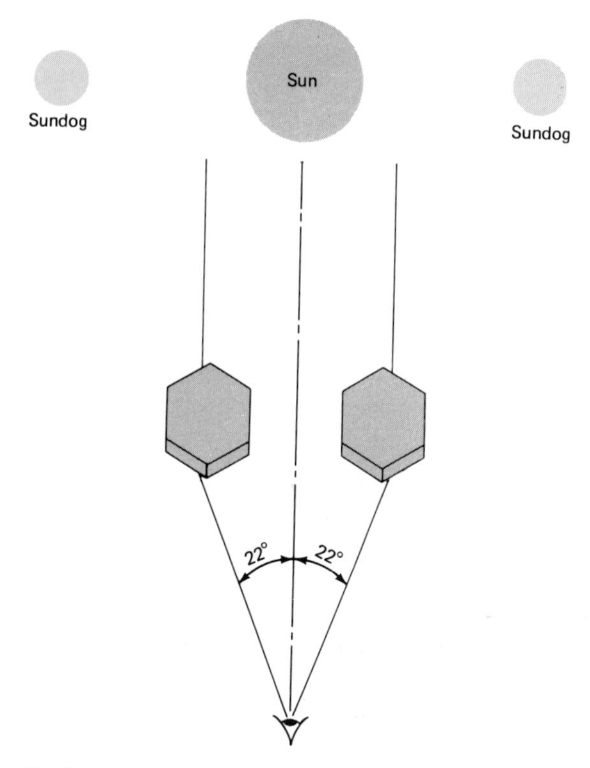

FIGURE 13-17 Sundogs form as sunlight passes through cirrus clouds containing ice crystal plates. The horizontally oriented plates cause the sunlight to be refracted at 22°, creating images of the sun 22° outward from the sun.

SUN PILLARS

Cirrus clouds also produce an optical effect called a **sun pillar.** This is also caused by hexagonal plates; in this case sunlight reflects off the top or bottom of the plates near the time of sunrise or sunset. A variety of approximately horizontal tilt angles of the hexagonal plates produces a pillar of light above or below the sun, depending on the sun angle and tilt angle of the ice crystals (Fig. 13-18).

ATMOSPHERIC ACOUSTICS

If you have ever been fishing on a large lake at dusk you have probably noticed that you can hear and understand voices more than a kilometer away when the winds are light. Or you may have heard a train whistle at dusk from so far away that it was unexpected. Sound waves are generated as the air expands and contracts around vibrating objects. These waves then spread outward in all directions. The air temperature and winds affect the spread of sound waves. The speed of sound waves in air depends on air temperature. The speed of sound at 0°C is 331 m/sec, but it travels more than 0.5 m/sec faster for each increase in temperature of 1°C.

The increased range of sound waves at dusk is related to the vertical temperature gradient (Fig. 13-19). A surface inversion deflects sound waves downward while a large lapse rate deflects them upward. As an inversion develops at dusk, the sound waves are kept near the earth's surface, thus increasing their range. A smooth water surface does not interfere with the sound waves as much as a rough surface would as they travel over it.

On a windy day sound travels further downwind than upwind, not so much because of the displacement of the sound waves by the wind, but because of the downward deflection of sound waves as they travel downwind (Fig. 13-19). Sound waves traveling upwind are soon deflected upward so much that they cannot be heard near the ground.

Calls for help in a snow storm are not carried very far by the atmosphere—not because of the cold temperature, but because of the sound-absorbing nature of soft snow. You may have noticed the silence after a new snowfall, as the snow blankets the ground much as a builder covers a ceiling with acoustical tile composed of porous material.

You may have wondered why the winds howl as they blow through pine trees, across wires, or over rooftops. Obstructions in the air stream create small vortices as the air passes behind them. These

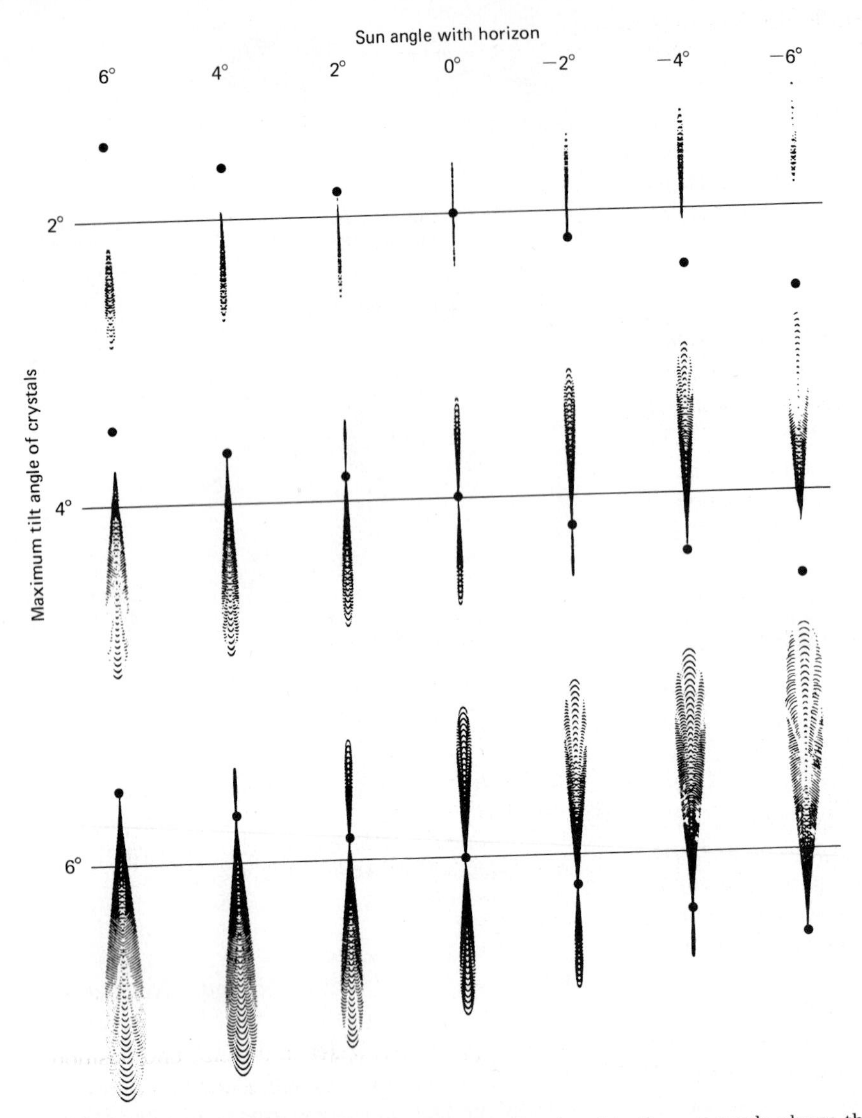

FIGURE 13-18 The appearance of sun pillars is related to the sun angle above the horizon and the tilt angle of the ice crystal. These simulated sun pillars show how they would appear as the sun sets. Each of three sequences is for a different ice crystal orientation from 2° to 6°. (From *Am. Sci.*, vol. 60, No. 3, 1972.)

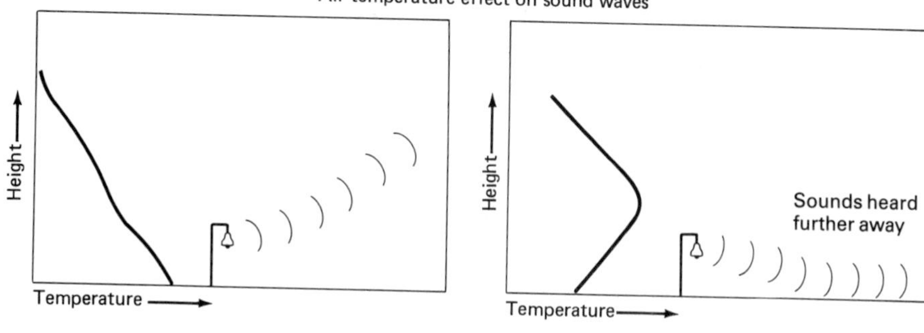

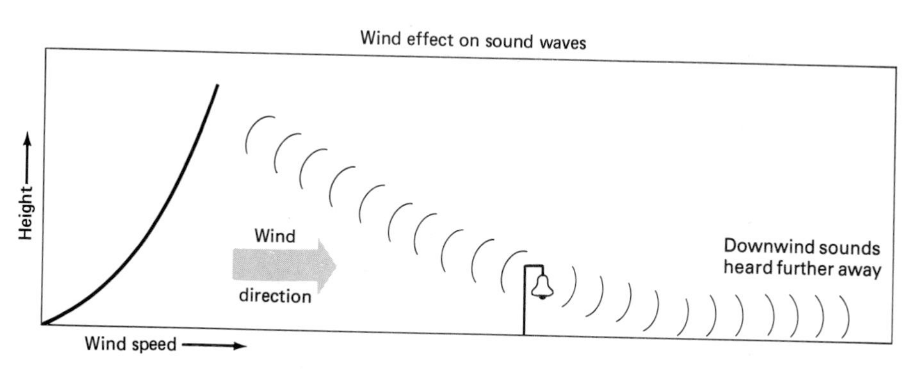

FIGURE 13-19 Decreases in temperature with height cause sound waves to be deflected upward since the speed of the waves is 5 m/sec slower for every 10°C decrease in air temperature. Sounds can be heard further through inversions as the sound waves are deflected downward by the temperature gradient. The typical wind profile with stronger winds at greater altitudes favors the propagation of sound waves along the downwind direction, but deflects sound waves upward, out of hearing range, in an upwind direction.

cause pressure differences and waves that reach our ears as sound. Such vortices and pressure differences created by the winds blowing past a telephone wire may cause it in turn to vibrate enough to produce sound waves in the same manner as a stringed musical instrument.

SUMMARY

A large number of interesting atmospheric displays and illusions occur due to the interaction of sunlight with the atmosphere as well as with ice crystals and raindrops. Specific atmospheric lapse rates produce apparent changes in the size and position of objects. Atmospheric optical effects produced as sunlight passes through ice crystals include haloes, sundogs, and sun pillars. Rainbows and coronas are produced as sunlight is affected by spherical raindrops. The reflection, refraction, and diffraction of light interacting with the atmosphere, liquid raindrops, or ice crystals explains these various atmospheric optical effects.

Light rays are bent or refracted as they pass from a substance of one density to another of differing density. Therefore atmospheric layers that are different in density affect the bending of light rays. Atmospheric density decreases with height produce bending light rays with a curvature in the same direction as the earth's curvature, while atmospheric density increases with height produce bending of light rays in the opposite direction. Such bending of light can cause objects to appear to be larger, smaller, or in a different position from their actual location. The atmospheric optical effects of looming and sinking occur as objects appear to be higher or lower than their actual position. Towering and foreshortening are optical effects produced as objects appear to be taller or shorter than they really are. Specific atmospheric lapse rates are required to produce each of these optical effects.

Desert mirages and highway mirages are common as the sun warms a thin atmospheric layer, producing a very large lapse rate near the earth's surface. This causes strong bending of light rays in a direction opposite to the earth's curvature and gives an image of water on a desert floor or highway from the sky above, with heat waves producing a shimmering effect.

Atmospheric bending of light rays also causes the stars, sun, and moon to be displaced from their actual position. The sun rises earlier and sets later than would occur without bending of the sunlight. The green flash is produced as various colors within sunlight are bent by slightly different amounts. The shortest wavelengths are bent most, with the blue light being scattered more readily; this makes green the last color to be seen at sunset, and it produces a rare green light called the green flash.

The formation of rainbows occurs as light is returned to the observer from liquid cloud droplets at an angle of about 42° for the primary rainbow. Separation of colors occurs as sunlight is refracted with one internal reflection inside raindrops to produce an angle of 42° 18′ for red light and 40° 36′ for violet light, with the other colors of the rainbow in between. The secondary rainbow is produced in a similar manner except that two internal reflections occur inside the raindrop, producing an angle of about 50° with the observer. Supernumerary arcs occur on the lighter, inside area of the primary rainbow because of diffraction of light.

Coronas are frequently observed around the moon and occasionally around the sun as liquid water droplets diffract sunlight into alternating circles of blue- and red-colored light.

Haloes occur from ice crystals within cirrus clouds, as light is refracted by 60° prisms at an angle of 22°. A 46° halo is also possible because of refracted light through 90° prisms. The longer wavelength red light is bent less to form the inner edge of haloes. Sundogs are also the result of refraction of light through 60° ice prisms; these are small images of the sun at 22° to the right or left of the sun. Sun pillars occur as the sunlight bounces off the top or bottom of the hexagonal plates, producing a pillar of light above or below the sun.

The propagation of sound waves is considerably faster through warmer air. This causes sounds to be held close to the ground as an inversion develops at night, which extends the hearing distance considerably. Sound waves traveling downwind under very windy conditions are affected in the same way, as the sound is carried faster by the stronger winds above the surface, thus deflecting them downward. Snow covers the earth with an acoustical layer that dampens sounds, while a smooth water surface allows sound waves to travel much further than normal.

STUDY AIDS

1. List various atmospheric optical effects that are produced by ice crystals, and make a separate list of those produced by raindrops.

2. Explain the difference among reflection, refraction, and diffraction of light.

3. Why are light rays bent with a different curvature if the lapse rate is greater than 0.3°C/10m as opposed to less than 0.3°C/10m?

4. Compare looming and sinking and the lapse rate that produces these atmospheric phenomena.

5. Explain towering and foreshortening and the atmospheric conditions that give rise to these optical effects.

6. Explain the desert and highway mirages.

7. Explain the occurrence of the green flash.

8. Describe the various features of the primary and secondary rainbows together with the formation mechanism.

9. How are haloes and sundogs formed?

10. Explain some atmospheric effects on sound.

TERMINOLOGY EXERCISE

Check the glossary if you are unsure of the meaning of any of the following terms used in Chapter 13.

Supernumerary arcs
Reflection
Refraction
Diffraction
Refractive index
Autoconvective lapse rate
Looming
Sinking
Towering
Foreshortening
Desert mirage
Highway mirage
Green flash
Rainbow
Secondary rainbow
Corona
Glory
Halo
Sundog
Sun pillar

THOUGHT QUESTIONS

1. Assume that an object is lying on the highway a great distance ahead of you on a warm sunny day. Describe the many possible changes in the appearance of this object as you approach it.

2. Explain the circumstances under which it is possible to see a rainbow in the east on one day and a rainbow in the west on another day.

3. List and explain the time of day and seasons of the year when you would be most likely to observe such atmospheric phenomena as sundogs and sun pillars.

4. The green flash is more commonly observed over the ocean than over land. Can you explain why this is so?

5. Explain what you would see as you watched a person walking into a desert mirage.

SPECIAL TOPIC 16

BLUE SKY, RED SKY, AND BLUE MOONS

The composition of the atmosphere changes because of the amount of dust and water vapor present. Sunlight and moonlight traveling through the atmosphere are affected by these. You may have noticed that the sky is very blue just after the passage of a cold front. As the air mass changes from mT to cP, a change occurs in the blue of the sky because the water vapor present in mT air masses causes a hazy appearance even when no clouds are present. Within a cold air mass the sky is very blue due to increased visibility through the dry atmosphere. The blue color exists because the shorter blue wavelengths in sunlight are scattered more by the atmosphere than the longer orange and red wavelengths. It is mainly the longer wavelengths that are able to penetrate the greater atmospheric thickness that sunlight must penetrate near the times of sunrise and sunset. Thus the sun appears red at these times.

Volcanic ash and smoke particles can cause some unusual atmospheric effects. Red sunsets were common for at least two years after the volcano Krakatoa in West Indonesia erupted in 1883. Unusually red sunsets have also been observed in the 1980s because of volcanic dust added by the eruption of Mt. St. Helens and other volcanoes. Because of the additional dust particles after volcano eruptions, the atmosphere itself takes on a red color in the absence of cirrus clouds, which normally create beautiful sunsets and sunrises.

Dust particles that are all about the same size may affect a particular wavelength of sunlight or moonlight. The familiar expression "Once in a blue moon" comes from the rare observation of a blue-colored moon caused by particles in the atmosphere selectively scattering the moonlight.

CHAPTER FOURTEEN
AIR POLLUTION METEOROLOGY

CONCERN ABOUT AIR POLLUTION

The problems associated with air pollution have become a major concern to inhabitants of many large cities in the United States. Los Angeles, California, for example, frequently has a pollution problem as sunlight causes chemical reactions between nitrogen dioxide and hydrocarbons, resulting in the formation of ozone and other toxic gases. This is called photochemical smog. In other cities the combustion of fossil fuels releases significant quantities of sulfur dioxide and particulates into the atmosphere. The automobile is the source of almost half the pollutants in the United States, contributing carbon monoxide, nitrogen dioxide, and hydrocarbons to the atmosphere.

The Clean Air Act was passed by Congress and signed into law in 1970. This act established Primary and Secondary Standards (Table 14-1) for air pollution concentrations based on available information concerning concentrations of pollutants that become dangerous to human health and vegetation. Since 1970 automobile manufacturers, electric power companies, and industry have been spending a significant amount of their budget installing and operating pollution control equipment to meet air pollution standards supervised by the Environmental Protection Agency.

The air pollution meteorology problem is a two-way street. The presence of pollution in the atmosphere may affect the weather and climate. At the same time the meteorological conditions greatly affect the concentration of pollutants at a particular location, as well as the rate of **dispersion of pollutants,** as we shall see in this chapter.

TABLE 14-1

National ambient air quality standards. Analytical determinations as documented in appendices of Federal Register. Secondary standard for sulfur oxides revised in Federal Register on September 13, 1973.

Contaminant	Primary	Secondary
Sulfur oxides	80 μg/m³ (0.03 ppm) annual arithmetic mean	Same
	365 μg/m³ (0.14 ppm) 24-hour concentration[a]	Same
		1300 μg/m³ (0.5 ppm) 3-hour concentration[a]
Particulate matter	75 μg/m³ annual geometric mean	Same
	260 μg/m³ 24-hour concentraton[a]	150 μg/m³ 24-hour concentration[a]
Carbon monoxide	10 mg/m³ (9 ppm) max. 8-hour concentration[a]	Same
	40 mg/m³ (35 ppm) max. 1-hour concentration[a]	Same
Oxidants	240 μg/m³ (0.12 ppm) max. 1-hour concentration[a]	Same
Nitrogen dioxide	100 μg/m³ (0.05 ppm) annual arithmetic mean	Same
Lead	1.5 μg/m³ quarterly arithmetic mean	Same

Source: Published in Federal Register, vol. 36, No. 84, April 30, 1971.

[a]Maximum value not to be exceeded more than once per year.

Pollution Sources Contaminants get into the atmosphere from a variety of sources. These can be categorized as **fixed sources** and **mobile sources.** Examples of mobile pollution sources are automobiles and aircraft. Pollutants are generated by the combustion process within engines. Large quantities of carbon monoxide are produced by gasoline engines, which also produce many other gases such as nitrogen dioxide and hydrocarbons in lesser quantities. Even water vapor produced by jet airplane engines may be considered a pollutant when delivered to the stratosphere where clouds are produced easily in the low temperatures and where the water content is ordinarily small. In recognition of the pollution problem, control devices have been placed on automobiles sold since 1968, and these have decreased the emission of pollutants significantly.

Fixed sources such as factories, homes, and office buildings may also add pollutants to the atmosphere. The burning of fossil fuels is typically accompanied by the release of sulfur dioxide and nitrogen dioxide. The location of industrial centers in cities, along with atmospheric conditions, dictates which geographic locations are exposed to the largest concentrations of pollutants.

In addition to the fixed and mobile sources of atmospheric pollution, another source is chemical reactions within the atmosphere. Sulfur dioxide turns into sulfur trioxide, which combines with water vapor to become sulfuric acid. Nitrogen dioxide combines with hydrocarbons, while sunlight furnishes the energy for further reactions to produce photochemical smog. Such reactions occur most commonly where concentrations of the constituent gases are great.

ATMOSPHERIC CONDITIONS FOR POLLUTION EPISODES

High-Pressure Effects on Pollution Concentration Particular meteorological conditions are conducive to increasing pollution concentrations. First of all, the type of atmospheric pressure system, high or low pressure, is important in determining the mixing of pollutants. The structure of an

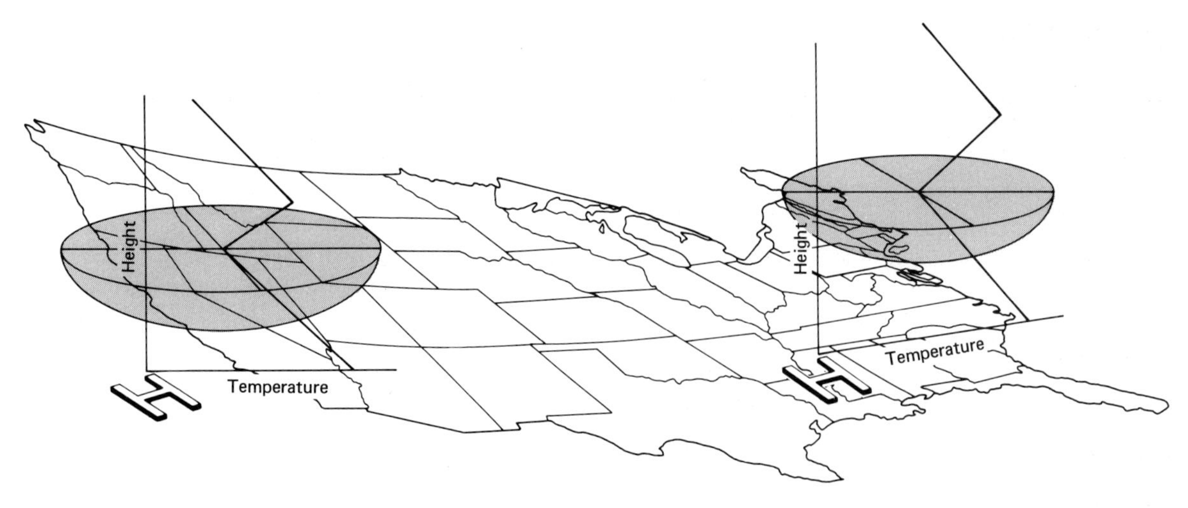

FIGURE 14-1 Large anticyclones or high-pressure systems are usually accompanied by upper air inversions. Such layers are very stratified since the density decreases rapidly with height through an inversion. Thus convection and atmospheric mixing is almost completely absent, causing pollutants from the surface to accumulate in the layer from the surface up to the inversion layer.

anticyclonic high-pressure system is such that descending air exists above the center of the high pressure and this leads to greater stability. Less mixing of air occurs in a region of high pressure, in contrast to the much better mixing that occurs in a low-pressure area from the ascending air and the unstable nature of the air associated with these atmospheric systems. The nature of airflow and the stability associated with high-pressure systems cause greater pollution problems, therefore, in these high-pressure areas.

In addition, the high-pressure system frequently develops an upper air inversion layer that is very stable. As the air subsides above the center of a high-pressure area it heats at the dry adiabatic lapse rate rather than at the measured lapse rate, as previously shown in Figure 5-10. This develops an upper air inversion that is very stable and that frequently traps pollutants below (Fig. 14-1). Therefore, pollution concentrations may continue to increase near the ground from continuous pollution sources since the air is not mixing within and above the inversion layer.

Jet Stream and Upper Atmospheric Effects The jet stream and upper atmospheric airflow patterns are important in determining the mixing of pollutants in the atmosphere. The presence of high wind speeds is an influencing factor on pollution levels. Greater wind speeds are associated with more turbulence and disperse pollutants to a greater extent than light winds. The wind flow characteristics in the upper atmosphere, at the jet stream level, frequently consist of much higher wind speeds around an upper air low-pressure system than around an upper air high. The greater pressure gradient around the low-pressure area produces greater wind speeds and better pollution dispersion. Therefore upper air ridges tend to be areas where less atmospheric mixing occurs because the winds are lighter (Fig. 14-2).

Lighter winds not only carry the pollutants a smaller distance in the same time interval, but they are also associated with much less turbulence, which contributes to less lateral and vertical dispersion of pollutants. Light winds in the upper atmosphere are an important factor in allowing

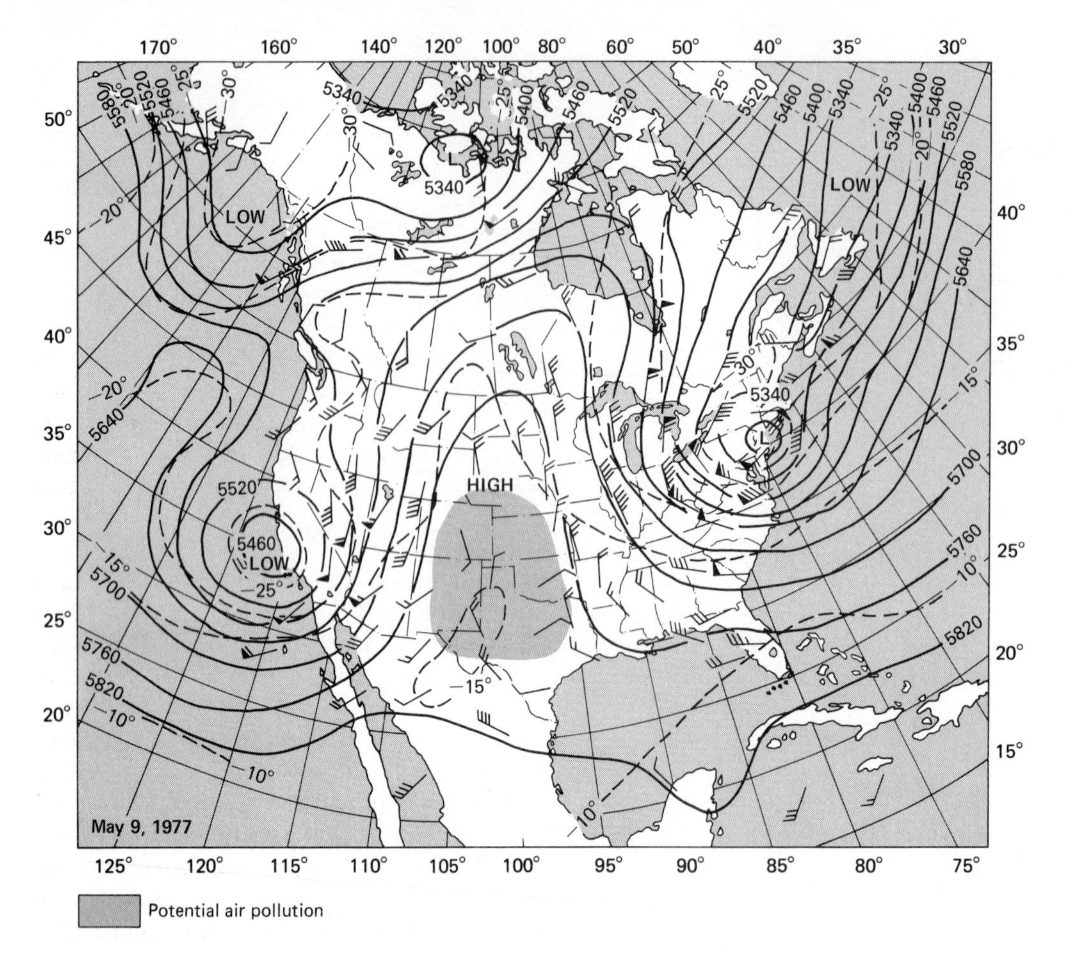

FIGURE 14-2 High pollution episodes are most likely when the upper atmospheric winds are light since strong winds in the upper atmosphere are necessary to initiate turbulence and carry pollutants away. Light winds with very little atmospheric mixing frequently occur over large areas, as in this example. Upper atmospheric ridges are ordinarily accompanied by light winds, while upper atmospheric troughs are accompanied by strong winds. Therefore upper air ridges can be used to identify geographical areas where potential air pollution problems exist.

unusual increases in the concentration of pollutants in the lower atmosphere. Measurements have verified the direct relationship between the concentration of pollutants in the lower atmosphere and wind speeds in the upper atmosphere.

Atmospheric Stability Effects The atmospheric lapse rate is very important in determining air pollution concentrations. The temperature profile governs the stability of the atmosphere, and this is related to vertical currents that cause more rapid mixing of pollutants in the atmosphere. Pollution concentrations are therefore related to lapse rates that may change hourly from a positive lapse rate with decreasing temperatures with height during the daytime to a negative lapse rate or inversion at night. Greater atmospheric stability is associated with less mixing of pollutants in the atmosphere.

Absolute stability of the atmosphere exists when the measured lapse rate is less than both the

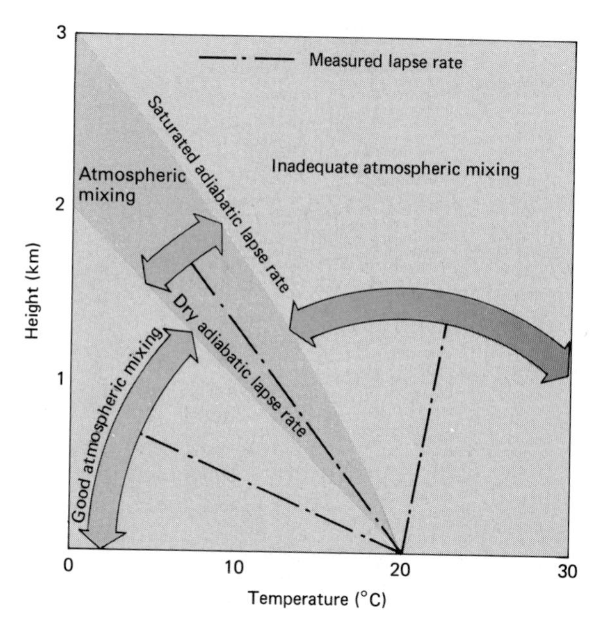

FIGURE 14-3 Turbulence and vertical mixing of the atmosphere are related to the particular existing lapse rate. Inversions and all lapse rates that are less than the saturated adiabatic lapse rate signify inadequate atmospheric mixing and cause an increase in the concentration of pollutants. If the lapse rate is greater than the dry adiabatic lapse rate, very good atmospheric mixing occurs, as convection and vertical motion in the atmosphere disperse pollutants. An atmospheric lapse rate between the saturated and dry adiabatic lapse rates signifies atmospheric mixing.

saturated and dry adiabatic lapse rates, producing a very stable atmosphere with practically no vertical currents and very little mixing in the atmosphere (Fig. 14-3). Conditional stability exists when the measured or environmental lapse rate is between the dry and saturated lapse rates. In this case atmospheric mixing occurs particularly in the afternoon when surface heating may be sufficient to contribute to the mixing of the atmosphere. The most unstable atmospheric condition, with very good mixing, occurs with absolute instability as the measured environmental lapse rate becomes greater than either the dry or the saturated lapse rate. The atmosphere is completely unstable with this lapse rate, allowing pollutants to readily disperse through the atmosphere.

The effects of inversions in the production of greater concentrations of air pollution are well known. The increase in temperature with height in an inversion layer causes a very stable atmospheric condition with very little atmospheric mixing, as previously illustrated in Figure 5-8. Convective activity is dampened with a reduction in the ability of the atmosphere to disperse pollutants.

Surface Inversion Effects Surface temperature inversions are common and frequently extend upward at least 100 m in the atmosphere. They can develop from several different causes. The usual temperature profile through the troposphere shows decreasing temperature with height. However, infrared radiation loss from the ground during the nighttime also cools the lower atmosphere, creating a surface temperature inversion. Such nighttime radiation-induced inversions are more likely with clear skies than with cloudy skies. It is obvious that cloud cover reduces the amount of radiation from the sun during the daytime, but it is much less obvious that the amount of radiation lost from earth at night is also reduced by cloud cover, thereby decreasing the development of inversions.

Cold climates of the world are more likely to have prolonged periods of time with surface inversions because of greater radiation loss than gain near the earth's surface. For this reason inversions are frequent near the North and the South Poles. Fortunately these are not very industrialized regions, so the atmosphere there becomes much less polluted than it might.

Diurnal measurements of the temperature profile in the lower atmosphere typically show an inversion that hampers pollution dispersion during the nighttime and shortly after sunrise (Fig. 14-4). The absorption of solar radiation on clear days causes heating at the surface with atmospheric

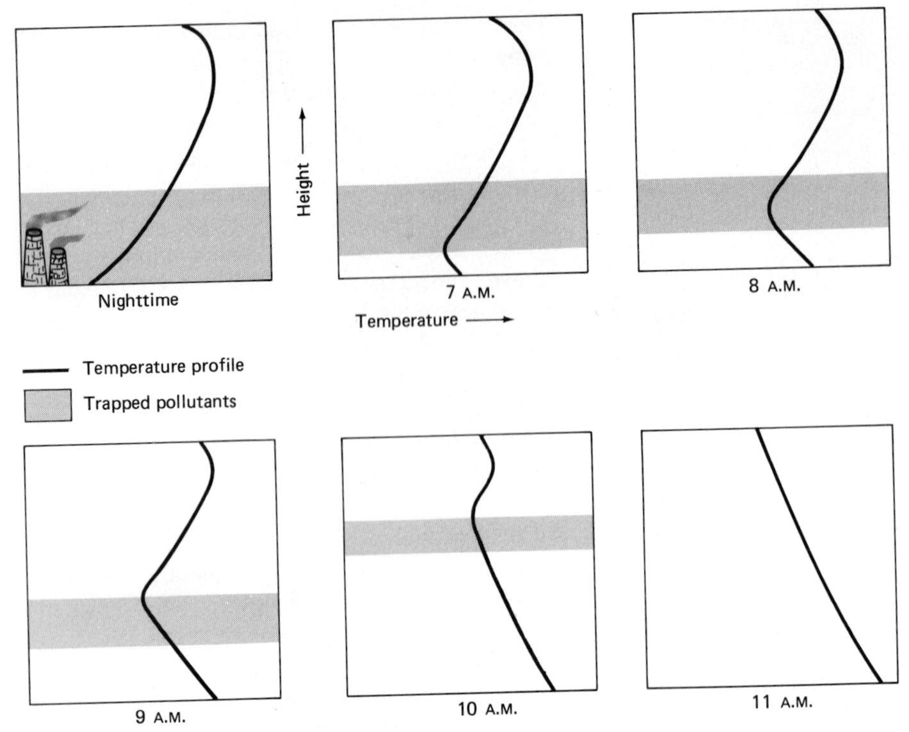

FIGURE 14-4 The common nighttime inversion traps pollutants within the layer if the height of smokestacks supplying the pollutants is lower than the top of the inversion layer. The absorption of sunlight at the earth's surface after sunrise normally causes the inversion to be eliminated before noon. Since pollutants are trapped by the inversion layer, it is frequently possible to observe this process as it occurs in the atmosphere.

mixing, leaving an inversion containing the night's accumulation of pollutants. The inversion layer becomes thinner and more concentrated with pollutants as the morning progresses but is normally eliminated by 10 or 11 A.M. The lapse rate continues to increase until the time of maximum temperature, when the greatest decrease in temperature with height occurs with associated maximum dispersion of atmospheric pollutants. A surface inversion develops again shortly after sunset as the atmosphere starts to cool from radiational losses and the accumulation of pollutants within it begins again.

Some measurements of the frequency of inversions have been compiled because of the great influence of inversions on pollution distributions. The frequency, expressed as the percentage of total hours that has inversions in the winter time, varies with location and is about 40% in the central United States (Fig. 14-5). Such surface inversions occur predominantly at night since they are influenced by radiation. Southern California and much of Arizona have inversions about 55% of the time. Parts of some of the Rocky Mountain states—Wyoming and Utah, for example—also have a high frequency of inversions. This is a major factor in air pollution

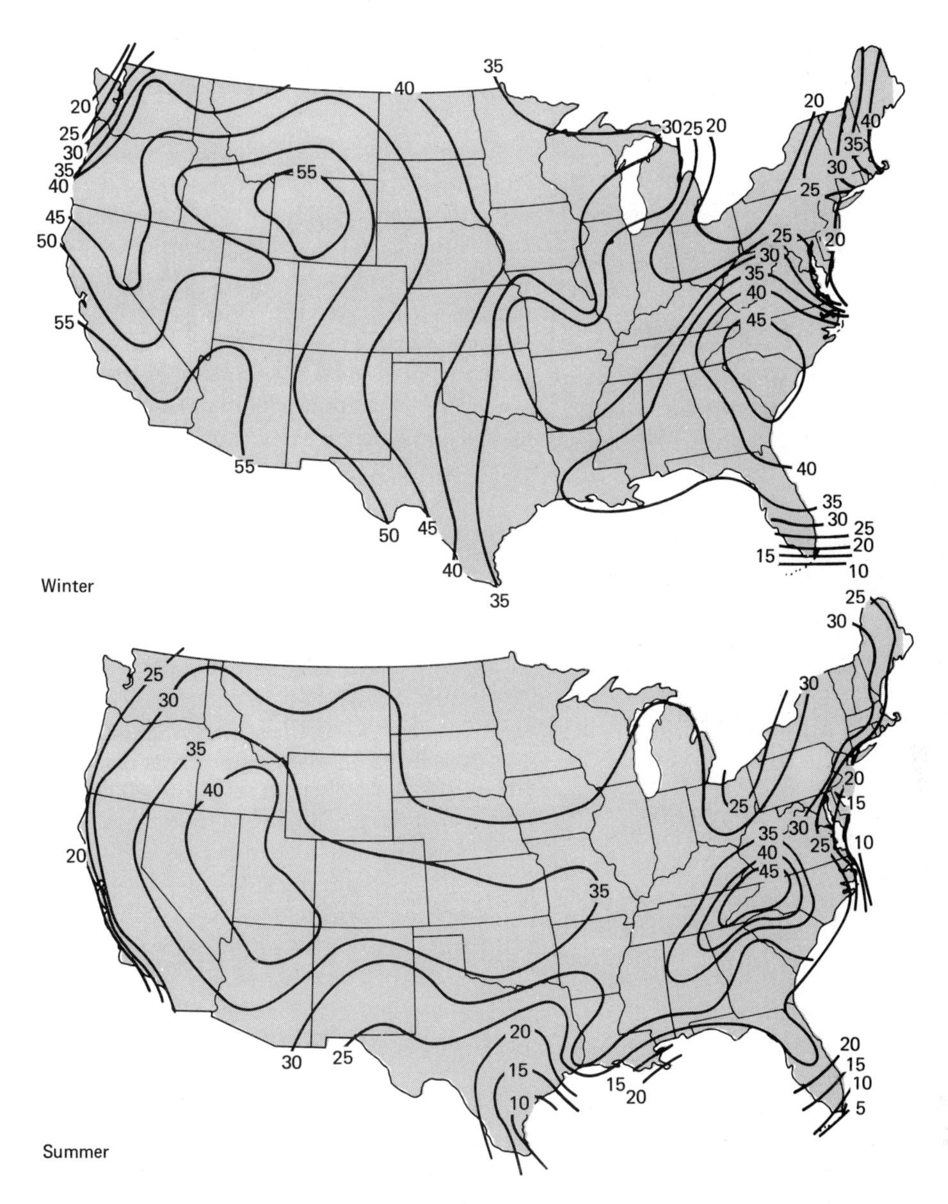

FIGURE 14-5 The occurrence of inversions below 1.5 km during the winter and summer varies with geographical location. The numbers represent the percentage of total hours when such inversions are present. Inversions are present more than 50% of the time during the winter over most of the western United States except for the extreme northwestern coastal areas. Another area of frequent inversions is the southern United States north of Florida. The largest variation with seasons occurs in southern California, where inversions are less common during the summer than during the winter. (From Hosler, 1961, *Monthly Weather Review.*)

FIGURE 14-6 Comparison of the visibility over San Diego, California, on two different days. The mountains are visible in the top photograph but completely disappear from view in the bottom photograph. (Courtesy of Philip R. Pryde.)

problems in these states. Another area with frequent surface inversions is in the southeastern United States, particularly the Carolinas, where surface inversions are present in summer and winter about 45% of the time. Southern Florida has very few surface inversions, with a frequency of less than 20% of the hours.

In general, fewer surface inversions occur in the warm months of the year. The central United States has inversions only about 35% of the time in summer. Fewer inversions also occur along the west coast in the summer. The extreme east coast has inversions only 15% or 20% of the time. This climatology of surface inversions is helpful in locating those areas of the United States where atmospheric mixing is the least and where more problems in dispersing pollutants are likely to occur.

Upper Air Inversion Effects Upper air inversions (1 or 2 km high) can have even greater undesirable effects on pollution concentrations than surface inversions since they are more persistent and last for a longer period of time. Upper air inversions frequently last for days rather than a few hours. Inversions develop in the atmosphere above the surface as a result of several different physical processes.

One important developmental mechanism is associated with the subsiding air above a high-pressure area, as already described. Subsiding air is heated by compression, with the top of atmospheric layers descending further and heating more, giving rise to upper air inversions. Therefore regions of the world that have frequent high-pressure areas also have upper air inversions.

The upper air inversion and the subtropical high-pressure area that is frequently near southern California and that lasts for extended periods of time are two of the reasons why California has extreme air pollution problems. The inversion gives subsiding air and causes the development of an upper air inversion that is very stable. The inversion layer is slightly lower in elevation than the mountain ranges to the east. Thus the high pressure places a lid on convective activity and the topography provides the walls of the container, preventing atmospheric pollutants from being carried away by the surface winds from over the ocean. This effectively traps pollutants in the lower atmosphere where the continuous supply from automobiles and industry causes persistently high concentrations (Fig. 14-6).

Upper air inversions also develop from air mass changes. Sources of maritime tropical air exist in the Atlantic and Pacific near 30° latitude. The air there is very warm and humid and frequently invades the United States, causing a change in the air mass type. Continental polar and continental arctic air masses located to the north of the United States frequently move southward to engulf the country in cold air. As these two major air masses meet, the leading edge of the cold air mass moves under the warm air mass, producing a cold front and precipitation as the warmer air is lifted. This process also produces upper air inversions (Fig. 14-7).

As the cold air moves under the warmer air, the resulting temperature profile includes an upper air inversion at the boundary between the two air masses; this is another mechanism for developing an inversion. These generally do not cause serious air pollution buildups unless the air masses stall and leave the upper air inversion to affect one locality for a longer period of time.

Upper air inversions are also produced by the diurnal heating and cooling cycle. As heating occurs during the day, followed by cooling at night, a deep inversion layer may develop if cloud cover interferes with daytime heating more than nighttime cooling. Heating of the earth's surface normally warms the lower atmosphere during the daytime to eliminate the inversion that formed there during the night. If insufficient heating occurs during the day to offset the cooling at night an upper air inversion is left in the atmosphere (Fig. 14-8).

An upper air inversion, regardless of its development history, is very influential in determining the dispersion of pollutants. The inversion layer is a very stable layer without atmospheric currents through it, which would help disperse pollutants below. The most detrimental upper air inversions are those that persist for longer periods of time and allow air pollution levels to continue to increase.

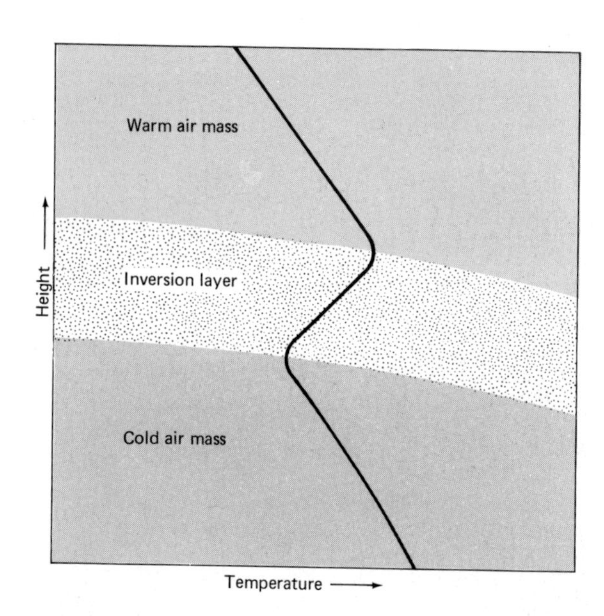

FIGURE 14-7 Upper air inversions are common after the passage of a cold front since a cold front represents the leading edge of a thin wedge of cold air. An inversion occurs at the boundary between the cold air and the warm air mass above the surface.

MIXING DEPTHS

The lapse rate can be used in another way to obtain important information on expected pollution concentrations. The atmospheric mixing depth for pollutants from the surface can be calculated from measurements of the lapse rate through the atmosphere. If a thick mixing depth exists in the atmosphere, pollutants are dispersed through the thicker layer, thereby reducing the concentration at any locality. With a shallow mixing depth, the pollutants are more concentrated in the thinner layer.

The lapse rate from radiosonde data obtained at 6 A.M. is normally used to calculate the expected mixing depth for the afternoon. This is accomplished by using the forecasted maximum temperature for the afternoon and assuming that the air of this temperature rises dry adiabatically and does not mix with the surrounding air. Thus the mixing depth is obtained by following a dry adiabatic lapse rate from the forecasted maximum surface temperature until it intersects the measured temperature profile (Fig. 14-9). That layer below the intersection represents the maximum mixing depth for that afternoon. This is called the calculated **afternoon mixing depth.**

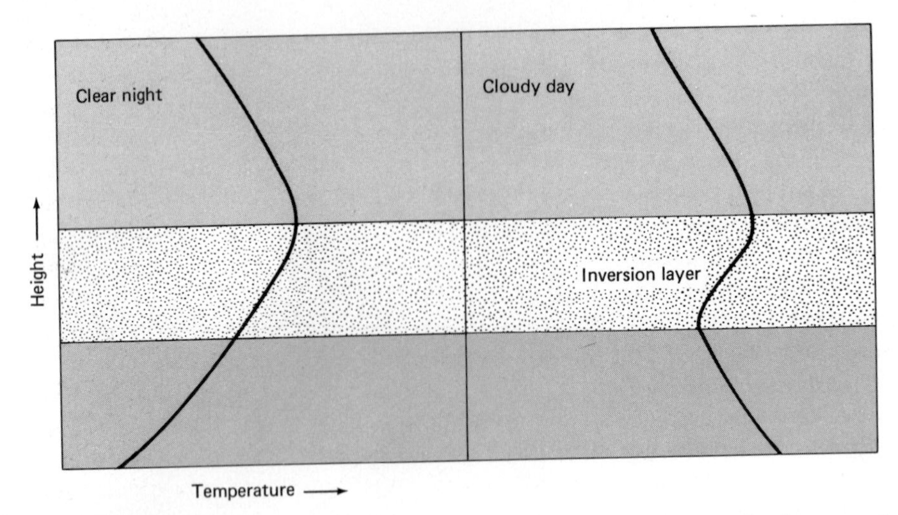

FIGURE 14-8 Another mechanism for developing an upper air inversion is unequal heating of the earth's surface between night and day. If the skies are clear, the earth's surface loses radiation to space during the night, creating a surface inversion. If cloud cover develops before the earth can be heated during the day, an upper air inversion layer may result.

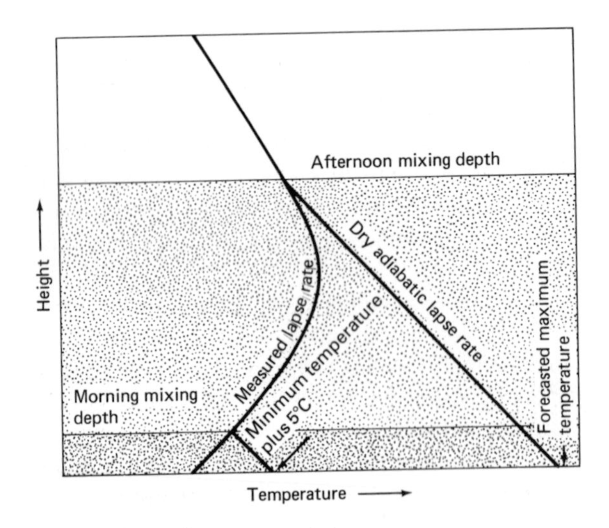

FIGURE 14-9 The morning and afternoon mixing depth can be calculated from the measured lapse rate at 6:00 A.M. combined with the minimum temperature and the forecasted maximum temperature for the day. The top of the afternoon mixing depth is determined by following a dry adiabatic lapse rate upward from the forecasted maximum surface temperature. The intersection of this dry adiabatic lapse rate with the measured atmospheric lapse rate represents the top of the afternoon mixing layer. This gives the afternoon mixing depth for pollutants originated at the surface, since surface heating will cause daytime mixing of the atmosphere to this height. The morning mixing depth is normally much more shallow and is determined by adding 5°C to the measured minimum temperature and again following a dry adiabatic lapse rate until it intersects the measured atmospheric lapse rate. Such calculations show that it is much better to release pollutants into the atmosphere during the day than at night.

The atmosphere is able to mix through a greater depth during the daylight hours than during nighttime hours because of more stable atmospheric conditions at night. The increased heating of urban areas (urban heat island) affects the minimum mixing depth that occurs during the night. The **morning mixing depth** is calculated from the measured lapse rate by adding 5°C to the minimum surface temperature to compensate for the urban heat island within the heart of urban areas and assuming that the air rises dry adiabatically without mixing. The level in the atmosphere corresponding to the intersection of this dry adiabatic lapse rate with the measured temperature profile gives the morning mixing depth. The nature of the atmosphere dictates that a very shallow mixing depth normally exists at night with a much greater mixing depth during the daylight hours.

The average morning and afternoon mixing depths vary considerably over the United States (Fig. 14-10). The morning mixing depth is very shallow, averaging only about 0.4 km. It is less than this over the Rocky Mountain states and is greatest along the Gulf and east coasts, where it may approach 1 km in depth. The afternoon mixing depth averages about 1.5 km and is thickest in the desert Southwest as a result of solar heating, and is shallowest along the coastlines. Regions with greater afternoon mixing depths have better air quality since this is the time when the atmosphere is most likely to cleanse itself by the mixing and dispersing of pollutants.

LOCAL WIND EFFECTS

Topography and local wind circulations are important in determining the dispersion of pollutants. Onshore and offshore breezes near water bodies should be taken into consideration more often in planning industrial locations near a shoreline. Many of the industrial plants located along the shoreline of Lake Michigan affect the air quality of Chicago, as well as of Gary and La Porte, Indiana. The lake breeze in the middle of the day frequently brings pollutants from these plants back over the city (Fig. 14-11). However, the air over the water is normally purer than that over the land, which results in lower concentrations of pollutants during the onshore lake breeze.

Local topography also affects pollution concentrations. In areas with significant topographic relief, winds tend to drain down the side slopes and main slopes at night to produce noticeable breezes. During the daylight hours as the land heats and warms the air above, winds blow up the slopes carrying pollutants with them (Fig. 14-12).

These local winds affect such cities as Denver, Colorado. In the early morning hours pollution from the city is visible as it flows down the hills carried by cold air drainage. The air has warmed sufficiently by about 10 A.M., however, to reverse direction and start moving back up the slopes, bringing the same pollutants back over the city. Thus, local wind systems generated by a particular topography can be important in determining air pollution concentrations.

URBAN HEAT ISLAND EFFECTS

The urban heat island (area with warmer air temperatures) influences the concentration of pollutants. The central regions of cities are normally warmer than the surrounding countryside. Part of the reason for this is related to the structural materials of cities since brick, concrete, and asphalt retain heat at night longer than natural surfaces, thus decreasing the cooling rate. However, cities are also warmer during the daylight hours. This means that an urban heat island exists during both the day and night. Pollution concentrations may affect the urban radiation balance and contribute to the development of a heat island. Greater concentrations of particulates and haze over cities result in a smaller amount of radiation loss at night. Therefore pollution concentration as well as surface materials contribute to the generation of the urban heat island.

The urban heat island is measurable even in small cities. The magnitude of the heat island has

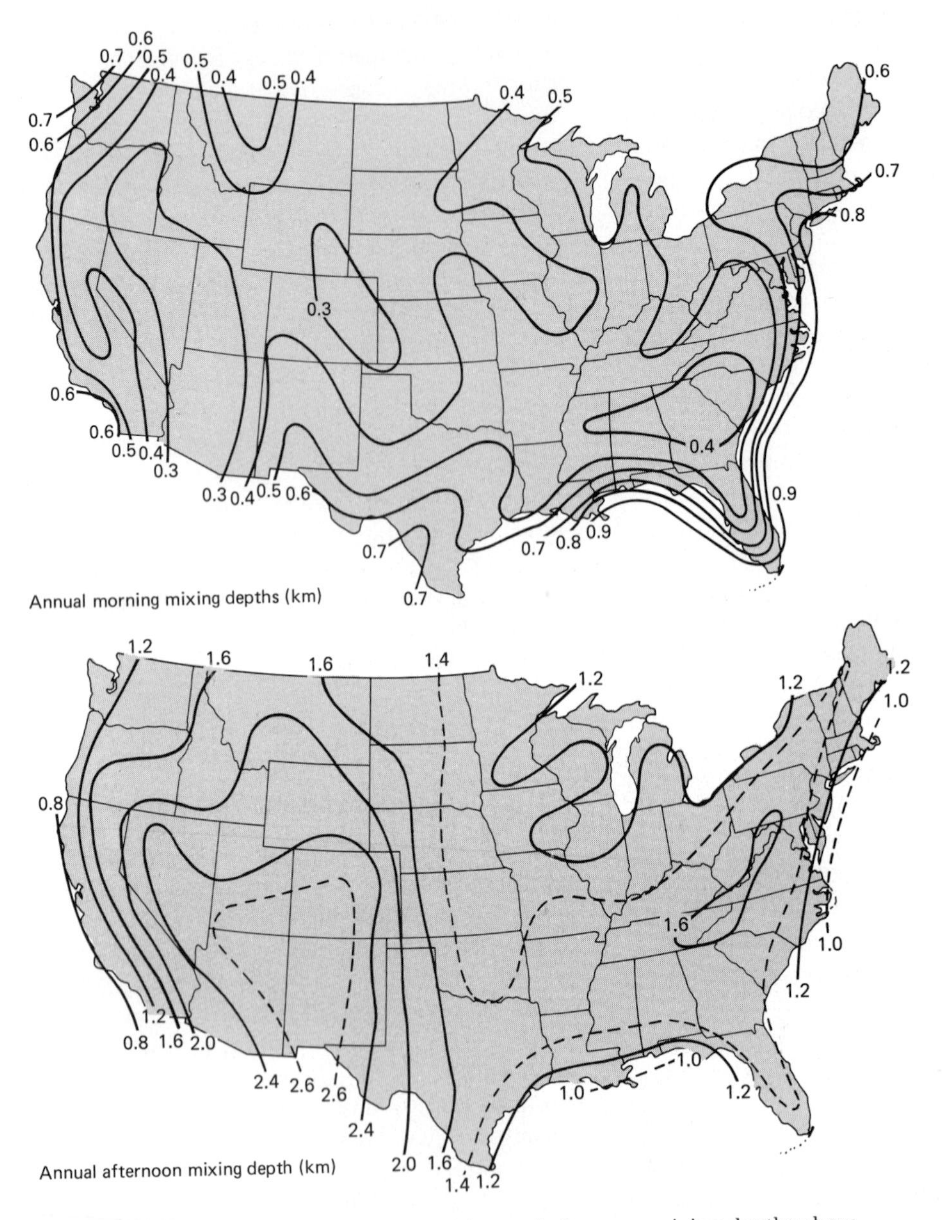

FIGURE 14-10 The average annual morning and afternoon mixing depths show variations with geographical location as well as between night and day. The morning mixing depth varies from 0.3 to 0.9 km and is smallest in the central United States and greatest in coastal areas. The afternoon mixing depth varies from 0.8 to 2.6 km and is greatest in the desert Southwest and least along coastal areas. (From G. C. Holtzworth, "Mixing Heights, Wind Speeds and Potential for Urban Air Pollution," Environmental Protection Agency, Research Triangle Park, North Carolina, 1972.)

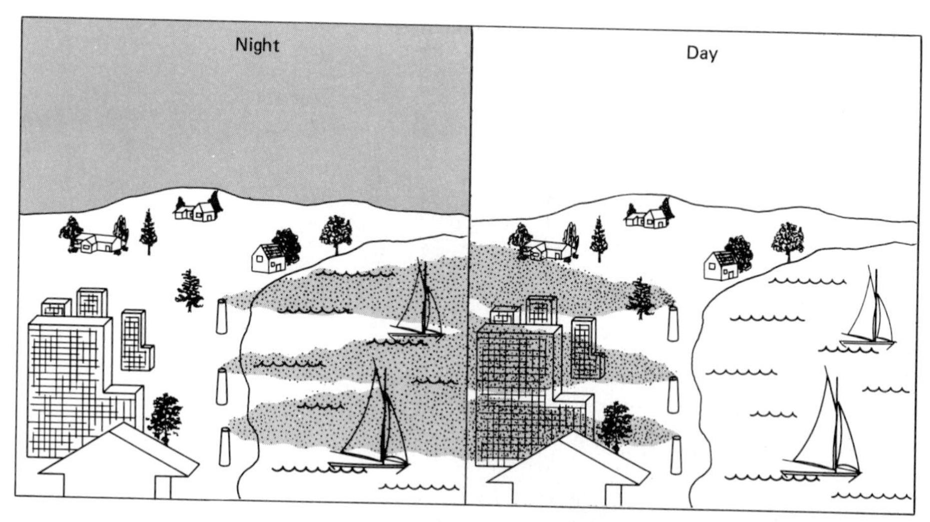

FIGURE 14-11 Local wind systems such as the lake breeze may be important in determining the spread of pollutants. The land breeze will carry pollutants from industrial plants or other sources along the shoreline over the water surface during the night while it will bring the pollutants over the land during the day. It is important to consider such local wind systems when establishing zoning regulations and doing urban planning.

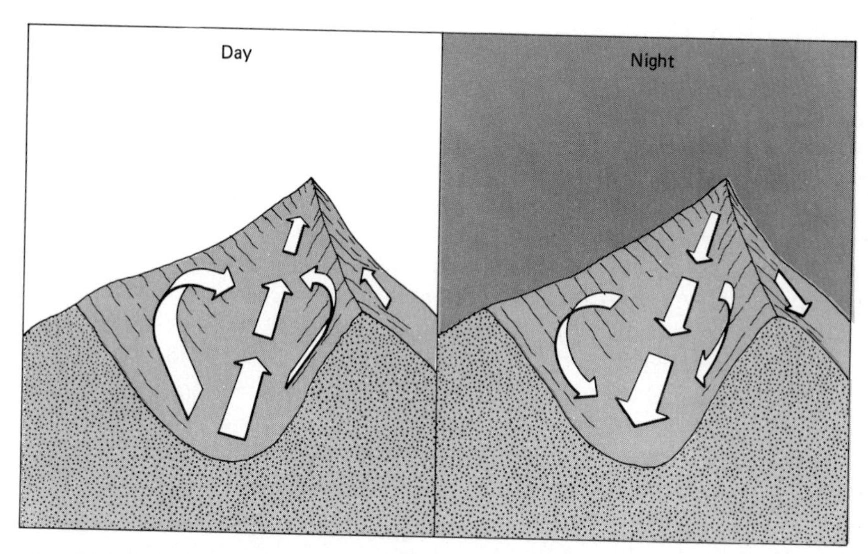

FIGURE 14-12 Another local wind system that is important in determining the spread of pollutants is that which occurs in valleys. During the day, surface heating causes the air to flow up the slopes because of warming of the surface air. This airflow will carry pollutants along in the same direction. During the night the airflow is down the slopes toward the valley floor, as the earth's surface cools the air in contact with it.

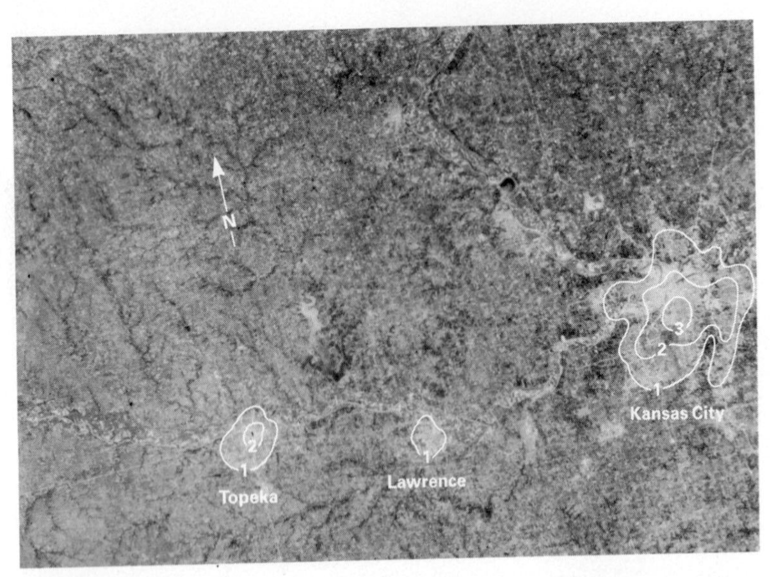

FIGURE 14-13 An urban heat island is common over cities of even fairly small size. The air temperature distribution is shown here for a specific day in Kansas. The increase in temperature is 3°C in central Kansas City, 2°C in central Topeka, and 1°C in Lawrence. The location and intensity of these heat islands is related to the degree of alteration of the earth's surface, as seen from satellite photographs such as this ERTS photograph.

been measured in Kansas City, Topeka, and Lawrence, Kansas, to determine the relationship between the urban heat island and the size of cities, as well as to determine other effects due to the presence of such heat islands. These three cities had populations of about 1,102,000, 132,000, and 46,000, respectively, in 1970.

These heat islands were measured in all seasons of the year, during both the day and night, with a mobile fast response thermistor. Typical heat islands in Kansas City and other cities, on a specific day, are shown in Figure 14-13. The highest temperatures occurred in the downtown business district with decreases outward. Numerous measurements revealed that the maximum heat island measured in Lawrence was 4.5°C, while the average was 2°C. The distribution of measured heat islands in three cities are shown in Figure 14-14. The average heat island measured in Topeka was about 3°C with a maximum measured value of 4.5°C. The average for Kansas City was 4°C. The greatest heat island measured in Kansas City was 7°C and the lowest measured was 1.5°C.

The magnitude of the heat island is related to the amount of disturbed urbanized area, as shown on satellite photographs (Fig. 14-15). A larger city generates a much larger heat island. Some of the accompanying local climatic changes will be discussed in Chapter 15, since it is apparent that human beings are affecting the local climate of urbanized areas. Part of the change is due to air pollution and part is due to other factors related to urban developments.

The urban heat island has several effects on pollution levels in cities. One of the effects is the development of a thin atmospheric layer near the earth's surface over cities, where mixing occurs during the night (Fig. 14-16). If this layer extends

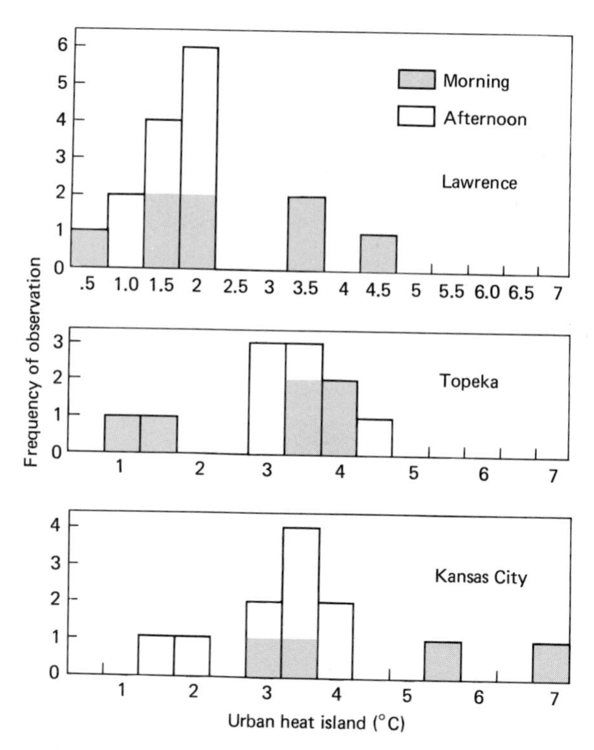

FIGURE 14-14 Comparison of the magnitude of the urban heat island in three different cities shows a distribution of heat islands for each, with the larger cities tending toward larger heat islands than smaller ones. Slightly larger heat islands were measured in the morning hours before sunrise although considerably intense heat islands were also measured in the daylight hours in the afternoon. The largest heat island measured was 7°C in Kansas City. (From *Atmospheric Environment*, vol. 8, J. R. Eagleman, "A Comparison of Urban Climatic Modifications in Three Cities," © 1972, Pergamon, New York.)

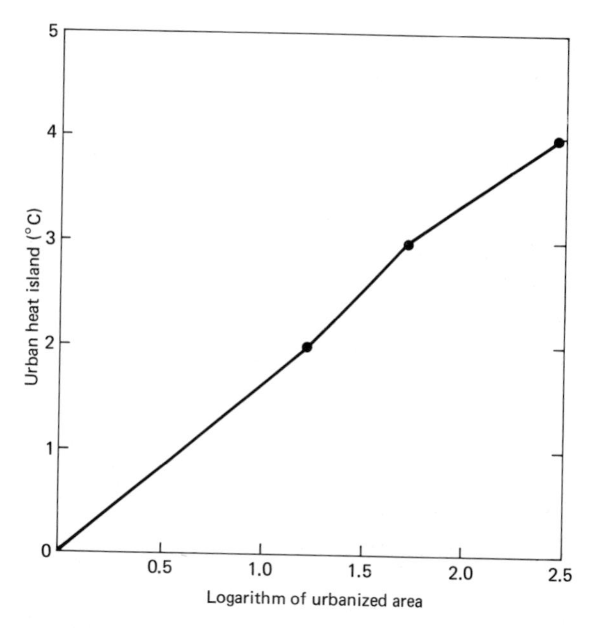

FIGURE 14-15 The magnitude of the urban heat island is related to the amount of urbanized area. The heat island increases according to the logarithm of the urbanized area, as shown on satellite photographs. (From J. R. Eagleman, "A Comparison of Urban Climatic Modifications in Three Cities," *Atmospheric Environment*, vol. 8, © 1972, Pergamon, New York.)

above the height of smokestacks in industrial plants, pollutants are mixed through the surface layer, thus increasing surface concentrations. The usual inversion during the night prevents the mixing of pollutants above this thin surface mixing layer.

Another effect of the heat island on pollution concentrations is related to the wind currents that may be produced by the heat island, which acts as a miniature thermal low-pressure area. There is some indication that light breezes blow toward the center of cities because of the heat island, much like the thermal low and the onshore breezes that blow during the daytime as land surfaces warm more than water. This creates a **country breeze** from the surrounding countryside toward the heart of the city (Fig. 14-16). Such a breeze may intensify air pollution problems since it is most noticeable

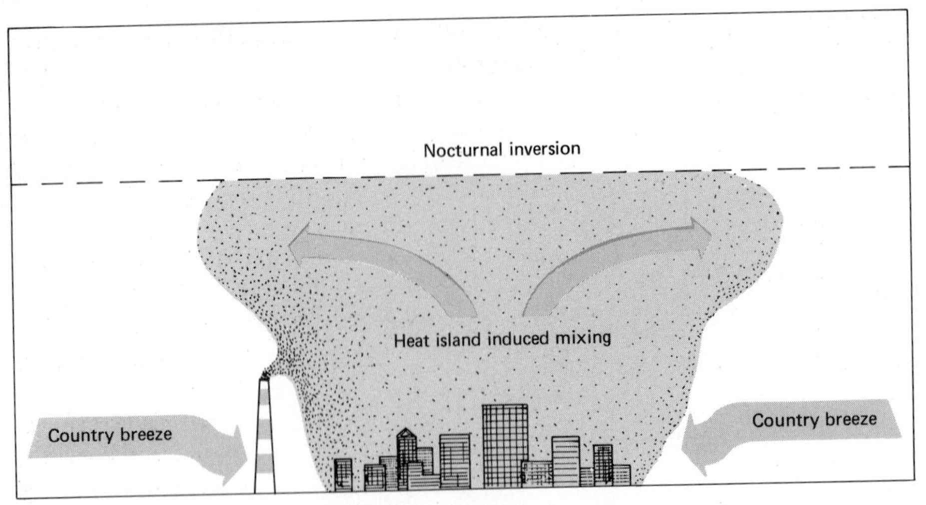

FIGURE 14-16 The heat island affects the distribution and mixing of pollutants since it induces mixing during the nighttime by creating a very shallow mixing layer within the nocturnal inversion. The heat island may also contribute to the development of a country breeze because of the warm temperatures over the city.

when the major wind systems are calm and air pollution problems more likely. The common location of industrial plants on the outskirts of cities combined with a country breeze that carries all pollutants toward the heart of the city may also increase air pollution levels over cities.

DISPERSION ESTIMATES

The concentration of pollutants over cities can be measured by instruments or it can be calculated from meteorological data. Calculations are preferable in some cases to determine the geographical distribution of pollutants since a large number of expensive measurements would be required to obtain the same information. The concentration of pollutants from a point source can be used with statistical equations that have been adapted for calculating the dispersion of pollutants from that point. The most widely used estimate is the **Gaussian plume dispersion model** (Fig. 14-17).

This model (equation) is based on a normal, or bell-shaped, distribution curve. With the concentration known at the source, it is assumed that dispersion of the pollutants downwind from this point follows the normal bell-shaped curve. This assumption makes it possible to calculate the spread of pollutants under particular meteorological conditions since the equation describing such a bell-shaped curve is known and can be used for calculations at any location downwind.

PLUME SHAPES

The shape of the **plume** from a particular smoke stack is influenced by the specific atmospheric conditions. Various plume shapes and associated temperature-height diagrams were illustrated in

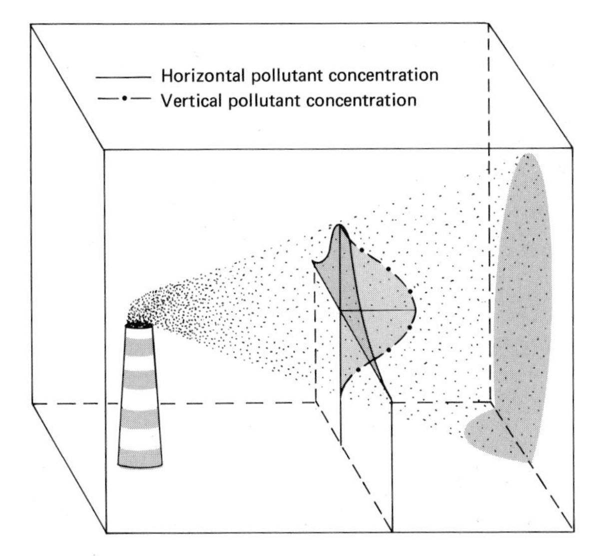

FIGURE 14-17 The concentration of pollutants downwind from a source can be calculated by assuming that the concentration follows the normal bell-shaped curve both vertically and horizontally. Since the equations for these curves are known, the concentrations at various distances downwind from the source can be calculated. This is called the Gaussian plume dispersion model.

Figure 5-14. **Lofting** occurs with a surface inversion and a decrease in temperature above the inversion. The plume appearance shows mixing upward but not downward into the surface inversion layer.

Coning is a typical plume appearance that occurs with neutral stability. If the lapse rate decreases at about the same rate as the dry adiabatic lapse rate, a coning plume is produced with some spreading of the pollutants farther away from the stack.

If the atmosphere is very unstable, with a lapse rate much greater than the dry adiabatic lapse rate, the plume has a **looping** appearance. The large parcels of air that are rising and descending in the atmosphere cause the plume to describe a looping pattern.

An inversion just above the top of the stack may result in **fumigation,** where pollutants are mixed in a shallow layer below the stack. This causes high concentrations of pollutants in a thin layer close to the surface.

A thicker inversion layer from the surface upward allows very little vertical mixing and also traps the pollutants. At the same time, horizontal fluctuations may spread the pollutants sideways. This is called **fanning.** If the wind is greater and some mixing occurs because of fluctuations of the wind, then concentrations on the ground are affected. Fanning normally results in a narrow layer of pollutants that goes in a direction governed by the wind.

The different shapes of plumes allow information on the atmosphere to be obtained by visual observation. They also provide information on expected mixing with different atmospheric conditions.

EXAMPLE OF A POLLUTION EPISODE

A few major air **pollution episodes** in the past were very dramatic and helped call attention to the air pollution problem. One of these occurred in Donora, Pennsylvania, in 1948. Donora had a populaton of about 14,000 at that time. In October of 1948 most of the people became ill and 20 people died because of the heavy concentrations of air pollution. The valley location of Donora contributed to the pollution episode.

The particular weather that occurred in late October 1948 is typical of atmospheric conditions during extreme atmospheric pollution. The upper air conditions showed a ridge located over Pennsylvania with wind speeds less than 20 km/h at 500 mb (Fig. 14-18). This was one of the factors that allowed the pollution concentrations to increase.

The October 26, 1948, surface map showed a large high-pressure area located over Pennsylvania. The high surface pressure lasted for four days, producing calm or very light surface winds. As a result of these synoptic conditions practically no atmospheric mixing occurred and the concentrations of pollutants from local industrial activity increased to harmful levels.

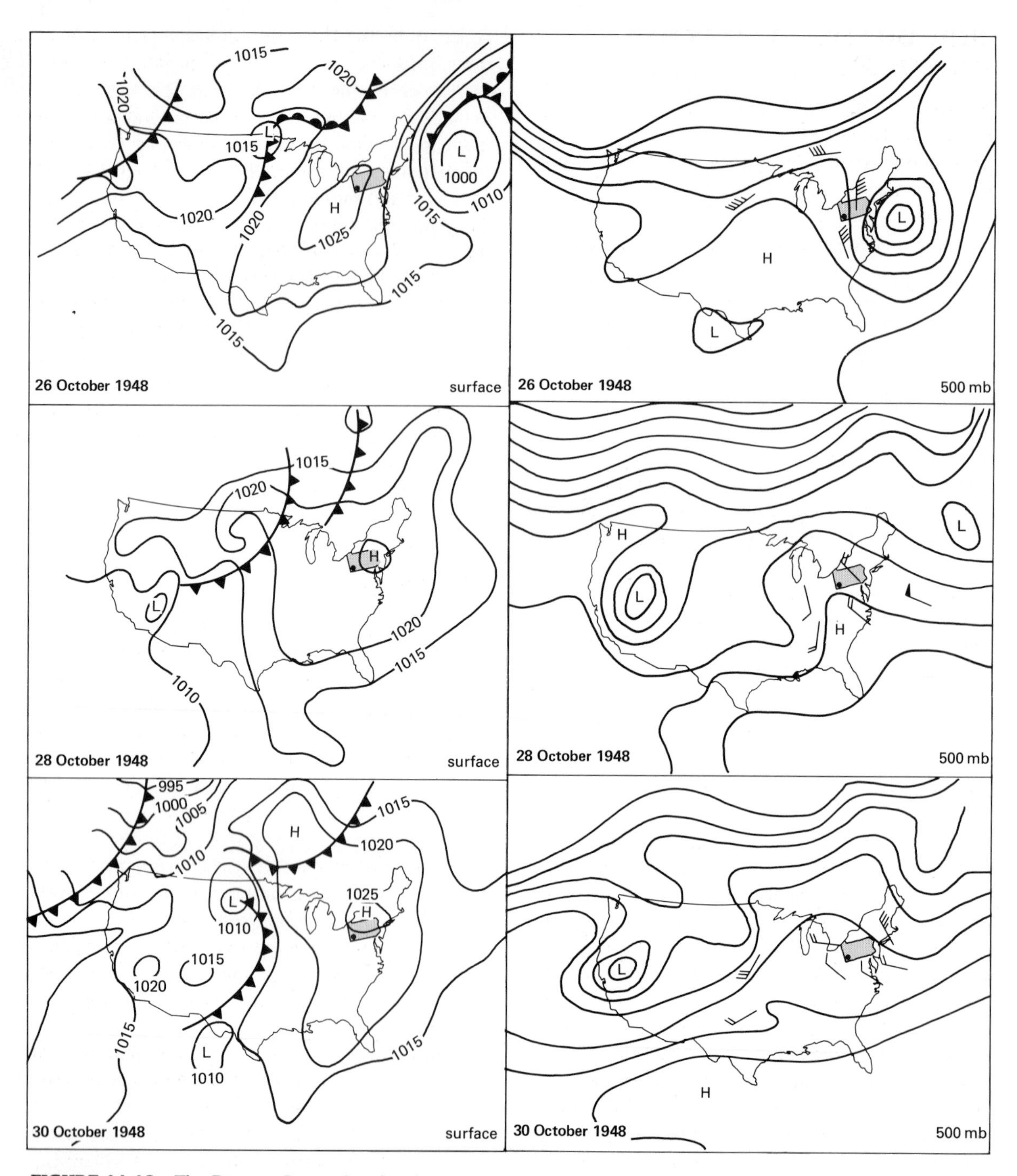

FIGURE 14-18 The Donora, Pennsylvania, air pollution episode in 1948 occurred as the upper atmospheric patterns and surface atmospheric conditions allowed pollution levels to build up to dangerous concentrations. The upper atmospheric winds and surface pressures are shown here, with a persistent ridge providing only very light winds over southwestern Pennsylvania. These light winds, both at the surface and above, were factors in the development of the pollution episode.

HIGH AIR POLLUTION POTENTIAL ADVISORIES

The ability of the atmosphere to mix is now forecasted, and High Air Pollution Potential Advisories are issued every day. Each advisory covers an area about the size of Oklahoma, and forecasts the atmospheric conditions that are likely to persist in that area for the next 36 hours, such as very light surface winds and no precipitation.

Specific criteria are placed on the wind speeds and the afternoon mixing depth. An index called the **afternoon ventilation** is obtained by multiplying the mixing depth by the average wind speed through this depth. If this value is less than 6000 m^2/sec, air pollution problems are indicated. Another indicator is simply the average surface wind speed, which must be less than 4 m/sec to indicate problems. If these conditions are forecasted to persist, then a High Air Pollution Advisory is issued. The advisory is based on the mixing ability of the atmosphere rather than on measurements of concentrations of pollutants. This information, when combined with known locations of pollution sources, gives a good indication of where pollution problems can be expected.

EMERGENCY ACTION PLANS

Most cities have, or are developing, emergency action plans. These are based primarily on measurements of pollutants, as commonly obtained in many cities. Enough measurements are available for many locations to give trends of pollution levels through time (Fig. 14-19). In general, as a result of the efforts of industry, automobile manufacturers, and government regulations to control atmospheric pollution, the trends show improving air quality. Emergency action plans are designed to deal with short-time pollution problems, and identify different stages that are based on well-defined pollution concentrations. Some cities have vigorous plans of action. For example, Philadelphia, Pennsylvania, can completely close traffic, industry, and outdoor activity if critical levels of pollution are reached.

Based on the High Air Pollution Potential Advisory program, the first stage is a **pollution forecast.** This indicates that the atmospheric conditions are such that pollution concentration increases are likely. The next stage is a **pollution alert.** In this stage the sulfur dioxide levels are at least 0.3 parts per million (ppm) or oxidants have reached 0.1 ppm. Any one of these can result in an alert situation. A **pollution warning** is issued if sulfur dioxide has reached 1.6 ppm for a 24-hour average or carbon monoxide has reached 30 ppm or oxidants have reached 0.4 ppm. The **pollution emergency** status is reached, with danger to human health, if the sulfur dioxide level reaches 1 ppm, carbon monoxide reaches 50 ppm on an 8-hour average, or if oxidants reach 0.4 ppm on a 4-hour average. Specific criteria designed to deal with the air pollution problem are defined for the different stages.

SUMMARY

Air pollution is bothersome to people in many large cities in the United States. Los Angeles and New York are especially bothered by air pollution and in many other cities there is considerable concern for air quality. Meteorological factors can be used to estimate the dispersion of atmospheric pollution and to forecast geographical areas that are likely to have greater air pollution problems.

The air circulation within a high-pressure system, with calmer winds and subsiding air, contributes to an environment where air pollution levels can increase. The jet stream and upper atmospheric conditions that are conducive to air pollution problems consist of light winds in the upper atmosphere that commonly occur within upper air ridges.

Atmospheric stability is another feature that must be considered in air pollution studies since an unstable atmosphere contains more vertical air currents and therefore disperses pollution much more readily. Surface inversions are extremely stable and allow very little mixing of pollutants within them. Surface inversions frequently occur at night because of radiation loss.

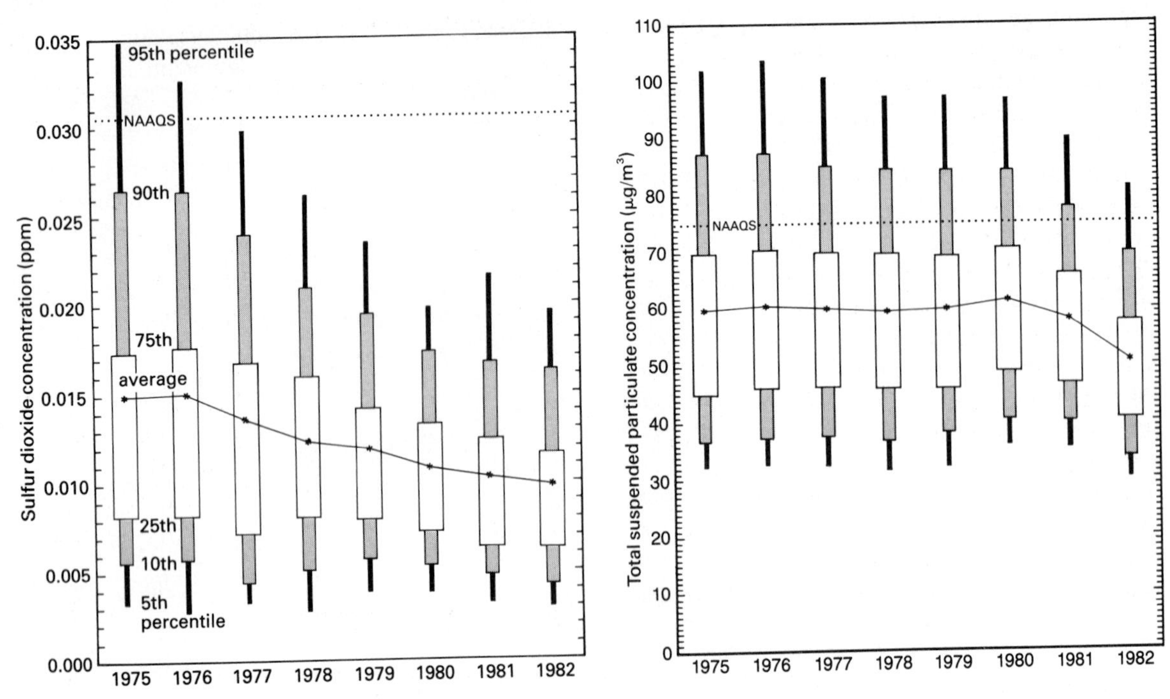

FIGURE 14-19 Trends of air pollution levels are shown here for total suspended particulate concentration and sulfur dioxide levels from 1975 through 1982, based on measurements at more than 300 sites. The value of the National Ambient Air Quality Standard (NAAQS) is shown for comparison. (From National Air Quality and Emission Trends Report, 1984 EPA, Research Triangle Park, N.C.)

Upper air inversions also affect the concentration of pollution in the lower atmosphere. These may develop from subsiding air above a high-pressure system, from air mass changes, or from an unequal heating and cooling daily cycle. Upper air inversions cause air pollution from the surface to be trapped within the layer between the surface and the upper air inversion, contributing to large increases in air pollution levels.

The atmospheric mixing depth is the calculated depth of mixing of pollutants in the lower atmosphere. It is calculated on the basis of the dry adiabatic lapse rate and radiosonde data.

Air pollution levels may be affected by local winds such as the lake breeze or mountain breeze. The urban heat island represents significant temperature increases in cities and may cause atmospheric mixing of pollutants through a thin layer at night when an inversion would otherwise occur. It may also contribute to local wind currents toward the center of the city that bring pollution from the surrounding suburbs.

Air pollution levels can be measured or calculated from meteorological data. The Gaussian plume dispersion model is a method that can be used for calculating the spread of pollutants from a point source. The shape of plumes from smokestacks is related to meteorological conditions. The particular plume shapes of coning, looping, fumigation, and fanning all occur under specific meteorological conditions.

Pollution episodes are most likely when atmospheric conditions contribute to small amounts of mixing. A surface high-pressure area and upper air ridge with an upper air inversion occurred over Donora, Pennsylvania, in October 1948. This allowed air pollution levels to increase to a point that became dangerous to people, with many becoming ill and 20 deaths occurring.

High Air Pollution Potential Advisories are now a part of the National Weather Service forecasting efforts. These are issued when the atmospheric conditions are appropriate for increasing the concentration of air pollution. Emergency action plans have been developed by most cities—often very vigorous plans of action. Emergency action plans are based on stages of pollution, with the first, based on the High Air Pollution Potential Advisory program, being a forecast. The next stage, based on particular concentrations of air pollution, is an alert. Advanced stages are the warning and emergency. These allow specific criteria to be established for dealing with the air pollution problem.

STUDY AIDS

1. Do you think people are more or less concerned about air pollution than they were ten years ago? Give reasons for your answer.

2. What effects do high-pressure areas have on air pollution concentrations?

3. What effects do the jet stream and upper atmospheric conditions have on air pollution levels?

4. Discuss the effects of atmospheric stability on air pollution levels.

5. Discuss the development of surface and upper air inversions and their effects on air pollution levels.

6. Explain the morning and afternoon mixing depth and describe how they are estimated.

7. Discuss the urban heat island and its effect on air pollution levels in cities.

8. Discuss the shape of particular atmospheric plumes as affected by the atmosphere.

9. Describe the High Air Pollution Potential Advisory program.

10. Discuss emergency action plans for air pollution episodes.

TERMINOLOGY EXERCISE

Check the glossary if you are unsure of the meaning of any of the following terms used in Chapter 14.

Dispersion of pollutants
Fixed source
Mobile source
Afternoon mixing depth
Morning mixing depth
Country breeze
Gaussian plume dispersion model
Plume
Lofting
Coning
Looping
Fumigation
Fanning
Pollution episode
Afternoon ventilation
Pollution forecast
Pollution alert
Pollution warning
Pollution emergency

THOUGHT QUESTIONS

1. Explain why some regions of the United States are more susceptible to air pollution problems than other locations. Should this affect the way air pollution problems are handled?

2. List and comment on some of the trade-offs between strict environmental laws and industrial development.

3. How successful do you think past efforts at air pollution control have been? Explain your answer.

4. Do you feel that the acid rain problem is likely to increase or decrease? Why? (See Special Topic 7.)

5. Name and identify the source of as many atmospheric pollutants as possible for your particular location.

CHAPTER FIFTEEN
CLIMATOLOGY

CLIMATE AND WEATHER

Climate is more than just average weather. Climate includes all the various weather events that occur over an extended period of time. Frequently the cold winter followed by a very mild one, or the summer drought followed by a very wet summer, is more descriptive of a region than the long-term average winter temperature or summer rainfall. A realistic description of the climate of a region must, therefore, include the extremes as well as the average weather.

As we have noted previously, predicting the behavior of the atmosphere more than a few hours in advance is a complex problem involving accurate measurements of the current state of the atmosphere and accurate equations to project atmospheric conditions into the future. Both these aspects of the problem contribute to weaker weather forecasts as the time projection increases. Since the climate of a particular location is governed by the long-term atmospheric conditions above the earth's surface, predictions of future climatic conditions are even more difficult than long-range weather forecasts. However, a nation's climate is one of its most important natural resources. Therefore our best guesses of future climatic trends are better than no information at all.

CLIMATIC CHANGE

As we contemplate such weather fluctuations as the record-breaking severe winters of the late 1970s, with the coldest average temperatures of this century and much greater snow cover than normal, we wonder whether an ice age is imminent, as some authors maintain. Our climate may

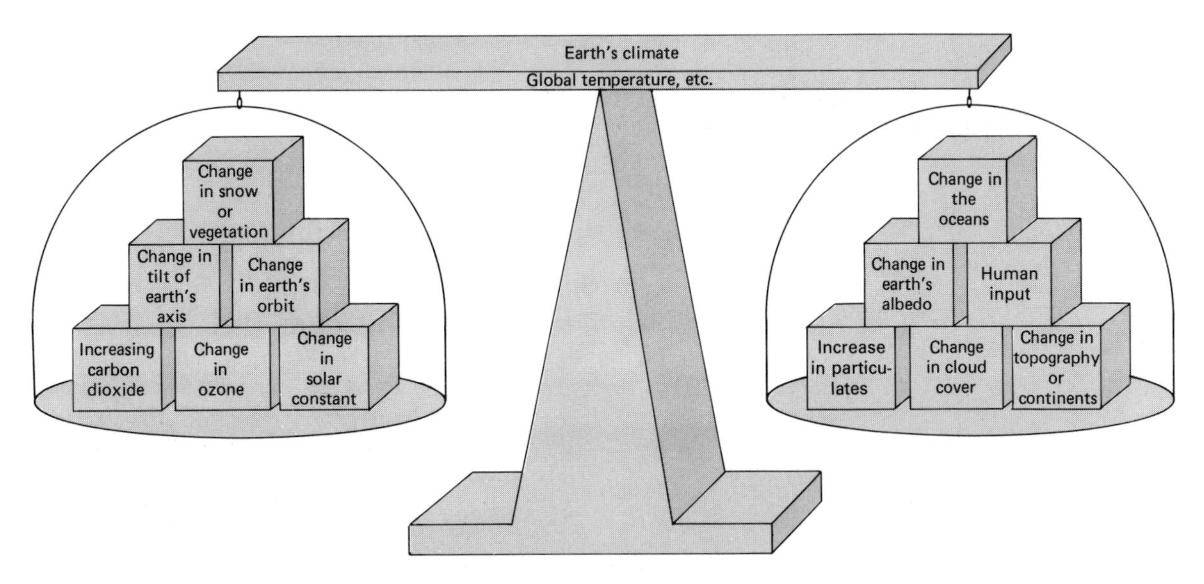

FIGURE 15-1 The earth's climate is in a delicate state of equilibrium; small changes in the various factors shown can perhaps tip the scales toward long-term cooling or warming.

change if one or more of a large number of influencing factors is altered. The primary **climatic control** is the incoming and outgoing radiation balance. This in turn is affected by a wide variety of factors, some of which are in a very delicate state of equilibrium (Fig. 15-1).

For example, an increase in the average global temperature of 1°C will probably occur by the year 2000 due to carbon dioxide, which continues to increase due to consumption of fossil fuels, even if no other climatic controls change (Fig. 15-2). Additional carbon dioxide in the atmosphere enhances the greenhouse effect and traps more heat within the earth-atmosphere system. However, the projected temperature increase may not occur since the particulate concentration of the atmosphere is also increasing. This reduces the incoming solar radiation more than the outgoing infrared radiation, perhaps counterbalancing the effect of increased carbon dioxide.

Another factor to be considered is small changes in cirrus clouds, which are very influential on the earth's energy budget. It has been calculated that an increase in cirrus clouds of less than 1% would offset the effects of increased carbon dioxide to the year 2000. There is some indication that the cirrus cloud cover is increasing at some locations, perhaps due to high altitude aircraft (Fig. 15-3).

As shown in Figure 3-15, the average temperature of the northern hemisphere increased from 1880 to 1940. Since 1940 a definite cooling trend has occurred of more than 0.5°C. This indicates that the projected warming from carbon dioxide increases is more than offset by other factors. Therefore, rather than creating undesirable effects, it is quite likely that carbon dioxide increases have had a beneficial effect in preventing even more rapidly decreasing global temperatures. It is not very reassuring, however, to realize that climatic change may respond to a factor, or combination of factors, that we are not able to pinpoint precisely.

As shown in Figure 15-1, our climate is delicately balanced, with small changes—originating outside the atmosphere, within the atmosphere, or

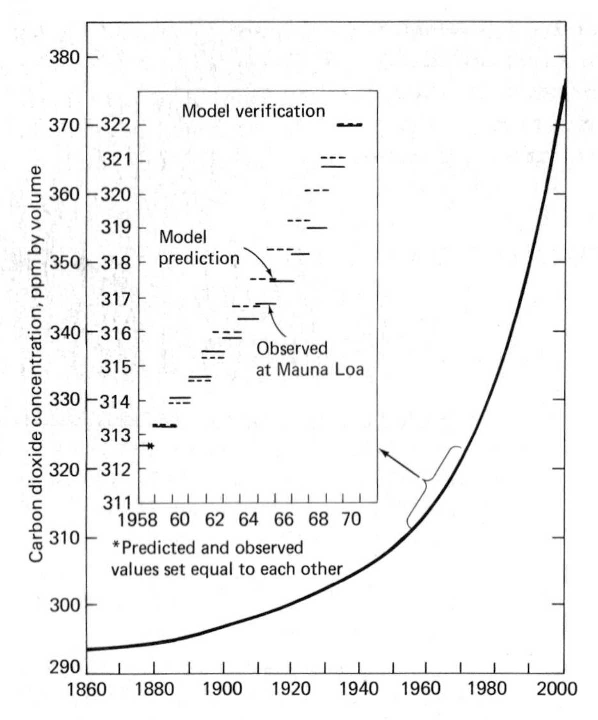

FIGURE 15-2 The amount of carbon dioxide in the atmosphere is increasing, as shown from these measurements and calculations. Projections into future years on the basis of these measurements furnish one piece of the jigsaw puzzle of future climatic trends. (From SMIC, *Inadvertent Climate Modification,* © The M.I.T. Press, 1970.)

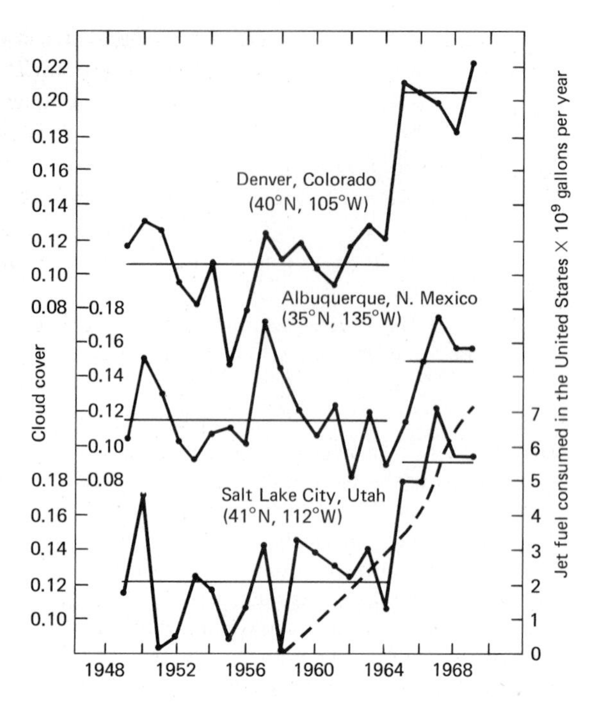

FIGURE 15-3 Measurements of the cirrus cloud cover at the indicated locations show a trend toward slightly more cloud cover. Changes in cloud cover may be an important aspect of climatic change. (From SMIC, *Inadvertent Climate Modification,* © The M.I.T. Press, 1970.)

on the earth's surface—tipping the scales toward a warmer or colder climate. We do not have to look back further than a few centuries to find considerable climatic fluctuation. The **little ice age** occurred from about 1400 to 1800, created from an average temperature change of less than 2°C (Fig. 15-4).

The year 1816 is famous as the "year without a summer." A reduction in temperature of about 1°C below normal throughout midlatitudes of the northern hemisphere followed the eruption of Tambora in Indonesia in 1815 and tons of volcanic dust were injected into the stratosphere. As a result, killing frosts occurred every month of the year, and there were food riots in parts of Europe. Within the United States the New England region was greatly affected, as snow fell in June and frosts during the summer resulted in complete crop failures as far south as the Carolinas. The widespread effects of the eruption of a single volcano emphasize the delicate nature of our climatic environment.

Thus, we should view with extreme caution such ideas as spreading carbon dust over the north polar ice cap to initiate melting, or diverting rivers away from the Arctic. If the northward flowing rivers of Siberia were diverted, as has been proposed, to provide water for the arid lands of central Asia, the loss of fresh water in the Arctic could result in

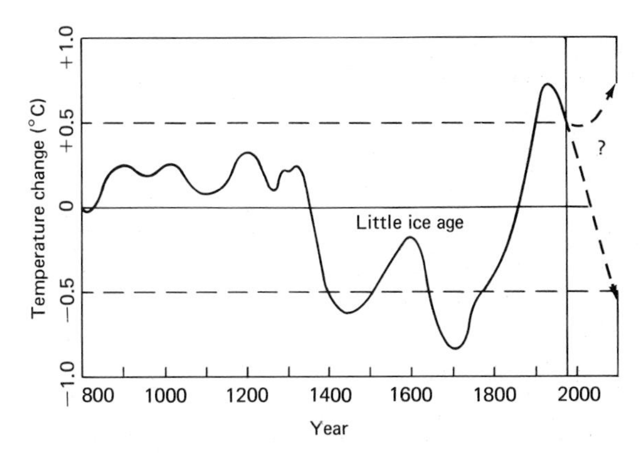

FIGURE 15-4 The average temperature of the northern hemisphere during the last 1000 years has fluctuated about 1.5°C. This temperature variation was accompanied by drastically altered climates in many regions. Although future trends are uncertain, a peak in temperature was reached in 1940, and a cooling trend is now in effect. (From *Understanding Climatic Change,* p. 130, with the permission of the National Academy of Sciences, Washington, D.C.)

a large-scale climatic change. The rate of freezing and evaporation of the saltier water would affect the albedo, surface temperatures, and precipitation over much larger areas.

If we consider that temperatures over the past several centuries have been cooler than current normal temperatures (recent 30-year averages) for over 90% of the time, the evidence indicates continued cooling of the climate in the future. Lowell Ponte* suggests that a cooling of our planet is so certain that governments must immediately consider the implications. Since the United States produces more than two-thirds of all food available for international exchange, and Canada supplies more than half the remainder, a shift of climatic conditions toward colder temperatures could swing the balance of economic power dramatically toward the United States, just as the oil embargo switched the economic power to the nations with surplus oil production. On this basis, Ponte envisions a whole new political arena called **climatocracy.** Since climate is a major factor in the economic strength of a nation, some nation may try to shift the global climate toward more favorable weather for their particular location, if this were deemed possible.

EXTERNAL CLIMATIC INFLUENCES

External factors such as variations in solar radiation have been suggested as a possible cause of climatic change. Solar radiation just outside the earth's atmosphere normally varies by about 7% during a year, as the earth revolves around the sun in its elliptical orbit. However, measurements indicate that the variation from year to year is less than 1%, probably too small for initiating climatic change. The shape of the earth's orbit varies from almost circular to elliptical, with an oscillation of about 96,000 years. The earth also wobbles on its axis, with one complete cycle of the earth's tilt requiring about 21,000 years. Such long-term variations are not very useful in explaining shorter climatic variations, however.

Another solar variation, the 22-year sunspot cycle, may initiate climatic changes through its effect on the earth's magnetic field, or by some other mechanism not yet discovered. Sunspots are seen as dark storms moving across the sun in pairs. They are darker because they have a cooler temperature than their surroundings and are accompanied by slightly more radiation in the x-ray wavelengths than the normal sun. Since their total area is small, they do not influence the quantity of radiation emitted by the sun. Sunspot cycles do not always coincide with observed weather events, however, although the very dry year of 1976 corresponded to a sunspot minimum, as described in Chapter 2, and part of the little ice age was also a period of prolonged minimum sunspot activity.

*Lowell Ponte, *The Cooling,* Englewood Cliffs, N.J.: Prentice-Hall, 1976.

ATMOSPHERIC COMPOSITION INFLUENCES

Climatic influences may arise from the ozone layer, which determines the temperature of the stratosphere. The use of freons (chlorofluorocarbons) as propellants in aerosal spray cans was curtailed not long ago because of the discovery that it is possible for them to reach the stratosphere, where they are turned into fluorine and chlorine atoms by the action of sunlight. In this form they are active in the destruction of ozone.

Ozone goes through a natural cycle of formation by sunlight action on oxygen atoms to destruction, perhaps 18 months later, through interaction with nitric oxide or other chemicals from the earth's surface. Thus, other gases besides fluorine and chlorine may affect the ozone layer, including nitrous oxide produced by fertilizer plants and carbon monoxide from cars. Estimates indicate, however, that the destruction of ozone from such pollutants is many times less than the natural variability of ozone.

Carbon dioxide and particulates may also affect atmospheric composition. Even though particulates may counterbalance the warming of the lower atmosphere due to the increased greenhouse effect of the carbon dioxide, the concentration of particulates may be great enough to overcompensate and cause a cooling of the earth. Heavy particulate concentrations arise from volcanic eruptions on the other side of the world, or from a variety of other causes such as strong winds that pick up dry soil particles from neighboring states.

A cooling of the atmosphere causes climatic changes that occur suddenly rather than gradually. One reason for this is related to the jet stream and its surrounding airflow. As the jet stream and surrounding westerly winds circle the world in a looping path, they may have three, four, five or more wavelengths since they form a continuous stream of air. If global temperatures are cooler, the jet stream shifts southward in the northern hemisphere, just as it shifts southward every winter, as previously illustrated in Figure 4-24. As its circumference increases since it must take a more southerly route during both summer and winter due to cooler global temperatures, the sudden shift to a different number of wavelengths will cause the climate to be drastically altered to colder and wetter conditions in locations where the jet stream influence changes from ridge to trough (Fig. 15-5). A shift toward a much drier climate will be experienced by regions that come under the influence of a ridge in the jet stream. The increased variability of weather associated with regions dominated by the jet stream will be extended southward into new regions.

A cooling of global temperatures will also have effects in tropical regions. The intertropical convergence zone will not move as far northward during the summer in the northern hemisphere since the jet stream and surrounding westerly winds will be moving further southward and limiting their northward movement. Thus the monsoons will produce less rain in India and the Sahel of tropical Africa. In fact, this effect has already occurred, with drought and famine resulting, as the Sahara desert has moved southward into the semiarid Sahel of Africa. The result has been mass starvation, peaking in 1973, that has taken the lives of more than 100,000 people, with millions more saved by food that was flown in from other nations, including the United States.

SYNERGISTIC CLIMATIC EFFECTS

Since the earth's climate is composed of a variety of interacting elements it should not be surprising that a number of **synergistic effects** (combined effects, which may be greater than the sum of the individual components) are possible. Some of these can be illustrated by speculation of carbon dioxide climatic effects. If the amount of carbon dioxide in the atmosphere increased sufficiently, the temperature would increase because of the greenhouse effect. A significant temperature increase would melt the floating polar ice cap. This would not change the ocean level and would be easier to melt than the much thicker glaciers in Greenland. These glaciers are so thick that their extension into the

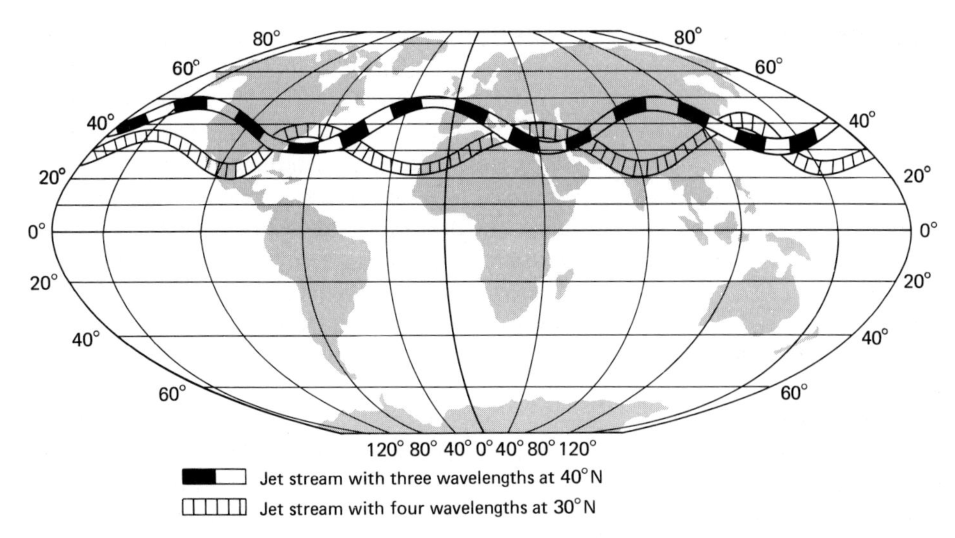

FIGURE 15-5 Comparison of a jet stream centered at 40° N latitude that has three wavelengths, with a jet steam at 30° N latitude that has four wavelengths. In each case the north–south variation of the looping pattern encompasses 20° latitude. Such a shift in the jet stream due to cooling global temperatures, with an associated change in predominant number of waves, would drastically alter the climate of most midlatitude locations.

cooler air above helps prevent their melting. If the melting of these glaciers were initiated, however, once half their volume was lost, they would continue to melt, as the glacier tops became exposed to warmer and warmer temperatures at lower elevations.

As the ocean levels increased from the melting of such a huge volume of ice, all of Florida would be under water, with the ocean waters backed up the Mississippi River as far as Missouri, covering all low-lying areas. Similar events would occur around the world. The greater surface area of the ocean would absorb more carbon dioxide from the air, reducing the greenhouse effect and eventually cooling the atmosphere. Glaciers would again begin to advance southward, causing a drop in ocean levels as the water became tied up in the snow and ice. With a sufficient decrease in the surface area of the ocean, excess carbon dioxide would be released to the atmosphere and the cycle would repeat itself.

A similar scenario can be developed by emphasizing the role of cloud cover and the large albedo of snow-covered surfaces. If a few years are colder than normal because of natural climatic fluctuations, less snow is melted from the polar ice cap during the summer. Thus the jet stream and midlatitude cyclone activity is displaced southward earlier in the fall, with a blanket of snow covering Canada and the northern United States earlier in the winter. The jet stream also reaches even further southward, causing a greater contrast with the warm equatorial air and resulting in increased cloud cover and more frequent frontal cyclones. As the extended snow cover reflects most of the solar radiation, less warming occurs and cold north winds are allowed to reach even further southward. Thus spring arrives later each year, as more sunlight is reflected and is also used to melt the snow before it can warm the earth.

As the earth cools, however, less evaporation occurs, with less atmospheric moisture for cloud formation and snowfall to feed the glaciers for continued growth. Also, less clouds and more sunny days cause greater warming, with more energy for melting snow and ice, thus causing continued change toward a warmer climate where the process can be repeated.

TERRESTRIAL CLIMATIC INFLUENCES

Topography and Oceans The physical effects of mountain chains built up by volcanic activity influence the climate. In midlatitudes, troughs in the jet stream are more frequent on the leeward side of a north–south mountain chain such as the Rocky Mountains. Sudden shifts of the jet stream from one side to the other of east–west mountain chains, such as the Himalayas, are typical. Such changes profoundly affect the climate. Some suggest that the ice ages of the past were caused by volcanic mountain building.

Ocean currents can be drastically altered by changes in the ocean level caused by the freezing and thawing of glaciers. The effects of an equatorial passage between North and South America, for example, could change the ocean currents. If the passage were large, the temperature of currents in both the Atlantic and Pacific oceans would be affected. Since warm and cool ocean currents interact with the air above, they could influence large climatic changes.

Potential Evapotranspiration Although oceans supply most of the water in the atmosphere, the return of moisture to the atmosphere from land surfaces is not insignificant. The **potential evapotranspiration** represents the water loss rate from land surfaces that have complete vegetative covers that are actively growing with plenty of water in the soil. The type and amount of vegetation not only changes the appearance of the terrain, but also influences the climate through its effect on the albedo, on water absorption, and on decreased dust content of the air above. The natural vegetation, as well as the particular crops that are grown, with related agricultural practices, are determined primarily by climatic factors—precipitation and potential evapotranspiration. Therefore the potential evapotranspiration rate will be considered in more detail.

The atmospheric variables that are most influential in determining the potential evapotranspiration rate are the temperature and relative humidity. Water is added to the atmosphere over land primarily through the leaves of vegetation. **Stomata** (small pores) on the lower side of leaves open during the day and close at night, allowing water to escape from vegetation during the day. Some plants are more adapted to dry weather since they can close their stomata whenever a slight water stress develops. Most vegetation is able to decrease the amount of water lost to the atmosphere by partially closing their stomata during very dry weather. For this reason the potential evapotranspiration is defined as the maximum water loss rate that occurs with plenty of water in the soil. This rate thus depends on the evaporative demand of the atmosphere and can be calculated from meteorological data.

An equation developed by Thornthwaite for calculating potential evapotranspiration used only mean monthly temperature data with adjustments for latitude. A more complicated equation by Penman is based on four variables: temperature, humidity, wind velocity, and net radiation. Although Penman's equation gives more accurate results, the data for the four different variables are frequently not available.

A method of calculating potential evapotranspiration was also developed by the author, based on the two most influential atmospheric variables—temperature and relative humidity—since these are more readily available. The distribution of annual potential evapotranspiration determined by this method is shown in Figure 15-6. The extreme southwestern United States could evaporate more than 200 cm/yr if the water were available. The potential water loss rate in southern Florida is 175 cm/yr. The potential evapotranspiration decreases northward to a rate of less than 75 cm/yr in the northern United States.

The potential evapotranspiration will be used in following sections to help characterize the existing climate of different regions.

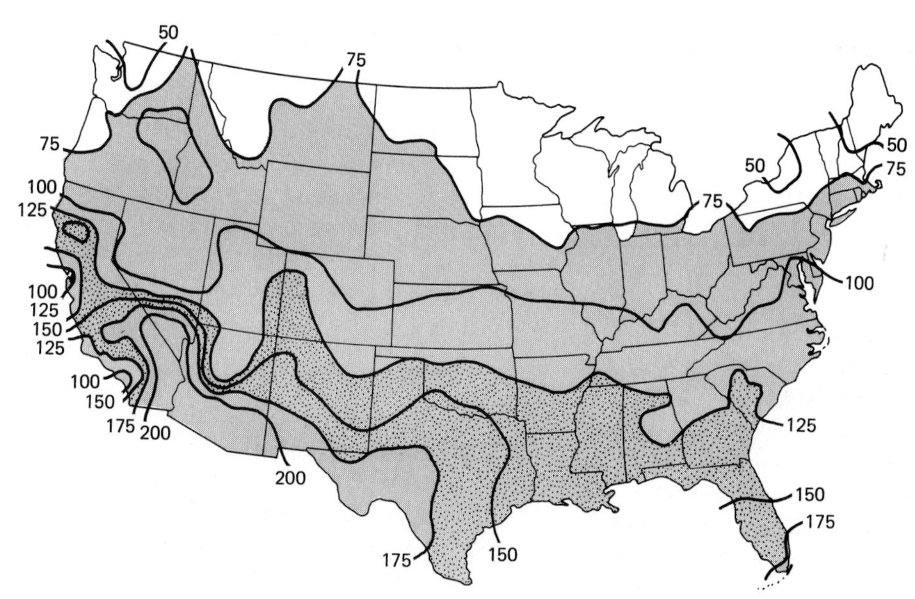

FIGURE 15-6 The annual potential evapotranspiration (in cm) in the United States is greatest in the desert Southwest, where more than 200 cm of water could be evaporated if it were available. In the northern United States only 50 cm/yr can be lost in some locations. (Modified from J. R. Eagleman, *The Visualization of Climate*, Lexington, Mass.: Lexington Books, D.C. Heath, 1976.)

Soil Moisture The amount of water in the soil is the major difference between the soil on Mars and Earth, and this difference allows global dust storms on Mars. Low soil moisture was of primary importance in generating the "dust bowl" conditions on Earth during the 1930s (Fig. 15-7). In addition to holding the soil particles together, soil moisture is essential for vegetation. A certain amount of water in the soil is unavailable to vegetation since it is tightly held by the soil particles and plant roots can't extract it. The moisture content where this occurs is called the **wilting point** for vegetation. After a rain occurs, a certain amount of the rainwater soaks downward through the soil and some runs off and is not available to plants. The moisture content after gravity drains the excess from the soil is called the **field capacity.** The amount of water available for plant use is held in the soil between these two limits. The available water is held by the small pore spaces in the soil against the pull of gravity and may remain there until it is utilized by plants.

The distribution of soil moisture at any location and time depends on the rainfall, moisture-holding characteristics of the soil, and potential evapotranspiration. Monthly variations in the amount of stored soil moisture at various locations across the United States are shown in Figure 15-8. Higher soil moisture content is more likely to occur in the winter, as soil moisture reserves are replenished. The soil moisture supply decreases during the summer, as the evaporative demand becomes large. The soil moisture content is very important for agriculture, the construction industry, and a wide variety of other activities.

Short-term variations in the soil moisture content of midlatitudes hypothetically affect short-term precipitation cycles (Fig. 15-9). A high soil moisture content allows more solar radiation to be used for evaporating water and less in heating the

FIGURE 15-7 The dust bowl of the 1930s resulted from extremely low soil moisture. Blowing soil changed the landscape in some locations. (Reproduced from the collections of the Library of Congress.)

land and lower atmosphere. As the soil surface dries it heats much more, thus warming the lower atmosphere. The importance of warmer surface temperatures in determining atmospheric stability and thunderstorm development have already been described. It does not, therefore, seem too unrealistic to suggest a **feedback mechanism,** where the cyclic variations in soil moisture content send messages to the atmosphere for more or less rain. The formation of rainfall, of course, would depend not only on this message but also on the presence of the appropriate atmospheric conditions for rainfall.

The soil moisture content is important in such a variety of applications that techniques are being developed to measure this variable from space. An experiment was designed by the author and conducted from the Skylab satellite to measure soil moisture by remote sensing techniques. The amount of radiation emitted by the earth at wavelengths of 21 cm is highly related to soil moisture content. This radiation is quite useful since it is affected very little by the presence of vegetation and clouds. Radiation from earth of 21-cm wave length was measured by Skylab instruments and correlated with the known soil moisture content in different areas of the United States to obtain a relationship between radiation and soil moisture. Based on this relationship, soil moisture can be obtained in other areas from radiation measurements alone. Figure 15-10 shows the soil moisture con-

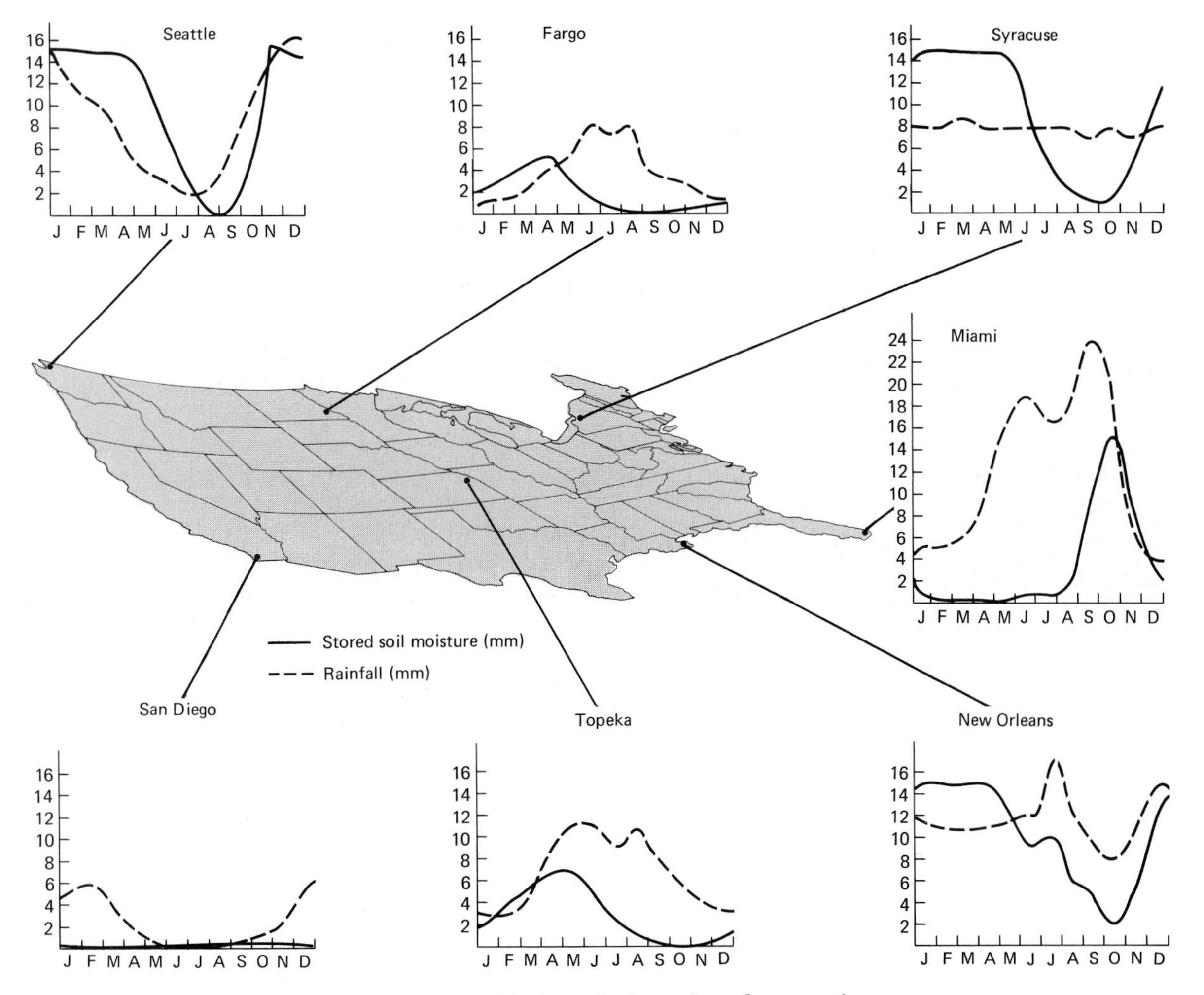

FIGURE 15-8 The amount of moisture stored in the soil, shown here for several locations, is influenced by the amount of rainfall and by water lost by evapotranspiration. Soil moisture reserves are normally depleted during the summer months and early winter because of the large demand for water during hot weather. The rainfull pattern of the extreme southwestern United States causes the soil moisture reserves to be depleted much earlier in the year.

tent for three strips of land, as determined from Skylab data obtained on three separate passes of the satellite.

There are many advantages of such remote sensing techniques, including almost instantaneous measurements and no destruction of the sample. A disadvantage is the broad area covered by satellite data, but this may represent an advantage for some applications.

Actual Evapotranspiration The **actual evapotranspiration** rate is usually less than the potential evapotranspiration rate since soil moisture is not always available in abundant quantities. Therefore

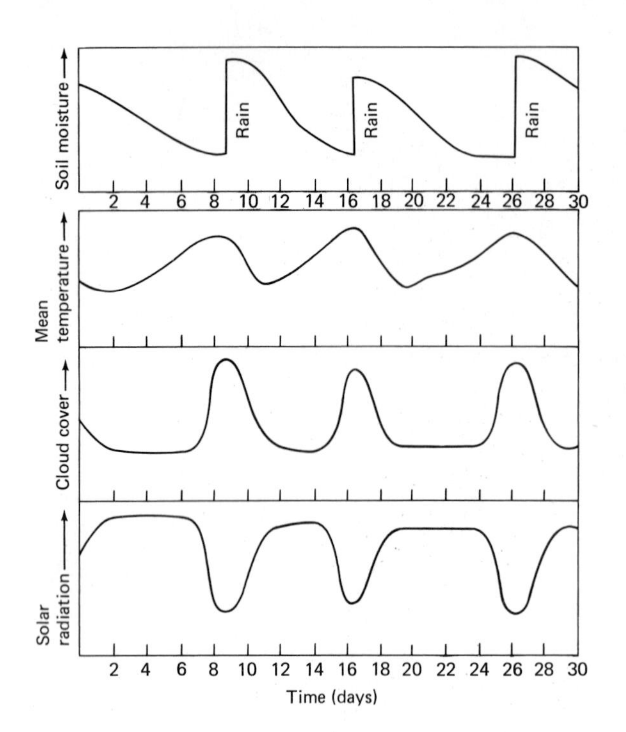

FIGURE 15-9 The soil moisture content affects various other climatic elements and may even initiate weather cycles, as illustrated here. As soil moisture reserves are depleted seven days after a rain in the summertime, for example, the mean air temperature starts to increase as less solar radiation is used for evaporating water at the surface. Warmer air temperatures near the ground may initiate convective activity and thunderstorm development because of the buoyancy of the warm air. Cloud development then affects the amount of solar radiation received, as it reflects much of the radiation. As rainfall occurs, soil moisture reserves are replenished and the mean air temperature decreases because radiation is used for evaporating water at the ground. The cycle is repeated as soil moisture reserves are again utilized. These hypothetical cycles can occur only if sufficient water vapor is in the atmosphere for the development of rainfall.

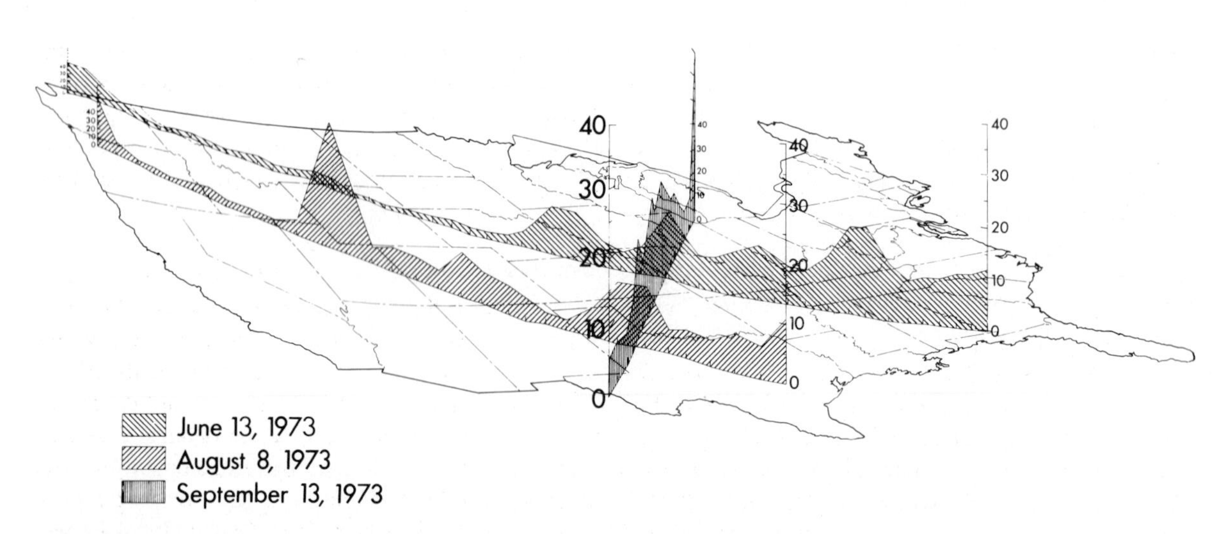

FIGURE 15-10 The distribution of soil moisture (percent by weight) was determined from radiation instruments on Skylab. Three different satellite passes were used to determine the soil moisture content for the three strips shown. (Reprinted by permission of the publisher, from J. R. Eagleman, *The Visualization of Climate,* Lexington, Mass.: Lexington Books, © D.C. Heath, 1975.)

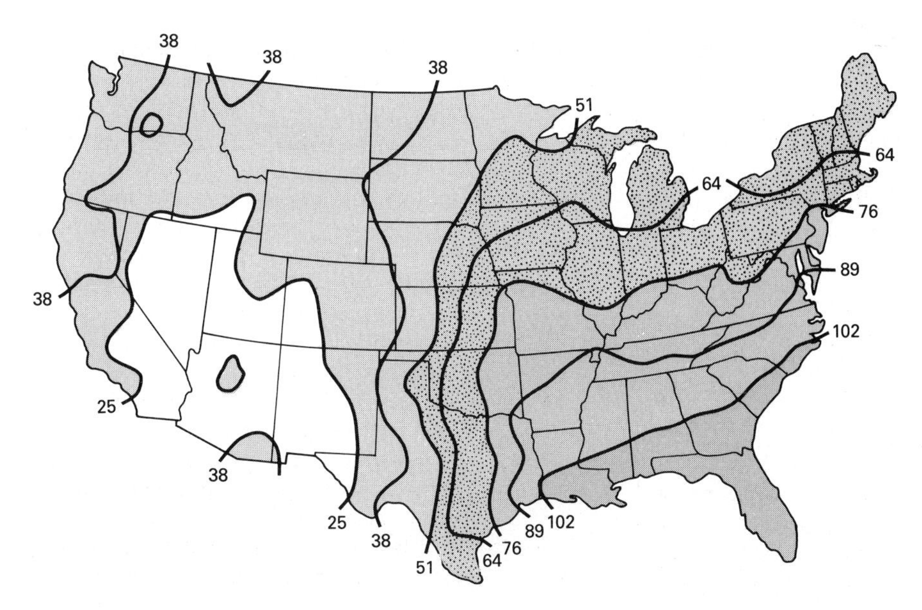

FIGURE 15-11 The annual calculated actual evapotranspiration (in cm) across the United States depends on the potential evapotranspiration and on the amount of water held in the soil to be lost to the atmosphere. The total amount of water lost to the atmosphere is least in the southwestern states. (Adapted from J. R. Eagleman, *The Visualization of Climate,* Lexington, Mass.: Lexington Books, © D.C. Heath, 1975.)

the actual evapotranspiration rate depends on the amount of water in the soil in addition to the evaporative demand of the atmosphere that determines the potential evapotranspiration rate. The actual evapotranspiration rate can also be calculated. This involves calculating the potential evapotranspiration first, and then making calculations based on information concerning the soil moisture content. The distribution of mean annual actual evapotranspiration is shown in Figure 15-11.

The actual rate is very different from the potential rate and shows water loss rates of less than 25 cm/yr in the southwestern United States and of over 100 cm/yr in the southeastern United States. The actual loss rate can be no more than the amount of water received as precipitation at any location. The lowest precipitation amount occurs at Yuma, Arizona (8 cm/yr). The actual evapotranspiration rate there is only a few centimeters per year in spite of the fact that more than 200 cm could be lost if it were available. In some localities the actual evapotranspiration may exceed the amount of rainfall for a short time as groundwater reserves are used to supply irrigation water. This cannot continue indefinitely, however, without generating critical water problems.

The Water Balance The **water balance** is a useful way of visualizing the climate of a particular location. The water balance is evaluated by comparing the potential evapotranspiration rate and the actual evapotranspiration rate with the amount of precipitation. An example of the water balance is shown in Figure 15-12. Surplus water occurs in months when the rainfall amount is higher than the potential evapotranspiration rate. During the summer months when the loss rate is greater than the rainfall rate, water stored in the soil is utilized. This allows the actual evapotranspiration to exceed the amount of precipitation and remain as great as the potential rate for a short time. The area between the actual and the potential evapotranspiration gives the amount of moisture deficiency. The area between the rainfall rate and the actual evapotranspiration rate, when it is higher, represents soil moisture utilization. A period of soil moisture recharge occurs when the rainfall rate is greater than the potential evapotranspiration rate.

The water balance can be used to compare the climate of different locations, as shown in Figure 15-13. The monthly rainfall distribution in Seattle, Washington, causes a very wet season during the winter months, with a slight moisture deficiency during the summer. Southward, at Red Bluff, California, the temperature and lower amount of rainfall combine to give an extremely dry climate during the summer months. In Los Angeles no surplus water occurs during the winter and the potential evapotranspiration is not as great during the summer. Yuma, Arizona, has only continual moisture deficits.

In the north-central United States, Duluth, Minnesota, has low potential evapotranspiration, with a monthly distribution very similar to the monthly precipitation. No moisture deficiency occurs there in an average year. Southward, at Des Moines, Iowa, the summers are warmer, with greater potential evapotranspiration and very little moisture deficit. At Topeka, Kansas, the moisture deficiency is greater during the summer, but the monthly rainfall distribution is similar in shape, providing adequate moisture for growing winter crops such as wheat. At Dallas, Texas, the moisture deficit during the summer is much greater.

In the northeastern United States, at Caribou, Maine, the average monthly precipitation is always greater than the potential evapotranspiration, thus producing a continuous surplus of water. At Rochester, New York, the monthly precipitation is more evenly distributed than at any other location in the United States, producing a unique water balance diagram. Southward, at Atlanta, Georgia, monthly precipitation amounts are higher, creating a water surplus during the winter months with some deficit during the summer. At Miami, Florida, the amount of rainfall is much greater during the summer and fall, with deficits only during the winter months. Rainfall amounts during the fall season in Miami, as well as in most other surrounding coastal areas, are influenced by the frequent passage of hurricanes with their accompanying large amounts of rainfall.

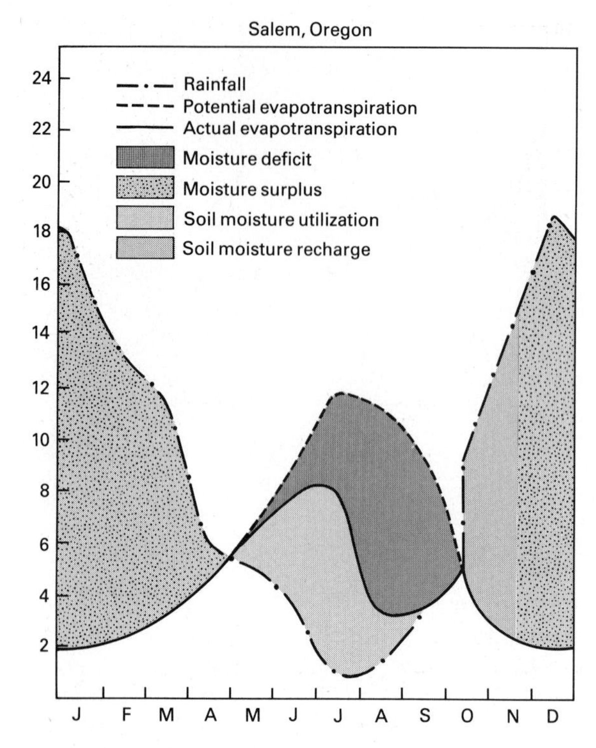

FIGURE 15-12 The water balance is determined by the amount of rainfall in comparison to the potential and actual evapotranspiration. A surplus of moisture exists for months when the rainfall rate is greater than the potential evapotranspiration. When the potential evapotranspiration is greater than the amount of rainfall, soil moisture utilization and moisture deficits occur. As the rainfall rate again rises above the potential evapotranspiration rate, periods of soil moisture recharge occur, followed by periods of surplus moisture as the soil becomes saturated. The water balance can therefore be used as a visual expression of the climate of a particular location.

Urban Climates In urban areas the various **inadvertent climate modifications** are of more concern

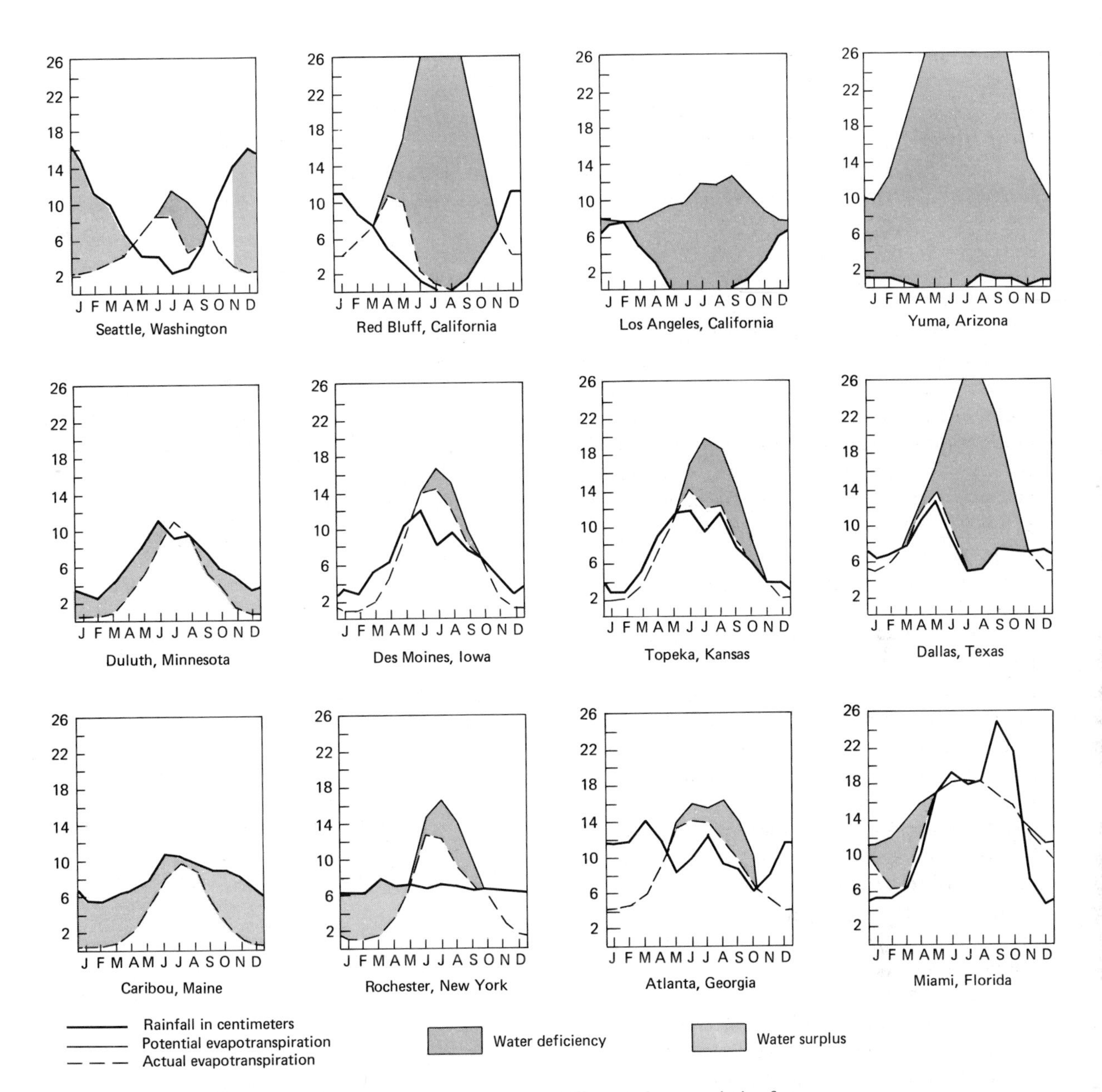

FIGURE 15-13 The water balance diagrams show very different characteristics for different locations. Large deficiencies of moisture exist at Yuma, Arizona; Red Bluff, California; Los Angeles, California; and various other locations. A continual surplus of water is common at Caribou, Maine. Other locations show oscillations between a surplus and a deficiency of water. (Adapted from J. R. Eagleman, *The Visualization of Climate,* Lexington, Mass.: Lexington Books, © D.C. Heath, 1975.)

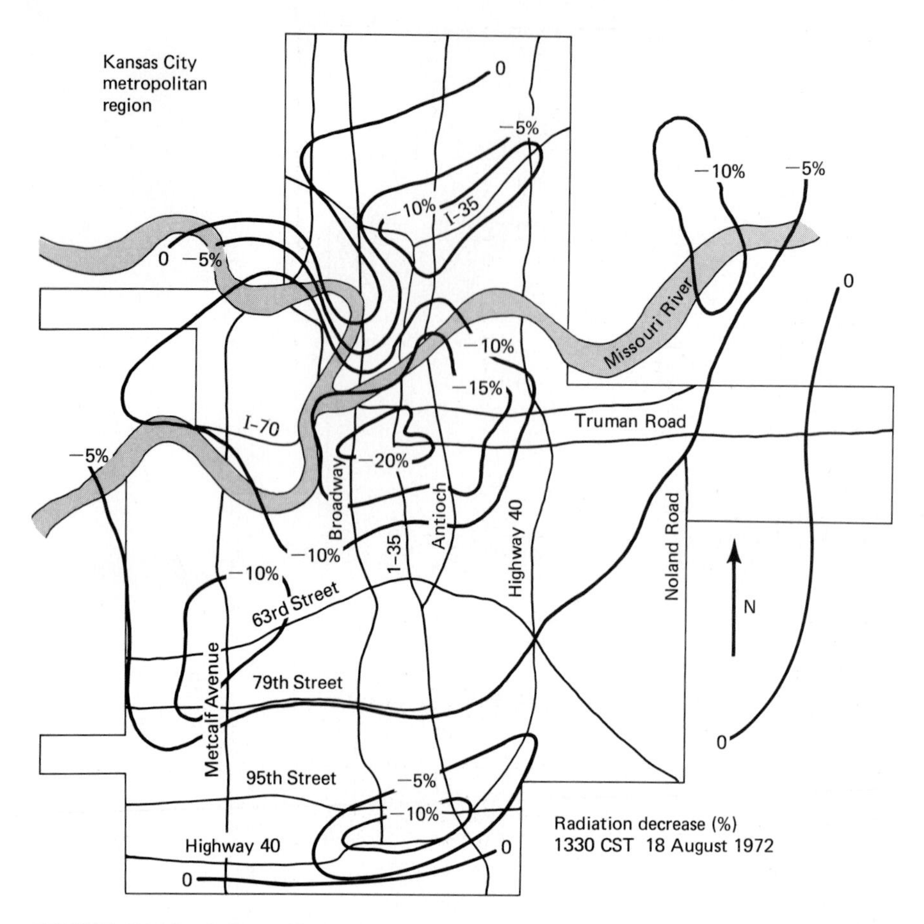

FIGURE 15-14 Solar radiation is normally decreased in urban areas. Solar radiation decreases of 20% were measured in downtown Kansas City on August 18, 1972. These radiation decreases are related to the poorer air quality in urban locations.

than the vegetation-oriented water balance. As seen in Chapter 14 a heat island is common in urban areas. Greater pollution of the urban atmosphere combined with heat-storing concrete, and multiple reflections between buildings, cause the urban climate to be different from the surrounding rural area. The temperature is not the only atmospheric property that is affected by man's activities. The lower visibility of the urban atmosphere decreases the amount of solar radiation. Measurements of this reduction are commonly as much as 20% (Fig. 15-14).

The relative humidity of urban areas is also different from surrounding areas. This is related to the temperature increase since higher temperatures reduce the relative humidity. The reduction in relative humidity and other climatic modifications in three cities are shown in Figure 15-15. The

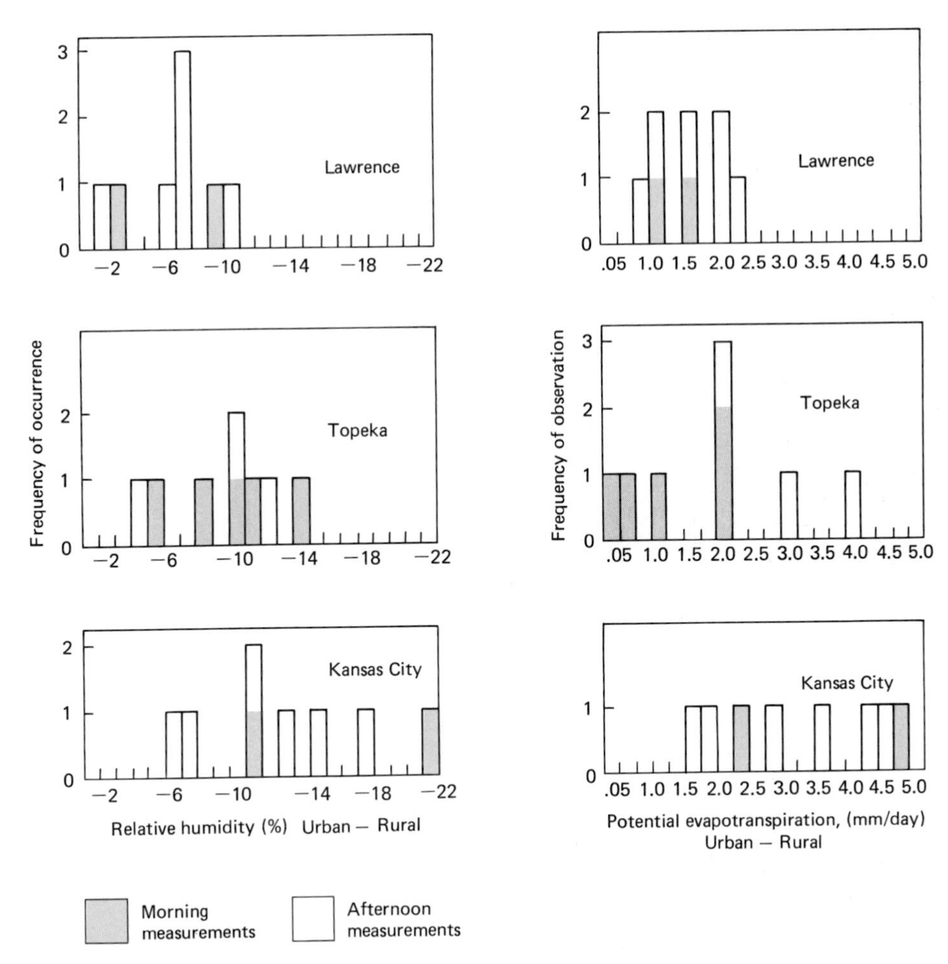

FIGURE 15-15 Measurements of the relative humidity and potential evapotranspiration in three cities of various size show considerable decreases in relative humidity, depending on city size, and increases in the potential evapotranspiration rate in urban areas. (Reprinted with permission from J. R. Eagleman, "A Comparison of Urban Climatic Modification in Three Cities," *Atmospheric Environment,* vol. 8, New York: Pergamon Press, 1974.)

greatest reduction in relative humidity was 22% measured in Kansas City, with an average reduction of about 13%. The average daytime reduction in relative humidity was about 9% in Topeka and 6% in Lawrence, Kansas.

The potential evapotranspiration rate is increased in urban areas since the temperature is greater and the relative humidity is less. The same atmospheric variables are important for human comfort since people are uncomfortable if the temperature and the humidity are too high (Fig. 15-16). This causes urban areas to be more uncomfortable during hot weather than the surrounding rural area.

The urban climate is a little more comfortable during cold weather. Wind speeds in cities are commonly reduced by the sheltering effects of large buildings by at least 5%. The amount of

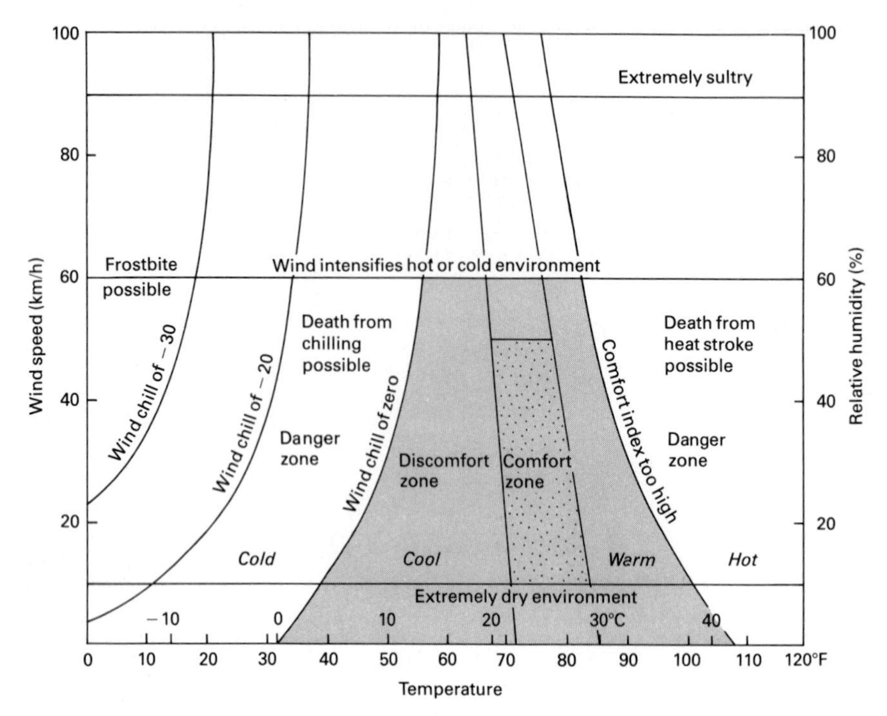

FIGURE 15-16 The human relevance of climatic ranges showing atmospheric comfort, discomfort, and danger for inhabitants. Within the comfort zone limited by temperature, wind, and relative humidity, a body feels comfortable at elevations less than 300 m with customary indoor clothing and performing sedentary or light work. Outside these limits corrective measures are necessary to restore the feeling of comfort. Outside the shaded discomfort zone the environment is dangerous and continued exposure without appropriate corrective measures may lead to injury or death.

reduction is greater with strong winds than for light winds. The wind speed is an important part of the **wind chill** factor, Table 15-1. Strong winds cause greater chilling and produce effective temperatures that are much colder than would occur with light winds. The urban heat island and lighter winds in cities combine to produce a better effective temperature for human activity during cold weather, but they also produce a more uncomfortable environment during hot weather.

The air quality of urban areas is commonly poorer than in surrounding rural areas because of greater industrial activity and more automobiles with discharges of pollutants into the atmosphere. This contributes to more frequent fogs over cities. It may also contribute to more or less precipitation in urban areas. For example, poor air quality may increase the amount of precipitation in some areas, such as in La Porte, Indiana (Fig. 15-17), while analysis of precipitation records in other locations indicate a reduction in rainfall amounts even though fog is more common. The effect of the city climate on rainfall may be similar to cloud modification by seeding, with the amount and type of cloud seeding by the pollutants determining whether more fog and tiny cloud droplets, with less rainfall, are produced or whether more large raindrops with precipitation increases are produced.

TABLE 15-1

Wind chill equivalent temperatures for various wind speeds and air temperatures

Dry bulb temperature (°C)	Wind speed (km/h)										
	6	10	20	30	40	50	60	70	80	90	100
20	20	18	16	14	13	13	12	12	12	12	12
16	16	14	11	9	7	7	6	6	5	5	5
12	12	9	5	3	1	0	0	−1	−1	−1	−1
8	8	5	0	−3	−5	−6	−7	−7	−8	−8	−8
4	4	0	−5	−8	−11	−12	−13	−14	−14	−14	−14
0	0	−4	−10	−14	−17	−18	−19	−20	−21	−21	−21
−4	−4	−8	−15	−20	−23	−25	−26	−27	−27	−27	−27
−8	−8	−13	−21	−25	−29	−31	−32	−33	−34	−34	−34
−12	−12	−17	−26	−31	−35	−37	−39	−40	−40	−40	−40
−16	−16	−22	−31	−37	−41	−43	−45	−46	−47	−47	−47
−20	−20	−26	−36	−43	−47	−49	−51	−52	−53	−53	−53
−24	−24	−31	−42	−48	−53	−56	−58	−59	−60	−60	−60
−28	−28	−35	−47	−54	−59	−62	−64	−65	−66	−66	−66
−32	−32	−40	−52	−60	−65	−68	−70	−72	−73	−73	−73
−36	−36	−44	−57	−65	−71	−74	−77	−78	−79	−79	−79
−40	−40	−49	−63	−71	−77	−80	−83	−85	−86	−86	−86
−44	−44	−53	−68	−77	−83	−87	−89	−91	−92	−92	−92
−48	−48	−58	−73	−82	−89	−93	−96	−98	−99	−99	−99
−52	−52	−62	−78	−88	−95	−99	−102	−104	−105	−105	−105
−56	−56	−67	−84	−94	−101	−105	−109	−111	−112	−112	−112
−60	−60	−71	−89	−99	−107	−112	−115	−117	−118	−118	−118

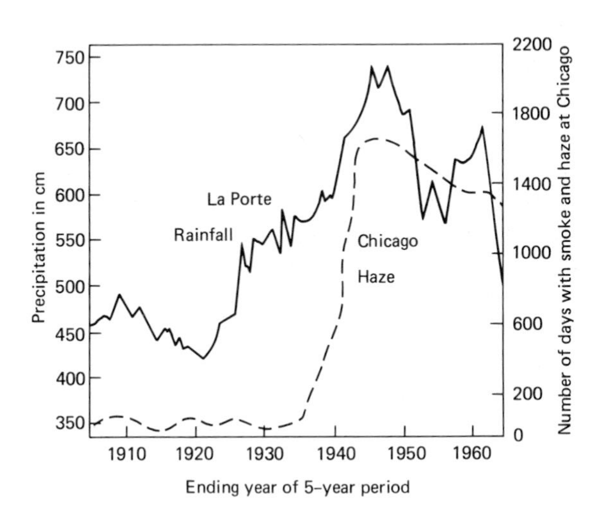

FIGURE 15-17 The effects of the urban climate on the amount of rainfall are, in general, hard to evaluate. La Porte, Indiana, is one of the few locations where rainfall amounts indicate an increase in precipitation because of urban climatic modification. (From Stanley A. Changnon, Jr., "The La Porte Weather Anomaly: Fact or Fiction?" from *Bul. of Am. Met. Soc.*, XLIX, No. 1, 1968.)

TABLE 15-2

Koeppen's climatic classification

First two symbols	Type	Criteria for first symbol
Af	Tropical rain forest	$T > 18°C$ in coldest month
Am	Tropical monsoon	
Aw	Tropical savanna	
BS	Semiarid, steppe	$R < 0.44\,(1.8T + 32 - N)$
BW	Arid, desert	$R < \frac{1}{2}\{0.44\,[1.8T + 32 - N]\}$
Cf	Humid subtropical	$T > -3°C$ but $< 18°C$ in coldest month
Cs	Summer-dry subtropical	
Cw	Winter-dry subtropical	
Df	Cold humid climate	$T > -3°C$ for coldest month and $T > 10°C$ for warmest month
Dw	Winter-dry cold climate	
ET	Tundra climate	$T < 10°C$ for warmest month
EF	Frost climate	
Second symbol	**Criteria**	
f	No dry season—$R > 3$ cm[a] all months ($N = 8.5$)	
s	Summer dry—one month $R < 3$ cm[a] ($N = 14$)	
w	Winter dry—one month $R < 0.1$ (wettest month) ($N = 3.5$)	
m	Rainfall of driest month greater than a; $a = 3.94 - R/10$	
T	$T > 0°C$ for warmest month	
F	$T < 0°C$ for warmest month	
Third symbol	**Criteria**	
a	Warmest month over 22°C	
b	Warmest month below 22°C	
c	Warmest month below 22°C; less than 4 months above 10°C	
d	Coldest month below −38°C	
h	Average annual temperature above 18°C	
k	Average annual temperature below 18°C	

[a]The value is 6 for A-type climates.

CLIMATIC CLASSIFICATION

Several different **climatic classification** schemes have been developed to group the many possible local climates into similar regions. Koeppen used monthly and annual temperature and precipitation in combination with vegetation boundaries to develop twelve different climatic types that were considered to be uniform regions. These twelve types are listed in Table 15-2 with the criteria for determining each type. The basic climatic types are distinguished by a combination of two letters. A third letter is used with some of the climatic types as a further description. The tropical climates (A) are determined by a temperature of more than 18°C in the coldest month. The desert climate (B) is determined by the annual amount of rainfall in comparison to the annual temperature. The subtropical climates (C) are determined by the temperature of the coldest month which has to be greater than −3°C but less than 18°C. The cold climates (D) are determined by a temperature less than −3°C in the coldest month and greater than 10°C in the warmest month. The tundra and frost climates (E) are determined by temperatures of less than 10°C in the warmest month.

The distribution of climates according to Koeppen's system is shown in Figure 15-18. General climatic types are frequently useful for transferring information or experience with one climatic type to

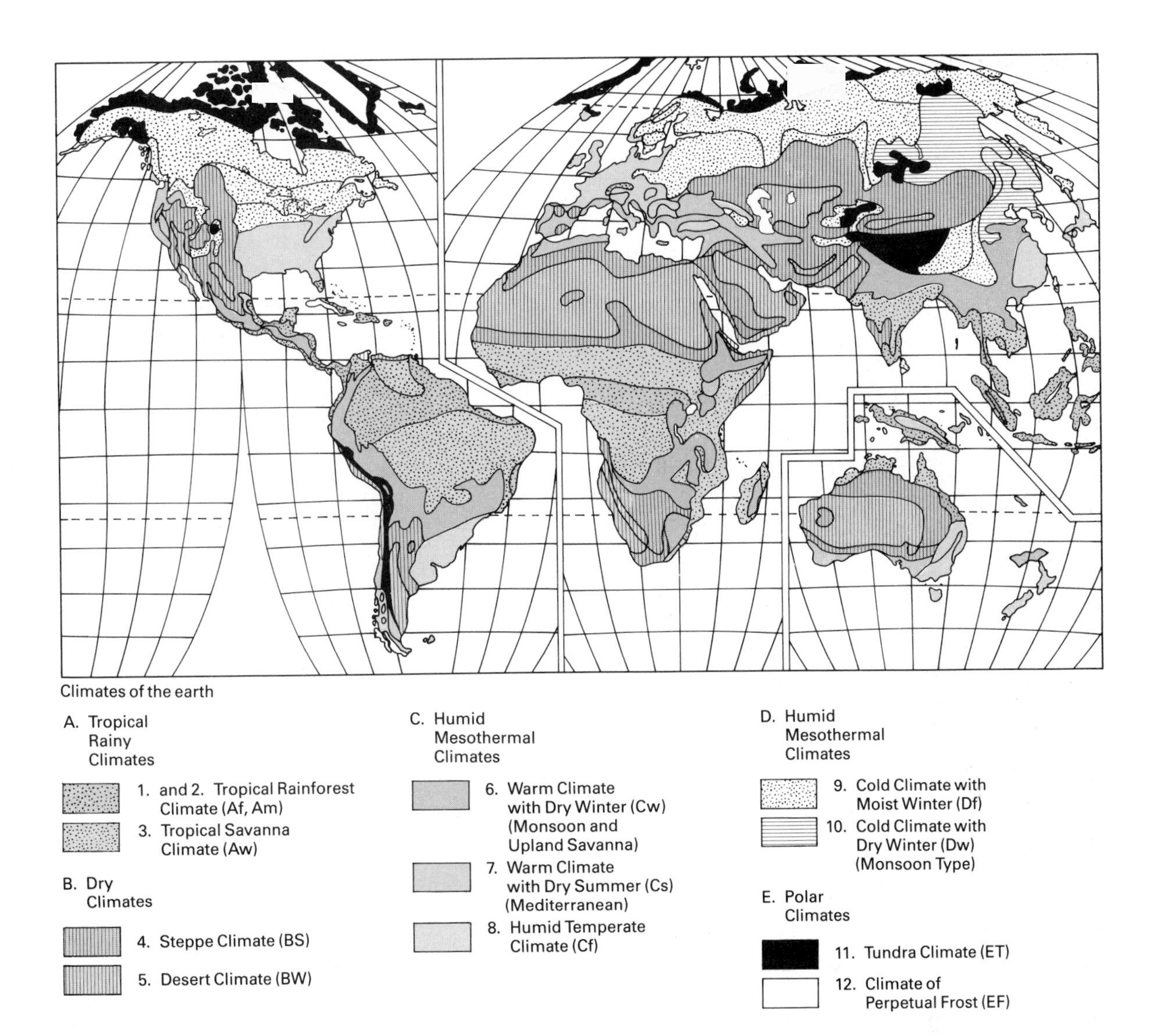

FIGURE 15-18 Climatic regions of the world, as developed by Koeppen.

different areas of the world with a similar type of climate.

A more recent climatic classification system, developed by the author, is based on a similar approach, as shown in Table 15-3. The temperature of the coldest month is used to determine whether the climate is tropical (T), warm (W), or mild (M). If the average temperature of the coldest month is less than 0°C, the temperature of the warmest month is used to specify the climate as either oscillating (O), cold (C), frigid (F), or polar (P). An index based on the potential and actual evapotranspiration is used to determine four different aridity–humidity regions. Various climatic regions of the United States according to the system are shown in Figure 15-19.

TABLE 15-3

Climatic classification system based on temperature and aridity

Temperature regions		Criteria	
Symbol	Type	Coldest month	Warmest month
T	Tropical	≥18°C	—
W	Warm	≥10°C	—
M	Mild	≥0°C	—
O	Oscillating	<0°C	≥20°C
C	Cold	<0°C	≥10°C
F	Frigid	<0°C	≥0°C
P	Polar	<0°C	<0°C

Aridity–humidity regions		Aridity–humidity index
Symbol	Type	(AHI)[a]
h	Humid	0%–25%
m	Moist	26%–50%
d	Dry	51%–75%
a	Arid	76%–100%

Rainfall distribution	
Symbol	Description
s	Summer-dry; rainfall for one month is less than 6 cm for tropical climates or 3 cm for all others.
w	Winter-dry; rainfall for one month is less than one-tenth of the wettest month.

Source: From J. R. Eagleman, *The Visualization of Climate* (Lexington, Mass.: Lexington Books, 1976).

[a]AHI = 100 (1 − AE/PE); AE = actual evapotranspiration; PE = potential evapotranspiration.

The central United States, for example, has an oscillating moist (Om) climate bordered by a mild moist (Mm) climate to the south and a cold moist (Cm) climate to the north. The eastern United States has climates varying northward from Florida with a tropical humid (Th) to a warm humid (Wh), mild humid (Mh), oscillating humid (Oh) to a cold humid (Ch) climate. Along the west coast the climate varies northward from a warm arid (Wa) to warm dry (Wd), mild dry (Md), mild moist (Mm) to mild humid (Mh).

Another approach to climatic classification was used by Terjung. This regionalization is related to human comfort through such factors as the comfort index. The comfort index is determined by the temperature and the humidity, as measured by the wet bulb temperature. The temperature and humidity were graphed and categories were assigned (Fig. 15-20). Numbers were then transferred to word definitions (given in Table 15-4), with symbols such as EH for extremely hot conditions, and so on. The other factor used by Terjung besides the comfort index was a wind chill index factor that places a number on the cooling power of the atmosphere. If it is cold and windy a person feels colder than if the temperature is the same with no wind blowing. So the cooling power of the air is important in the response of a person to the climate.

Using the two factors, comfort index and wind chill index, Terjung developed annual physioclimatic extremes, by combining climatic data for January and July. Figure 15-21 shows the distribution of annual physioclimatic extremes developed by using 20 categories. Although this is an interesting method of determining climatic regions, it has limitations. One of the problems with such a classification is that people respond differently to the same environment since response depends on the age of the individual. Older people with blood circulation problems are colder in the same environment where younger people are comfortable. Response also depends on activity since an active person will be warmer than one who is sedentary.

The climate may vary over short distances, causing any general classification system to contain considerable variation within each large climatic region. A mountain, for example, has large climatic changes with small increases in elevation. In fact, tall mountains, even in tropical regions, have rings of different climates where coffee and rice can be grown from about 0.5 to 1 km, sugar from 1 to 2 km, with maize and beans growing up to 3 km, and the snowline above 4 km. Climatic classifications of the United States are very generalized in mountain regions because the climatic change is so extreme over very short distances.

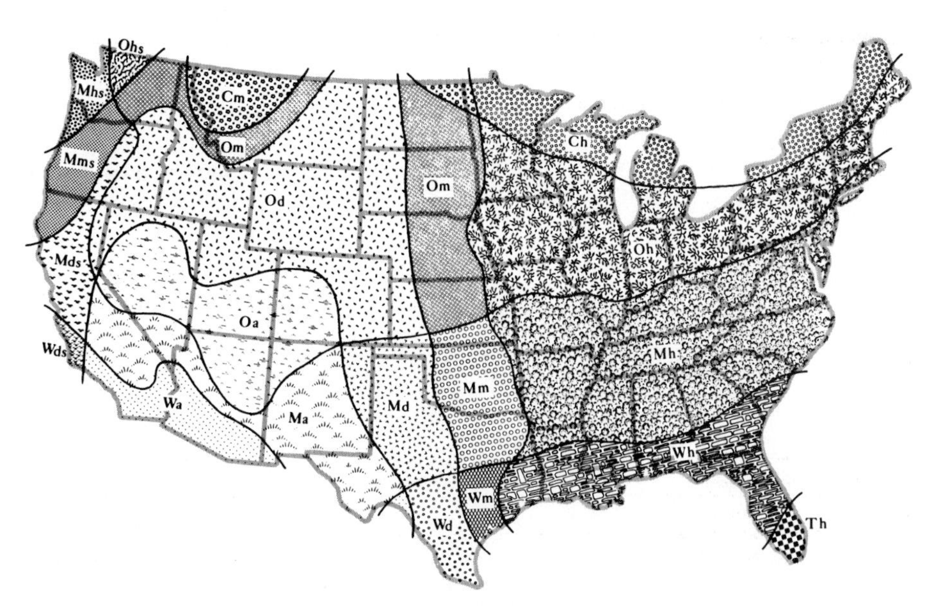

FIGURE 15-19 Climatic regions in the conterminous United States determined from temperature, aridity, and rainfall distribution. (Reprinted by permission of the publisher, from J. R. Eagleman, *The Visualization of Climate*, Lexington, Mass.: Lexington Books, © D. C. Heath, 1975.)

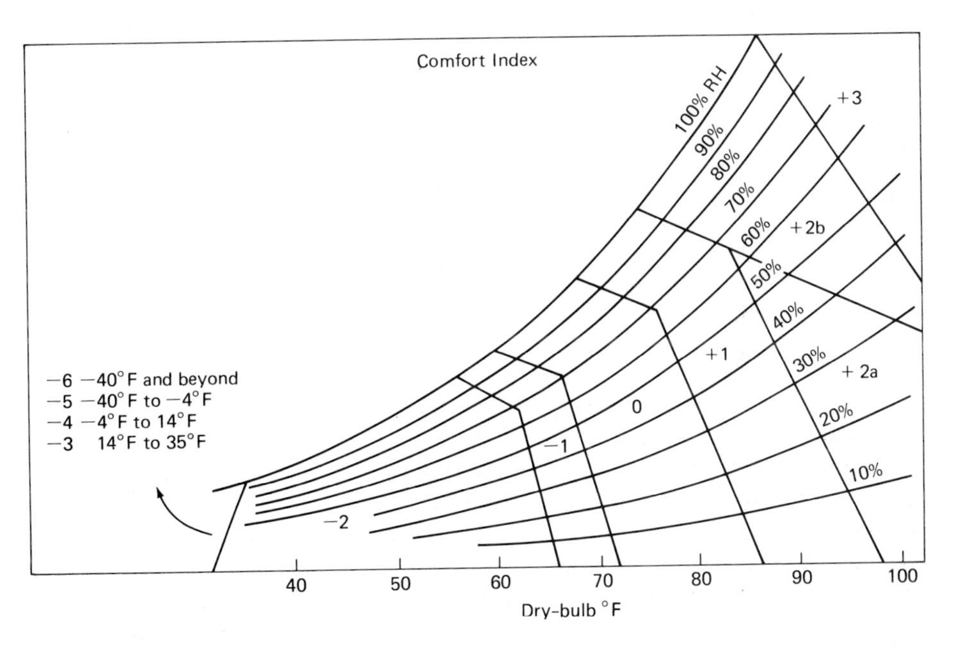

FIGURE 15-20 The categories used by Terjung are related to the comfort index and are used to develop a climatic classification system. (Reproduced by permission from the Ann. Assoc. Am. Geographers, vol. 56, 1966.)

TABLE 15-4

Terjung's climatic classification of 1966

Symbol	Type	Comfort index category (from Figure 15-20)
EH	Extremely hot	+3
S	Sultry	+2b
H	Hot	+2a
W	Warm	+1
M	Moderate	0
C	Cool	−1
K	Keen	−2
CD	Cold	−3
VC	Very cold	−4
EC	Extremely cold	−5
UC	Ultracold	−6

Nevertheless, they may be useful for transferring information or experience of the climate of one location to the climate of other parts of the world.

THE FUTURE

In the future, our understanding, and perhaps our control, of the atmosphere will further our attempts to produce ideal weather and climate. Weather forecasts will become more accurate and will be able to be made farther in advance as computer models and atmospheric measurements continue to improve. We may someday be able to plan our activities with full confidence that the weather next week will be exactly as described by the forecaster on TV. However, many new discoveries concerning the behavior of our atmosphere must be made before this will be possible.

If we do not like the forecast for tomorrow's weather, perhaps we will someday be able to order our own variety of weather. Current weather modification gives us only a hint of this possibility. Before it can become a reality, new techniques and a greater understanding of the atmosphere must be reached.

In the near future advances will be made in meteorology as the many different aspects of the atmosphere, such as the separate types of storms, the jet stream, and the intertropical convergence zone, are studied and understood more thoroughly. Further advances will be made as the separate parts of the jigsaw puzzle are fit together to form the complete picture.

Perhaps you will be one of those to discover some new secret of our atmosphere, hidden throughout the ages, that will unlock previously closed doors.

SUMMARY

The climate is the composite of all weather events that occur over an extended time period in a particular region. The climate changes due to our ever-active atmosphere. Future changes are uncertain, but since 1940 they have included temperature decreases of more than 0.5°C.

The climate of a location is determined by external influences including solar radiation, tilt of the earth's axis, and shape of the earth's orbit.

The composition of the atmosphere influences the climate through the greenhouse effect of carbon dioxide and the cooling effect of particulates in the atmosphere. Changes in the jet stream pattern occur rapidly as it shifts from one number of wave-

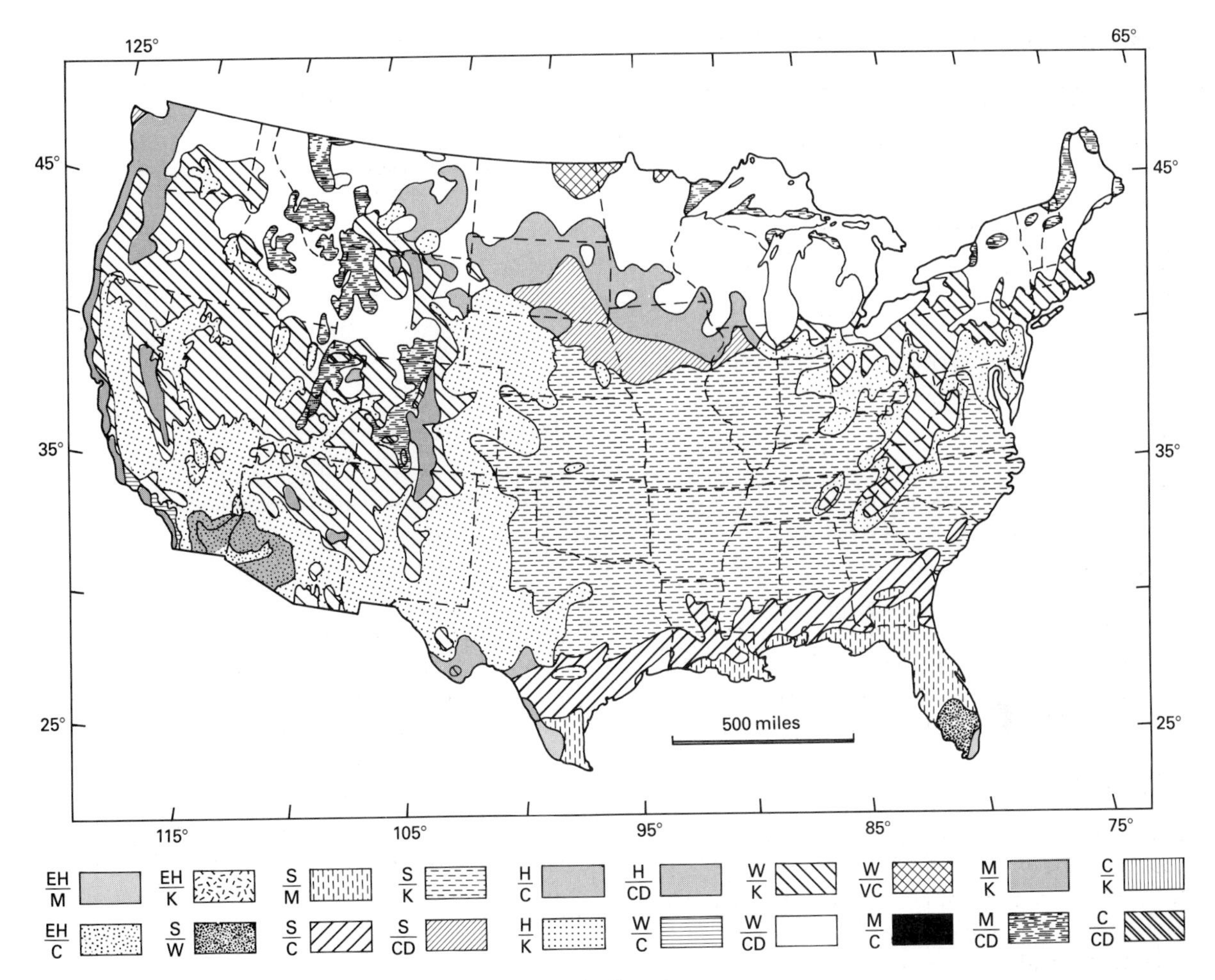

FIGURE 15-21 Terjung's distribution of annual physioclimatic extremes. (Reproduced by permission from the Ann. Assoc. Am. Geographers, vol. 56, 1966.)

lengths to another. Net cooling causes the jet stream to shift southward in the northern hemisphere, resulting in large climatic changes along its boundary zone.

Terrestrial influences on climate include the topography, especially mountain chains and ocean currents. The water balance, which includes the potential and actual evapotranspiration along with the amount of precipitation, is a very important climatic factor. The soil moisture and vegetation are important in holding soil particles on the ground, in addition to creating climatic effects.

The climate of urban areas is altered by human activities. Some of the urban climatic modifications are increased temperature, reduction in amount of solar radiation, a decrease in the relative humidity, increase in potential evapotranspiration, an increase in human discomfort during hot weather, a decrease in human discomfort during cold weather, lower air quality, and more frequent fog.

General climatic classification schemes have been developed to group the many local natural climates into similar climatic regions. The classification system developed by Koeppen is based on vegetation boundaries, with monthly and annual temperature and precipitation determining the climatic region. Another system, developed by the author, is based on critical temperatures of the warmest and coldest months, along with an index based on the potential and actual evapotranspiration. A third approach to climatic classification is that of Terjung, which is based on human comfort using the comfort index and wind chill index. Such climatic classifications are very generalized and are most useful for transferring climatic information from one location to a distant location.

STUDY AIDS

1. How does the climate differ from the weather of your county.
2. Discuss your views of future climatic trends.
3. Describe possible causes and effects of global atmospheric cooling.
4. Explain potential evapotranspiration and its evaluation.
5. Discuss the amount of available soil moisture for vegetation.
6. Discuss the importance of soil moisture and its measurement.
7. How does the actual evapotranspiration rate differ from the potential rate?
8. Explain the water balance and discuss its use in visualizing climate.
9. Discuss some of the inadvertent modifications of urban climate.
10. Briefly compare the three different climatic classification systems described and discuss their possible uses.
11. Select a topic that you feel is an important unanswered climatic question and describe how you would obtain information to answer the question.

TERMINOLOGY EXERCISE

Check the glossary if you are unsure of the meaning of any of the following terms used in Chapter 15.

Climatic control
Little ice age
Climatocracy
Synergistic effects
Potential evapotranspiration
Stomata
Wilting point
Field capacity
Feedback mechanism
Actual evapotranspiration
Water balance
Inadvertent climate modification
Wind chill
Climatic classification

THOUGHT QUESTIONS

1. What is the particular climatic type for your location? Explain what this tells you.
2. Do you feel that past climatic trends are a useful indication of the type of climate to be expected in the future? Give reasons for your answer.
3. List and comment on as many different climatic factors as possible that are influenced by human activity.
4. Do you expect significant changes in the climate of the near future because of volcanic eruptions? Explain your answer.
5. Do you feel that most people could agree on what the ideal climate should be? Discuss the ramifications of your answer.

SPECIAL TOPIC 17

SCIENTIST CLOCKS SEVENTEENTH-CENTURY SOLAR WIND

A puzzling seventeenth-century gap in solar activity may hold clues to predicting the earth's climate.

A scientist with the National Oceanic and Atmospheric Administration (NOAA) has found that in the late seventeenth-century the solar wind, the stream of energetic particles that the sun spits out, slowed to a comparative breeze, and may have been linked to the occurrence on earth of the "little ice age."

Such a pause in solar activity—and its attendant effects on earth—could happen again at any time, according to Dr. Stephen Suess, a researcher with the Commerce Department agency's Space Environment Laboratory.

For 70 years, from about 1645 to 1715, the 11-year sunspot cycle considered as normal seems to have disappeared. The few sunspots that did appear were small and short-lived. Displays of the aurora on earth, which are tied to sunspots, faded, and the solar corona, the halo of glowing gas that becomes visible round the sun during eclipses, was smaller and duller than it is today. This period of solar quiet is called the Maunder Minimum, after the English astronomer who first postulated it.

At the same time, the earth was suffering through the coldest part of the little ice age, a period of long, severe winters that lasted from about 1450 to 1850. Suess is one of a growing number of scientists who believe this was no coincidence.

Evidence is accumulating that solar activity affects earth's weather and climate. With the solar wind an obvious intermediary between the sun and earth, understanding what this wind was like during the Maunder Minimum could give clues as to how changes on the sun affect earth.

The solar wind normally races outward from the sun at a rate of 700 to 800 km/sec (a million and a half miles an hour) in some places. In other places it may idle along at a mere 300 km/sec (670,000 mi/h).

Suess, after studying a variety of evidence, including data gathered by Dr. John Eddy of the National Center for Atmospheric Research, has concluded that the solar wind during the Maunder Minimum was comparatively thin, slow, and featureless.

In this same period, other data indicates the bombardment of the earth by cosmic rays was more intense than at any time since. When solar activity is high, according to Suess, cosmic ray intensity decreases. The dearth of sunspots and the disappearance of an apparent corona led him to infer that the magnetic field of the solar corona was weakened, affecting the interplanetary magnetic field (the extension of the solar field that reaches to earth and beyond) in such a way as to admit more cosmic radiation.

The aurora on earth not only are tied to sunspots, but are directly related to the speed of the solar wind. The slower the wind, the fewer auroral displays at lower latitudes. Historical records suggest that the zones where

(Continued on next page)

aurora are visible shrank closer to the poles during the Maunder Minimum, disappearing from the skies in temperate latitudes, where their appearance probably would have been noted and recorded.

Everything points, the Commerce Department scientist believes, to a low-velocity solar wind and a faint corona with a weak magnetic field. The wind would have been essentially featureless as well as slow—a homogeneous flow blowing outward with equal force in all directions.

These results are of more than historical interest, according to Suess. "The sun can no longer be thought of as a non-varying star on time scales even as short as decades," he said. "It is not unlikely that even over the coming 50 years there will be significant changes in average properties of the solar wind. Observing and understanding these changes will make it possible to better understand past phenomena such as the Maunder Minimum and, more importantly, to anticipate these changes and gain the benefits of better preparation for their consequences."

From Louise Purrett-Carroll, *U.S. Dept. of Commerce News.*

APPENDIX A EQUATIONS

Hydrostatic Equation The hydrostatic equation relates atmospheric pressure to height. The pressure at any height in the atmosphere is equal to the weight of the atmosphere above that height. In the absence of vertical motion the downward forces must be equal to the upward forces. We can use this fact to calculate the relationship between height and pressure. Consider a column of air, or unit cross section, extending from the earth through the entire atmosphere.

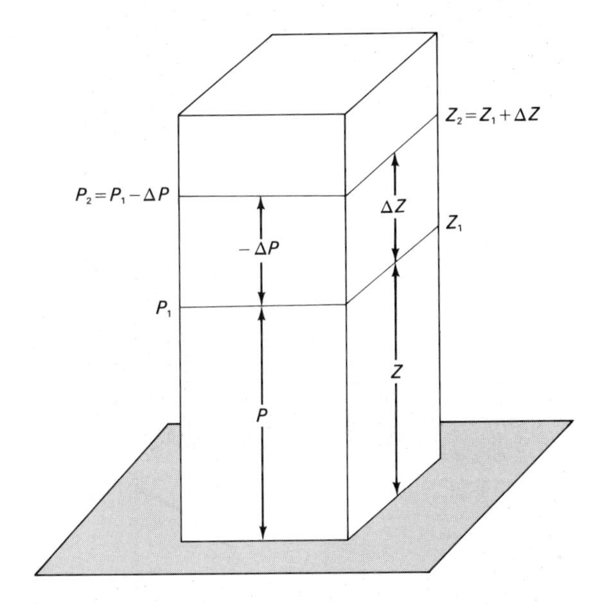

The downward force is the gravitational pull on the atmospheric mass and the upward force arises from the decreasing pressure upward. Assume the air has a density (ρ) defined as the amount of mass per unit volume. The mass of air between heights Z_1 and Z_2 in the column can be expressed as the product of the density (ρ) and ΔZ since we are using a cross-sectional area of unity. ΔZ indicates a small difference in height. The downward force on the air at height Z is the acceleration of gravity (g) multiplied by the mass, or $g\rho\Delta Z$. At equilibrium this downward force must be balanced by the upward force arising from the pressure decrease, $-\Delta P$. This balance of forces means that

$$-\Delta P = g\rho\Delta Z \text{ or } \Delta P/\Delta Z = -g\rho$$

This is the **hydrostatic equation** that is very useful in atmospheric pressure and height relationships. It tells us that for constant density an exact relationship exists between height and pressure. As height increases pressure decreases.

Example: Calculate the height of the 500-mb level if the average density of the air from the surface through the layer to 500 mb is 0.001 g/cm³. The acceleration of gravity is 980 cm/sec². The height of the 500-mb level (ΔZ) is 500,000 dyn/cm² ($-\Delta P$) divided by the product of 980 cm/sec² (g) multiplied by 0.001 g/cm³ (ρ). Since 1 dyne equals 1 gcm/sec², the height of the 500-mb level is 510,200 cm, 5102 m, or 5.102 km. The assumption of an average density from the surface to 500 mb is, of course, an extreme simplification, since density decreases rapidly with height.

Radiation Laws The wavelength (λ) is a basic property of radiation. It is related to the amount of oscillation or frequency (f) by the speed of light (c):

$$c = \lambda f$$

where $c = 3 \times 10^{10}$ cm/sec.

Objects that are perfect emitters of radiation, that is, that behave as black bodies, absorb all the energy falling on them and radiate energy at a rate (E) determined by the fourth power of their absolute temperature (T):

$$E = \sigma T^4$$

Where σ is a constant equal to 5.67×10^{-8} W/m² deg⁴ and the rate of radiation is in watts per square meter. This equation is called the **Stefan-Boltzmann law.**

Since the sun and earth both emit radiation approximately as black bodies, this law can be applied to their radiation.

At a particular wavelength the **emissivity** (ϵ)—amount of energy emitted by an object in comparison to black body emission—may be less than unity. In this case the gray body emission is

$$E = \epsilon\sigma T^4$$

Another radiation law, **Kirchhoff's law,** states that the emissivity of radiation at a particular wavelength must be equal to the absorptivity (a)—the fractional amount of energy absorbed in comparison to a black body. Thus strong absorbers are also strong emitters at that particular wavelength:

$$a_\lambda = \epsilon_\lambda$$

The wavelength of maximum emission (λ_{max}) for a black body can be calculated in micrometers from **Wien's law** if the temperature (T) in degrees absolute is known:

$$\lambda_{max} = 2897/T$$

Thus hotter objects have greatest emission of radiation at a shorter wavelength than colder objects.

Equation of State Some of the characteristics of gases were discovered by Boyle, who found in 1662 that at constant temperature (T), the volume (V) per unit mass (M) of gas is inversely proportional to the pressure (P). Since density (ρ) is the mass per unit volume, the pressure varies directly with density ($P = C\rho$), where C is a constant. More than a century later, in 1787, J. A. C. Charles discovered that at constant pressure, the volume of a fixed mass of any gas increases by the same amount for every degree rise in temperature. Expressed in terms of density, an increase in temperature results in a decrease in density of the gas.

Boyle's law and **Charles' law** can be combined to form the **ideal gas law** or **equation of state:**

$$P\,V/M = RT \quad \text{or} \quad P = \rho RT$$

where R is a constant value of 2.87×10^6 erg/g K in the cgs (centimeter–gram–second) system, or 287 J/kg K in the mks (meter–kilogram–second) system. The equation of state expresses quantitatively the relationship between the various properties of gases and allows the calculation of any one variable of the equation if the other two are known.

Conservation of Absolute Vorticity The vorticity of air is the amount of circulation in a given area. The total or absolute vorticity is the combined circulation of air with respect to the earth plus the amount of spin or circulation of the earth. The principle of conservation of absolute vorticity (V_{abs}) states that the total vorticity of an air mass remains constant even though the air mass vorticity (V_{air}) or the earth's vorticity (V_{earth}) may change:

$$V_{abs} = V_{air} + V_{earth}$$

Since the earth has greater vorticity at the poles than at the equator, a moving air mass may have considerably different vorticity with respect to the earth while its absolute vorticity remains constant. This concept leads to an explanation of the long waves in the air currents.

Pressure Gradient Acceleration The atmospheric pressure is determined by the gravitational pull on the mass of air above a particular location. Since the atmosphere is a gas, the resulting pressure at the earth's surface acts equally in all directions. If the mass of air above two locations is different then a horizontal pressure difference exists and a greater force may be exerted on one side of an air parcel than on the other. The magnitude of the acceleration resulting from this force can be calculated by the following consideration.

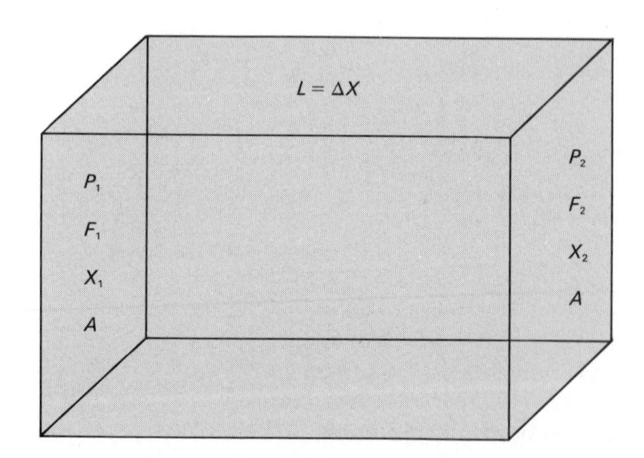

Assume that a given chunk of the atmosphere of cross-sectional area (A) and density (ρ) is affected by forces F_1 and F_2 to produce different pressures P_1 and P_2. Since pressure is force per unit area, $P_1 = F_1/A$ and

$P_2 = F_2/A$. The force per unit mass in the atmosphere is directly proportional to the pressure difference ($P_2 - P_1$) and inversely proportional to the distance between them (L), the density, and the area:

$$(F_2 - F_1)/A\rho L = (P_2 - P_1)/\rho L$$

The pressure gradient is $(P_2 - P_1)/L$ or $\Delta P/\Delta X$ since it measures the difference in pressure per unit distance. The acceleration of the air due to the pressure gradient (A_{pg}) must be equal to the force per unit mass and is

$$A_{pg} = -\frac{1}{\rho}\frac{\Delta P}{\Delta X}$$

The minus sign indicates that the horizontal acceleration is opposite in direction to the direction of increasing pressure, that is, from higher to lower pressure. If the pressure gradient were the only factor causing acceleration of the air, movement would be perpendicular to the isobars on a weather map, with winds stronger where the isobars are closer together.

Coriolis Acceleration Air does not travel directly from high to low pressure because of the influence of the earth's rotation. At the North and South Pole the spin about a vertical axis (Ω) is one rotation per day or 7.29×10^{-5} rad/sec, while the spin about a vertical axis is zero at the equator. The rotation about a vertical axis at any latitude (ϕ) is $\Omega \sin\phi$. G. G. de Coriolis (1792–1843) used mathematics to describe accelerations on the rotating earth and showed that the acceleration also depends on the velocity (V) of the wind. The **Coriolis acceleration** (A_{Cor}) is

$$A_{Cor} = 2V\Omega \sin\phi$$

The magnitude of the Coriolis acceleration is thus determined by the wind velocity, latitude, and rotational speed of the earth. The direction of the acceleration is always at right angles to the path of motion, and is directed to the right of a moving air stream (facing the direction the air is going) in the northern hemisphere and to the left in the southern hemisphere.

Centrifugal Acceleration Newton's first law of motion tells us that the velocity of a moving object does not change as long as there are no unbalanced forces acting on it. Air parcels that follow a curved path are changing direction and must therefore be experiencing acceleration. For a ball rotating on a string, a centripetal acceleration supplied by the string holds the ball against the **centrifugal acceleration** that would cause the ball to travel in a straight path if the string should break. The centrifugal acceleration (A_{cent}) is directed outward from the center of rotation with a magnitude directly related to the velocity (V) of the object and inversely related to the radius of curvature (r):

$$A_{cent} = V^2/r$$

Thus the centrifugal acceleration can be calculated and is small for a large radius of curvature (such as for the midlatitude cyclone) but becomes more important than the Coriolis acceleration as the radius of curvature decreases to that of the hurricane and especially to that of a tornado.

Equation of Motion Newton's second law tells us that the total acceleration (A_{total}) of an object must equal the forces applied per unit mass. The horizontal acceleration of air results from the combined effect of the pressure gradient acceleration, Coriolis acceleration, centrifugal acceleration, and frictional deceleration (A_f).

$$A_{total} = A_{pg} + A_{Cor} + A_{cent} + A_f$$

Since the pressure gradient, Coriolis, and centrifugal accelerations have been quantified (in preceding sections), the equation of motion for a pressure gradient along a distance ΔX becomes

$$A_{total} = -\frac{1}{\rho}\frac{\Delta P}{\Delta X} + 2V\Omega \sin\phi + V^2/r - A_f$$

The geostrophic wind can be expressed quantitatively by assuming that the only accelerations are due to the pressure gradient and the earth's rotation and that these two accelerations balance. The gradient wind can be expressed quantitatively by assuming that the pressure gradient acceleration, Coriolis acceleration, and centrifugal acceleration balance. Cyclostrophic flow is quantified if the pressure gradient acceleration equals the centrifugal acceleration and these accelerations balance. In each case the equation is solved for the velocity (V) to give the particular wind.

First Law of Thermodynamics When a quantity of heat (ΔH) is supplied to a gas certain physical changes take place. These changes are expressed by altered temperature, pressure, or volume of the gas. The magnitude of the change is related to a constant called the specific heat of air at constant pressure (C_p) or at constant volume (C_v). If α is the volume per unit mass, then heat added to a gas increases the internal energy ($C_v \Delta T$) and does work by expansion ($P\Delta\alpha$):

$$\Delta H = C_v \Delta T + P\Delta\alpha$$

If the gas is not free to expand, then volume changes are replaced by pressure changes:

$$\Delta H = C_p \Delta T - \alpha \Delta P$$

These two equations are both forms of the **first law of thermodynamics.** Since temperature and pressure changes can be measured in the atmosphere much more easily than volume changes, the second equation is more useful for atmospheric applications.

Dry Adiabatic Lapse Rate If a gas undergoes physical changes by a process where no heat is added to it or removed from it, the process is called adiabatic. An example of such a process is the change in temperature of a rising or descending parcel of unsaturated air that does not mix or exchange heat with its surroundings. The first law of thermodynamics and the hydrostatic equation can be used to calculate the value of the expansion cooling rate for rising air or compression heating rate for descending air. This rate of temperature change is called the **dry adiabatic lapse rate.** If no heat is added or lost, $\Delta H = 0$ and the first law of thermodynamics becomes

$$C_p \Delta T = \alpha \Delta P$$

This tells us that the air cools if the pressure is lowered and heats if the pressure rises. We can convert the pressure change to height change from the hydrostatic equation, since $\Delta P = -\rho g \Delta Z$. Since $\alpha\rho = 1$, we can rearrange the expression into a form to give the lapse rate ($-\Delta T/\Delta Z$) denotes by Γ:

$$\Gamma = \frac{g}{C_p} = \frac{9.80 \text{ m/sec}^2}{1004 \text{ J/°C kg}} = 9.8\text{°C/km}$$

The dry adiabatic lapse rate (Υ) thus has a value of 9.8°C/km or approximately 1 Celsius degree for each 100 m change in height.

Height of Cloud Base Measurement of the surface air temperature (T) and dew point temperature (D) can be used to obtain a good estimate of the height of the cumulus cloud base. Rising unsaturated air will cool at 1°C/100 m, so the temperature at the cloud base (T_c) of height (h) in meters will be:

$$T_c = T - \left[\frac{1}{100}\right] h$$

The dew point temperature decreases with height at a constant rate of 0.2°C/100 m. Thus the dew point at the cloud base (D_c) of height (h) is

$$D_c = D - \left[\frac{0.2}{100}\right] h$$

Cumulus clouds will form in rising air currents when the air cools to saturation, that is, the temperature equals the dew point temperature. Thus T_c must equal D_c at the cloud base:

$$T - \left[\frac{1}{100}\right] h = D - \left[\frac{0.2}{100}\right] h$$

This can be reduced to:

$$h = 125\,(T - D)$$

This equation can then be used to calculate the height of the cloud base in meters from surface temperature and dew point measurements.

Snell's Law When light passes from one medium to another, the rays are usually bent at the surface separating the two mediums. Consider a flat water surface with light rays making an angle i (angle of incidence) with an imaginary line perpendicular to the water surface. After entering the water, the light will travel a new direction, making a different angle (j) with the same perpendicular line. **Snell's law** states that the sine of the two angles must be equal:

$$\frac{\sin i}{\sin j} = n$$

where n is a constant for a given material and is called the index of refraction. This constant is equal to 1.33 for water. The index of refraction of light is related to the density of the two mediums and is slightly different for different wavelengths. The value for air is approximately unity, with significant variations for air of varying density.

APPENDIX B
UNITS AND CONVERSION FACTORS

The metric system of measurement has been the accepted system throughout the world except in English-speaking countries, but now these too are slowly converting to this system. The basic unit of length is the meter while the liter is used for volume and gram for mass. Prefixes are used to designate powers of 10 for the different basic units. For example, a kilometer is 1000 meters and a millimeter is one-thousandth of a meter. These and other prefixes are listed here.

METRIC UNITS

p	pico	one-trillionth	10^{-12}
n	nano	one-billionth	10^{-9}
μ	micro	one-millionth	10^{-6}
m	milli	one-thousandth	10^{-3}
c	centi	one-hundredth	10^{-2}
d	deci	one-tenth	10^{-1}
da	deka	ten	10
h	hecto	one hundred	10^{2}
k	kilo	one thousand	10^{3}
M	mega	one million	10^{6}
G	giga	one billion	10^{9}
T	tera	one trillion	10^{12}

CONVERSION FACTORS

Multiply	by	to obtain
acres	0.40469	hectares
atmospheres	76.0	centimeters of mercury
atmospheres	1.0133×10^{6}	dynes centimeter^2
atmospheres	29.921	inches of mercury
atmospheres	101.33	kilopascals
atmospheres	1,013.3	millibars
atmospheres	14.696	pounds/ inches^2
calories	4.186	watt-seconds or joules
calories	1.1628×10^{-3}	watt-hours

(Continued on next page)

CONVERSION FACTORS *(Continued)*

Multiply	by	to obtain
centimeters	0.3937	inches
centimeters	1×10^4	micrometers
centimeters of mercury	13.333	millibars
feet	30.48	centimeters
feet	0.3048	meters
grams	0.035274	ounces
inches	2.540	centimeters
inches	25,400	micrometers
inches of mercury	33.867	millibars
Joules	0.2389	calories
kilowatts	0.23889	kilocalories/second
kilometers	0.62137	miles
knots	1.1516	miles/hour
knots	1.8533	kilometers/hour
liters	0.26418	gallons
meters	3.2808	feet
meters	1.0936	yards
meters/second	2.237	miles/hour
micrometers	1×10^{-4}	centimeters
micrometers	3.937×10^{-5}	inches
micrometers	1×10^{-6}	meters
miles, U.S., statute	1.6093	kilometers
miles (statute)/hour	0.44704	meters/second
millibars	0.02953	inches of mercury
millibars	0.0750	centimeters of mercury
pounds	453.59	grams
pounds	0.45359	kilograms
square centimeters	0.155	square inches
square kilometers	247.10	acres
square miles	2.59	square kilometers
yards	0.9144	meters

APPENDIX C
WEATHER MAP SYMBOLS AND ENTRIES

Weather data are sent to other locations in coded form. The symbolic form of the message and a sample coded message, which is a series of five numbers, is illustrated here for one weather station (Washington, D.C., Number 405). The numbers, indicating sky cover, wind direction, and so on, are decoded and plotted in specific locations on a weather map, as indicated in the sample plotted report for Washington, D.C. The symbolic station model shows the locations and symbolic representation of the various weather data. For example, *WW* stands for *present weather.* From the table for present weather, in this appendix, it can be seen that the coded message (71 in the example; row 70, column 1 in the table for present weather) indicates continuous snow. This is plotted with the symbol for snow in the sample plotted report. Any one of the 100 symbols shown in the table for present weather can be plotted in this location. Other tables in this appendix give detailed explanations for other symbols shown in the symbolic station model.

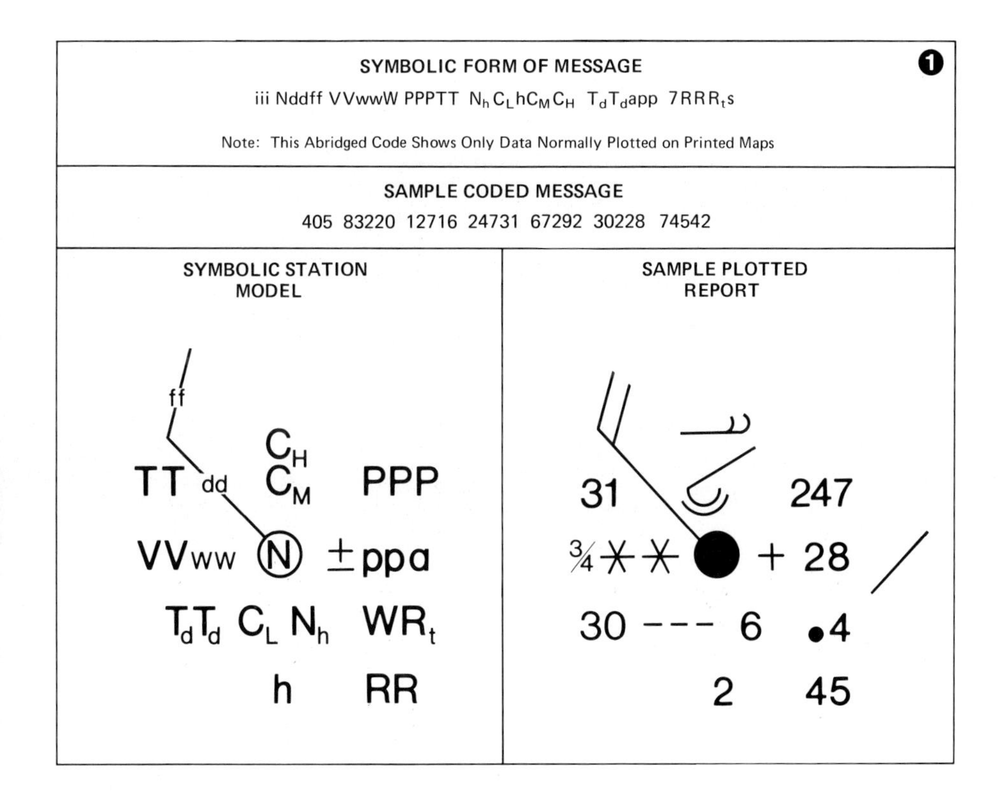

Front symbols are given below:

▲▲▲▲	Cold front (surface)	▼●▼●	Stationary front (surface)
●●●●	Warm front (surface)	◠◠◠◠	Warm front (aloft)
●▲●▲	Occluded front (surface)	△△△△	Cold front (aloft)

A front which is disappearing or is weak and decreasing in intensity is labeled "Frontolysis."

A front which is forming is labeled "Frontogenesis."

A "squall line" is a line of thunderstorms or squalls usually accompanied by heavy showers and shifting winds, and is indicated as

—— ·· —— ·· —— ·· ——

The paths followed by individual disturbances are called storm tracks and are shown as ➝ ➝. The symbols ☒ indicate past positions of the low pressure center at 6-hour intervals.

"HIGH" (H) and "LOW" (L) indicate the centers of high and low barometric pressure.

EXPLANATION OF SYMBOLS AND MAP ENTRIES ❷

Symbols in order as they appear in the message	Explanation of symbols and decode of example above	Remarks on coding and plotting
iii	Station number 405 = Washington	Usually printed on manuscript maps below station circle. Omitted on Daily Weather Map in favor of printed station names.
N	Total amount of cloud 8 = completely covered	Observed in tenths of cloud cover and coded in Oktas (eighths) according to code table in block ❻. Plotted in symbols shown in same table.
dd	True direction from which wind is blowing 32 = 320° = NW	Coded in tens of degrees and plotted as the shaft of an arrow extending from the station circle toward the direction from which the wind is blowing.
ff	Wind speed in knots 20 = 20 knots	Coded in knots (nautical miles per hour) and plotted as feathers and half-feathers representing 10 and 5 knots, respectively, on the shaft of the wind direction arrow. See block ❾.
VV	Visibility in miles and fractions 12 = 12/16 or 3/4 miles	Decoded and plotted in miles and fractions up to 3 1/8 miles. Visibilities above 3 1/8 miles but less than 10 miles are plotted to the nearest whole mile. Values higher than 10 miles are omitted from the map.
ww	Present weather 71 = continuous slight snow	Coded in figures taken from the "ww" table (block ❽) and plotted in the corresponding symbols same block. Entries for code figures 00, 01, 02, and 03 are omitted from this map.
W	Past weather 6 = rain	Coded in figures taken from the "W" table (block ⓫) and plotted in the corresponding symbols same block. No entry made for code figures 0, 1, or 2.
PPP	Barometric pressure (in millibars) reduced to sea level 247 = 1024.7 mb.	Coded and plotted in tens, units, and tenths of millibars. The initial 9 or 10 and the decimal point are omitted.
TT	Current air temperature 31 = 31°F.	Coded and plotted in actual value in whole degrees F.

EXPLANATION OF SYMBOLS AND MAP ENTRIES ❷ (continued)		
Symbols in order as they appear in the message	Explanation of symbols and decode of example above	Remarks on coding and plotting
N_h	Fraction of sky covered by low or middle cloud 6 = 0.7 or 0.8	Observed and coded in tenths of cloud cover. Plotted on map as code figure in message. See block ❼.
C_L	Cloud type 7 = Fractostratus and/or Fractocumulus of bad weather (scud)	Predominating clouds of types in C_L table (block ❸) are coded from that table and plotted in corresponding symbols.
h	Height of base of cloud 2 = 300 to 599 feet	Observed in feet and coded and plotted as code figures according to code table in block ❺.
C_M	Cloud type 9 = Altocumulus of chaotic sky	See C_M table in block ❸.
C_H	Cloud type 2 = Dense cirrus in patches	See C_H table in block ❸.
T_dT_d	Temperature of dew point 30 = 30° F.	Coded and plotted in actual value in whole degrees F.
a	Characteristic of barograph trace 2 = rising steadily or unsteadily	Coded according to table in block ❿ and plotted in corresponding symbols.
pp	Pressure change in 3 hours preceding observation 28–2.8 millibars	Coded and plotted in units and tenths of millbars.
7	Indicator figure	Not plotted.
RR	Amount of precipitation 45 = 0.45 inch	Coded and plotted in inches to the nearest hundredth of an inch.
R_t	Time precipitation began or ended 4 = 3 to 4 hours ago	Coded and plotted in figures from table in block ❹.
s	Depth of snow on ground	Not plotted.

CLOUD ABBREVIATION
St or Fs-Stratus or Fractostratus
Ci-Cirrus
Cs-Cirrostratus
Cc-Cirrocumulus
Ac-Altocumulus
As-Altostratus
Sc-Stratocumulus
Ns-Nimbostratus
Cu or Fc-Cumulus or Fractocumulus
Cb-Cumulonimbus

C_L		DESCRIPTION (Abridged From W.M.O. Code)	C_M		DESCRIPTION (Abridged From W.M.O. Code)	C_H		DESCRIPTION (Abridged From W.M.O. Code) ❸
1		Cu of fair weather, little vertical development and seemingly flattened	1		Thin As (most of cloud layer semi-transparent)	1		Filaments of Ci, or "mares tails," scattered and not increasing
2		Cu of considerable development, generally towering, with or without other Cu or Sc bases all at same level	2		Thick As, greater part sufficiently dense to hide sun (or moon), or Ns	2		Dense Ci in patches or twisted sheaves, usually not increasing, sometimes like remains of Cb; or towers or tufts
3		Cb with tops lacking clear-cut outlines, but distinctly not cirriform or anvil-shaped; with or without Cu, Sc, or St	3		Thin Ac, mostly semitransparent; cloud elements not changing much and at a single level	3		Dense Ci, often anvil-shaped, derived from or associated with Cb
4		Sc formed by spreading out of Cu; Cu often present also	4		Thin Ac in patches; cloud elements continually changing and/or occurring at more than one level	4		Ci, often hook-shaped, gradually spreading over the sky and usually thickening as a whole
5		Sc not formed by spreading out of Cu	5		Thin Ac in bands or in a layer gradually spreading over sky and usually thickening as a whole	5		Ci and Cs, often in converging bands, or Cs alone; generally overspreading and growing denser; the continuous layer not reaching 45° altitude
6		St or Fs or both, but no Fs of bad weather	6		Ac formed by the spreading out of Cu	6		Ci and Cs, often in converging bands, or Cs alone; generally overspreading and growing denser; the continuous layer exceeding 45° altitude
7		Fs and/ or Fc of bad weather (scud)	7		Double-layered Ac, or a thick layer of Ac, not increasing; or Ac with As and/or Ns	7		Veil of Cs covering the entire sky
8		Cu and Sc (not formed by spreading out of Cu) with bases at different levels	8		Ac in the form of Cu-shaped tufts or Ac with turrets	8		Cs not increasing and not covering entire sky
9		Cb having a clearly fibrous (cirriform) top, often anvil-shaped, with or without Cu, Sc, St, or scud	9		Ac of a chaotic sky, usually at different levels; patches of dense Ci are usually present also	9		Cc alone or Cc with some Ci or Cs, but the Cc being the main cirriform cloud

R_t	TIME OF PRECIPITATION ❹	h	HEIGHT IN FEET (Rounded Off)	HEIGHT IN METERS (Approximate) ❺	N	SKY COVERAGE (Total Amount) ❻	N_h	SKY COVERAGE (Low and/or Middle Clouds) ❼
0	No Precipitation	0	0-149	0-49		No clouds	0	No clouds
1	Less than 1 hour ago	1	150-299	50-99		Less than one-tenth or one-tenth	1	Less than one-tenth or one-tenth
2	1 to 2 hours ago	2	300-599	100-199		Two-tenths or three-tenths	2	Two-tenths or three-tenths
3	2 to 3 hours ago	3	600-999	200-299		Four-tenths	3	Four-tenths
4	3 to 4 hours ago	4	1,000-1,999	300-599		Five-tenths	4	Five-tenths
5	4 to 5 hours ago	5	2,000-3,499	600-999		Six-tenths	5	Six-tenths
6	5 to 6 hours ago	6	3,500-4,999	1,000-1,499		Seven-tenths or eight-tenths	6	Seven-tenths or eight-tenths
7	6 to 12 hours ago	7	5,000-6,499	1,500-1,999		Nine-tenths or overcast with openings	7	Nine-tenths or overcast with openings
8	More than 12 hours ago	8	6,500-7,999	2,000-2,499		Completely overcast	8	Completely overcast
9	Unknown	9	At or above 8,000, or no clouds	At or above 2,500, or no clouds		Sky obscured	9	Sky obscured

	0	1	2	3
00	Cloud development NOT observed or NOT observable during past hours	 Clouds generally dissolving or becoming less developed during past hour	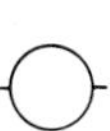 State of sky on the whole unchanged during past hour	 Clouds generally forming or developing during past hour
10	Light fog	Patches of shallow fog at station, NOT deeper than 6 feet on land	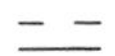 More or less continuous shallow fog at station, NOT deeper than 6 feet on land	 Lightning visible, no thunder heard
20	 Drizzle (NOT freezing and NOT falling as showers) during past hour, but NOT at time of observation	 Rain (NOT freezing and NOT falling as showers) during past hour, but NOT at time of observation	 Snow (NOT falling as showers) during past hour, but NOT at time of observation	 Rain and snow (NOT falling as showers) during past hour, but NOT at time of observation
30	 Slight or moderate dust storm or sand storm, has decreased during past hour	 Slight or moderate dust storm or sand storm, no appreciable change during past hour	 Slight or moderate dust storm or sand storm, has increased during past hour	 Severe dust storm or sand storm, has decreased during past hour
40	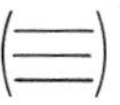 Fog at distance at time of observation, but NOT at station during past hour	Fog in patches	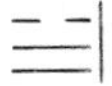 Fog, sky discernible, has become thinner during past hour	 Fog, sky NOT discernible, has become thinner during past hour
50	 Intermittent drizzle (NOT freezing) slight at time of observation	Continuous drizzle (NOT freezing) slight at time of observation	 Intermittent drizzle (NOT freezing) moderate at time of observation	 Intermittent drizzle (NOT freezing), moderate at time of observation
60	 Intermittent rain (NOT freezing), slight at time of observation	 Continuous rain (NOT freezing), slight at time of observation	 Intermittent rain (NOT freezing), moderate at time of observation	 Continuous rain (NOT freezing), moderate at time of observation
70	 Intermittent fall of snowflakes, slight at time of observation	 Continuous fall of snowflakes, slight at time of observation	 Intermittent fall of snowflakes, moderate at time of observation	 Continuous fall of snowflakes, moderate at time of observation
80	 Slight rain shower(s)	 Moderate or heavy rain showers(s)	 Violent rain shower(s)	 Slight shower(s) of rain and snow mixed
90	 Moderate or heavy shower(s) of hail, with or without rain or rain and snow mixed, not associated with thunder	 Slight rain at time of observation; thunderstorm during past hour, but NOT at time of observation	 Moderate or heavy rain at time of observation, thunderstorm during past hour, but NOT at time of observation	 Slight snow or rain and snow mixed or hail at time of observation, thunderstorm during past hour, but not at time of observation

4	5	6	7
 Visibility reduced by smoke	 Haze	 Widespread dust in suspension in the air, NOT raised by wind, at time of observation	 Dust or sand raised by wind at time of observation
 Percipitation within sight, but NOT reaching the ground	 Precipitation within sight, reaching the ground but distant from station	 Precipitation within sight, reaching the ground, near to but NOT at station	 Thunder heard, but no precipitation at the station
 Freezing drizzle or freezing rain (NOT falling as showers) during past hour, but NOT at time of observation	 Showers of rain during past hour, but NOT at time of observation	 Showers of snow, or of rain and snow, during past hour, but NOT at time of observation	 Showers of hail, or of hail and rain, during past hour, but NOT at time of observation
 Severe dust storm or sand storm, no appreciable change during past hour	 Severe dust storm or sand storm, has increased during past hour	 Slight or moderate drifting snow, generally low	 Heavy drifting snow, generally low
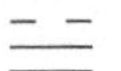 Fog, sky discernible, no appreciable change during past hour	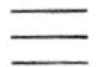 Fog, sky NOT discernible, no appreciable change during past hour	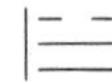 Fog, sky discernible, has begun or become thicker during past hour	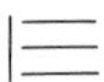 Fog, sky NOT discernible, has begun or become thicker during past hour
 Intermittent drizzle (NOT freezing), thick at time of observation	 Continuous drizzle (NOT freezing), thick at time of observation	 Slight freezing drizzle	 Moderate or thick freezing drizzle
 Intermittent rain (NOT freezing), heavy at time of observation	 Continuous rain (NOT freezing), heavy at time of observation	 Slight freezing rain	 Moderate or heavy freezing rain
 Intermittent fall of snowflakes, heavy at time of observation	 Continuous fall of snowflakes, heavy at time of observation	 Ice needles (with or without fog)	 Granular snow (with or without fog)
 Moderate or heavy shower(s) of rain and snow mixed	 Slight snow shower(s)	 Moderate or heavy snow shower(s)	 Slight shower(s) of soft or small hail with or without rain or rain and snow mixed
 Moderate or heavy snow, or rain and snow mixed or hail at time of observation; thunderstorm during past hour, but NOT at time of observation	 Slight or moderate thunderstorm without hail, but with rain and or snow at time of observation	 Slight or moderate thunderstorm, with hail at time of observation	 Heavy thunderstorm, without hail, but with rain and/or snow at time of observation

8	9

	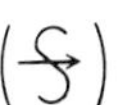
Well developed dust devil(s) within past hour	Duststorm or sandstorm within sight of or at station during past hour

Squall(s) within sight during past hour	Funnel cloud(s) within sight during past hour
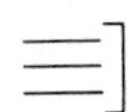	
Fog during past hour, but NOT at time of observation	Thunderstorm (with or without precipitation) during past hour, but NOT at time of observation

Slight or moderate drifting snow, generally high	Heavy drifting snow, generally high

Fog, depositing rime, sky discernible	Fog, depositing rime, sky NOT discernible

Drizzle and rain, slight	Drizzle and rain, moderate or heavy

Rain or drizzle and snow, slight	Rain or drizzle and snow, mod. or heavy

Isolated starlike snow crystals (with or without fog)	Ice pellets (sleet, U.S. definition)

Moderate or heavy shower(s) of soft or small hail with or without rain or rain and snow mixed	Slight shower(s) of hail, with or without rain or rain and snow mixed, not associated with thunder

Thunderstorm combined with dust storm or sand storm at time of observation	Heavy thunderstorm with hail at time of observation

(Descriptions Abridged from W.M.O. Code)

ff	Miles (Statute) Per Hour	Knots	⑨ Kilometers Per Hour
	Calm	Calm	Calm
	1-2	1-2	1-3
	3-8	3-7	4-13
	9-14	8-12	14-19
	15-20	13-17	20-32
	21-25	18-22	33-40
	26-31	23-27	41-50
	32-37	28-32	51-60
	38-43	33-37	61-69
	44-49	38-42	70-79
	50-54	43-47	80-87
	55-60	48-52	88-96
	61-66	53-57	97-106
	67-71	58-62	107-114
	72-77	63-67	115-124
	78-83	68-72	125-134
	84-89	73-77	135-143
	119-123	103-107	144-198

Code Number	a	Barometric Tendency (10)	
0	⁄\	Rising, then falling	
1	⁄‾	Rising, then steady; or rising, then rising more slowly	Barometer now higher than 3 hours ago
2	/	Rising steadily, or unsteadily	
3	✓	Falling or steady, then rising; or rising, then rising more quickly	
4	—	Steady, same as 3 hours ago	
5	\⁄	Falling, then rising, same or lower than 3 hours ago	
6	_	Falling, then steady; or falling, then falling more slowly	Barometer now lower than 3 hours ago
7	\	Falling steadily, or unsteadily	
8	^\	Steady or rising, then falling; or falling, then falling more quickly	

Code Number	W	Past Weather (11)	
0		Clear or few clouds	Not plotted
1		Partly cloudy (scattered) or variable sky	
2		Cloudy (broken) or overcast	
3	S/+	Sandstorm or duststorm, or drifting or blowing snow	
4	≡	Fog, or smoke, or thick dust haze	
5	,	Drizzle	
6	•	Rain	
7	*	Snow, or rain and snow mixed, or ice pellets (sleet)	
8	∇	Shower(s)	
9	⊤ʀ	Thunderstorm, with or without precipitation	

BIBLIOGRAPHY AND RESOURCE MATERIALS

BOOKS

Anthes, R. A., H. A. Panofsky, J. J. Cahir, and A. Rango. *The Atmosphere.* Columbus, Ohio: Charles E. Merrill, 1978.

Bach, W. *Atmospheric Pollution.* New York: McGraw-Hill, 1972.

Barry, R. G., and R. J. Chorley. *Atmosphere, Weather, and Climate.* New York: Holt, Rinehart and Winston, 1970.

Battan, L. T. *Harvesting the Clouds.* Garden City, N.Y.: Anchor Books, Doubleday, 1969.

Bentley, W. A., and W. J. Humphreys. *Snow Crystals.* New York: Dover, 1962.

Chalmers, J. A. *Atmospheric Electricity.* New York: Pergamon Press, 1968.

Dunn, G. E., and B. I. Miller. *Atlantic Hurricanes.* Baton Rouge, La.: Louisiana State University Press, 1964.

Eagleman, J. R. *Severe and Unusual Weather.* New York: Van Nostrand Reinhold, 1983.

Eagleman, J. R. *The Visualization of Climate.* Lexington, Mass.: Lexington Books, D. C. Heath, 1976.

Eagleman, J. R., V. U. Muirhead, and N. Willems. *Thunderstorms, Tornadoes and Building Damage.* Lexington, Mass.: Lexington Books, D. C. Heath, 1975.

Edinger, J. G. *Watching the Winds.* Garden City, N.Y.: Doubleday, 1967.

Fleagle, R. G., J. A. Crutchfield, R. W. Johnson, and M. F. Abdo. *Weather Modification in the Public Interest.* Seattle: University of Washington Press, 1974.

George, J. J. *Weather Forecasting for Aeronautics.* New York: Academic Press, 1960.

Gokhale, N. R. *Hailstorms and Hailstone Growth.* Albany: State University of New York Press, 1975.

Goody, M., and C. G. Walker. *Atmospheres.* Englewood Cliffs, N.J.: Prentice-Hall, 1972.

Huschke, E., editor. *Glossary of Meteorology.* Boston, Mass.: American Meteorological Society, 1959.

Inadvertent Climate Modification. Report of the Study of Man's Impact on Climate (SMIC). Cambridge, Mass.: M.I.T Press,1971.

International Cloud Atlas (complete and abridged editions). World Meteorological Organization, Geneva, Switzerland, 1956.

LaChapelle, E. R. *Field Guide to Snow Crystals.* Seattle: University of Washington Press, 1969.

Landsberg, H. E. *Weather and Health.* Garden City, N.Y.: Anchor Books, Doubleday, 1969.

Lehr, P. E., R. W. Burnett, and H. S. Zim. *Weather.* New York: Golden Press, 1957.

Lowry, W. P. *Weather and Life, An Introduction to Biometeorology.* New York: Academic Press, 1969.

Ludlam, F. H. *Clouds and Storms: The Behavior and Effect of Water in the Atmosphere.* State College: Pennsylvania State University Press, 1980.

Mason, J. *Clouds, Rain and Rainmaking.* New York: Cambridge University Press, 1962.

Mather, J. R. *Climatology Fundamentals and Applications.* New York: McGraw-Hill, 1974.

Maunder, W. J. *The Value of Weather.* London: Methuen and Co., 1970.

Middleton, W. E. K. *The Invention of the Meteorological Instruments.* Baltimore: The Johns Hopkins University Press, 1969.

Miller, A., and J. C. Thompson. *Elements of Meteorology.* Columbus, Ohio: Charles E. Merrill, 1975.

Neiburger, M., J. G. Edinger, and W. D. Bonner. *Understanding Our Atmospheric Environment.* San Francisco: W. H. Freeman, 1973.

Reiter, R. *Jet-Stream Meteorology.* Chicago: The University of Chicago Press, 1963.

Riehl, H. *Introduction to the Atmosphere.* New York: McGraw-Hill, 1972.

Schonland, B. *The Flight of the Thunderbolts.* Oxford: Clarendon Press, 1964.

Stern, A. C., H. C. Wohlers, R. W. Boubel, and W. P. Lowry, *Fundamentals of Air Pollution.* New York: Academic Press, 1973.

U. S. Weather Bureau–Federal Aviation Agency. *Aviation Weather.* Washington, D. C., 1965.

Wallace, J. M., and P. V. Hobbs. *Atmospheric Science: An Introductory Survey.* New York: Academic Press, 1977.

Widger, K., Jr. *Meteorological Satellites.* New York: Holt, Rinehart and Winston, 1966.

PERIODICALS (NONTECHNICAL)

Weatherwise, published bimonthly by Heldref Publications, 400 Albemarle St., N.W., Washington, D.C., 20016.

Weather, published monthly by the Royal Meteorological Society, London, England.

U.S. Department of Commerce, NOAA, Washington, D.C.: Average Monthly Weather Outlook, Climates of the States, Climatic Charts for the United States, Climatological Data for the U.S. by States, Daily Weather Map, Monthly Climatic Data for the World.

For other government pamphlets write for PL48 (Price List 48-Weather, Astronomy, Meteorology), Superintendent of Documents, Government Printing Office, Washington, D.C. 20402.

FILMS (16 MM COLOR AND SOUND)

Many excellent films are available that are appropriate for an introductory meteorology class. Some of these are listed next, together with the lending agency.

Motion Picture Service
Department of Commerce—NOAA
12231 Wilkins Av.
Rockville, Md. 20852
Tel. (301) 443-8411
- Floods (15 min)
- Hurricane Decision (14 min)
- Tornado (15 min)
- Twister (27 min)
- Neosho (Tornado Preparedness) (24 min)
- GATE to World Weather (28 min)

Modern Learning Aids
P.O. Box 302
Rochester, N.Y. 14603
Tel. (716) 467-3400

or

Universal Education and Visual Arts
100 Universal City Plaza
Universal City, Calif. 91608
Tel. (213) 985-4321
- Above the Horizon (20 min)
- Formation of Raindrops (20 min)
- Solar Radiation I (20 min)
- Solar Radiation II (20 min)
- Atmospheric Electricity (20 min)
- Convective Clouds (20 min)
- Sea Surface Meteorology (20 min)
- Planetary Circulation of the Atmosphere (20 min)
- It's an Ill Wind (air pollution) (20 min)

McGraw-Hill Text Films
330 West 42nd St.
New York, N.Y.
- The Inconstant Air (27 min)
- The Flaming Sky
- Can We Control the Weather? (25 min)

Films available through the Local Civil Defense
- It Happened in Texas (Hurricane Beulah) (10 min)
- A Lady Called Camille (24 min)
- Storm (Hurricane Agnes) (28 min)
- Survival In a Winter Storm (27 min)
- Tornado (17 Min)
- Your Chance to Live Series (14 min each)
  - Hurricane
  - Pollution
  - Flood
  - Tornado
  - Heatwave
  - Winter Storm
  - Psychological Response

Film Librarian
Aetna Life and Casualty
151 Farmington Av.
Hartford, Conn. 06115
Tel. (203) 275-0123
Hurricane (27 min)

Learning Corporation of America
717 Fifth Av.
New York, N.Y. 10022
Tel. (212) 397-9360
Pollution of the Upper and Lower Atmosphere (20 min)
Urban Impact on Weather and Climate (20 min)

Encyclopaedia Britannica Educational Corp.
425 N. Michigan Av.
Chicago, Ill. 60611
Storms—The Restless Atmosphere (20 min)
Weather Forecasting (20 min)
What Makes Rain (20 min)

Environmental Films, Inc.
Box 302
St. Johnsbury, Ver. 05819
The Tornado, Approaching the Unapproachable (23 min)

NASA films can be obtained from
Lyndon B. Johnson Space Center
Photographic Technical Laboratory
Audiovisual Office, Code JL-13
Houston, Tex. 77058
Hurricane Below, HQ233 (14 min)
Tornado Below, HQ246 (14 min)
Pollution Below, HQ247 (14 min)
Flood Below, HQ249 (16 min)
Research in the Atmosphere, HQa180 (25 min)

Many of these and other films are available from
National Audiovisual Center (GSA)
Film Order Desk
National Archives and Records Service
Washington, D.C. 20409

GLOSSARY/INDEX

Absolute instability A general stability category that exists when the measured lapse rate is greater than the dry adiabatic lapse rate. 122, 317

Absolute stability A general stability category that exists when the measured lapse rate is less than the saturated lapse rate. 122, 316, 317

Absorption bands Narrow bands in the electromagnetic spectrum where radiation is almost completely absorbed. 70

Absorptivity, 66

Acid rain, 151

Actual evapotranspiration The rate of water lost from vegetation and soil, ordinarily at a slower rate than the potential rate. 343, 344

Adiabatic, 116

Advection fog Fog formed by winds that contrast in temperature with the earth's surface, producing saturated air with the formation of liquid droplets. 147

AFOS An acronym for automation of field operations and services. It consists of minicomputers linked in a nationwide network to weather service offices and observing stations. 57, 59

Afternoon mixing depth The depth of the lower atmosphere where mixing of the air and pollutants occurs as the ground warms. 321

Afternoon ventilation An air pollution index related to the mixing depth and wind speed. 331

Air discharge A lightning discharge that occurs as the electrical charge is dissipated in the air before reaching the ground. 229

Air mass A large expanse of air having similar temperature and humidity at any given height. 6, 17, 155, 200

Air mass thunderstorm A thunderstorm that forms in a warm moist air mass. 199

Albedo The percentage of electromagnetic radiation that is reflected by an object. 66, 71–74

Aleutian low, 100

Anemometer An instrument that measures wind speed. 6, 38

Anticyclone Region of high atmospheric pressure, with winds around it in a clockwise direction in the northern hemisphere. 14, 164, 315, 316

Anticyclonic Clockwise airflow around high-pressure centers in the northern hemisphere. 14, 315

Anvil The cirrus outflow forming the top of a large thunderstorm and giving it an anvil shape. 26, 131, 134, 201

Aristotle, 3, 4, 5

Atmosphere The gaseous fluid surrounding a planet. The earth's atmosphere is composed primarily of nitrogen and oxygen. 7–12

Atmospheric acoustics, 307

Atmospheric density, 77, 78

Atmospheric displays, 294

Atmospheric electricity, 243

Atmospheric lapse rate Change in air temperature with height. See **lapse rate.** 113

Atmospheric models The atmosphere may be modeled (simulated) by mathematical equations or by physical models; colored water may be used to simulate fluid aspects of air, for example. 212, 214, 217

Atmospheric pressure The downward force of the air per unit area, averaging 101.3 kPa at sea level. 12, 42, 99–101

Atmospheric processes, 3

Atmospheric science, 3

Atmospheric window A region of the electromagnetic spectrum from 8 to 12 μm where the atmosphere is transparent to radiation. 70

Auroras Atmospheric displays caused by excited gases in the earth's magnetic field. 8, 10

Autoconvective lapse rate A lapse rate of 34.2°C/km (0.3°C/10 m) that represents constant density with altitude. 296

Backing Counterclockwise shift in wind direction. 86

Ball lightning An unusual form of lightning that consists of a ball of electricity or charged air. 229

Banding features The nature of the cloud bands of hurricanes that are useful in satellite techniques for hurricane study. 194

Barogram An atmospheric pressure trace on a chart produced by a self-registering barometer (barograph). 38

Beneficial competition A hail formation concept in which the number of hail embryos may be increased by cloud seeding to increase the number and reduce the size of hail. 287

Bergeron-Findeisen precipitation process The raindrop formation process, including growth of ice crystals at the expense of liquid water droplets. 141

Blackbody An object or substance that emits the maximum possible amount of radiation. 66, 362

Blackbody emission Maximum amount of radiation emission possible for a given temperature. 66

Blizzard A very cold wind with blowing dry snow. Freezing winds are at least 60 km/h in speed. 3, 171–74

Bolt from the blue A lightning discharge that occurs as the conducting channel extends as far as 15 km horizontally away from the thunderstorm. 229

Boyle's law A basic law of chemistry discovered by Sir Robert Boyle that relates the pressure of a gas to its volume at constant temperature. *See also* **ideal gas law** and **equation of state.** 362

Buys-Ballot law A law that says if you stand with your back to the wind in the northern hemisphere, the area of lower pressure is to your left. 13

Carbon dioxide cloud Clouds formed of carbon dioxide or dry ice, such as those on Mars. 32

Cascade effect Effect caused as large raindrops break apart as their size increases; further growth of these separate drops may cause them to split again and again producing a cascade effect. 143

Ceilometer An instrument for measuring the height of cloud base by means of a light beam. 128

Celsius scale The standard temperature scale in the metric system, with the freezing point at 0 and boiling point of water at 100. 6

Centrifugal acceleration The outward acceleration of a rotating body that would cause it to go in a straight line if the force holding the body in a circular path were eliminated. 87, 89

Chinook wind A very hot dry wind that forms as air descends on the leeward side of mountains; also called *foehn* in some areas. 135, 136

Cirrus High clouds composed of ice crystals. 23

Climatic classification A grouping of similar local climates to form a larger climatic region. 352–57

Climatic control A factor or activity that causes the climate to have its particular characteristics. 385

Climatocracy The political and economic aspects of a nation's climate. 337

Climatological data, 40

Climatological forecast Forecasts of weather conditions or climate expected to occur more than a month in the future. 49

Climatology A branch of meteorology that deals with the long-term characteristics of weather. 334–60

Cloud base The lowest portion of a cloud. 128

Cloud cover The amount of the sky obscured by clouds when observed at a particular location. 128

Cloud deck The top of a cloud layer, usually viewed from an aircraft. 280

Cloud genera The ten main groupings of clouds used for assigning names to the various cloud formations. 23, 24, 129, 130, 133

Cloud seeding The introduction of artificial agents (usually dry ice or silver iodide) into a cloud to modify it or increase rainfall. 7, 277

Coalescence The process whereby large cloud drops falling faster than smaller ones cause collisions and mergers that contribute to cloud drop growth. 140, 281, 282

Cold core high-pressure system An anticyclone that can be located beneath an upper air low-pressure area because of the cold air and greater mass associated with the core of the high-pressure area. 94

Cold front The leading edge of a cold air mass. 20, 21, 43, 156

Collar cloud The larger cyclonic circulation at the base of a thunderstorm that surrounds the tornado. It is a part of the mesocyclone. 257

Colliding clouds, 259

Colorado River Basin Project A weather modification project sponsored by the Bureau of Reclamation. 285

Colored precipitation, 146

Compression heating The increase in temperature of a gas that occurs simply by decreasing the volume that the gas can occupy, or increasing the pressure on it. 135, 136, 315

Computers Modern electronic computers are instruments that are able to store and manipulate millions of numbers per second. 4, 47, 56, 217

Condensation The process whereby water vapor is transferred to liquid form. 17, 281

Condensation nuclei Small particles in the atmosphere that serve as the core of tiny condensing cloud droplets. These may be dust, salt, or other material. 137, 281, 282

Conditional stability A general stability category that exists when the measured lapse rate is between the dry and moist adiabatic lapse rates. 122, 317

Coning Slight dispersion of a pollution plume upward, downward, and horizontally. 329

Conservation of absolute vorticity, 362

Contrails Trail of clouds, usually cirrus, forming behind aircraft as they travel through the atmosphere. 265

Convection Rising air currents due to heating as sunlight warms the ground and lower atmosphere. Warm air rises because it is less dense after it expands. 134

Convective thunderstorm An ordinary thunderstorm formed from surface heating and convection currents in moist air. 199

Conversion factors, 365

Coriolis acceleration An effect caused by the earth's rotation that causes moving objects, unattached to the earth, to accelerate to the right (left) of their path of motion in the northern (southern) hemisphere. 87–89, 162, 180

Corona A circle of light around the sun or moon produced by diffraction of light by liquid water droplets in clouds. 305

Country breeze A breeze from the surrounding country toward the heart of a city in response to the urban heat island. 327

Cumulonimbus cloud Cloud with vertical growth and precipitation; a thunderstorm. 24, 134, 199–223

Cumulus Clouds that exhibit vertical growth. 24, 25, 133, 200

Curvature effect The increase in vapor pressure of small cloud droplets with greater surface curvature that causes more rapid evaporation than that from larger drops with less curvature. 138, 139

Cyclogenesis Deepening of a low-pressure system with increasing development of a midlatitude cyclone. 155, 162, 165

Cyclone An area of low pressure with counterclockwise winds around it in the northern hemisphere and clockwise winds in the southern hemisphere. 14

Cyclonic Counterclockwise airflow around a low-pressure center in the northern hemisphere; clockwise in southern. 14, 154, 177

Cyclonic storms, 6, 154, 177, 245

Cyclonic vorticity Rotation in a counterclockwise (cyclonic) direction. 160, 162

Cyclostrophic wind Wind in a circular path, in which the Coriolis acceleration is negligible compared to the centrifugal and pressure gradient accelerations. 363

Dalton's law A law that says the total pressure of a mixture of gases is the sum of the partial pressures, where the partial pressure is the pressure each gas would exert if it alone were present. 12

Damage path The path or strip covered by a tornado where various types of damage occur. 217, 255

Damaging wind Wind of 97 km/h, or strong enough to cause damage to trees and property. 200

Density The amount of mass per unit volume. 12

Deposition The transition process of water from vapor to ice without passing through the liquid stage. 17, 129, 281

Deposition nuclei Tiny particles in the atmosphere that serve as the core of tiny ice crystals as water vapor changes to the solid form; also called *ice nuclei.* 137

Desert mirage An optical effect that produces inverted or partial images over hot surfaces. The bending of light rays gives the appearance of water on the desert. 300

Desert temperature, 76

Dew point temperature The temperature of the air if it is cooled to saturation. 206, 207

Diffraction Streams of brighter light produced as light passes objects having a diameter similar to the wavelength of light. 295, 305

Dispersion of pollutants The rate at which impurities in the air are decreased in concentration by atmospheric mixing. 313

Displacement (sun), 302

Diurnal temperature change, 114

Diurnal temperature range The temperature difference between the minimum at night and the maximum during the day. 76

Doppler radar Radar that employs the shift in frequency of radiowaves returned from moving objects to give the objects' speed. 214, 216, 218, 264

Double vortex thunderstorm The internal structure of a severe thunderstorm that consists of two counterrotating vortices located side by side, with the cyclonic vortex to the south. 214, 215

Drainage winds, 96

Drifting snow Term used in weather forecasts to indicate that strong winds will pile snow into drifts that may become quite deep. 173

Drought, 52

Drought warning program, 293

Dry adiabatic lapse rate The rate of expansion cooling (10°C/km) as a parcel of air rises, if no other heat additions occur, or the rate of compression heating as it descends (also 10°C/km). 115

Dry line The leading edge of an advancing hot dry air mass of continental tropical origin, desert Southwest for the United States. 202, 211

Dual Doppler radar Two Doppler radar sets aimed at the same cloud provide three-dimensional information on the winds and precipitation distribution within the cloud. 54

Dust bowl, 341, 342

Dust devil A small surface-generated vortex storm that occurs in hot deserts and in other locations when the ground is hot. 82

Forked lightning Cloud-to-ground lightning with branched channels. 228

Franklin, 6

Freezing rain Rain that turns to ice as it strikes cold objects or the cold ground. 143

Friction-layer wind Within the layer of air below 1 km, the earth and objects on it exert friction on wind currents that change the direction and reduce the speed of the wind. 92

Frictional deceleration The decrease in wind speed because of frictional drag from the earth's surface. 87, 89, 90

Front, 6, 19–22

Frontal analysis A weather map analysis involving the location of fronts on a weather map. 43

Frontal cyclone A low-pressure center with weather fronts (cold and/or warm) associated with it. Same as *midlatitude* or *wave cyclone*. 154–77

Frontal fog Fog formed near a weather front. 147

Frost, 152

Fumigation An increased concentration of atmospheric pollutants, usually associated with an inversion, that prevents atmospheric mixing. 329

Funnel cloud A rotating vortex extending from the base of a thunderstorm that would become a tornado if it touched the ground. 207

Galileo, 5

Gaussian plume dispersion model An equation used for calculating the concentration of pollutants downwind from a known point source. 328

Geostrophic wind The calculated or measured wind when the pressure gradient acceleration equals the Coriolis acceleration. 90

Glaciers, 339

Global heat budget, 71–73

Global temperature, 335

Glory Rings of light surrounding the shadow of your head when dew is on the ground, or around the shadow of an airplane as it falls on a cloud. 305

Gradient wind Wind blowing along curved isobars with a balance of centrifugal, Coriolis, and pressure gradient accelerations. 91, 92

Graupel A pellet formed from packed snow flakes or an ice crystal after riming has given it the appearance of small soft hail. 145

Green flash A very brief flash of green light as the sun rises or sets, caused by the bending of light by the atmosphere. 303

Greenhouse effect Lower atmospheric warming by atmospheric trapping of infrared radiation while solar radiation passes through. 75

Greenwich Mean Time (GMT) Since the prime meridian of 0° longitude passes through Greenwich, England, this location is used for a universal time throughout the world. Each of the full time zones are measured outward 15° from the prime meridian. Thus 1400 (2:00 P.M.) GMT is 0900 EST, 0800 CST, and so on. GMT − 5 = EST, GMT − 6 = CST, and so on. Also called *Zulu (z) time*. 36

Gust front The miniature cold front created by the downdraft of a large thunderstorm as the evaporation-cooled air spreads outward beneath the thunderstorm. 211, 216

Gustiness of wind, 115

Hadley cell Atmospheric circulation from the equator to 30° latitude that forms a cell from the thermal circulation. 98

Hail Solid precipitation in the form of chunks or balls of ice with diameters greater than 5 mm. The stones fall from cumulonimbus clouds. 145
- deaths, 231
- distribution, 235
- drifts, 233
- embryos, 239
- generation, 216

Hailstones, 234, 235, 237

Hailstorm A thunderstorm (cumulonimbus cloud) that produces hail. 231

Hail suppression The artificial reduction of hail damage from a thunderstorm by cloud seeding or other techniques. 287, 291

Hail swath The localized strip beneath a thunderstorm that is subjected to hail as the thunderstorm moves along. 234

Little ice age The period from about 1400 to 1800 when global temperatures were about 1.5°C colder, winters were much more severe, and glaciers advanced. 336, 337

Lofting The process in which a pollution plume mixes upward but not downward. 329

Long wave Another name for the meanders or Rossby waves in the airflow patterns at the jet stream level. 7, 79

Long-wave cyclone A type of surface frontal cyclone that is generated beneath an upper air low-pressure center within a long wave of the jet stream. 167, 169

Looming An optical effect that causes objects to appear higher than their actual location. 298

Looping The appearance of a pollution plume undergoing rapid dispersion as large air currents carry it up and down, in addition to dispersing it at the edges. 329

Low-level jet A stream of air moving northward from the Gulf of Mexico, generally located over the surface moisture tongue. This air current is lower in altitude (about 2 km) and contains lower wind speeds than the polar jet stream. 207

Low-pressure system A region of lower atmospheric pressure in midlatitudes, generally accompanied by cloudy, more windy weather, and precipitation. 44

Macroscale A scale used in meteorology extending from 100 to 1000 km. This is the normal synoptic scale for obtaining weather data. 47

Mammatus cloud Sometimes called a tornado sky, since this pouchy-type cloud formation indicates turbulence within a thunderstorm and frequently accompanies tornadoes. 263

Mechanical lifting The forced uplift of air as it flows over topographic barriers. 132

Mercury barometer An instrument used for measuring atmospheric pressure using a column of mercury to balance the weight of the atmosphere. 5

Mesocyclone The cyclonic rotating part of a large thunderstorm. 216

Mesoscale A scale used in meteorology extending from 1 to 100 km. 47

Mesosphere The atmospheric layer, above the stratosphere, extending from about 50 to 90 km. 11

Meteorological forecast Forecasts of weather events for six hours to five days ahead. 46

Meteorological satellites Satellites placed in orbit specifically for meteorological functions; these include the TIROS, ESSA, ATS, NOAA, SMS, GOES, and Nimbus weather satellites. 54, 55

Microscale A scale used in meteorology extending from a centimeter to a kilometer.

Millibar A unit of pressure in the metric system. One atmosphere equals 1013 millibars (mb). 12

Mixing ratio The mass of water vapor per unit mass of dry air. 17

Mobile source A moving source of air pollution, such as an automobile. 314

Moist adiabatic lapse rate The expansion cooling rate of rising saturated air (4 to 10°C/km). 120, 121

Moisture tongue A stream of warm moist air flowing northward across the United States from the Gulf of Mexico. 207

Monsoon Seasonal rains that are produced as winds from the ocean provide a continual moisture supply. 96, 107

Morning mixing depth The depth of the lower atmosphere where mixing of air and pollutants occur during the early morning hours. 323

Mountain breeze The downslope winds that occur particularly at night as the air cools along mountain slopes. 95

Multicell thunderstorms Thunderstorms that form in a group. A single mature cell may have a dissipating cell on its leeward side and a developing cell on its windward side. 211, 212

Nacreous clouds "Mother of pearl" clouds that form in the stratosphere at about 25 km and may be seen after sunset because of their height. 131

National ambient air standards, 314

National Climatic Center The organization responsible for storing and publishing weather data gathered in the United States. It is located in Asheville, North Carolina. 40

National Hurricane Center The government agency, located in Miami, Florida, that is responsible for tracking hurricanes and issuing hurricane warnings. 53

National Meteorological Center The organization responsible for receiving and analyzing synoptic weather data to produce various weather maps and forecasts for public use. It is located near Washington, D.C., at Suitland, Maryland. 40, 49, 53, 62

National Severe Storms Forecast Center The government agency, located in Kansas City, Missouri, that is responsible for forecasting tornadoes and severe thunderstorms. 54

National Weather Service The organization responsible for measuring, analyzing, forecasting, and informing the public of weather conditions. 46

Neutral stability, 116

Newton's second law This law states that the forces on an object will be equal to the acceleration multiplied by the mass of the object. 87

Newton's third law of motion The law of motion that states that for every force acting on a body, there is an equal and opposite reactive force.

NHRE The National Hail Research Experiment conducted by the National Center for Atmospheric Research. 287

Noctilucent clouds Thin clouds that form high in the mesosphere above 75 km. These are visible after sunset because of their height. 131

North American high, 100

Numerical cloud model A cloud simulated by mathematical and numerical methods, usually requiring the use of a computer. 217, 220

Numerical weather prediction Weather forecasting accomplished by the use of mathematical equations, manipulated by computers, using meteorological measurements to supply the initial numbers. 7, 46–47, 62

Oblate spheroid A common shape of large hail; egg- or doorknob-shaped rather than circular. 234

Occluded front The front that is formed as a cold front overtakes a warm front. 20, 22

Occlusion The process in which the cold front moves around a frontal cyclone faster than the warm front and overtakes it. 165, 167

Ocean currents, 79

Ocean swells Long, regular undulations of the ocean level, as wave motion from a hurricane, for example, is propagated hundreds of kilometers ahead of the storm. 189

Operational cloud seeding Cloud seeding performed to eliminate drought or to modify the weather, rather than for research purposes. 285

Ozone A form of oxygen composed of three atoms. Ozone in the upper atmosphere absorbs ultraviolet radiation from the sun. 9

Parcel stability method A means of determining atmospheric stability by considering a displaced parcel of air. If it returns to its original position the atmosphere is stable; if it accelerates away from its original position, the atmosphere is unstable. 116

Pascal, 5, 6

Pedestal cloud, 238, 240

Persistence forecast A simple weather forecast made by considering the past history of weather events and projecting them into the future without change. These are good for only a few hours at most. 46

Photochemical smog, 313

Plume The visible effluent from a smokestack. 328

Polarized light, 64

Polar jet stream The stream of air blowing from a

westerly direction at a height of about 12 km, over midlatitude locations. It generally separates polar air masses from warmer air. 98, 100–107

Polar molecule A molecule with a positive and negative pole that causes intermolecular attractions. 138–39

Pollution alert Issued when air pollution levels reach a specific value: 0.3 ppm of SO_2, for example. 331

Pollution emergency Issued when air pollution levels reach a specific value: 1 ppm of SO_2 for an 8-hour average, for example. 331

Pollution episode A buildup of pollution concentrations to the level where they are harmful to plants and animals. 316, 329, 331

Pollution forecast A forecast that pollution levels will increase because of weak atmospheric mixing. 331

Pollution warning Issued when air pollution levels reach a specific value: 1.6 ppm of sulfur dioxide for a 24-hour average, for example. 331

Potential evapotranspiration The maximum water loss rate from vegetation and soil for a particular atmospheric condition. 340

Precipitable water vapor The depth of water that would result if all the vapor in the atmosphere above a location were condensed into liquid water.

Precipitation enhancement, 281

Precipitation management The concept of controlling the amount and distribution of precipitation through cloud seeding techniques. 227

Precipitation probability The degree of certainty or chance of rainfall at a given location. 47, 49

Prefrontal wave An atmospheric wave that may be propagated ahead of an advancing cold front in the warm air mass. 202, 203

Pressure, average, 99–101

Pressure gradient force The force exerted on air by a difference in pressure between two points. One of the forces that causes wind. 87

Pressure tendency Change in pressure over the last three hours, used in weather analysis. 42

Prevailing westerlies A name eroneously applied to the surface winds in the belt from 30° to 60°N latitude. This region is, however, dominated by "prevailing westerlies" in the upper troposphere. 98

Project Stormfury The governmental hurricane modification program. 285

Quantitative precipitation forecast, 47

Radar An instrument useful for remote sensing of meteorological phenomena. It operates by sending radio waves and monitoring those returned by such reflecting objects as raindrops within clouds. 7, 54, 55, 65

Radiation Electromagnetic energy emitted by all objects in amounts related to the temperature of the object. 64

Radiation fog Fog formed as objects on the ground cool by losing radiation. 147

Radiosonde The instrument package carried by weather balloons to measure the temperature, humidity, and pressure of the atmosphere. 39

Rain shaft A streak of rain falling from a cloud. 247

Rainbow The display of colors as raindrops separate light according to wavelength, 274, 303

Raindrop growth, 141, 281

Rawinsonde An instrument carried by weather balloons to measure the temperature, humidity, pressure, and winds of the atmosphere. 39, 116

Red rain, 146

Red sky, 312

Redfield, 6

Reflection Light bounced off an object at the same angle and wavelength as that of the incident light ray. 295

Reflectivity, 66

Refraction The bending of light at a particular angle as it passes through a transparent object. 295

Refractive index The ratio of the speed of light in a vacuum to its speed in a transparent medium. 295

Relative humidity The amount of water vapor in the air divided by the amount of water vapor at saturation expressed as a percentage. 17

Releaser cloud A cloud composed of ice crystals that naturally seed a cloud containing supercooled liquid water. 283

Remote sensing Any of a variety of instruments and techniques that allow the collecting of data from a distance. 194

Return streamer The lightning stroke involving the flow of charges from the ground to the cloud. This follows the stepped leader. 229

Ribbon lightning A lightning discharge with a ribbon appearance, as strong winds move the conducting channel between successive strokes. 229

Richardson, L. F., 7

Ridge A region of anticyclonic curvature of winds in the atmosphere located around a high-pressure area. 14, 27, 44–46, 162

Riming The process where tiny liquid droplets collect on ice crystals or other objects and freeze. 143

Roll cloud A cloud formation associated with the leading edge of a thunderstorm. It has the appearance of a horizontal roll. 216

Rossby wave Waves in the airflow patterns of the jet stream and upper atmosphere of long wavelength. 7

Safety:
- blizzard, 171–74
- hurricane, 194
- tornado, 266–70

Sahelian drought, 107, 338

St. Elmo's Fire A visible glow from objects as a charge is induced from a thunderstorm overhead. 227

Santa Ana wind Strong downslope winds through mountain passes in southern California. 136

Satellites, 7

Saturation The condition when the maximum amount of water vapor that can exist in the air is present. 17

Schaefer, Vincent, 7

Sea breeze The onshore winds that develop during the day as the land heats faster than water. 95

Secondary rainbow A second rainbow that is fainter than the primary rainbow and appears higher in the sky. 303

Severe thunderstorm Thunderstorms with lightning and damaging winds or hail. 199–223

Sferics A contraction of *atmospherics:* electrical activity from thunderstorms. 264

Sheet lightning A lightning discharge from distant thunderstorms that lights up the evening sky. 228

Shelf cloud, 238, 240

Short-wave cyclone A type of surface frontal cyclone that is generated by a short-wave disturbance superimposed on the jet stream flow patterns. 167, 170

Siberian high, 100

Silver iodide A chemical that is burned to produce particles similar enough to ice crystals to serve as deposition nuclei for the artificial stimulation of ice crystal growth. 277, 279, 280

Sinking An optical effect that causes objects to appear lower than their actual location. 298

Skipping tornado A tornado that touches the ground, then lifts back into the air, followed by another touchdown that may be repeated. This produces an intermittent damage path. 266

Sleet, 143

Smoke plumes, 122, 123

Snell's law An equation relating the amount of bending of light to the angles before and after bending. 295, 364

Snow Solid precipitation in the form of minute ice flakes that occur below 0°C. 3, 4, 145, 173

Snow pellets Precipitation of white or opaque spherical ice of 5 mm diameter or less. 145

Snow roller Rolls of snow produced as the wind forms melting snow into a roll. 147, 167

Soaring, 127

Soil moisture, 341

Solar constant The radiation from the sun on a surface perpendicular to the sun's rays and located just outside the earth's atmosphere

Solar constant (continued)
(1.38 kW/m²). 63

Solar flare A storm on the sun that may change the amount or type of radiation on earth. 243

Solstice Either of two times of the year when the sun is the greatest distance from the celestial equator, occurring about June 22 and December 22. 68, 69

Solute effect The reduction of vapor pressure of small concentrated cloud droplets that prevents their rapid evaporation and may allow their growth when pure droplets are evaporating. 138, 139

Source region The large geographical region where an air mass acquires its temperature and humidity characteristics. 17–19

Specific humidity The mass of water vapor in the air per unit mass of moist air. 17

Spender cloud A cloud with numerous supercooled liquid water droplets that is seeded with ice crystals from a natural cloud (**releaser cloud**). 283

Sponge effect The misconception that cloud seeding in one area robs another area because the atmosphere contains a finite amount of water vapor. 284

Squall line A line of thunderstorms. 132, 203, 209, 210

Stable atmosphere The atmospheric condition when little or no vertical mixing occurs due to the nature of the temperature change with height. 117, 118

Stationary front A surface boundary between air masses that have stalled. 20, 22

Stefan-Boltzmann law An equation that relates the amount of energy emitted to the temperature of the emitter. 361

Stomata Small pores on the lower side of plant leaves that allow water to escape to the atmosphere. 340

Stratosphere The layer of the atmosphere above the troposphere, defined on the basis of constant or increasing temperature with altitude. 10

Stratus Low clouds usually having a stratified form and containing liquid water droplets. 24, 130, 131

Streamline Line on a weather map showing the path a parcel of air would follow as it moves in the atmosphere. 44, 177

Streamline analysis Method of drawing streamlines on a weather map by using wind direction measurements plotted on the map. 44

Streamline confluence A region where the paths taken by parcels of air (streamlines) come closer and closer together. Also called *streamline convergence*. 205

Streamline diffluence A region where the paths taken by parcels of air (streamlines) spread apart. Also called *streamline divergence*. 205

Sublimation The transition process of water from ice to vapor without passing through the liquid stage. 16

Subtropical jet stream A stream of air from the west at about 14 km height and about 30° latitude. This jet stream is weaker than the polar jet stream and usually separates equatorial air masses from cooler air to the north. 98, 100

Sundog An image of the sun at 22° or 46° from it, produced by ice crystals. 306

Sun pillar A pillar of light above or below the sun near the time of sunrise or sunset, produced by ice crystals. 307

Sunspots Storms that occur in pairs on the sun at intervals with period peaks in activity. These affect radiation from the sun and may affect weather on earth. 51, 244

Supercell thunderstorms The largest thunderstorms; these approach steady-state internal structure for a few hours and are able to produce hail and/or tornadoes. 207

Supercooled droplets Liquid water droplets between 0 and −40°C that are unfrozen and would immediately freeze if particles were present to start the solidifying process. 129

Supernumerary arcs Bands of color inside the primary rainbow caused by light diffraction. 295

Swells Ocean waves that move out of a strong wind environment where they were generated become ocean swells as their crests become rounded and lower. 189

Synergistic effects The effects of two or more factors on climate that produce a result greater than either factor could produce separately. 338

Synoptic Dealing with the gathering of meteorological data over a large area at a specified instant of time, for the purpose of projecting the data into the future to give weather forecasts. 36

Synoptic chart Weather map produced by plotting synoptic meteorological data. 40

Thermal circulation Airflow resulting from heating the air. 97

Thermal equator The intertropical convergence zone where the northeasterly trade winds of the northern hemisphere meet the southeasterly trade winds of the southern hemisphere. 107

Thermal low An area of low atmospheric pressure produced from continued surface heating, as in the desert Southwest. 96, 100

Thermal updraft The rising air current in a large thunderstorm that is buoyant because of its warmer temperature compared to surrounding air. 214, 239

Thermodynamic diagram A chart useful for determining atmospheric stability, since the dry and moist adiabatic lapse rates are already plotted. 119, 121

Thermoelectric effect The process whereby a temperature gradient in ice crystals produces a separation of electrical charges, with positive charges in colder portions. 226

Thermogram A temperature trace on a chart produced by a self-registering thermometer called a thermograph.

Thermosphere The atmospheric layer above the mesosphere. It extends from 90 km to outer space. 11

Thunder The audible sound waves produced as a lightning discharge expands a conducting channel of air. 224, 226

Tornado The most violent vortex storm on earth. It originates within a thunderstorm. 257
- control, 265
- damage, 266–70
- days, 247
- deaths, 268
- detection, 262, 263
- distribution, 247, 250
- forecasts, 7, 54
- formation, 254
- generation, 216, 217
- measurements, 273
- multiple, 217
- pressure, 260
- safety, 266–70

Tornado day A day when one or more tornadoes occur somewhere in the United States. 247

Tornado detector An electronic device that is used on an experimental basis for detecting tornadoes. 263

Tornadolike vortex A vortex created in the laboratory to simulate a tornado. 265

Towering An optical effect that causes objects to appear taller than they actually are. 299

Trade wind inversion A stable inversion associated with sinking air above high-pressure regions in the trade winds. 177

Trade winds The name of the northeasterly (southeasterly) winds that predominate from the equator to about 30° N latitude (30° S latitude). 98

Trajectory (hail) Trajectory path an object takes. *See* **hail trajectory.** 240

Transmissivity, 66

Travelers' advisory Issued by the National Weather Service when blizzards or ice storms are forecast to make traveling dangerous. 173

Tropical depression A stormy region in the tropics more intense than a tropical disturbance but less intense than a tropical storm. 178

Tropical disturbance A region of cloudy weather in the tropics. These sometimes develop into a hurricane. 178

Tropical storm A vortex storm that occurs over the tropical ocean. In a more specific sense it refers to a stage in hurricane development where the intensity is slightly less than a hurricane. 178

Troposphere The lowest layer of the atmosphere where the temperature decreases with altitude. 10

Trough A region of cyclonic curvature of winds in the atmosphere, located around a low-pressure area. 14, 27, 44–46, 162–67, 249

Trough cyclone The type of frontal cyclone that is generated beneath the trough of the jet stream. 165, 168

Turbulence Very unorganized atmospheric motion including gusts and lulls in the wind. 83, 84

Typhoon The name applied to tropical storms in some parts of the Pacific Ocean; similar storms are called hurricanes in the Atlantic Ocean. 178

Ultraviolet radiation Electromagnetic energy with wavelengths just shorter than visible light. 9, 66, 71

Unstable atmosphere The atmospheric condition when vertical mixing occurs due to the temperature profile that creates buoyant parcels of air. 114, 116, 118

Upper air divergence Outflow of air in the upper troposphere created by horizontal divergence. 160, 205

Upper air inversion An inversion occurring above the surface, usually associated with a high-pressure area or a cold front. 118, 320

Upper air support Strength provided to surface low-pressure areas by horizontal divergence in the upper air, or to surface high-pressure areas by horizontal convergence in the upper atmosphere. 164

Upslope fog Fog formed as air flows up the slope of a topographic barrier. 147

Urban heat island A region in urban areas of increased temperatures compared with surrounding rural areas. 323

Vapor pressure The pressure exerted by water vapor on its surroundings. It is a force due to molecular action per unit area of surface. 17, 134, 138, 281

Veer Clockwise shift in wind direction. 86

Visible light That portion of the electromagnetic spectrum from 0.4 to 0.7 μm wavelengths that is visible. 65, 67

Vorticity The amount of rotation of the air; circulation per unit area. 47, 160–62, 206

Wake capture The collection of small cloud droplets as they are pulled into the wake and collide with the backside of larger falling drops. 141

Warm front The leading edge of a warm air mass. 20, 21, 43

Warning A warning issued by the National Weather Service to say that a storm has formed and is approaching. Warnings are issued for blizzards, tornadoes, hurricanes, floods, and severe thunderstorms. 172, 194

Watch A watch issued by the National Weather Service means a storm has formed or conditions are right for its formation. Watches are issued for blizzards, tornadoes, hurricanes,

WEATHER MAP SYMBOLS FOR PRESENT WEATHER

(Continued from front endpapers)

	6	7	8
00	Widespread dust in suspension in the air, NOT raised by wind, at time of observation	Dust or sand raised by wind at time of observation	 Well developed dust devil(s) within past hour
10	Precipitation within sight, reaching the ground, near to but NOT at station	Thunder heard, but no precipitation at the station	 Squall(s) within sight during past hour
20	Showers of snow, or of rain and snow, during past hour, but NOT at time of observation	Showers of hail, or of hail and rain, during past hour, but NOT at time of observation	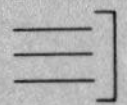 Fog during past hour, but NOT at time of observation
30	Slight or moderate drifting snow, generally low	Heavy drifting snow, generally low	 Slight or moderate drifting snow, generally high
40	Fog, sky discernible, has begun or become thicker during past hour	Fog, sky NOT discernible, has begun or become thicker during past hour	 Fog, depositing rime, sky discernible
50	Slight freezing drizzle	Moderate or thick freezing drizzle	 Drizzle and rain, slight
60	Slight freezing rain	Moderate or heavy freezing rain	 Rain or drizzle and snow, slight
70	Ice needles (with or without fog)	Granular snow (with or without fog)	 Isolated starlike snow crystals (with or without fog)
80	Moderate or heavy snow shower(s)	Slight shower(s) of soft or small hail with or without rain or rain and snow mixed	 Moderate or heavy shower(s) of soft or small hail with or without rain or rain and snow mixed
90	Slight or moderate thunderstorm, with hail at time of observation	 Heavy thunderstorm, without hail, but with rain and/or snow at time of observation	 Thunderstorm combined with dust storm or sand storm at time of observation